# ENCYCLOPEDIA OF SURFACTANTS

## *VOLUME I*

*Compiled by*
**Michael and Irene Ash**

*Chemical Publishing Co.*
*New York, N.Y.*

# Encyclopedia of Surfactants
# Volume 1



ISBN: 978-0-8206-0041-3

Chemical Publishing Company:
www.chemical-publishing.com
www.chemicalpublishing.net

First Edition:

© **Chemical Publishing Company, Inc.** - New York 1980

Second Impression:

**Chemical Publishing Company, Inc.** - 2011

Printed in the United States of America

# PREFACE

This comprehensive, modern encyclopedia of surfactant products has been compiled with the cooperation of the major national and international surfactant manufacturers in order to give the user quick and thorough information on each tradename product that is included. Information from thousands of brochures, technical bulletins, and data sheets has been gleaned in order to make a complete profile on each surfactant product available from one source—this three volume compendium. Though sometimes limited by the information provided by the manufacturers, in most instances, the chemical ingredients, properties, form, applications, etc. of each product are discussed in detail.

The use of surfactants pervades almost every chemical industrial concern—including textiles, paints, pesticides, foods, cosmetics, pharamceuticals, etc. We hope that because of the universality of surfactants, this reference work will be able to serve as a useful tool to the many chemical engineers and salespeople here and abroad.

We want to extend our thanks to Roberta Dakan for her dedication and tireless efforts in helping make this encyclopedia as accurate and consistent as possible.

When the temperature at which properties such as viscosity, density, surface tension, solubilities, etc. is not included, a standard temperature of 25 C is to be assumed. The information in this publication is reliable to the best of our knowledge. We would be glad to be informed of any errors or omissions so that these can be integrated into subsequent editions of this encyclopedia.

M. and I. Ash

# CONTRIBUTORS

Aceto Chem. Co., Inc.
126-02 Northern Blvd.
Flushing, N.Y. 11368

Ajinomoto USA, Inc.
9 West 57th St.
New York, N.Y. 10019

Akzo Chemie bv
Stationsstraat 48
P.O. Box 247
3800 AE-Amersfoort
The Netherlands

Akzo Chemie Italia SpA.
Via Vismara
20020 Arese
Milano, Italy

Albright & Wilson Ltd.
Detergent Sector
P.O. Box 15
White Haven, Cumberland
CA 28 9QQ, England

Alcolac Inc.
3440 Fairfield Rd.
Baltimore, Md. 21226

Alconox Inc.
215 Park Ave. So.
New York, N.Y. 10003

Alkaril Chemicals Ltd.
3265 Wolfedale Road
Mississauga, Ont., Canada

Allied Colloids Inc.
One Robinson Lane
Ridgewood, N.J. 07450

Alox Corp.
3934 Buffalo Ave.
P.O. Box 517
Niagara Falls, N.Y. 14302

Amerchol Corp.
(Unit of CPC International Inc.)
P.O. Box 351
Talmadge Road
Edison, N.J. 08902

American Color & Chem. Corp.
11400 Westinghouse Blvd.
Charlotte, N.C. 28232

American Cyanamid Co.
Fine Chemicals Dept.
Pearl River, N.Y. 10965

American Cyanamid Co.
Industrial Chemicals Div.
Berdan Ave.
Wayne, N.J. 07470

American Cyanamid Co.
Organic Chemicals Div.
Bound Brook, N.J. 08805

American Hoechst Corp.
Industrial Chemicals Div.
Rt. 202-206 North
Somerville, N.J. 08876

American Lecithin Co., Inc.
32-34 61st St.
Woodside, L.I., N.Y. 11377

Apex Chem. Co., Inc.
200 South First St.
Elizabethport, N.J. 07206

Arizona Chemical Co.
840 Valley Brook Ave.
Lyndhurst, N.J. 07071

Arjay Inc.
P.O. Box 45045
Houston, Texas 77045

Arkansas Co., Inc.
185 Foundry St.
P.O. Box 210
Newark, N.J. 07101

Armak–Industrial Chemicals
P.O. Box 1805
Chicago, Ill. 60690

Armstrong Chem. Co., Inc.
1530 South Jackson St.
Janesville, Wis. 53545

Arol Chem. Prod. Co.
649 Ferry St.
Newark, N.J. 07105

Atlas Chem.
See ICI United States Inc.

Atlas Refinery, Inc.
142 Lockwood St.
Newark, N.J. 07105

BASF Wyandotte Corp.
Industrial Chemicals Group
1609 Biddle Ave.
Wyandotte, Mich. 48192

BASF Wyandotte Corp.
Textile Colors and Chem. Dept.
100 Cherry Hill Road
Parsippany, N.J. 07054

BASF Aktiengesellschaft
Carl-Bosch-Str. 36
6700 Ludwigshafen
Germany

Berol Kemi AB
S-444 01 Stenungsund 1
Sweden

Bostik South
Div. of USM Corp.
P.O. Box 5695
Greenville, S.C. 29606

Carson Chemicals, Inc.
2779 East El Presidio
Long Beach, Calif. 90810

Catawba-Charlab, Inc.
P.O. Box 948
Charlotte, N.C. 28231

Central Soya Co., Inc.
Chemurgy Div.
1300 Fort Wayne National Bank
Bldg.
Fort Wayne, Ind. 46802

Chemform Corp.
Vertac Bridgeport Ind. Park
P.O. Box 1745
West Memphis, AR 72301

Ciba-Geigy Corp.
Dyestuff & Chemicals Div.
P.O. Box 11422
Greensboro, N.C. 27409

Cincinnati Milacron Chem. Inc.
West St.
Cincinnati, Ohio 45215

Cities Service Co.
ICD, Cities Service Bldg.
3445 Peachtree Rd. N.E.
Atlanta, Georgia 30326

Clintwood Chemical Co.
4342 S. Wolcott Ave.
Chicago, Ill. 60609

Conoco Chemicals
Continental Oil Co.
5 Greenway Plaza East
P.O. Box 2197
Houston, TX 77001

Croda Chemicals Ltd.
Cowick Hall
Snaith Goole
North Humberside DN14 9AA
England

Croda Inc.
51 Madison Ave.
New York, N.Y. 10010

Cyclo Chemicals Corp.
7500 N.W. 66th St.
Miami, Fla. 33166

Diamond Shamrock Corp.
Process Chemicals Div.
350 Mt. Kemble Ave.
Morristown, N.J. 07960

Dixo Co., Inc.
158 Central Ave.
Rochelle Park, N.J. 07662

Dow Chemical Europe S.A.
Leland I. Doan Str. 3 CH-8810
Horgen
Switzerland

Dow Corning Corp.
Midland, Mich. 48640

Du Pont de Nemours, E.I. & Co.
Specialty Chemicals & Products
Nemours Bldg.
Wilmington, DE 19898

Durkee Industrial Foods Group/
SCM Corp.
900 Union Commerce Bldg.
Cleveland, Ohio 44115

Dutton & Reinisch Ltd.
Crown House, London Road
Morden, Surrey
SM45DU, England

Emery Industries, Inc.
1501 W. Elizabeth Ave.
Linden, N.J. 07036

Emkay Chemical Co.
319-325 Second St.
Elizabeth, N.J. 07206

Emulsion Systems Inc.
58 Metropolitan Ave.
Brooklyn, N.Y. 11211

Essential Chemicals Corp.
28391 Essential Rd.
Merton, Wis. 53056

Exxon Company, USA
P.O. Box 2180
Houston, Texas 77001

GAF Corp.
Chemical Products
140 W. 51st St.
New York, N.Y. 10020

General Electric Co.
Silicone Prod. Dept.
Waterford, N.Y. 12188

General Mills Chemicals, Inc.
See Henkel Inc.
Chemical Specialties Div.

Georgia-Pacific Corp.
P.O. Box 1236
Bellingham, Wash. 98225

Glyco Chemicals, Inc.
51 Weaver St.
Greenwich, Conn. 06830

W.R. Grace & Co.
Organic Chemicals Div.
55 Hayden Ave.
Lexington, MA 02173

Graden Chemical Co., Inc.
426 Bryan St.
Havertown, Pa. 19083

Grindsted Products
Grindstedvaerket A/S
38 Edwin Rahrs Vej
DK-8220 Brabrand
Denmark

C.P. Hall Co.
7300 S. Central Ave.
Chicago, Ill. 60638

Hart Chemical Ltd.
Guelph, Canada N1H 6K8

Henkel Chemicals (Canada) Ltd.
9550 Blvd. Ray Lawson
Ville d'Anjou (Montreal)
P.Q. Canada

Henkel Inc.
Chemical Specialties Div.
1301 Jefferson St.
Hoboken, N.J. 07030

Hercules Inc.
Organics Dept.
Hercules Tower
910 Market St.
Wilmington, Del. 19894

Hilton-Davis Chemical Co.
2235 Langdon Farm Rd.
Cincinnati, Ohio 45237

ICI United States Inc.
New Murphy Rd. & Concord Pike
Wilmington, Del. 19897

IMC Chem. Group–NP Div.
4415 W. Harrison St.
Hillside, Ill. 60162

Lankro Chemicals Ltd.
Eccles
Manchester M30 OBH
England

Monsanto Co.
800 N. Lindbergh Blvd.
St. Louis, Mo. 63166

NL Industries
NL Baroid
P.O. Box 1675
2404 Southwest Freeway
Houston, TX 77001

NL Industries
Industrial Chemicals Div.
P.O. Box 700
Hightstown, N.J. 08520

Onyx Chemical Co.
Millmaster Onyx Group
190 Warren St.
Jersey City, N.J. 07302

Proctor & Gamble Co.
Sharonwood Technical Center
Cincinnati, Ohio 45217

PVO International Inc.
416 Division St.
Boonton, N.J. 07005

Reilly-Whiteman Inc.
Washington, & Righter Sts.
Conshohocken, Pa. 19428

Rewo Chemicals, Inc.
125 Wilbur Pl.
Bohemia, N.Y. 11716

Sandoz Colors & Chemicals
Hanover, N.J. 07936

Sandoz Ltd.
CH-4002
Basle, Switzerland

Shamton Lubricants & Chem. Ltd.
Tel Aviv, Israel

Shell Chemical Co.
One Shell Plaza
Houston, TX 77002

Sherex Chemical Co., Inc.
P.O. Box 646
Dublin, Ohio 43017

Standard Chemical Co.
Mill Lane
Cheadle, Chesire SK8 2NX
England

Stepan Chemical Co.
Edens & Winnetka Rds.
Northfield, Ill. 60093

Swastik Household & Industrial Products Ltd.
Shaghibag House
13 Walchand Hirachand Marg
Bellard Estate
Bombay 400 038 India

Swift Chemical Co.
383 Orenda Road
Bramaleg, Ontario, Canada

Thompson-Hayward Chem. Co.
5200 Speaker Rd.
Kansas City, Kansas 66106

3M Co.
3M Center
Commercial Chemical Div.
St. Paul, MN 55105

Toho Chemical Industry Co., Ltd.
1-14-9 Nihonbashi
Kakigara-cho, Chuo-ku
Tokyo, Japan

Van Schuppen Chemie
Nieuweweg, Veenendaal
Netherlands

R.T. Vanderbilt Co., Inc.
30 Winfield St.
Norwalk, Conn. 06855

Witco Chemical Corp.
Organics Div. & Sonneborn Div.
277 Park Ave.
New York, N.Y. 10017

Witco Chem. Corp.
Ultra Div.
2 Wood St.
Paterson, N.J. 07524

# ABBREVIATIONS

| | |
|---|---|
| aq. | aqueous |
| BHA | butylated hydroxyanisole |
| B.P. | boiling point |
| C | degrees Centigrade |
| cc | cubic centimeter(s) |
| CC | Closed cup |
| cm | centimeter(s) |
| COC | Cleveland open cup |
| conc. | concentration, concentrated |
| cps | centipoise(s) |
| cs or cst | centistoke(s) |
| disp. | dispersion, dispersible |
| dist. | distilled |
| EO | ethylene oxide |
| esp. | especially |
| ETO | ethylene oxide |
| F | degrees Fahrenheit |
| FDA | Food & Drug Administration |
| F.P. | freezing point |
| g | gram(s) |
| gal | gallon(s) |
| HLB | hydrophilic lipophilic balance |
| ingred. | ingredient(s) |
| IPA | isopropyl alcohol |
| IPM | isopropyl myristate |
| kg | kilogram(s) |
| l | liter(s) |
| lb | pound(s) |
| max. | maximum |
| MEK | methyl ethyl ketone |
| min. | minimum |
| ml | milliliter(s) |
| mm | millimeter(s) |
| M.P. | melting point |
| m.w. | molecular weight |
| NF | National Formulary |

| | |
|---|---|
| no. | number |
| O/W | oil in water |
| PEG | polyethylene glycol |
| pH | hydrogen ion concentration |
| PMCC | Pensky-Martens closed cup |
| POE | polyoxyethylated |
| pt. | point |
| PVC | polyvinyl chloride |
| quat. | quaternary |
| ref. | refractive |
| sapon. | saponification |
| sol. | soluble |
| solid. | solidification |
| sol'n. | solution |
| sp. gr. | specific gravity |
| SSU | Saybolt Seconds Universal |
| Stod. | Stoddard |
| SUS | Saybolt Universal Seconds |
| TCC | Taggart closed cup |
| tech. | technical |
| T/L | tank load |
| TOC | Taggart open cup |
| T/T | tank truck |
| USP | United States Pharmacopeia |
| uv | ultraviolet |
| visc. | viscosity |
| W/O | water in oil |
| wt | weight |
| % | percent |
| $\approx$ | approximately |
| $<$ | less than |
| $>$ | greater than |
| $\leqslant$ | less than or equal to |
| $\geqslant$ | greater than or equal to |
| $\gg$ | much larger than |
| $\ll$ | much smaller than |
| # | number |

The following products, which are included in this volume, are registered with the U.S. Patent Office.

Abex®
Actiflo®
Acto®
Adsee®
Aerosol®
Agrimul®
Ahco®
Alacsan®
Alcojet®
Alconox®
Alfol®
Alkaterge®
Alipal®
Amacarrier®
Amasoft®
Amawet®
Amerchol®
Ameroxol®
Aminogen®
Antara®
Antaron®
Antarox®
Arlacel®
Arlasolve®
Arlatone®
Arlex®
Armac®
Armeen®
Armol®
Armohib®
Arquad®
Atlas®
Atlox®
Atmer®
Atmos®
Atmul®
Atpet®
Atplus®
Atsurf®
Bio Soft®
Blancol®
Brij®
Casul®
Carbonox®
CC-16®
Centrol®
Centrolene®
Centrolex®
Centromix®
Centrophil®
Chemcol®
Cremophor®
Cutina®
Cyncal®
Darvan®
Daxad®
Deceresol®
Det-O-Jet®
Dow Corning®
Dresinate®
Drewmulse®
Duomac®
Duomeen®
Duoquad®
Duponol®
Emerest®
Emery®
Emid®
Emulphogene®
Emulphor®
Ethoduomeen®
Ethofat®
Ethomeen®
Ethoquad®
Flexricin®
Fluorad®

# A

**A-641.** ICI Unites States Inc.

*Chemical Description:* Noncrystallizing sorbitol sol'n.
*Category:* Polyhydric alcohol (Polyol)
*Form:* Clear viscous liquid; Color: Water white to pale yellow; Odor: None; Taste: Pleasantly sweet
*Properties:* Ref. Index: 1.455–1.465; Sp. gr.: ≈ 1.31; Visc.: ≈ 140 cps
*Composition:* 53% sorbitol content

**AA-10.** Cities Service Co.

*Category:* Detergent, wetting agent, emulsifier
*Applications:* Emulsifier
*Form:* Liquid, flake; Color: White
*Properties:* Odor: None; Solubility: Highly soluble in water; Density: 2 lb/gal; Sp. gr.: 1/5 $H_2O$; Stability: Solid
*Composition:* 98% active ingredient
*Reaction:* Neutral
*Toxicity:* Nontoxic
*Containers:* Fiber drums

**Abex 18-S.** Alcolac

*Category:* Emulsifier, anionic
*Applications:* Polymer emulsion as binders for coated paper and paperboard; in paint and floor coatings; in polymerization of vinyl acetate and acrylics
*Form:* Liquid; Color: Pale yellow
*Properties:* Density: 8.5 lb/gal; Visc.: @ 15 C, 200 cps;Stability: Stable for broad range of pH and temperature; Surface Tension: 1.0% sol'n.–48 dynes/cm; 0.1% sol'n.–45–47 dynes/cm
*Composition:* 35% active ingredient

**Abex 22-S.** Alcolac

*Category:* Emulsifier, anionic
*Applications:* In emulsion polymer latices for floor polish, latex paints, paper coatings, pigment binders and adhesives
*Form:* Liquid; Color: Pale yellow
*Properties:* Density: 8.7 lb/gal; Visc.: @ 25 C, 1000 cps; Stability: Good; Surface Tension: 40 dynes/cm at critical micelle concentration

*Composition:* 25% active ingredient

**Abex 26-S.** Alcolac
*Category:* Emulsifier, anionic surfactant
*Applications:* Emulsifier for acrylics, vinyl acetates and other vinyl monomers
*Form:* Liquid; Color: Pale yellow
*Properties:* Density: 8.8 lb/gal; Visc.: @ 25 C, 100 cps; Stability: Good in presence of multivalent salts; Surface Tension: 55 dynes/cm at critical micelle concentration
*Composition:* 33% active ingredient

**Abex JKB.** Alcolac
*Category:* Emulsifier, anionic surfactant
*Applications:* Emulsifier for highly carboxylated acrylic monomers; latex thickeners, alkali-soluble resins; metalized floor finishes; textile sizing agents and handmodifiers
*Form:* Liquid; Color: Clear
*Composition:* 30% active ingredient

**Abex VA-40.** Alcolac
*Category:* Emulsifier, anionic surfactant
*Applications:* Emulsifier for high solids vinyl acetates
*Form:* Liquid; Color: Pale yellow
*Properties:* Visc.: @ 25 C, 600 cps
*Composition:* 40% active ingredient

**Abex VA-50.** Alcolac
*Category:* Emulsifier, anionic surfactant
*Applications:* Emulsifier for high solids vinyl acetates
*Form:* Liquid; Color: Pale yellow
*Properties:* Visc.: @ 25 C, 400 cps; Surface Tension: 47.4 dynes/cm (0.01% sol'n.)
*Composition:* 46% active ingredient

**Absorption bases.** Croda Chem.
*Chemical Description:* Cholesterol rich extracts of lanolin compounded into inert hydrocarbon bases
*Category:* Emollient
*Applications:* Cosmetics, especially emollient skin creams, lotions; baby preparations; make up
*Form:* Solid, liquid

**Accobetaine CL.** Armstrong Chem.
*Chemical Description:* Complex coco betaine
*Category:* Detergent, wetting agent, emulsifier, high foaming agent, solubilizer
*Applications:* Household and cosmetics uses (hair shampoos, liquid hand cleaners, dishwashing compounds, rug shampoo, etc.)
*Form:* Liquid; Color: Yellow; Odor: Pleasant
*Properties:* Solubility: Good in water; F.P.: 0 C
*Composition:* 35% active ingredient
*Toxicity:* Nontoxic

**Accomeen C2.** Armstrong Chem.
*Chemical Description:* Ethoxylated coco amine
*Category:* Emulsifier, antistat, nonionic surfactant

*Form:* Liquid; Color: 4 (Gardner); Odor: Amine
*Properties:* Solubility: Organic solvent; insoluble water; Density: 7.25 lb/gal; Sp. gr.: 0.87
*Composition:* 99% active ingredient; 2 moles ethylene oxide on primary coco amine
*Reaction:* Acid
*Warnings and Caution:* Corrosive

**Accomeen C5.** Armstrong Chem.
*ChemicalDescription:* Ethoxylated coco amine
*Category:* Emulsifier, nonionic surfactant
*Applications:* Emulsifier, antistat
*Form:* Liquid; Color: 4 (Gardner); Odor: Amine
*Properties:* Solubility: Organic solvents and water; Density: 8.15 lb/gal; Sp. gr.: 0.98
*Composition:* 99% active ingredient; 5 moles ethylene oxide on primary coco amine
*Reaction:* Alkaline
*Warnings and Caution:* Corrosive

**Accomeen C10.** Armstrong Chem.
*Chemical Description.* Ethoxylated coco amine
*Category:* Emulsifier, nonionic surfactant
*Applications:* Emulsifier, antistat
*Form:* Liquid; Color: 6 (Gardner); Odor: Amine
*Properties:* Solubility: Organic solvents and water; Density: 8.3 lb/gal; Sp. gr.: 1.0
*Composition:* 99% active ingredient; 10 moles ethylene oxide on primary coco amine
*Reaction:* Acid
*Warnings and Caution:* Corrosive

**Accomeen C15.** Armstrong Chem.
*ChemicalDescription:* Ethoxylated coco amine
*Category:* Emulsifier, nonionic surfactant
*Applications:* Emulsifier, antistat, textile spinning aid
*Form:* Liquid; Color: 6 max (Gardner); Odor: Mild amine
*Properties:* Density: 8.7 lb/gal; Sp. gr.: 1.04
*Composition:* 99% active ingredient; 15 moles ethylene oxide on primary coco amine
*Reaction:* Acid
*Toxicity:* Corrosive
*Warnings and Caution:* Corrosive

**Accomeen S2.** Armstrong Chem.
*Chemical Description:* Ethoxylated soya amine
*Category:* Emulsifier, nonionic surfactant
*Applications:* Emulsifier, antistat
*Form:* Liquid; Color: 5 (Gardner); Odor: Amine-type
*Properties:* Solubility: Organic solvents, insoluble water; Density: 7.6 lb/gal; Sp. gr.: 0.91; Surface Tension: 31.3 @ 0.1% sol'n.
*Composition:* 99% active ingredient; 2 moles ethylene oxide on primary soya amine
*Reaction:* Acid
*Warnings and Caution:* Corrosive

**Accomeen S5.** Armstrong Chem.
*ChemicalDescription:* Ethoxylated soya amine
*Category:* Emulsifier, nonionic surfactant

*Applications:* Emulsifier, antistat
*Form:* Liquid; Color: 10 (Gardner); Odor: Amine-type
*Properties:* Solubility: Organic solvents; disp. in water; Density: 7.9 lb/gal; Sp. gr.: 0.95
*Composition:* 99% active ingredient; 5 moles ethylene oxide on primary soya amine
*Reaction:* Acid
*Warnings and Caution:* Corrosive

**Accomeen S10.** Armstrong Chem.
*Chemical Description:* Ethoxylated soya amine
*Category:* Emulsifier, antistat, nonionic surfactant
*Form:* Liquid; Color: 10 (Gardner); Odor: Amine
*Properties:* Solubility: Organic solvents and water; Density: 8.5 lb/gal; Sp. gr.: 1.02; Surface Tension: 43 @ 0.1%
*Composition:* 99% active ingredient; 10 moles ethylene oxide on primary soya amine
*Reaction:* Acid
*Warnings and Caution:* Corrosive

**Accomeen T2.** Armstrong Chem.
*Chemical Description:* Ethoxylated tallow amine
*Category:* Emulsifier, antistat, nonionic surfactant
*Form:* Paste; Color: 3 (Gardner); Odor: Amine type
*Properties:* Organic solvents, insoluble in water; Density: 7.7 lb/gal; Sp. gr.: 0.92
*Composition:* 99% active ingredient; 2 moles ethylene oxide on primary tallow amine
*Reaction:* Acid
*Warnings and Caution:* Corrosive

**Accomeen T15.** Armstrong Chem.
*Chemical Description:* Ethoxylated tallow amine
*Category:* Emulsifier, nonionic surfactant
*Applications:* Emulsifier dispersant
*Form:* Liquid; Color: 5 (Gardner); Odor: Amine type
*Properties:* Refractive Index: 1; Solubility: Organic solvents, insoluble in water; Density: 8.1 lb/gal; Sp. gr.: 0.97
*Composition:* 99% active ingredient; 5 moles ethylene oxide on primary tallow amine
*Reaction:* Acid
*Warnings and Caution:* Corrosive

**Accomid C.** Armstrong Chem.
*Chemical Description:* Coconut oil diethanolamide
*Category:* Detergent, stabilizer, nonionic surfactant
*Applications:* Shampoo, bubble baths, diswashing detergent
*Form:* Liquid; Color: 4 (Gardner); Odor: Mild
*Properties:* Solubility: Completely soluble in water at low concentration; Density: 8.3 lb/gal; Sp. gr.: 0.99
*Composition:* 98% active ingredient

**Acconon 300 MO.** Armstrong Chem.
*Chemical Description:* Polyethylene glycol 300 monooleate
*Category:* Emulsifier, nonionic surfactant
*Applications:* Surface activity; cosmetics, food, agriculture, plastics

*Form:* Liquid; Color: 6 (Gardner)
*Properties:* Solubility: Organic solvents; Density: 8.3 lb/gal; Sp. gr.: 0.99; M.P.: <–5 C
*Composition:* 99% active ingredient; 6.8 moles ethylene oxide added to oleic acid
*Containers:* Drums

**Acconon 400 MO.** Armstrong Chem.
*Chemical Description:* Polyethylene glycol 400 monooleate
*Category:* Emulsifier, nonionic surfactant
*Applications:* Surface activity, lubricant, chemical intermediary; cosmetics, food, agriculture, plastics
*Form:* Liquid; Color: 6 (Gardner)
*Properties:* Solubility: Organic solvents; Density: 8.4 lb/gal; Sp. gr.: 1.01; M.P.: < 10 C; Biodegradable
*Composition:* 99% active ingredient; 9.1 moles ethylene oxide added to oleic acid

**Acconon 1300.** Armstrong Chem.
*Chemical Description:* Polyethoxylated-polypropoxylated $C_{12-20}$ alcohols
*Form:* Liquid; Color: Light yellow, 3 max (Gardner)
*Properties:* Density: 8.35–8.45 lb/gal; Sp. gr.: 1.016–1.019; Cloud Pt.: °F, 135–145

**Acconon TGH.** Armstrong Chem.
*Chemical Description:* Complex tallow glyceride ethoxylate
*Category:* Wetting agent, foaming agent, nonionic surfactant
*Applications:* personal care and household products
*Form:* Liquid; Color: Yellow; Odor: Mild

**Accoquat 2C-75.** Armstrong Chem.
*Chemical Description:* Dicoco dimethyl ammonium chloride
*Category:* Emulsifier, coupling agent, cationic
*Applications:* Car spray waxes, dust control oil, removal of spots
*Form:* Liquid; Color: Amber; Odor: Mild
*Properties:* Solubility: Solvent is isopropanol; Density: 7.2 lb/gal; Sp. gr.: 0.86
*Composition:* 75% active ingred.
*Warnings and Caution:* Flammable liquid

**Accoquat 2C-75–H.** Armstrong Chem. Co.
*Chemical Description:* Dicoco dimethyl ammonium chloride–high flash
*Category:* Emulsifier, cationic
*Applications:* Car spray waxes, dust control oils, removal of spots
*Form:* Liquid; Color: Amber; Odor: Mild
*Properties:* Solubility: Solvent is hexylene glycol; Flash Pt.: High
*Composition:* 75% active ingred.

**Accosoft 550.** Armstrong Chem.
*Chemical Description.* Tallow amido amine quaternary compound
*Category:* Cationic surfactant
*Applications:* Household and industrial fabric softener, dispersion for household fabric softener, fabric antistat agent
*Form:* Liquid, gel; Color: 6 (Gardner)

*Properties:* Solubility: Solvent—Isopropanol; Density: 8.0; Sp. gr.: 0.96; Biodegradable
*Composition:* 75 or 90% active ingredient
*Toxicity:* Nontoxic
*Warnings and Caution:* Flammable, supports bacterial growth

**Accosoft 620.** Armstrong Chem.
*Chemical Description:* Tallow amido amine quaternary compound
*Category:* Cationic surfactant
*Applications:* Household or industrial fabric softener, fabric antistat agent
*Form:* Liquid; Color: 6 (Gardner)
*Properties:* Solubility: Solvent—IPA; Density: 8.0; Sp. gr.: 0.96; Biodegradable
*Composition:* 75 or 90% active ingredient
*Toxicity:* Nontoxic
*Warnings and Caution:* Flammable, supports bacterial growth

**Accosoft 707.** Armstrong Chem.
*Chemical Description.* Dimethyl dihydrogenated tallow ammonium chloride
*Category:* Cationic surfactant
*Applications:* Household or industrial fabric softener, fabric antistat agent
*Form:* Paste; Color: 3 (Gardner)
*Properties:* Solubility: Solvent—IPA, water; Density: 7.3; Sp. gr.: 8.7
*Composition:* 75% active ingredient
*Toxicity:* Nontoxic
*Warnings and Caution:* Flammable, supports bacterial growth

**Accosoft 748.** Armstrong Chem.
*Chemical Description.* Dimethyl dihydrogenated tallow ammonium metho sulfate
*Category:* Cationic surfactant
*Applications:* Household or industrial fabric softener or antistat particularly suited to drier use
*Form:* Paste, solid; Color: 4 (Gardner)
*Properties:* Solubility: Solvent—IPA, water, M.P.: 125–135 C; Biodegradable
*Composition:* 90% active ingredient
*Toxicity:* Nontoxic
*Containers:* Tankwagons, RR tank cars and drums
*Warnings and Caution:* Flammable, supports bacterial growth

**Accosoft A-155.** Armstrong Chem.
*Chemical Description.* Tallow amido amine quaternary compound
*Category:* Cationic surfactant
*Applications:* Household fabric softener base; dispersions for fabric softener; fabric antistat agent
*Form:* Gel; Color: 7 (Gardner)
*Properties:* Biodegradable
*Composition:* 75 or 90% active ingredient
*Toxicity:* Nontoxic
*Containers:* 4000–6000 gal stainless steel tankwagons
*Category:* Emulsifier, solubilizer
*Form:* Liquid; Color: Yellow
*Properties:* Sp. gr.: 1.1; Visc.: 360–400 cps; HLB: 16.7; Sapon. Value: 40–52; Acid No.: 2.0 max.; Hydroxyl No.: 97–113
*Composition:* 98% active ingred.
*Containers:* 55 gal lined bung

drums

**Accosperse 20.** Armstrong Chem.
*Chemical Description:* POE-20 sorbitan monolaurate

**Accosperse 60.** Armstrong Chem.
*Chemical Description:* POE-20 sorbitan monostearate
*Category:* Emulsifier
*Form:* Liquid, gel; Color: Amber
*Properties:* Sp. gr.: 1.1; Visc.: 550–600 cps; HLB: 14.9; Sapon. Value: 45–55; Acid No.: 2.0 max.; Hydroxyl No.: 80–96
*Composition:* 98% active ingred.

**Accosperse 80.** Armstrong Chem.
*Chemical Description:* POE-20 sorbitan monooleate
*Category:* Emulsifier
*Form:* Liquid; Color: Amber
*Properties:* Sp. gr.: 1.1; Visc.: 400–450; HLB: 15.0; Sapon. Value: 45–55; Acid No.: 2.0 max.; Hydroxyl No.: 65–81
*Composition:* 98% active ingred.
*Containers:* 55 gal lined bung drums

**Accosperse TGH.** Armstrong Chem.
*Chemical Description:* Alkoxylated mono-, di-glyceride
*Category:* Emulsifier, foaming and wetting agent, nonionic
*Applications:* Personal care and cosmetic products, conditioners, cleansing creams; industrial emulsions
*Form:* Liquid; Color: Yellow; Odor: Mild
*Properties:* F.P.: 30 F; Cloud Pt.: 140–146 F (1% in dist. water); Biodegradable
*Composition:* 98–100% active
*Reaction:* Neutral

**Acetoquat CPB.** Aceto Chem.
*Chemical Description:* Cetyl pyridinium bromide
*Category:* Germicide, cationic
*Applications:* Germicide, sanitizing agent
*Form:* Powder
*Composition:* 95% active ingredient

**Acetoquat CPC.** Aceto Chem.
*Chemical Description:* Cetyl pyridinium chloride
*Category:* Germicide, cationic
*Applications:* Germicides, sanitizing agent
*Form:* Powder
*Composition:* 100% active ingredient

**Acetoquat CTAB.** Aceto Chem.
*Chemical Description:* Cetyl trimethyl ammonium bromide
*Category:* Germicide, cationic
*Applications:* Germicide, sanitiz-agent
*Form:* Powder
*Composition:* 95% active ingredient

**Acidan.** Grindsted Products
*Chemical Description:* Citric acid esters of monoglycerides
*Category:* Emulsifier, anionic surfactant
*Applications:* Food industry in coffee whiteners, frying margarine, meat emulsions to improve dispersion, whitening and emulsification

*Form:* Powder, flakes
*Composition:* 100% active ingredient

**ACL 56.** Monsanto
*Chemical Description:* Sodium dichloroisocyanurate dihydrate
*Category:* Bleaching compound, sanitizer
*Applications:* Detergents, bleaches, pool disinfection
*Composition:* 56% available chlorine

**ACL 59.** Monsanto
*Chemical Description:* Potassium dichloroisocyanurate
*Category:* Bleaching compound, sanitizer
*Applications:* Pool disinfectant and sanitizer; oxidizer in dishwashing compounds
*Composition:* 59% available chlorine

**ACL 60.** Monsanto
*Chemical Description:* Sodium dichloroisocyanurate
*Category:* Bleaching compound, sanitizer
*Applications:* Detergent, bleach, pool disinfectant
*Composition:* 63% available chlorine

**ACL 66.** Monsanto
*Chemical Description:* (monotrichloro) tetra-(monopotassium dichloro)-penta–isocyanurate
*Category:* Bleaching compound, sanitizer
*Applications:* sanitizers, detergent
*Composition:* 66.5% available chlorine

**ACL 85.** Monsanto
*Chemical Description:* trichloroisocyanuric acid
*Category:* Bleaching compound, sanitizer
*Applications:* Pool disinfectant in compacted forms; detergent
*Composition:* 90% available chlorine

**Aconol X10.** Hart Chem.
*Chemical Description:* Tall oil ethoxylate
*Category:* Emulsifier, nonionic surfactant
*Applications:* Low-foaming nonionic for Built detergent systems
*Form:* Liquid; Color: Clear
*Properties:* Refractive Index: 1.4655; Solubility: Clear soluble; Density: 1.039; Visc.: 140 cps
*Composition:* 95% active ingredient
*Containers:* Tankcars, drums
*Warnings and Caution:* Avoid contact with skin and eyes

**Acrawax.** Glyco Chem.
*Applications:* Synthetic waxes and plastic lubricants
*Form:* Hard brittle solid; Color: Tan
*Properties:* Sp. gr.: 1.04 @ 20 C; M.P.: 90–96 C; Flash Pt.: ≈ 205 C

**Acrawax B.** Glyco Chem.
*Form:* Hard solid; Color: Light brown
*Properties:* Sp. gr.: 0.97 @ 20 C;

M.P.: 82–90 C; Flash Pt.: ≈ 200 C

**Acrawax C.** Glyco Chem.
*Form:* Hard solid; Color: Light tan
*Properties:* Sp. gr.: 0.97 @ 20 C; M.P.: 142–145 C; Flash Pt.: ≈ 285 C

**Acrawax HM-S.** Glyco Chem.
*Form:* Coarse powder; Color: Tan
*Properties:* Sp. gr.: 0.97 @ 20 C; M.P.: 243–246 C; Flash Pt.: ≈ 298 C

**Actiflo 68, 70.** Central Soya Co.
*Chemical Description:* Natural lecithin
*Category:* Emulsifier, wetting agent, amphoteric
*Applications:* Emulsifier, dispersant, wetting agent; food, cosmetics, textiles, plastics and rubber
*Form:* Liquid
*Composition:* 66–72% active ingredient

**Acto 450.** Exxon Corp.
*Chemical Description:* Alkylaryl sodium sulfonate
*Category:* Detergent, wetting agent, emulsifier, anionic
*Applications:* Detergents for textiles, skins, leather, dyeing, mines and minerals, paints; Wetting agents for dyeing, rewetting dried skins, tanning, insecticides, herbicides, fungicides; Emulsifiers in fat liquoring, cutting and soluble oils, dispersing waxes, pigments, resins; printing, preparation of emulsified paints; making of bitumen emulsions; rust preventative
*Form:* Liquid; Color: Reddish-amber; Odor: Alcohol
*Properties:* m.w.: 470; Solubility: oil-soluble, slight sol. in water but readily dispersible; Density: 8.24 lb/gal; Sp. gr.: 0.99; Stability: Good < 180 F
*Composition:* 44% active ingredient
*Toxicity:* Causes eye burn on contact; may cause mild skin irritation
*Warnings and Cautions:* Flammable
*References:* Data sheet DG-4F

**Acto 500.** Exxon Corp.
*Chemical Description:* Alkylaryl sodium sulfonate
*Category:* See Acto 450
*Applications:* See Acto 450
*Form:* Viscous liquid; Color: Amber; Odor: Slight bland petroleum oil
*Properties:* m.w.: 460; Solubility: oil-soluble; slight sol. in water but readily dispersible; Density: 8.33 lb/gal; Sp. gr.: 1.00; Stability: Good < 180 F
*Composition:* 50% active ingredient
*Toxicity:* See Acto 450
*References:* Data sheet DG-4F

**Acto 630.** Exxon Corp.
*Chemical Description:* Alkylaryl sodium sulfonate
*Category:* See Acto 450
*Applications:* See Acto 450
*Form:* Viscous liquid; Color: Amber; Odor: Slight bland petroleum oil odor
*Properties:* m.w.: 470; Solubility:

oil-soluble, slight sol. in water but readily dispersible; Density: 8.41 lb/gal; Sp. gr.: 1.01; Stability: Good < 180 F
*Composition:* 63% active ingredient
*Toxicity:* See Acto 450
*References:* Data sheet DG-4F

**Acto 632.** Exxon Corp.
*Chemical Description:* Alkylaryl sodium sulfonate
*Category:* See Acto 450
*Applications:* See Acto 450
*Form:* See Acto 450
*Properties:* m.w.: 430; Solubility: oil-soluble, slight sol. in water but readily dispersible; Density: 8.41 lb/gal; Sp. gr.: 1.01; Stability: Good < 180 F
*Composition:* 63% active ingredient
*Toxicity:* See Acto 450
*References:* Data sheet DG-4F

**Acto 636.** Exxon Corp.
*Chemical Description:* Alkylaryl sodium sulfonate
*Category:* See Acto 450
*Applications:* See Acto 450
*Form:* Viscous liquid; Color: Orange; Odor: Slight bland petroleum oil odor
*Properties:* m.w.: 470; Solubility: oil-soluble, slight sol. in water but readily dispersible; Density: 8.5 lb/gal; Sp. gr.: 1.02; Stability: Good < 180 F
*Composition:* 63% active ingredient
*Toxicity:* See Acto 450
*References:* Data sheet DG-4F

**Acto 639.** Exxon Corp.
*Chemical Description:* Alkylaryl sodium sulfonate
*Category:* See Acto 450
*Applications:* See Acto 450
*Form:* See Acto 500
*Properties:* m.w.: 520; Solubility: oil-soluble, slight sol. in water but readily dispersible; Density: 8.5 lb/gal; Sp. gr.: 1.02; Stability: Good < 180 F
*Composition:* 63% active ingredient
*Toxicity:* See Acto 450
*References:* Data sheet DG-4F

**Acylan.** Croda Chem.
*Chemical Description:* Acetylated lanolin
*Category:* Emollient, conditioner
*Applications:* Cosmetics, especially hypoallergenic and baby cosmetics
*Form:* Paste; Color: Yellow
*Composition:* 100% active ingred.
*Reaction:* Nonionic

**Acylglutamate CS-11.** Ajinomoto
*Chemical Description:* Monosodium N-cocoyl-L-glutamate
*Category:* Anionic detergent
*Applications:* Personal care; high detergency and foaming power; emollient for skin; bacteriostatic effect
*Form:* powder
*Properties:* Stability: Good in hard water; Biodegradable
*Composition:* 100% active ingredient
*Toxicity:* Nonskin irritation even for infants and eczema cases

**Acylglutamate CT-12.** Ajinomoto

*Chemical Description:* Monotriethanolamine N-cocoyl-L-glutamate
*Category:* See Acylg. CS-11.
*Applications:* See Acylg. CS-11.
*Form:* Liquid
*Properties:* See Acylg. CS-11.
*Compositon:* 20% active ingred.
*Toxicity:* See Acylg. CS-11.

**Acylglutamate GS-11.** Ajinomoto
*Chemical Description:* Monosodium N-mixed fatty acid acyl glutamate
*Category:* See Acylg. CS-11.
*Applications:* See Acylg. CS-11.
*Properties:* See Acylg. CS-11.
*Toxicity:* See Acylg. CS-11.

**Acylglutamate HS-11.** Ajinomoto
*Chemical Description:* Monosodium N-hydrogenated tallowyl-L-glutamate
*Category:* See Acylg. CS-11.
*Applications:* See Acylg. CS-11.
*Form:* Powder
*Properties:* See Acylg. CS-11.
*Composition:* 100% active ingred.
*Toxicity:* See Acylg. CS-11.

**Acylglutamate LS-11.** Ajinomoto
*Chemical Description:* Monosodium N-lauroyl-L-glutamate
*Category:* See Acylg. CS-11.
*Applications:* See Acylg. CS-11.
*Form:* Powder
*Properties:* See Acylg. CS-11.
*Composition:* 100% active ingred.
*Toxicity:* See Acylg. CS-11.

**Adinol T.** Croda Chem.
*Chemical Description:* Sodium N methyl N oleoyl taurate
*Category:* Detergent, wetting agent, emulsifier, dispersant, foaming agent
*Applications:* Shampoo, foam bath and skin cleansing compositions
*Form:* Liquid, paste, gel, powder;
*Properties:* Solubility: Good in hard water

**Admex 433.** Sherex
*Chemical Description:* Low-viscosity polyester
*Category:* Plasticizer
*Applications:* Foam, extrusion, calendered films, coated fabric; solvent resistant, compatible in high humidity

**Admex 515.** Sherex
*Chemical Description:* Low-viscosity polyester
*Category:* Plasticizer
*Applications:* Calendered and coated fabric, vinyl foam, plastisols, organosols, pigment grinding, wire coating

**Admex 522.** Sherex
*Chemical Description:* Low-viscosity polyester
*Category:* Plasticizer
*Applications:* Extrusion, calendered film and sheet, wire coating, organosols, plastisols, flooring
*Properties:* Stability: Stable to light and heat

**Admex 523.** Sherex
*Chemical Description:* Medium-viscosity polyester
*Category:* Plasticizer
*Applications:* Transfer hose and tubing, roof flashing and edging,

vinyl foams and gaskets requiring gas and oil resistance

**Admex 525.** Sherex
*Chemical Description:* Primary polyester
*Category:* Plasticizer
*Applications:* Coating fabrics, paper coatings, pigment dispersions, plastisols, organosols, tapes, vinyl foams, extruded profiles
*Properties:* Visc.: Low; Good low temperature flexibility and pigment wetting

**Admex 529.** Sherex
*Chemical Description:* General purpose polyester
*Category:* Plasticizer
*Applications:* Calendered and coated fabric, wire coatings, flooring

**Admex 710.** Sherex
*Chemical Description:* Epoxidized soybean oil
*Category:* Plasticizer
*Applications:* Plasticizer/stabilizer for PVC, food wrappings, beverage transfer tubing, floor tile, baby and nursery goods, nitro-cellulose and butyrate coatings; reactive intermediate

**Admex 711.** Sherex
*Chemical Description:* Epoxidized soybean oil
*Category and Applications:* See Admex 710

**Admex 746.** Sherex
*Chemical Description:* Epoxidized tallate ester
*Category:* Plasticizer, stabilizer
*Applications:* Calendered film and sheeting; PVC compounds; plastisol viscosity depressant

**Admex 752.** Sherex
*Chemical Description:* High molecular weight ester
*Category:* Plasticizer
*Applications:* Upholstery sheeting, coated fabric, headliner stock, side panel trim, armrest fabric, crash-pad molding
*Properties:* Excellent fog resistance, low temperature flexibility, low viscosity

**Admex 760.** Sherex
*Chemical Description:* High molecular weight polyester
*Category:* Plasticizer
*Applications:* Pressure-sensitive tape, electrical sleeving, organosols, vinyl solution coatings, drycleanable film

**Admex 761.** Sherex
*Chemical Description:* Medium molecular weight polyester
*Category:* Plasticizer
*Applications:* 105 C wire coating, electrical tape and sleeving, tape and mattress covers

**Admex 770.** Sherex
*Chemical Description:* Medium molecular weight polyester
*Category:* Plasticizer
*Applications:* Refrigerator gaskets, baby pants, nursery goods, high-clarity film and sheeting

**Admex 775.** Sherex
*Chemical Description:* Medium molecular weight polyester

*Category:* Plasticizer
*Applications:* Refrigerator and freezer gaskets, coated fabrics, saturated paper coatings, infant wear, tapes

**Admex 890.** Sherex
*Chemical Description:* Medium molecular weight polyester
*Category:* Plasticizer
*Applications:* Refrigerator and freezer gaskets, coated fabric, appliance wire coating, automotive gasketing, tape applications

**Admex ELO.** Sherex
*Chemical Description:* Epoxidized linseed oil
*Category:* Stabilizer, plasticizer
*Applications:* Food wrap, beverage transfer tubing, PVC, film

**Adogen 415.** Sherex
*Chemical Description:* Trimethyl soya quaternary ammonium chloride
*Category:* Antistat, emulsifier, dispersant, cationic
*Form:* Liquid
*Composition:* 50% active ingred.

**Adogen 432.** Sherex
*Chemical Description:* Dialkyl dimethyl ammonium chloride
*Category:* Fabric softener, cationic
*Applications:* Retail, commercial laundry and textile mill formulations and paper
*Form:* Liquid
*Composition:* 75% solids content

**Adogen 432CG.** Sherex
*Chemical Description:* Dialkyl dimethyl ammonium chloride
*Category:* Creme rinse concentrate, cationic
*Applications:* Hair creme rinse
*Form:* Liquid
*Composition:* 75% solids content

**Adogen 442.** Sherex
*Chemical Description:* Dimethyl dihydrogenated tallow ammonium chloride
*Category:* Fabric softener, cationic
*Applications:* Home and commercial launderies, textile processing
*Form:* Paste
*Composition:* 75% solids content

**Adogen 462.** Sherex
*Chemical Description:* Dimethyl dicoco ammonium chloride
*Category:* Antistat, emulsifier, flocculating agent, cationic
*Applications:* Dispersant for oilfield chemicals
*Form:* Liquid
*Composition:* 75% quat. content

**Adogen 464.** Sherex
*Chemical Description:* Methyl tri ($C_8$–$C_{10}$) ammonium chloride
*Category:* Flotation agent, cationic
*Applications:* Solvent extraction of metal
*Form:* Liquid
*Composition:* 90% quat. content

**Adogen 471.** Sherex
*Chemical Description:* Trimethyl tallow ammonium chloride
*Category:* Emulsifier, dispersant, antistat, cationic
*Form:* Liquid
*Composition:* 50% quat. content

**Adogen 477.** Sherex
*Chemical Description:* N-tallow pentamethyl propane diammonium dichloride
*Category:* Emulsifier, dispersant, cationic
*Form:* Liquid
*Composition:* 50% quat. content

**Adsee 775.** Witco Chem. Corp., Organics Div.
*Chemical Description:* Blend of polyoxyethylene ethers and special resins
*Category:* Agricultural surfactant, nonionic
*Applications:* Spreader sticker and penetrant; adjuvant for spray-tank mixtures
*Form:* Liquid
*Composition:* 100% conc.

**Adsee 799.** Witco Chem. Corp., Organics Div.
*Chemical Description:* Alkylpolyoxyethylene ether
*Category:* Agricultural surfactant, nonionic
*Applications:* Soil penetrant
*Form:* Liquid
*Composition:* 100% conc.

**Adsee 801.** Witco Chem. Corp., Organics Div.
*Chemical Description:* Surfactant blend
*Category and Applications:* See Adsee 799

**Adsee 2096.** Witco Chem. Corp., Organics Div.
*Chemical Description:* Surfactant blend
*Category:* Agricultural surfactant, nonionic
*Applications:* Spreader penetrant for wetting of pesticide formulations; adjuvant for spray-tank mixtures

**Adsee 2141.** Witco Chem. Corp., Organics Div.
*Chemical Description:* Surfactant blend
*Category:* Agricultural surfactant, anionic/nonionic
*Applications:* Premix surfactant for high- and low-content monosodium methylarsonate (MSMA) formulations
*Form:* Liquid
*Composition:* 100% conc.

**Adsee AC62-38.** Witco Chem. Corp., Organics Div.
*Chemical Description:* Surfactant blend
*Category:* Agricultural surfactant, anionic/nonionic
*Applications:* Premix surfactant for high- and low-content monosodium methylarsonate (MSMA) formulations

**AE-1214/3.** Procter & Gamble Co.
*Chemical Description:* Ethoxylated linear alcohols, nonionic
*Category:* Detergent, emulsifier
*Form:* Liquid
*Composition:* 100% active ingred.

**AE-1214/6.** Procter & Gamble Co.
*Chemical Description:* Ethoxylated linear alcohols, nonionic
*Category:* Lubricant, wetting agent
*Applications:* Textiles

*Form:* Solid
*Composition:* 100% active ingred.

**AEPD.** IMC Chem.–NP Div.
*Chemical Description:* 2-amino-2-ethyl-1,3-propanediol
*Category:* Emulsifier, catalyst, anionic
*Applications:* As raw materials in polymer applications, e.g. oxazoline structures (as corrosion inhibitors, antifoam compounds, synthetic waxes); in coatings as pigment dispersants; in emulsions for cosmetics, cleaners, industrial insecticides, asphalt emulsions; in boiler-water treatment; as catalysts; in formaldehyde scavenging
*Form:* Color: 2 (Gardner)
*Properties:* m.w.: 119.17; Density: 9.15 lb/gal; Sp. gr.: 1.101 @ 40/40 C; M.P.: 38 C; B.P.: 152 C
*Toxicity:* Causes severe eye burns
*Containers:* Bulk or 450 lb drums
*References:* TDS No. 10

**Aerosol 18.** American Cyanamid
*Chemical Description:* Disodium N-octadecyl sulfosuccinamate
*Category:* Emulsifier, dispersant, foaming agent; anionic surfactant
*Applications:* Cosmetics, foaming latexes; emulsion polymerization, lubrication; scouring powders and soaps; industrial cleaners; agricultural foams
*Form:* Creamy paste; Color: 10 (Gardner)
*Properties:* Solubility: Good in water; Density: ≈ 8.9 lb/gal; Sp. gr.: ≈ 1.07; Visc.: Variable, semi-liquid; F.P.: Solid paste < 21 C; Stability: Good in acid and alkaline media; Surface Tension: 41 dynes/cm; Biodegradable
*Composition:* 35% active ingred.
*References:* Circular PRC-32

**Aerosol 19.** American Cyanamid
*Chemical Description:* Disodium N-alkylsulfosuccinamate
*Category:* Emulsifier, foaming and wetting agent
*Applications:* Foaming agent in foaming latexes, wallboard, cement, plastics, carpet & upholstery shampoos, hand soaps; Emulsifier in emulsion polymerization
*Form:* Clear liquid; Color: 10 (Gardner)
*Properties:* Solubility: Good in water; Density: ≈ 8.9 lb/gal; Sp. gr.: ≈ 1.07; Stability: Good in acid and alkaline media; Surface Tension: 41 dynes/cm; Biodegradable
*Composition:* 35% active ingred.
*References:* Circular PRC-145

**Aerosol 22.** American Cyanamid
*Chemical Description:* Tetrasodium N-(1,2-dicarboxyethyl)-N-octadecyl sulfosuccinamate
*Category:* Emulsifier, dispersant, anionic surfactant
*Applications:* Agricultural emulsions; emulsion polymerization; industrial, household and cosmetic cleaners; alkaline metal cleaners; cosmetics; anti-gelling agent; latex stabilizer; textile nonwoven applications
*Form:* Clear to cloudy liquid; Color: 8 (Gardner)
*Properties:* Solubility: Excellent in water; Density: 9.4 lb/gal; Sp. gr.: 1.12; Visc.: at 25 C, 53 cps; F.P.: Separates below 10 C; M.P.: > 200 C; Surface Tension: 41 dynes/

cm; Biodegradable
*Composition:* 35% active ingred.
*References:* Circular PRC-90

**Aerosol 30.** American Cyanamid
*Chemical Description:* Cocoamidopropyl betaine
*Category:* Amphoteric surfactant; foam booster; viscosity builder
*Applications:* hair and bath products, skin cleansers, disinfectants
*Form:* Clear, yellow liquid; Color: 4 max (Gardner)
*Properties:* Solubility: Excellent in water; Sp. gr.: 1.04; F.P.: 0 C; Surface Tension: 35 dynes/cm; Partially biodegradable
*Composition:* 35% active ingred.
*Toxicity:* Very mild on skin, eyes
*References:* Circular PRT-160
*Patent:* 3,225,074 (U.S.)

**Aerosol 200.** American Cyanamid
*Chemical Description:* Disodium (ethoxylated alkylamide) sulfosuccinate
*Category:* Emulsifier, foaming agent, dispersant
*Applications:* Cosmetics: shampoos, shaving creams, liquid hand soaps, bubble baths; carpet and upholstery cleaners; vinyl acetate and acrylic emulsions; foaming applications
*Form:* Clear, yellow liquid; Color: 7 max (Gardner); Odor: Mild
*Properties:* Solubility: Excellent in water; Density: 8.93 lb/gal; Sp. gr.: 1.07; Visc.: 200 cps max (Brookfield, #1 spindle, 60 rpm); Cloud Pt.: $<0$ C; Surface Tension: 35 dynes/cm; Biodegradable
*Composition:* 30% active ingred.
*Toxicity:* Very mild
*References:* Circular PRT-153

**Aerosol 413.** American Cyanamid
*Chemical Description:* Disodium ethanol alkyl ($C_{18}$) amide sulfosuccinate
*Category:* Emulsifier, foaming agent, dispersant, wetting agent
*Applications:* Vinyl acetate and acrylate emulsions; shampoos; foaming applications
*Form:* Clear, yellow liquid; Color: 5–7 (Gardner)
*Properties:* Solubility: Excellent in water; Density: 8.95 lb/gal; Visc.: at 25 C, 70–90 cps; Stability: Good; Surface Tension: 38 dynes/cm; Biodegradable
*Composition:* 35% active ingred.
*References:* Circular PRT-154

**Aerosol 501.** American Cyanamid
*Chemical Description:* Disodium alkyl sulfosuccinate
*Category:* Emulsifier, wetting agent, dispersant, foaming agent, anionic-nonionic surfactant
*Applications:* Acrylic and vinyl acetate emulsions; self-crosslinking latexes; textile wetting and foaming applications
*Form:* Clear liquid; Color: 60 max (APHA)
*Properties:* Solubility: Excellent in water; Density: 9.66 lb/gal; Sp. gr.: 1.16; Visc.: 25 C, $\approx$ 260 cps; F.P.: $-3$ C; Stability: Good; Surface Tension: 28 dynes/cm
*Composition:* 50% active ingred.
*References:* Circular PRC-37
*Patent:* 3,947,400 (U.S.)

**Aerosol A-102.** American Cyanamid
*Chemical Description:* Disodium ethoxylated alochol half ester of sulfosuccinic acid
*Category:* Emulsifier, solubilizer, foaming agent, dispersant, anionic-nonionic surfactant
*Applications:* Emulsion polymerization of polyvinyl acetate/acrylics; textile industry—emulsions, wetting agent, pad-bath additive; solubilizing agent
*Form:* Clear liquid; Color: Colorless-to-light-yellow
*Properties:* Solubility: Excellent in water; Density: 9.01 lb/gal; Sp. gr.: 1.08; Visc.: @ 25 C, ≈ 40 cps; F.P.: −4 C; Stability: Excellent in acid, fair in base; Surface Tension: 29 dynes/cm
*Composition:* 30% active ingred.
*References:* Circular PRC-66

**Aerosol A-103.** American Cyanamid
*Chemical Description:* Disodium salt of half ester of sulfosuccinate acid derived from an ethoxylated nonyl phenol
*Category:* Emulsifier, wetting agent, anionic-nonionic surfactant
*Applications:* Polyvinyl acetate acrylic emulsions; Textiles—emulsions, pad-bath additive, wetting agent; solubilizer, surface tension depressant
*Form:* Liquid; Color: Colorless to light yellow
*Properties:* Solubility: Excellent in water; Density: ≈ 9.1 lb/gal; Sp. gr.: 1.09; Visc.: 25 C, 170–190 cps; F.P.: −9 C; Surface Tension: 34 dynes/cm
*Composition:* 35% active ingred.
*References:* Circular PRC-88

**Aerosol A-196.** American Cyanamid
*Chemical Description:* Sodium dicyclohexyl sulfosuccinate
*Category:* Emulsifier, dispersant, anionic surfactant
*Applications:* Sole emulsifier for modified styrene butadiene; post additive to stabilize latex and promote adhesion; potential low foaming surfactant
*Form:* White pellets, clear to cloudy liquid
*Properties:* Stability: Good mechanical stability; Surface Tension: 39 dynes/cm, high critical micelle concentration; Biodegradable
*Composition:* 85% active pellets, 15% water; 40% active solution in water (110 F)
*References:* Circular PRC-49

**Aerosol A-268.** American Cyanamid
*Chemical Description:* Disodium isodecyl sulfosuccinate
*Category:* Emulsifier, anionic surfactant
*Applications:* Sole emulsifier for PVC latexes—vinyl, vinylidene chloride, acrylics; surface tension depressant; solubilizer
*Form:* Clear liquid
*Properties:* Solubility: Excellent in water; Density: 9.96 lb/gal; Sp. gr.: 1.19; Visc.: 25 C, ≈ 150 cps; F.P.: −5 C; Stability: Good; Surface Tension: 28 dynes/cm
*Composition:* 50% active ingred.
*References:* Circular PRC-50

**Aerosol AY-65.** American Cyanamid
*Chemical Description:* Sodium diamyl sulfosuccinate
*Category:* Wetting agent, dispersant, anionic surfactant
*Applications:* Wetting agent esp. effective in conc. salt sol'ns.; agriculture; emulsion polymerization; electroplating; ore leaching; cleaning of porcelain, tile, brick, cement
*Form:* Clear liquid
*Properties:* Solubility: Solvent–water/ethanol; Density: 9.0 lb/gal; Sp. gr.: 1.081; F.P.: < 18 C; Surface Tension: 29 dynes/cm; Biodegradable
*Composition:* 65% active ingred.
*References:* Circular PRC-44

**Aerosol AY-100.** American Cyanamid
*Chemical Description:* See Aerosol AY-65
*Category:* See Aerosol AY-65
*Applications:* See Aerosol AY-65
*Form:* Hard waxy solid
*Properties:* Solubility: Sol. in water and organic solvents; Density: 10.0 lb/gal; Sp. gr.: 1.2; M.P.: 105–108 C; Stability: Good; Surface Tension: 29 dynes/cm; Biodegradable; good salt tolerance
*Composition:* 100% active ingred.
*References:* Circular PRC-44

**Aerosol C-61.** American Cyanamid
*Chemical Description:* Ethoxylated alkyl guanidine-amine complex
*Category:* Wetting agent, dispersant, settling agent, cationic surfactant
*Applications:* Textile softener, pigment dispersant, de-emulsifying agent, settling agent, antistatic agent for plastics, dye penetrant, foaming agent, acid cleaner formulations
*Form:* Creamy paste or semiliquid; Color: Light-tan
*Properties:* Density: ≈ 8 lb/gal; Sp. gr.: ≈ 1.0; F.P.: loses fluidity < 16 C; Stability: Good in acid and alkaline media; Surface Tension: 34 dynes/cm, Partially biodegradable
*Composition:* 70% active ingred.
*Toxicity:* Irritating to eyes, skin
*References:* Circular PRC-67

**Aerosol GPG.** American Cyanamid
*Chemical Description:* See Aerosol OT-70-PG
*Category:* General purpose grade OT surfactant; see Aerosol OT-70-PG
*Applications:* See Aerosol OT-70-PG
*Form:* Clear slightly viscous liquid
*Properties:* Solubility: Limited sol. in water, good organic solvents; Visc.: ≈ 200 cps; F.P.: –40 C; Surface Tension: 26 dynes/cm; Biodegradable
*Composition:* 70% active in water and alcohol
*References:* Circular PRT-73

**Aerosol IB-45.** American Cyanamid
*Chemical Description:* Sodium diisobutyl sulfosuccinate
*Category:* Emulsifier, wetting and dispersing agent, anionic surfactant
*Applications:* Emulsion polymerization of styrene butadiene; wet-

ting in conc. salt sol'ns.; dispersing dyes and pigments
*Form:* Clear liquid
*Properties:* Solubility: Sol. in water; F.P.: 20–21 C; Stability: Excellent in conc. salt sol'ns.; Surface Tension: 49 dynes/cm; Biodegradable; extremely hydrophilic
*Composition:* 45% active ingred.
*Toxicity:* May be irritating to eyes
*References:* Circular PRC-74

**Aerosol MA-80.** American Cyanamid
*Chemical Description:* Sodium dihexyl sulfosuccinate
*Category:* Wetting agent, emulsifier, dispersant, solubilizing agent, anionic surfactant
*Applications:* Emulsion polymerization; textile wetting applications; battery separators; electroplating; ore leaching; enhances germicidal activity
*Form:* Clear, slightly viscous liquid; Color: 50 max (APHA)
*Properties:* Solubility: Good in water, alcohol, and organic solvents; Density: ≈ 9.4 lb/gal; Sp. gr.: ≈ 1.13; F.P.: –28 C; M.P.: 199–292 C; Surface Tension: 28 dynes/cm; Slowly biodegradable
*Composition:* 80% active ingred.
*Toxicity:* Irritating to eyes, skin
*References:* Circular PRC-38

**Aerosol OS.** American Cyanamid
*Chemical Description:* Sodium isopropyl naphthalene sulfonate
*Category:* Emulsifier, wetting and dispersing agent, anionic surfactant
*Applications:* Pigment dispersant, wettable powders, agricultural herbicides, alkaline metal cleaners, industrial and consumer cleaners, etching solutions, emulsion polymerization
*Form:* White to light cream, free-flowing powder
*Properties:* Solubility: 50% in water; Density: 11.93 lb/gal; Sp. gr.: 1.43; M.P.: 205 C; Stability: excellent in concentrated acids and alkalis; Surface Tension: 37 dynes/cm; Slowly biodegradable
*Composition:* 75% active powder, 25% sodium sulfate
*Toxicity:* Irritating to eyes, skin; avoid dust inhalation
*References:* Circular PRC-122

**Aerosol OT-70 PG.** American Cyanamid
*Chemical Description:* Sodium dioctyl sulfosuccinate
*Category:* Wetting agent, surface tension depressant, emulsifier, anionic surfactant
*Applications:* Wetting agent in textiles, rubber, paper, petroleum, metal, paint, plastics and agriculture industries; antifog compositions; battery separators; coal dust control; cosmetics; dry cleaning detergents; emulsifying waxes; emulsion & suspension polymerization; industrial cleaners; paint production; pharmaceutical–topical; wettable powders
*Form:* Liquid; Color: Clear to slightly yellow
*Properties:* Solubility: limited water sol., good organic sol.; Solvent: Water, alcohol; Sp. gr.: 1.09; Visc.: 200–400 cps; Surface Tension: 26 dynes/cm; Biodegradable; Flash

Pt.: High
*Composition:* 70% active in water and propylene glycol
*References:* Circular PRT-73B

**Aerosol OT-75.** American Cyanamid
*Chemical Description:* See Aerosol OT-70-PG
*Category:* See Aerosol OT-70-PG
*Applications:* See Aerosol OT-70-PG
*Form:* Clear slightly viscous liquid
*Properties:* Solubility: Limited water sol., good organic solvents; Solvent–water, alcohol; Sp. gr.: 1.09; Visc.: ≈ 200 cps; F.P.: –40 C; Surface Tension: 26 dynes/cm; Biodegradable
*Composition:* 75% active ingred.
*References:* Circular PRT-73B

**Aerosol OT-100.** American Cyanamid
*Chemical Description:* See Aerosol OT-70-PG
*Category:* See Aerosol OT-70-PG
*Applications:* See Aerosol OT-70-PG
*Form:* Waxy solid
*Properties:* Solubility: Limited sol. in water, excellent for organic solvents; Sp. gr.: 1.1; M.P.: 153–157 C; Biodegradable
*Composition:* 100% active ingred.
*References:* Circular PRT-73B

**Aerosol OT-B.** American Cyanamid
*Chemical Description:* See Aerosol OT-70-PG
*Category:* See Aerosol OT-70-PG
*Applications:* See Aerosol OT-70-PG
*Form:* Granular solid; Color: White
*Properties:* Solubility: Limited sol. in water, insol. organic solvents; Solvent–Sodium benzoate; Sp. gr.: 1.1; M.P.: 300 C; Surface Tension: 26 dynes/cm; Biodegradable
*Composition:* 85% active in 15% sodium benzoate
*References:* Circular PRT 73-B

**Aerosol OT-S.** American Cyanamid
*Chemical Description:* See Aerosol OT-70–PG
*Category:* See Aerosol OT-70-PG
*Applications:* See Aerosol OT-70-PG
*Form:* Transparent liquid; Color: Light amber
*Properties:* Solubility: Limited sol. in water, good for organic solvents; Sp. gr.: 1.0; Visc.: 200–300 cps; Surface Tension: 26 dynes/cm; Biodegradable
*Composition:* 70% active in odorless mineral spirits
*References:* Circular PRT 73-B

**Aerosol TR-70.** American Cyanamid
*Chemical Description:* Sodium bistridecyl sulfosuccinate
*Category:* Emulsifier, Dispersant, anionic surfactant
*Applications:* Emulsion and suspension polymerization of vinyl chloride; emulsion polymerization of vinyl acetate; dispersing agent for resins, pigments, polymers and dyes in organic systems; pigment dispersing–printing inks; rust preventative
*Form:* Clear solution
*Properties:* Solubility: High sol. in organic media, limited water sol.; *Density:* ≈ 8.3; Sp. gr.: 0.995; Visc.: ≈ 110 cps; F.P. –40 C; Sta-

bility: Good; Surface Tension: 26 dynes/cm, low critical micelle concentration; Biodegradable
*Composition:* 70% active in water and alcohol
*References:* Circular PRC-66

**AF 60.** General Electric
*Chemical Description:* Dimethyl polysiloxane antifoam emulsion in water
*Category:* Defoaming agent
*Applications:* Industrial defoamer in aqueous systems; adhesive manufacture; ink manufacture; latex processing; soap manufacture; starch processing; paint (water based) manufacture
*Form:* Fluid; Color: White; Odor: Slight sorbic acid odor
*Properties:* Solubility: Good in water; Density: 8.4 lb/gal; Sp. gr.: 1.01; Visc.: 300 max. cps; Stability: Good
*Composition:* 30% active ingred.
*Toxicity:* May cause eye irritation
*Containers:* 5 gal cubitainers; 55 gal drums

**AF 66.** General Electric
*Chemical Description:* Filled polydimethy siloxane
*Category:* Antifoam agent
*Applications:* Antifoam agent in nonaqueous industrial applications; adhesive, ink and paint manufacture; resin polymerization; petroleum processing
*Form:* Fluid; Color: Grayish white Odor: None
*Properties:* Solubility: Sol. in aliphatic, aromatic, chlorinated solvents, negligible sol. in water; Density: 8.4 lb/gal; Sp. gr.: 1.01; Visc.: 500 max (Cstks.); Flash point: 600 F min. (open cup); Stability: Indefinite
*Composition:* 100% active ingred.
*Toxicity:* May cause eye irritation
*Containers:* 5 gal containers, 55 gal fiberpak drums

**AF 72.** General Electric
*Chemical Description:* Polydimethylsiloxane
*Category:* Antifoam agent, nonionic surfactant
*Applications:* Antifoam in food processing; adhesive, ink (water-based), paint (water-based), and soap manufacture; latex processing; manufacture or processing of fruit, instant coffee, instant foods, jams, jelly, juice, pickles, rice, soft drinks, sauces; sugar refining
*Form:* Fluid; Color: White; Odor: Slight sorbic acid odor
*Properties:* Solubility: Good in water; Density: 8.4 lb/gal; Sp. gr.: 1.01; Visc.: 1000 max. cps; Stability: Good
*Composition:* 30% active ingred.
*Toxicity:* May cause eye irritation
*Containers:* 5 gal cubitainers, 55 gal drums

**AF 75.** General Electric
*Chemical Description:* Polydimethylsiloxane
*Category:* Antifoam agent, nonionic surfactant
*Applications:* Food grade antifoam; chemical processing of adhesives, ink, latex, soap, starch; textiles and paper industry—paper

coating and manufacture, textile dyeing, finishing, sizing and wood pulping; processing or manufacture of fruit, instant coffee, jams, juice, pickles, rice, sauce, soft drinks, syrup, potato, sugar refining; leather finishing; metalworking
*Form:* Fluid; Color: White; Odor: Slight sorbic acid odor
*Properties:* Solubility: Good in water; Density: 8.4 lb/gal; Sp. gr.: 1.02; Visc.: 2500 cps; Stability: Good
*Composition:* 10% active ingred.
*Toxicity:* May cause eye irritation
*Containers:* 55 gal fiberpak drums

**Agrilan A.** Lankro Chem. Ltd.
*Chemical Description:* Blend oil soluble sulfonates and polyoxyethylene ethers
*Category:* Emulsifier, anionic/nonionic
*Applications:* Emulsifier for agricultural products, solvents, oils, etc.
*Form:* Viscous liquid; Color: Brown; Odor: Faint
*Properties:* Solubility: Turbid water-sol. @ 20 C; Sp. gr.: 1.062 @ 20 C; Visc.: 2400 cs @ 20 C; Flash Pt.: > 300 F (COC); Pour Pt.: < 0 C; pH: 4.5–6.5 (1% aq.)
*Composition:* 100% active ingred.
*Containers:* Store in SS, glass, resin or polythene-lined equipment
*Warnings and Caution:* Avoid prolonged contact with concentrated product; keep from strong oxidizing and reducing agents (potentially explosive)

**Agrilan AC.** Lankro Chem. Ltd.
*Chemical Description:* Blend oil soluble sulfonates and polyoxyethylene ethers
*Category:* Emulsifier, anionic/nonionic
*Applications:* Emulsifier for agricultural products (pesticides and herbicide systems), oils and aromatic solvents in degreaser formulations
*Form:* Viscous liquid; Color: Amber; Odor: Alcoholic
*Properties:* Solubility: Water-sol. @ 20 C; Sp. gr.: 1.052 @ 20 C; Visc.: 500 cs @ 20 C; Flash Pt.: > 120 F (Abel CC); Pour Pt.: 17 C; pH: 4.5–6.5 (1% aq.)
*Composition:* 90% active ingred.
*Toxicity:* Avoid prolonged contact with concentrated form
*Containers:* Store in SS, glass, resin, or polythene-lined equipment
*Warnings and Caution:* Contains flammable solvent; do not mix with strong oxidizing or reducing agents (potentially explosive)

**Agrilan BA.** Lankro Chem. Ltd.
*Chemical Description:* Blend oil soluble sulfonates and polyoxyethylene ethers
*Category:* Emulsifier, anionic/nonionic
*Applications:* Emulsifier for pesticides, herbicides, oils and aromatic solvents in degreaser formulations
*Form:* Viscous liquid; Color: Brown; Odor: Alcoholic
*Properties:* Solubility: Turbid wa-

ter-sol. @ 20 C; Sp. gr.: 1.034 @ 20 C; Visc.: 3100 cs @ 20 C; Flash Pt.: > 90 F (Abel CC); Pour Pt.: < 0 C; pH: 4.5–6.5 (1% aq.)
*Composition:* 80% active ingred.
*Toxicity:* Avoid prolonged contact with concentrated form
*Containers:* Store in SS, glass, resin, or polythene-lined equipment
*Warnings and Caution:* Contains flammable solvent; do not mix with strong oxidizing or reducing agents (potentially explosive)

**Agrilan BM.** Lankro Chem. Ltd.
*Chemical Description:* Blend of oil soluble sulfonates and polyoxyethylene ethers
*Category:* Emulsifier, anionic/nonionic
*Applications:* Emulsifier for pesticides, herbicides, oils and aromatic solvents in degreaser formulations
*Form:* Liquid; Color: Amber; Odor: Characteristic
*Properties:* Solubility: Turbid water-sol. @ 20 C; Sp. gr.: 1.022 @ 20 C; Visc.: 196 cs @ 20 C; Flash Pt.: 78 F (Abel CC); Pour Pt.: < 0 C; pH: 5.0–6.5 (1% aq.)
*Composition:* 70% active ingred.
*Toxicity:* Avoid prolonged contact with concentrated form
*Containers:* Store in SS, glass, resin, or polythene-lined equipment
*Warnings and Caution:* Contains flammable solvent; do not mix with strong oxidizing or reducing agents (potentially explosive)

**Agrimul 26-B.** Diamond Shamrock Process Chemicals Div.
*Chemical Description:* Nonionic surfactant
*Category:* Emulsifier, nonionic
*Applications:* Emulsifier for citrus and other agricultural spray oils
*Form:* Liquid; Color: Amber
*Properties:* Density: 8.3 lb/gal; Visc.: 95 cps; Pour Pt.: 15 C
*Composition:* 100% conc.

**Agrimul 70-A.** Diamond Shamrock Process Chemicals Div.
*Chemical Description:* Alkyl aryl polyether alcohol
*Category:* Emulsifier, nonionic
*Applications:* Agricultural emulsifier for chlorinated hydrocarbons
*Form:* Liquid; Color: Amber
*Properties:* Density: 8.8 lb/gal; Visc.: 8850 cps; Pour Pt.: 0 C
*Composition:* 100% conc.

**Agrimul A-300.** Diamond Shamrock Process Chemicals Div.
*Chemical Description:* Aromatic sulfonate oxide condensate blend
*Category:* Emulsifier, anionic/nonionic
*Applications:* Agricultural emulsifier for chlorinated and phosphated insecticide concentrates
*Form:* Liquid; Color: Amber
*Properties:* Density: 8.9 lb/gal; Visc.: 1050 cps; Pour Pt.: 5 C
*Composition:* 100% conc.

**Ahco A-117.** ICI United States
*Chemical Description:* Alkylaryl sulfonate

*Category:* Emulsifier, anionic surfactant
*Applications:* Emulsifier: dispersing waxes, pigments, resins; Detergent in textile, leather, laundering industries; Wetting agent for dyeing and finishing
*Form:* Liquid; Color: Amber; Odor: Faint
*Properties:* Solubility: Isopropanol @ 25 C; Sp. gr.: ≈ 1.0; Visc.: ≈ 7000 cps
*Composition:* 100% active ingred.
*Reaction:* Alkaline
*Toxicity:* Irritating to eyes, skin
*References:* TB 628

**Ahco AB-100.** ICI United States
*Chemical Description:* Formulated surfactant
*Category:* Emulsifier, anionic-nonionic blend
*Applications:* Dye carrier, solvent emulsifier
*Form:* Liquid; Color: Amber; Odor: Solvent
*Properties:* Solubility: Isopropanol @ 25 C; Sp. gr.: ≈ 1.04; Visc.: 1900 cs; Flash Pt.: ≈ 64 F; Pour Pt.: ≈ –15 F
*Composition:* Proprietary % active
*Toxicity:* Poisonous; strong irritant to eyes and skin; avoid inhalation
*Warnings and Cautions:* Poisonous if ingested; flammable
*References:* TB 633

**Ahco AB-118.** ICI United States
*Chemical Description:* Formulated surfactant; anionic-nonionic blend
*Category:* Emulsifier
*Applications:* Dye carrier; solvent emulsifier
*Form:* Liquid; Color: Amber; Odor: Solvent
*Properties:* Solubility: Isopropanol @ 25 C; Sp. gr.: ≈ 1.0; Visc.: 300 cs; Flash Pt.: ≈ 100 F; Pour Pt.: ≈ 20 F
*Composition:* Proprietary % active
*Reaction:* Alkaline
*Toxicity:* Poisonous; strong irritant to eyes and skin; avoid inhalation
*Warnings and Cautions:* Poisonous if ingested; flammable
*References:* TB 643

**Ahco AB-120.** ICI United States
*Chemical Description:* Formulated surfactant
*Category:* Emulsifier; anionic
*Applications:* Dye carrier; solvent emulsifier
*Form:* Liquid; Color: Amber; Odor: Solvent
*Properties:* Water @ 25 C; Sp. gr.: ≈ 1.0; Visc.: 600 cps; Flash Pt.: ≈ 80 F; Pour Pt.: ≈ 40 F; Stability: Good
*Composition:* Proprietary % active
*Reaction:* Alkaline
*Toxicity:* Poisonous; strong irritant to skin and eyes; avoid inhalation
*Warnings and Cautions:* Poisonous if ingested; flammable
*References:* TB 645

**Ahco AB-135.** ICI United States
*Chemical Description:* Formulated surfactant
*Category:* Emulsifier, anionic-nonionic blend
*Applications:* Dye carrier; solvent emulsifier
*Form:* Liquid; Color: Amber; Odor: Solvent

*Properties:* Sp. gr.: ≈ 1.0; Visc.: ≈ 100 cps; Flash Pt.: ≈ 90 F; Pour Pt.: ≈ 25 F; Stability: Good
*Composition:* Proprietary % active
*Reaction:* Alkaline
*Toxicity:* Poisionous if ingested; strong irritant to skin and eyes
*Warnings and Cautions:* Poisonous if ingested; flammable
*References:* TB 647

**Ahco AB-146.** ICI United States
*Chemical Description:* Formulated surfactant
*Category:* Emulsifier; anionic
*Applications:* Dye carrier; solvent emulsifiers
*Form:* Liquid; Color: Amber; Odor: Solvent
*Properties:* Solubility: Isopropanol @ 25 C; Sp. gr.: ≈ 1.0; Visc.: 1000 cps; Stability: Good
*Composition:* Proprietary % active
*Reaction:* Alkaline
*Toxicity:* Poisonous; strong irritant to skin and eyes; avoid inhalation
*Warnings and Cautions:* Poisonous if ingested; flammable
*References:* TB 648

**Ahco AB-160.** ICI United States
*Chemical Description:* Formulated surfactant
*Category:* Emulsifier; anionic
*Applications:* Dye carrier; solvent emulsifier
*Form:* Liquid; Color: Amber; Odor: Solvent
*Properties:* Solubility: Isopropanol @ 25 C; Sp. gr.: ≈ 1; Visc.: 750 cs; Stability: Good
*Composition:* Proprietary % active
*Reaction:* Alkaline
*Toxicity:* Poisonous; strong irritant to skin and eyes; avoid inhalation
*Warnings and Caution:* Poisonous if ingested; flammable
*References:* TB 644

**Ahco AB-228.** ICI United States
*Chemical Description:* Formulated surfactant
*Category:* Emulsifier; anionic
*Applications:* Dye carrier; solvent emulsifier
*Form:* Liquid; Color: Amber; Odor: Faint
*Properties:* Solubility: Water @ 25 C; Sp. gr.: ≈ 1.0; Visc.: 2600 cs; Stability: Good
*Composition:* 100% active ingred.
*Reaction:* Neutral
*Toxicity:* Strong irritant to eyes; irritant to skin
*Warnings and Cautions:* Avoid eye contact
*References:* TB 646

**Ahco AJ-110.** ICI United States
*Chemical Description:* Sulfated castor oil
*Category:* Wetting agent, lubricant, penetrant, dispersant, emulsifier, anionic surfactant
*Applications:* Dye bath agent; detergent–textile, leather industries; wetting agent in desizing/boiling off cottons, bleaching, re-wetting dried skins, tanning, finishing, sizing; fat liquoring
*Form:* Liquid; Color: Amber; Odor: Castor oil
*Properties:* Solubility: Water @ 25 C; Sp. gr.: ≈ 1.05
*Composition:* 50% active ingred., 50% water

*Reaction:* Alkaline
*Toxicity:* Eye irritant
*Warnings and Caution:* Avoid contact with eyes
*References:* TB 582

**Akwilox 133.** Swift Chem.
*Chemical Description:* Brominated oil
*Category:* Food additive
*Applications:* Viscosity adjustment for soft drinks
*Form:* Liquid; Color: Light amber; Odor: Bland; Taste: Bland
*Properties:* Sp. gr.: 1.33; Stability: 200 hr min.
*Composition:* 100% active ingred.
*Containers:* 600 lb drum; 65 lb pail

**Alacsan 7LUF.** Alcolac
*Chemical Description:* Modified alkyl dimethyl benzyl ammonium chloride
*Category:* Detergent; cationic
*Applications:* Toiletries, household cleaners, textile and plastics industries, diaper rash formulations, medicated soaps, protective hand creams, vaginal preparations, deodorants, after-shave lotions
*Form:* Viscous liquid; Color: 5 max (Gardner)
*Properties:* Strong attraction to electronegative surfaces; good solubilizing properties; high foam capacity
*Composition:* 50% active ingred.

**Albacer.** Glycol Chem.
*Applications:* Synthetic waxes and plastic lubricants
*Form:* Very hard solid; Color: Cream
*Properties:* Sp. gr.: 1.05 @ 20 C; M.P.: 75–85 C; Flash Pt.: ≈ 196 C

**Alcojet.** Alconox, Inc.
*Chemical Description:* Blend of sodium metasilicate, sodium carbonate, polyoxyethylene ester of mixed fatty and resin acids
*Category:* Detergent, nonionic
*Applications:* In mechanical washers; stain removal for laboratory and hospital glassware, metal instruments, etc.
*Form:* Powder
*Properties:* Solubility: Water; Biodegradable
*Containers:* 5 lb tubette; 25, 50 lb carton, 100, 300 lb drums

**Alcolec Series.** American Lecithin
*Chemical Description:* Phosphatides
*Category:* Wetting agent, emulsifier, release agent, nonionic
*Applications:* Foods, cosmetics, paints, pharmaceuticals
*Form:* Liquid, powder; Color: Amber
*Properties:* Solubility: Most fat solvents except acetone; Sp. gr.: 1.03; Visc.: 10,000 cps; Stability: Good
*Composition:* 60–100% active ingredient; other components: fatty oil
*Reaction:* pH 6.6
*Containers:* 460 lb drums

**Alcolec 439-C.** American Lecithin
*Chemical Description:* Water-dispersible lecithin
*Category:* Wetting agent, emulsifier, nonionic

*Applications:* Water-base paint emulsifier
*Form:* Liquid; Color: Amber; Odor: Bland
*Properties:* Solubility: Most fat solvents except acetone; Sp. gr.: 1.03; Visc.: 7000 cps; Stability: Good
*Composition:* 100% active ingred.
*Containers:* 460 lb drums

**Alcolec Granules.** American Lecithin
*Chemical Description:* Soy phosphatides, 95%
*Category:* Emulsifier, stabilizer, nonionic
*Applications:* Diet supplement; cosmetics; pharmaceuticals
*Form:* Beads; Color: Amber; Odor: Bland
*Properties:* Sp. gr.: 0.5
*Composition:* 95+% active ingred.
*Containers:* 200 lb drums

**Alcolec Powder.** American Lecithin
*Chemical Description:* Soy phosphatides, 95%
*Category:* Wetting agent, emulsifier, stabilizer, nonionic
*Applications:* Diet supplements, foods, cosmetics, pharmaceuticals
*Form:* Powder
*Properties:* Sp. gr.: 0.6
*Composition:* 100% active ingred.
*Containers:* 200 lb drums

**Alcolec Z-3.** American Lecithin
*Chemical Description:* Hydroxylated lecithin
*Category:* Wetting agent, emulsifier, nonionic
*Applications:* Foods
*Form:* Liquid
*Properties:* Solubility: Most fat solvents except acetone; Visc.: 8000 cps

**Alconate 80.** Witco Chem., Sonneborne Div.
*Category:* Emulsifier, wetting agent
*Form:* Viscous liquid; Color: Dark brown
*Properties:* Sp. gr.: 1.1 @ 60 F
*Composition:* 77% active ingred.; other components: mineral oil (15%), water (8.0%)
*Reaction:* Acid
*Toxicity:* Nontoxic with respect to Federal Hazardous Substance Labelling Act; may cause eye burn
*Containers:* Drum, T/W, T/C
*Warnings and Caution:* May cause eye burn

**Alconox.** Alconox, Inc.
*Chemical Description:* Blend of alkyl aryl sulfonates, lauryl alcohol sulfates, phosphates, carbonates
*Category:* Detergent with wetting, sequestering and synergistic agents, anionic
*Applications:* Detergent for removal of soil, grit, blood, fats from glass, metal, plastic, rubber; ultrasonic cleaners
*Form:* Powder; Color: Colorless; Odor: None
*Properties:* Solubility: Complete in water; Biodegradable; Anticorrosive properties
*Containers:* Drums, Cartons, boxes
*References:* Technical Bulletin #1

**Alcopol AH New.** Allied Colloids Ltd.
*Chemical Description:* Mono ethanolamine salt of fatty acid ester
*Category:* Dispersant, anionic
*Applications:* Dispersant esp. useful in nonaqueous systems; dry cleaning aid; good water carrying properties
*Form:* Liquid
*Composition:* 99% conc.

**Alcopol OD.** Allied Colloids Ltd.
*Chemical Description:* Sodium ditredecyl sulfosuccinate
*Category:* Dispersant, wetting agent, emulsifier, anionic
*Form:* Liquid
*Composition:* 60% conc.

**Alcopol OB.** Allied Colloids Ltd.
*Chemical Description:* Sodium deisobutyl sulfosuccinate
*Category and Applications:* See Alcopol FL
*Form:* Liquid
*Composition:* 45% conc.

**Alcopol OS.** Allied Colloids Ltd.
*Chemical Description:* Sodium dehexyl sulfosuccinate
*Category and Applications:* See Alcopol FL
*Form:* Liquid
*Composition:* 60% conc.

**Alcotab.** Alconox, Inc.
*Chemical Description:* Blend of alkyl aryl sulfonates, lauryl alcohol sulfates, phosphates, carbonates
*Category:* Detergent with wetting, sequestering and synergistic agents, anionic
*Application:* Cleaners in syphon-type pipette or test-tube rinsers
*Form:* Tablet
*Properties:* Biodegradable
*References:* Technical Bulletin #3

**Aldo 8.** Glyco Chem.
Same as Aldo TC in Cosmetic Grade

**Aldo 10.** Glyco Chem.
Same as Aldo TGTC in cosmetic grade

**Aldo 12.** Glyco Chem.
Same as Aldo TCL in cosmetic grade

**Aldo DGDO.** Glyco Chem.
*Chemical Description:* Decaglycerol decaoleate
*Form:* Liquid; Color: Amber
*Properties:* HLB: 3.0; Sap. Value: 165–185

**Aldo HGDS.** Glyco Chem.
*Chemical Description:* Hexaglycerol distearate
*Form:* Solid; Color: Tan
*Properties:* M.P.: 53–57 C; HLB: 4.0; Sap. Value: 105–125

**Aldo HMO.** Glyco Chem.
*Chemical Description:* Glycerol monooleate, high mono.
*Form:* Semi-liquid; Color: Yellow
*Properties:* M.P.: $< 20$ C; HLB: 3.0; Sap. Value: 145–155

**Aldo HMS.** Glyco Chem.
*Chemical Description:* Glycerol monostearate, high mono.
*Form:* White beads

*Properties:* M.P.: 61–68; HLB: 2.8; Sap. Value: 165–175

**Aldo LS.** Glyco Chem.
*Chemical Description:* Glycerol lactostearate
*Form:* Solid; Color: White
*Properties:* M.P.: 47–54 C; HLB: 9.0; Sap. Value: 245–260

**Aldo MC.** Glyco Chem.
*Chemical Description:* Glycerol monococoate
*Category:* Emulsifier, nonionic
*Form:* Soft solid; Color: Cream
*Properties:* Solubility: Sol. in ethanol, ethyl acetate, toluol, naphtha, mineral and vegetable oils; disp. in water; Sp. gr.: 0.97; M.P.: 20–26 C; HLB: 6.8; Sapon. Value: 180–190; Acid No.: < 6; Iodine Value: < 10; pH: 7.5–8.5 (5% aq.)
*Composition:* 100% conc.
*References:* Bulletin E-100

**Aldo MLD.** Glyco Chem.
*Chemical Description:* Glycerol monolaurate (disp.)
*Category:* Emulsifier, nonionic
*Form:* Soft solid; Color: Cream
*Properties:* Solubility: Sol. in ethanol, ethyl acetate, toluol, naphtha, mineral and vegetable oils; disp. in water; Sp. gr.: 0.97; M.P.: 21–26 C; HLB: 6.8; Sapon. Value: 185–195; Acid No.: < 5; Iodine Value: < 9; pH: 7.5–8.5 (5% aq.)
*Composition:* 100% conc.
*References:* Bulletin E-100

**Aldo MO.** Glyco Chem.
*Chemical Description:* Glycerol monooleate
*Category:* Emulsifier, nonionic
*Applications:* Baking, ice creams, cosmetics and industrial processes
*Form:* Soft solid; Color: Yellow
*Properties:* Solubility: Sol. in ethyl acetate; miscible in certain proportions with methanol, ethanol, acetone, toluol, naphtha, mineral and vegetable oils; Sp. gr.: 0.95; M.P.: < 25 C; HLB: 3.4; Sapon. Value: 170–180; Acid No.: < 5; Iodine Value: 68–73; pH: 4.5–6.5
*Composition:* 100% conc.
*References:* Bulletin E-100

**Aldo MOD.** Glyco Chem.
*Chemical Description:* Glycerol monooleate (disp.); available in animal, vegetable, food and Kosher grades
*Category:* Emulsifier, nonionic
*Applications:* Baking, ice creams, cosmetics, industrial processes
*Form:* Liquid; Color: Yellow
*Properties:* Solubility: Sol. in toluol, naphtha, mineral and vegetable oils; disp. in water; Sp. gr.: 0.95; M.P.: < 0 C; HLB: 5.0; Sapon. Value: 141–147; Acid No.: < 8; Iodine Value: 65–69; pH: 8.5–9.5 (5% aq.)
*Composition:* 100% conc.
*References:* Bulletin E-100

**Aldo MO Tech.** Glyco Chem.
*Chemical Description:* Glycerol monooleate, tech. grade
*Category:* Emulsifier, nonionic
*Form:* Soft solid; Color: Amber
*Properties:* Solubility: Sol. in ethyl acetate; miscible in certain proportions with methanol, ethanol, ace-

tone, toluol, naphtha, mineral and vegetable oils; Sp. gr.: 0.95; M.P.: < 25 C; HLB: 3.4; Sapon. Value: 170–180; Acid No.: < 5; Iodine Value: 68–73; pH: 4.5–6.5
*Composition:* 100% conc.
*References:* Bulletin E-100

**Aldo MR.** Glyco Chem.
*Chemical Description:* Glycerol monoricinoleate (disp.)
*Category:* Emulsifier, nonionic
*Form:* Liquid; Color: Yellow
*Properties:* Solubility: Sol. in methanol, ethanol, ethyl acetate, toluol, vegetable oil; disp. in water; Sp. gr.: 1.02; M.P.: < –8 C; HLB: 5.6; Sapon. Value: 122–132; Acid No.: < 4; Iodine Value: 66–69; pH: 8.3–9.3
*Composition:* 100% conc.
*References:* Bulletin E-100

**Aldo MS.** Glyco Chem.
*Chemical Description:* Glycerol monostearate
*Category:* Emulsifier, nonionic
*Applications:* Baking, ice creams, cosmetics, industrial processes
*Form:* Beads; Color: White
*Properties:* Solubility: Sol. hot in methanol, ethanol, toluol, naphtha, mineral and vegetable oils; disp. hot in water; Sp. gr.: 0.97; M.P.: 57–61 C; HLB: 3.6; Sapon. Value: 158–165; Acid No.: < 3; Iodine Value: < 5; pH: 7.6–8.6 (3% aq.)
*Composition:* 100% conc.
*References:* Bulletin E-100

**Aldo MS Industrial.** Glyco Chem.
*Chemical Description:* Glycerol monostearate (industrial grade)
*Category:* Emulsifier, nonionic
*Form:* Flakes; Color: Tan
*Properties:* Solubility: Sol. hot in methanol, ethanol, toluol, naphtha, mineral and vegetable oils; disp. hot in water; Sp. gr.: 0.97; M.P.: 55–62 C; HLB: 3.6; Sapon. Value: 157–177; Acid No.: < 5; Iodine Value: < 5; pH: 6.5–8.5 (5% aq.)
*Composition:* 100% conc.
*References:* Bulletin E-100

**Aldo MSA.** Glyco Chem.
*Chemical Description:* Glycerol monostearate, acid stable
*Form:* White beads
*Properties:* M.P.: 51–58 C; HLB: 11.0; Sap. Value: 90–100

**Aldo MSC.** Glyco Chem.
*Chemical Description:* Glycerol monostearate (cosmetic)
*Category:* Emulsifier, nonionic
*Form:* Flakes; Color: White
*Properties:* Solubility: Sol. hot in methanol, ethanol, toluol, naphtha, mineral and vegetable oils; disp. hot in water; Sp. gr.: 0.97; M.P.: 55–58 C; HLB: 3.5; Sapon. Value: 168–176; Acid No.: < 6; Iodine Value: < 3; pH: 4.0–7.0 (5% aq.)
*Composition:* 100% conc.
*References:* Bulletin E-100

**Aldo MSD.** Glyco Chem.
*Chemical Description:* Glycerol monostearate, disp.
*Category:* Emulsifier, nonionic

*Applications:* Baking, ice creams, cosmetics, industrial processes
*Form:* Beads; Color: White
*Properties:* Solubility: Sol. hot in methanol, ethanol, toluol, naphtha, mineral and vegetable oils; disp. hot in water; Sp. gr.: 0.97; M.P.: 56–60 C; HLB: 5.8; Sapon. Value: 140–150; Acid No.: < 3.5; Iodine Value: < 5; pH: 9.2–10.2 (3% aq.)
*Composition:* 100% conc.
*References:* Bulletin E-100

**Aldo MSLG.** Glyco Chem.
*Chemical Description:* Glycerol monostearate, low glycerine
*Form:* White beads
*Properties:* M.P.: 58–62 C; HLB: 3.3; Sap. Value: 165–175

**Aldo TC.** Glyco Chem.
*Chemical Description:* Liq. Fract. esters, edible coconut oil
*Form:* Straw. liquid
*Properties:* M.P./Solid Pt.: < – 5 C; HLB: 1.4; Sap. Value: 335–360

**Aldo TCL.** Glyco Chem.
*Chemical Description:* Liq. fract. esters, edible coconut oil
*Form:* Straw. liquid
*Properties:* M.P./Solid Pt.: < 0 C; HLB: 1.4; Sap. Value: 300–325

**Aldo TGMS.** Glyco Chem.
*Chemical Description:* Triglycerol monostearate
*Form:* Tan beads
*Properties:* M.P.: 54–58 C; HLB: 4.0; Sap. Value: 120–140

**Aldo TGTC.** Glyco Chem.
*Chemical Description:* Liq. fract. esters, edible coconut oil
*Form:* Straw. liquid
*Properties:* M.P./Solid Pt.: < 0 C; HLB: 4.0; Sap. Value: 250–275

**Aldo TO.** Glyco Chem.
*Chemical Description:* Glycerol trioleate
*Form:* Yellow liquid
*Properties:* M.P.: < 0 C; HLB: 0.8; Sap. Value: 188–195

**Aldosperse AA 40/60.** Glyco Chem.
*Chemical Description:* 40% Aldosperse MS-20, 60% Aldo HMS
*Form:* White beads
*Properties:* M.P.: 40–60 C; HLB: 8.0; Sap. Value: 120–130

**Aldosperse MS-20.** Glyco Chem.
*Chemical Description:* POE 20 glycerol monostearate
*Form:* Soft solid; Color: White
*Properties:* M.P.: 30–40 C; HLB: 13.1; Sap. Value: 65–75

**Aldosperse O-20.** Glyco Chem.
*Chemical Description:* 80% glycerol monostearate, 20% POE 20 sorbitan monooleate
*Form:* Straw. beads
*Properties:* M.P.: 55–65 C; HLB: 5.2; Sap. Value: 140–150

**Aldosperse TS-20.** Glyco Chem.
*Chemical Description:* 80% glycerol monostearate, 20% POE 20 sorbitan tristearate
*Form:* Straw. beads
*Properties:* M.P.: 55–65 C; HLB: 4.3; Sap. Value: 150–160

**Aldosperse TS-40.** Glyco Chem.
*Chemical Description:* 60% glycerol monostearate, 40% POE 20 sorbitan tristearate
*Form:* Straw. beads
*Properties:* M.P.: 40–60 C; HLB: 8.0; Sap. Value: 120–130

**Alfol 4 Alcohol.** Continental Oil Co.
*Applications:* Intermediate for amines, fatty acid esters
*Form:* Liquid; Color: Colorless; Odor: Characteristic, penetrating, sweet
*Properties:* Refractive Index: 1.3993; Solubility: Limited sol. in water; Density: 6.76; Sp. gr.: 0.809; Visc.: 3.4; M.P. −128 C; B.P.: 117.7 C
*Composition:* 99.3% active ingred.; 0.07% water

**Alfol 6 Alcohol.** Continental Oil Co.
*Form:* Liquid; Color: Colorless
*Properties:* Refractive Index: 1.4179; Density: 6.83; Sp. gr.: 0.823; Visc.: 5.5; M.P.: −49; B.P.: 315
*Composition:* 98.5% active ingred.; 0.15% max water

**Alfol 8 Alcohol.** Continental Oil Co.
*Category:* Wetting agent, emulsifier
*Applications:* Raw material for preparation of acetate esters, stabilizers, mercaptans, caprylic acid, octaldehyde; intermediate for surfactants
*Form:* Liquid; Color: Colorless; Odor: Mild aromatic
*Properties:* Solubility: Immiscible: water; Miscible: alcohol, chloroform, ether, benzene; Density: 6.86; Sp. gr.: 0.823; M.P. 5; B.P.: 383
*Composition:* 99.2% active ingred.; 0.03% water
*Containers:* SS, heresite-lined steel or aluminum or dry carbon steel tanks

**Alfol 10 Alcohol.** Continental Oil Co.
*Category:* Emulsifier
*Applications:* Chemical intermediate for surfactants, alcohol sulfates, ethoxylates, ether sulfates; foam builders
*Form:* Liquid; Color: Colorless; Odor: Sweet, fat-like
*Properties:* Solubility: Alcohol, ether, benzene, glacial acetic acid; insol. water; Density: 6.90; Sp. gr.: 0.828; M.P.: 44; B.P.: 448
*Composition:* 99% active ingred.; 0.04% water
*Containers:* Dry carbon steel; SS; heresite-lined steel or aluminum

**Alfol 12 Alcohol.** Continental Oil Co.
*Form:* Liquid; Color: Colorless
*Properties:* Sp. gr.: 0.831; M.P.: 75; B.P. 498
*Composition:* 99.1% active ingred.
*Containers:* Dry carbon steel; SS, epoxy or heresite-lined steel, aluminum
*Warnings and Caution:* Avoid inhaling vapors from the hot liquid; wash body parts contacting the alcohol with water; combustible

**Alfol 14 Alcohol.** Continental Oil Co.
*Form:* Solid, crystalline; Color: White; Odor: Typical fatty alco-

hol
*Properties:* Sp. gr.: 0.824; M.P.: 99; B.P. 552
*Composition:* 99% active ingred.
*Reaction:* Typical of compounds with the OH functional group
*Containers:* Dry carbon steel, SS, epoxy or heresite-lined steel, aluminum
*Warnings and Caution:* Avoid inhaling vapors from the hot liquid; wash body parts contacting the alcohol with water; combustible; treat with same precautions as with high boiling hydrocarbons

**Alfol 16 Alcohol.** Continental Oil Co.
*Form:* Waxy solid; Color: White; Odor: Typical fatty alcohol
*Properties:* Density: 6.77; Sp. gr.: 0.813; Visc.: 6.77; M.P. 117; B.P.: 604
*Composition:* 98.9% active ingred.
*Containers:* Dry carbon steel; aluminum, epoxy or heresite-lined steel
*Warnings and Caution:* Avoid inhaling vapors form the hot liquid; Wash body parts contacting the alcohol with water; combustible; treat with same precautions as high boiling hydrocarbons

**Alfol 18 Alcohol.** Continental Oil Co.
*Applications:* Surfactant range alcohol
*Form:* Waxy solid; Color: White; Odor: Typical fatty alcohol
*Properties:* Density: 6.71; Sp. gr.: 0.8075; Visc.: 13.5; M.P.: 135; B.P.: 640
*Composition:* 98.7% active ingred.
*Reaction:* Typical of compounds with the –OH functional group
*Containers:* Aluminum, SS, epoxy or heresite-lined steel. dry carbon steel
*Warnings and Caution:* Avoid inhaling vapors from the hot liquid; Wash body parts contacting the alcohol with water; combustible; treat with same precautions as high boiling hydrocarbons

**Alfol 20 Alcohol.** Continental Oil Co.
*Chemical Description:* Blend of $C_2O$ and higher even-carbon-number primary linear alcohols
*Category:* Lubricant, defoamer
*Applications:* Lubricants for plastics, textiles, metals; ethoxylation and sulfation to prepare emulsifier and surfactants; esters; defoamers for paper, latex, other aqueous slurries; waxes; chemical intermediate
*Form:* Solid; Color: Off white
*Properties:* Density: 6.80; Sp. gr.: 0.817; M.P.: 131

**Alfol 610 Alcohol.** Continental Oil Co.
*Category:* Intermediate
*Applications:* Intermediate for plasticizers and surfactants; chemical specialty as emulsifier and foam builders; mercerizing agent in textile industry; intermediate for aldehydes, acids, amines, halides and mercaptans, lubricating oil additive
*Form:* Liquid; Color: Colorless; Odor: Mild, sweet

*Properties:* Solubility: Insol. water, sol. alcohol, ether, benzene; Density: 6.92; Sp. gr.: 0.829; Visc.: 11.0; M.P.: –8; B.P.: 350
*Composition:* 99% active ingred.
*Warnings and Caution:* Avoid inhaling vapors from the hot liquid; Wash body parts contacting the alcohol with water; combustible; treat with same precautions as high boiling hydrocarbons

**Alfol 810 Alcohol.** Continental Oil Co.
*Category:* Intermediate
*Applications:* High-quality derivatives; intermediate for plasticizers, surfactants; chemical specialty as emulsifier and foam builder; mercerizing agent in textiles; lubricating oil additive; chemical intermediate for manufacture of halides, amines, aldehydes, acids and mercaptans
*Form:* Liquid; Color: Clear colorless; Odor: Mild, sweet
*Properties:* Solubility: Sol. alcohol, ether, benzene–almost insol. water; Density: 6.93; Sp. gr.: 0.831; Visc.: 8.9; M.P.: 7; B.P.: 400
*Composition:* 99% active ingred.
*Containers:* SS, aluminum, heresite-lined steel; dry carbon steel

**Alfol 1012 Alcohol.** Continental Oil Co.
*Category:* Intermediate
*Applications:* Surfactant intermediate
*Form:* Liquid; Color: Colorless; Odor: Sweet
*Properties:* Density: 7.0; Sp. gr.: 0.834; Visc.: 15.0; M.P.: 35–40; B.P.: 427
*Composition:* 99% active ingred.
*Reaction:* Typical of compounds with the –OH functional group
*Containers:* SS, aluminum, epoxy-lined or heresite-lined steel tanks; dry mild steel

**Alfol 1214 Alcohol.** Continental Oil Co.
*Category:* Surfactant intermediate
*Form:* Liquid; Color: Clear, colorless; Odor: Sweet, typical fatty alcohol
*Properties:* Density: 7.0; Sp. gr.: 0.838; Visc.: 14.3 (120); M.P.: 73–75; B.P.: 518
*Composition:* 99% active ingred.
*Reaction:* Typical of compounds with the –OH functional group
*Containers:* SS, aluminum, heresite-lined steel; dry carbon steel
*Warnings and Caution:* Avoid inhaling vapor from the hot liquid; wash body parts contacting the alcohol with water; combustible; treat with same precautions as high boiling hydrocarbons

**Alfol 1216 Alcohol.** Continental Oil Co.
*Category:* Surfactant intermediate
*Form:* Liquid; Color: Clear colorless; Odor: Typical fatty alcohol
*Properties:* Density: 7.0; Sp. gr.: 0.830; Visc.: 15.5 (120); M.P.: 64–70; B.P.: 514
*Composition:* 98.5% active ingred.
*Reaction:* Typical of compounds with the –OH functional group
*Containers:* See Alfol 1214 Alcohol

*Warnings and Caution:* See Alfol 1214 Alcohol

**Alfol 1218 Alcohol.** Continental Oil Co.
*Category:* Surfactant intermediate
*Form:* Solid; Color: White
*Properties:* Density: 6.85; Sp. gr.: 0.825; Visc.: 15.5 (120); M.P.: 99–101; B.P.: 524
*Composition:* 100% active ingred.
*Reaction:* Typical of compounds with the –OH functional group
*Containers:* See Alfol 1214 Alcohol
*Warnings and Caution:* See Alfol 1214 Alcohol

**Alfol 1412 Alcohol.** Continental Oil Co.
*Category:* Surfactant intermediate
*Form:* Soft, waxy solid; Color: White; Odor: Typical fatty alcohol
*Properties:* Density: 7.0; Sp. gr.: 0.838; Visc.: 16.7 (100); M.P. 72–75; B.P.: 560
*Composition:* 99.4% active ingred.
*Reaction:* Typical of compounds with the –OH functional group
*Containers:* See Alfol 1214 Alcohol
*Warnings and Caution:* See Alfol 1214 Alcohol

**Alfol 1618 Alcohol.** Continental Oil Co.
*Category:* Surfactant intermediate
*Form:* Waxy solid; Color: White; Odor: Typical fatty alcohol
*Properties:* Density: 6.81; Sp. gr.: 0.820; Visc.: 22 (120); M.P.: 110–115; B.P.: 609
*Composition:* 98.7% active ingred.
*Reaction:* Typical of compounds with the –OH functional group
*Containers:* See Alfol 1214 Alcohol
*Warnings and Caution:* See Alfol 1214 Alcohol

**Alfol 1620 Alcohol.** Continental Oil Co.
*Category:* Surfactant intermediate
*Form:* Waxy solid; Color: White; Odor: Typical fatty alcohol
*Properties:* Density: 6.81; Sp. gr.: 0.817; Visc.: 13.5 (160); M.P.: 110–113; B.P.: 605
*Composition:* 98.5% active ingred.
*Reaction:* Typical of compounds with the –OH functional group
*Containers:* See Alfol 1214 Alcohol
*Warnings and Caution:* See Alfol 1214 Alcohol

**Alfol C22-Alcohol.** Continental Oil Co.
*Chemical Description:* Blend of $C_{22}$ and higher even-carbon-number primary linear alcohols
*Category:* Lubricant, defoamer
*Applications:* Lubricant for plastics, textiles, metals; ethoxylation and sulfation to prepare emulsifiers and surfactants; esters; defoamers for paper, latex and other aqueous slurries; waxes; chemical intermediate
*Form:* Solid; Color: Off white
*Properties:* Solubility: nil in water; Density: 6.81; Sp. gr.: 0.817; M.P.: 132

**Alfonic 1012-60.** Conoco Chem.
*Chemical Description:* Ethoxylated

linear alcohols (60% E.O.)
*Category:* Detergent, wetting agent, emulsifier, foaming agent, intermediate, dispersant, nonionic
*Applications:* Industrial–metal cleaners, textile and leather processing agents, agricultural formulations, pulp and paper processing, maintenance cleaners, coatings; Household–light-duty dishwashing liquids. heavy-duty (liquid or dry) laundry products; all-purpose cleaners, car washing compounds
*Form:* Liquid; Color: 20 (APHA); Odor: Sweet
*Properties:* Density: 8.2; Sp. gr.: 0.98; Visc.: 20.5 cps @ 100; M.P.: 36; B.P.: 214; Stability: Good in presence of anionics, cationics, acids, alkalies, other builders; compatible with most oxiding and bleaching agents in textile industry; Surface Tension: 26 dynes/cm @ 0.1%; Biodegradable
*Composition:* 99.8% active ingred.
*Toxicity:* Relatively nontoxic; in conc. form may produce eye or skin irritation
*Containers:* aluminum, SS, phenolic coated tanks
*Warnings and Caution:* If it contacts the body, flush with water; combustible; treat with same precautions as hydrocarbon; foam or dry chemicals to extinguish fires

**Alfonic 1412-40 Ethoxylate.** Conoco Chem.
*Chemical Description:* Ethoxylated linear alcohols (40% E.O.)
*Category:* Nonionic
*Applications:* Raw material for manufacture of sodium, ammonium, potassium, and mono-, di-, or triethanolamine ethoxysulfates; can be further processed to yield ethoxyphosphates
*Form:* Liquid; Color: Clear; Odor: Slight
*Properties:* Density: 7.64; Sp. gr.: 0.91; Visc.: 52; M.P. 41; Biodegradable
*Composition:* 100% active ingred.
*Toxicity:* Relatively nontoxic; in conc. form may cause skin or eye irritation
*Warnings and Caution:* If contacts body, flush exposed area with water

**Alfonic 1412-60.** Conoco Chem.
*Chemical Description:* Ethoxylated linear alcohols (60% E.O.)
*Category:* Detergent, wetting agent, emulsifier, foaming agent, nonionic surfactant
*Applications:* Industrial–metal cleaners, textile and leather processing agents, agricultural formulations, pulp and paper processing, maintenance cleaners, coatings; Household–heavy-duty (liquid or dry) laundry products
*Form:* Liquid; Color: Cloudy; Odor: Mild, sweet; Viscous at room temperature
*Properties:* Density: 8.2; Sp. gr.: 0.98; Visc.: 33 cps @ 100; Stability: Good in presence of anionic and cationic compounds, acids, alkalies, other detergent builders; Biodegradable
*Composition:* 100% active ingred.
*Toxicity:* Relatively nontoxic; in conc. form could cause skin or eye irritation

*Warnings and Caution:* Should not be used with strong oxidizing agents such as conc. hydrogen peroxide (potentially explosive); if contacts body, flush exposed area with water

**Alfonic 1412-A Ether Sulfate.** Conoco Chem.
*Chemical Description:* Sulfated, ethoxylated alcohol derivatives
*Category:* Detergent, wetting agent, emulsifier, foaming agent
*Applications:* Household–light-duty dishwashing liquids, bubble baths, fine fabric cleaner, wool washing liquids, upholstery and rug cleaners, cosmetic cleansers, all-purpose cleaners; Industrial–emulsion polymerization, petroleum production, wallboard foaming agent, paper processing, agricultural formulations, maintenance cleaners
*Form:* Liquid; Color: Clear; Odor: Mild
*Properties:* Density: 8.4; Visc.: 76 cps @ 100; Biodegradable
*Composition:* 58% active ingred.; other components: water
*Toxicity:* Moderate irritation to eyes; nontoxic skin, oral, inhalation tests
*Containers:* 316 SS
*Warnings and Caution:* If contacts body, flush exposed area with water

**Alfonic 1412-S Ether Sulfate.** Conoco Chem.
*Chemical Description:* Sulfated ethoxylated alcohol derivatives
*Category:* Detergent, emulsifier, wetting and foaming agent, anionic
*Applications:* Household–light-duty dishwashing liquids, bubble baths, fine fabric and wool cleaners, upholstery and rug cleaners, cosmetic cleansers, all-purpose cleaners; Industrial–emulsion polymerization, petroleum production, wallboard foaming agent, paper processing, agricultural formulations, maintenance cleaners
*Form:* Liquid; Color: Clear; Odor: Mild
*Properties:* Density: 8.8; Visc.: 64 cps @ 100; Biodegradable
*Composition:* 38% active ingred.; 0.2% water; $Na_2CO_3$ added as buffer
*Toxicity:* Moderate eye irritation; nontoxic by skin, oral, inhalation tests
*Containers:* 316 SS
*Warnings and Caution:* If it contacts body, flush exposed area with water

**Alfonic 1618-65.** Conoco Chem.
*Chemical Description:* Ethoxylated linear alcohols (65% E.O.)
*Category:* Detergent, emulsifier, foaming and wetting agent; nonionic
*Applications:* Industrial–metal cleaners, textile and leather processing, agricultural formulations, pulp and paper processing, maintenance cleaners, coatings; Household–heavy-duty (liquid or dry) laundry products
*Form:* Solid; Color: Near white; Odor: Nil
*Properties:* Solubility: Good in water, ethanol, glycol ether (mon-

obutyl), isopropyl alcohol, xylene, perchlorethyl; slight in kerosene; insol. mineral oil; Density: 8.2; Sp. gr.: 0.98; Visc.: 81 cps @ 100; M.P.: 88; B.P.: 234; Biodegradable
*Composition:* 100% active ingred.
*Reaction:* Compatible and stable in presence of anionics and cationics, acids, alkalies, other detergent builders; compatible with most oxidizing and bleaching agents at normal use concentrations
*Toxicity:* Relatively nontoxic; may produce skin or eye irritation
*Containers:* Aluminum, SS, phenolic coated tanks
*Warnings and Caution:* If contacts body, flush exposed area with water; treat with same precautions as high boiling hydrocarbon; with strong oxidizing agents (e.g., conc. hydrogen peroxide) potential explosive mixtures can result

**Alipal CD-128.** GAF Corp.
*Chemical Description:* Sulfate-ester-type surfactant based on a linear hydrophobe
*Category:* Foaming agent, wetting agent, anionic
*Applications:* Use in all-purpose detergents and concentrates; foaming agent for air drilling of oil and gas wells; wetting agent for conc. caustic sol'ns.
*Form:* Liquid; Color: Straw
*Properties:* Solubility: Good in electrolyte systems; Sp. gr.: 1.056; Pour Pt.: –5 C; Stability: Good in high electrolyte, alkaline
*Composition:* 56% min. active ingredient
*References:* Data sheet

**Alipal CO-433.** GAF Corp.
*Chemical Description:* Sodium salt of a sulfated alkylphenoxypoly (ethyleneoxy)-ethanol
*Category:* Detergent, wetting, dispersing, emulsifying and foaming agents; anionic
*Applications:* Liquid all-purpose cleaners; light-duty dishwashing detergents; textile wet processing; cosmetics; emulsion polymerization and post-stabilization; pesticide emulsifiers
*Form:* Clear liquid; Color: 4 max (Varnish); Odor: Mild aromatic
*Properties:* Solubility: Water sol.; Density: 8.9 lb/gal; Sp. gr.: 1.065; Visc.: 2500 cps; Stability: Good; Surface Tension: 32 dynes/cm (1% conc.)
*Composition:* 28% active ingred.
*References:* Tech. Bulletin 6220-001

**Alipal CO-436.** GAF Corp.
*Chemical Description:* Ammonium salt of a sulfated alkylphenoxy-poly(ethyleneoxy)-ethanol
*Category and Applications:* See Alipal CO-433
*Form:* Clear liquid; Color: 5 max. (Varnish); Odor: Alcoholic
*Properties:* Solubility: Water sol.; Density: 8.9 lb/gal; Sp. gr.: 1.065; Visc.: 100 cps; Stability: Good; Surface Tension: 34 dynes/cm (1% sol'n.)
*Composition:* 58% active ingred.
*Toxicity:* Eye irritant

**Alipal EO-526.** GAF Corp.
*Chemical Description:* Sodium salt of a sulfated alkylphenoxypoly

(ethyleneoxy)ethanol
*Category and Applications:* See Alipal CO-433
*Form:* Hazy liquid; Color: 10 max (Varnish); Odor: Alcoholic
*Properties:* Solubility: Water sol.; Density: 9.3 lb/gal; Sp. gr.: 1.115; Visc.: 400 cps; Stability: Good; Surface Tension: 35 dynes/cm (1% sol'n.)
*Composition:* 58% active ingred.

**Alipal EP-110.** GAF Corp.
*Chemical Description:* Ammonium salt of sulfated alkylphenoxypoly (ethyleneoxy) ethanol
*Category:* Emulsifier
*Applications:* Emulsion polymerization yielding latices with outstanding multivalent ion stability, water-spot resistance, freedom from grain and coagulum, film clarity and uniformity
*Form:* Liquid (slight haze); Color: Pale yellow
*Properties:* Sp. gr.: 1.04; Visc.: 91.0 cks; Pour Pt.: 0 C; Surface Tension: 38.3 dynes/cm (1%)
*Composition:* 30% active ingred.
*Toxicity:* Moderate eye irritation

**Alipal EP-115.** GAF Corp.
*Chemical Description:* Ammonium salt of a sulfated alkylphenoxypoly (ethyleneoxy) ethanol
*Form:* Clear viscous liquid; Color: 3 max (VCS)
*Composition:* 30% active ingred.

**Alipal EP-120.** GAF Corp.
*Chemical Description:* Ammonium salt of sulfated alkylphenoxypoly (ethyleneoxy) ethanol
*Category:* Emulsifier
*Applications:* See Alipal EP-110
*Form:* Liquid (slight haze); Color: Pale yellow
*Properties:* Sp. gr.: 1.06; Visc.: 111 cks; Pour Pt.: 8 C; Surface Tension: 40 dynes/cm (1%)

**Alkadet LD-50.** Alkaril Chem.
*Chemical Description:* Blend of surfactants
*Category:* Detergent, anionic
*Applications:* Light-duty liquid detergent concentrate for dishwashing and general purpose, esp. scouring metal surfaces, all flooring surfaces
*Form:* Clear liquid; Color: Amber
*Properties:* Visc.: 500 cps; Cloud Pt.: $< 30$ F
*Composition:* 48% active ingred.

**Alkadet LD50-Q.** Alkaril Chem.
*Chemical Description:* Compounded liquid detergent concentrate
*Category:* Detergent, anionic
*Applications:* Liquid detergent concentrate for dishwashing and general purpose cleaning esp. flooring surfaces, lacquered, painted surfaces
*Form:* Clear liquid; Color: Green
*Properties:* Sp. gr.: 1.075; Visc.: 100–130 cps
*Composition:* 50% active ingred.

**Alkadet LD-60.** Alkaril Chem.
*Chemical Description:* Blend of surfactants
*Category and Applications:* See Alkadet LD-50
*Form:* Clear liquid; Color: Amber
*Properties:* Sp. gr.: 1.05; Visc.:

≈ 100 cps; Cloud Pt.: 7 C max.
*Composition:* 58–60% active

**Alkameen C-5.** Alkaril Chem.
*Chemical Description:* Tertiary amines; ethoxylates of primary tallow amines
*Category:* Emulsifier, detergent, wetting agent, cationic
*Applications:* Emulsifier–fat liquoring; making of cutting and soluble oils; dispersing waxes, pigments, resins; printing; Wetting agent–desizing and boiling off cottons; dyeing; tanning; insecticides, herbicides, fungicides; Detergents–textiles; skins; leather; dyeing
*Form:* Liquid; Color: Amber; Odor: Typical
*Properties:* Solubility: Insol. water
*Composition:* 100% active ingred.
*Reaction:* Alkaline

**Alkameen C-10.** Alkaril Chem.
*Chemical Description:* Tertiary amines; ethoxylates of primary tallow amines
*Category:* Emulsifier, detergent, wetting agent, cationic
*Applications:* Emulsifier–fat liquoring, making of cutting and soluble oils; Wetting Agents–pickling; electroplating; insecticides, herbicides, fungicides; Detergents–textiles; mines and minerals; cosmetics
*Form:* Liquid; Color: Amber; Odor: Typical
*Properties:* Solubility: Insol. water
*Composition:* 100% active ingred.
*Reaction:* Alkaline

**Alkameen HT-2.** Alkaril Chem.
*Chemical Description:* Tertiary amines; ethoxylates of hydrogenated primary tallow amines
*Category:* Emulsifier, cationic
*Applications:* Emulsifier–dispersing waxes, pigments, resins; Wetting agent–rotary washer, waterproofing; Detergents–textiles
*Form:* Solid; Color: Amber; Odor: Typical
*Properties:* Solubility: Insol. water
*Composition:* 100% active ingred.
*Reaction:* Alkaline

**Alkameen HT-5.** Alkaril Chem.
*Chemical Description:* See Alkameen HT-2
*Category:* Detergent, wetting agent, emulsifier, cationic
*Applications:* Emulsifier–fat liquoring; Wetting agent–insecticides, herbicides, fungicides; Detergent–textiles
*Form:* Solid; Color: Amber; Odor: Typical
*Properties:* Solubility: Insol. water
*Composition:* 100% active ingred.
*Reaction:* Alkaline

**Alkameen HT-12.** Alkaril Chem.
*Chemical Description:* See Alkameen HT-2
*Category:* Emulsifier, detergent, wetting agent, cationic
*Applications:* Emulsifier–fat liquoring; dispersing waxes, pigments, resins; Detergent–textiles
*Form:* Solid; Color: Amber; Odor: Typical
*Properties:* Solubility: Insol. water
*Composition:* 100% active ingred.
*Reaction:* Alkaline

**Alkameen HT-15.** Alkaril Chem.

*Chemical Description:* See Alkameen HT-2
*Category:* Detergent, emulsifier, wetting agent, cationic
*Applications:* Emulsifer–printing; Wetting agent–dyeing; Detergent–textiles, dyeing
*Form:* Solid; Color: Amber; Odor: Typical
*Composition:* 100% active ingred.
*Reaction:* Alkaline

**Alkameen HT-25.** Alkaril Chem.
*Chemical Description:* See Alkameen HT-2
*Category:* Detergent, wetting agent, emulsifier, cationic
*Applications:* Emulsifier–dispersing waxes, pigments, resins; preparation of emulsified paints; Detergents–textiles; mines and mining; paints
*Form:* Solid; Color: Yellow; Odor: Typical
*Composition:* 100% active ingred.
*Reaction:* Alkaline

**Alkameen HT-30.** Alkaril Chem.
*Chemical Description:* See Alkameen HT-2
*Category:* Emulsifier, wetting agent, cationic
*Applications:* Emulsifier–dispersing waxes, pigments, resins; preparation of emulsified paints; Detergents–textiles, mines and minerals, paints
*Form:* Solid; Color: Tan; Odor: Typical
*Composition:* 100% active ingred.
*Reaction:* Alkaline

**Alkamide 1002.** Alkaril Chem.
*Chemical Description:* 2:1 Coconut Alkanolamide
*Category:* Detergent, Nonionic-anionic surfactant
*Applications:* Hair shampoo, bubble bath, specialty detergents
*Form:* Liquid; Color: Straw-color
*Properties:* Solubility: Good in water; thickening and foam stabilizing properties
*Composition:* 100% active ingred.
*References:* Technical Data # 15

**Alkamide 2104.** Alkaril Chem.
*Chemical Description:* 2:1 Coconut Diethanolamide
*Category:* Detergent, foam booster, nonionic-anionic
*Applications:* Detergent–textiles, household, cosmetics, hair shampoos, bubble bath, cleaners of street, building, hospitals; Emulsifier–making of cutting and soluble oils
*Form:* Viscous liquid; Color: Amber; Odor: Low
*Properties:* Solubility: Good in water; thickening and foam stabilizing properties
*Composition:* 100% active ingred.
*References:* Technical Data # 11

**Alkamide 2106.** Alkaril Chem.
*Chemical Description:* 2:1 Coconut Alkanolamide
*Category:* Detergent, nonionic-anionic
*Applications:* Compounded with alkaline salts for industrial and household floor cleaning formulations
*Form:* Viscous liquid; Color: Amber

*Properties:* Water-soluble; thickening and foam stabilizing properties; good wetting and soil suspension properties
*Composition:* 100% active ingred.
*References:* Technical Data # 12

**Alkamide 2112.** Alkaril Chem.
*Chemical Description:* 2:1 Fatty acid alkanolamide, sulfonate blend
*Category:* Detergent, nonionic-anionic
*Applications:* Heavy-duty detergent base; base for highly alkaline wax strippers and degreasers; cleaners of street, building, hospitals
*Form:* Viscous liquid; Color: Brown; Odor: Low
*Properties:* Solubility: Good in water; Cloud Pt.: 45 F max.
*Composition:* 100% active ingred.
*Reaction:* Alkaline
*References:* Technical Data

**Alkamide 2122.** Alkaril Chem.
*Chemical Description:* 2:1 alkanolamide sulfonate concentrate
*Category:* Detergent, wetting, dispersing and air entraining agent, nonionic-anionic
*Applications:* Air entraining agent in construction materials, e.g., prestressed concrete, concrete brick, pipe and block; light-duty liquid detergents
*Form:* Viscous liquid; Color: Amber
*Properties:* Solubility: Good in water; Cloud Pt.: 45 F max.; High foam stabilizing properties
*Composition:* 100% active ingred.
*References:* Technical Data # 16

**Alkamide 2124.** Alkaril Chem.
*Chemical Description:* 2:1 Lauric diethanolamide
*Category:* Detergent, wetting agent, emulsifier, foaming agent, nonionic-anionic surfactant
*Applications:* Hair shampoo concentrate, liquid detergent, bubble bath formulations, specialty emulsifiers
*Form:* Viscous liquid; Color: Amber
*Properties:* Solubility: Good in water; Sp. gr.: 1.0–1.05; Good thickening and foam stabilizing action
*Composition:* 100% active ingred.
*References:* Technical Data # 13

**Alkamide 2204.** Alkaril Chem.
*Chemical Description:* 2:1 Coconut diethanolamide
*Category:* Detergent, nonionic-anionic
*Applications:* Heavy-duty detergent base, household use, cleaners of street, buildings, hospital; rust-inhibitor
*Form:* Viscous liquid; Color: Amber; Odor: Low
*Properties:* Solubility: Good in water
*Composition:* 100% active ingred.
*Containers:* Steel tanks or drums may be used due to anti-rusting properties
*References:* Technical Data # 14

**Alkamide CDE.** Alkaril Chem.
*Chemical Description:* Coconut diethanolamide
*Category:* Detergent, emulsifier, stabilizer, nonionic surfactant

*Applications:* Foam booster, viscosity builder; Detergent–textiles, skins, leather, laundering, household, cosmetics, street, buildings and hospital cleaners; Emulsifiers–making of cutting and soluble oils; printing
*Form:* Liquid; Color: Light amber; Odor: Low
*Properties:* Solubility: In detergent systems, aromatic and chlorinated aliphatic solvents, petroleum solvents, mineral oils; Sp. gr.: 1.04 @ 20 C
*Composition:* 90% active ingred.; other components: free diethanolamine (7% max.), free fatty acid (0.5% max.), and amide ester (4% max.)
*Reaction:* Alkaline
*References:* Technical Data # 4

**Alkamide CDO.** Alkaril Chem.
*Chemical Description:* Coconut diethanolamide
*Category:* Detergent, nonionic
*Applications:* Foam booster, viscosity builder; Detergent–textiles, skins, leather, laundering, household, cosmetics, street, building and hospital cleaners; Wetting agent–removal of spots, rotary washer; Emulsufier–making of cutting and soluble oils; dispersing waxes, pigments, resins
*Form:* Liquid; Color: Amber; Odor: Low
*Properties:* Solubility: In aromatic and chlorinated aliphatic solvents, most petroleum solvents, mineral oils; Sp. gr.: 1.04 @ 20 C
*Composition:* 82% active ingred.; other components: free diethanolamine (7.0% max.), free fatty acid (0.5% max.), amide ester (4% max.)
*References:* Technical Data # 3

**Alkamide CL63.** Alkaril Chem.
*Chemical Description:* Coconut-lauric diethanolamine
*Category:* Detergent, emulsifier, nonionic
*Applications:* Toiletry preparations–shampoos, bubble bath (as active skin conditioner and foam stabilizer); cleaning formulations–rug shampoo, all-purpose cleaners thickening agents
*Form:* Liquid; Color: Clear to cloudy at room temperature
*Properties:* Solubility: In detergent systems, aromatic and chlorinated aliphatic solvents, petroleum solvents, mineral oils; Sp. gr.: 1.03–1.05
*Composition:* 92% min. active ingred.; other components: free diethanolamine (7% max.); free fatty acid (0.5% max.), amide ester (4 % max.)
*References:* Technical Data # 2

**Alkamide CME.** Alkaril Chem.
*Chemical Description:* Coconut monoethanolamides
*Category:* Detergent, foam booster, viscosity builder, opacifier
*Applications:* Detergents–laundering, household, cosmetics; street, buildings, hospital cleaners; pharmaceuticals
*Form:* Waxy solid; Color: White; Odor: Low
*Properties:* Solubility: In detergent systems, aromatic or chlorinated

aliphatic solvents, petroleum solvents, mineral oils; M.P.: 56 C
*Composition:* 95% active ingred.; other components: free monoethanolamine (2% max.), free fatty acid (0.5% max.)
*Reaction:* Neutral
*References:* Technical Data # 5

**Alkamide CMO.** Alkaril Chem.
*Chemical Description:* Coconut monoethanolamide
*Category:* Detergent, nonionic
*Applications:* Detergent–laundering, household, cosmetics, pharmaceuticals
*Form:* Waxy solid; Color: Cream
*Properties:* Solubility: In detergent systems, aromatic or chlorinated aliphatic solvents, petroleum solvents, mineral oils; M.P.: 50 C
*Composition:* 90% active ingred.; other components: free monoethanolamine (2% max.), free fatty acid (1.0% max.)
*References:* Technical Data # 5

**Alkamide HTDE.** Alkaril Chem.
*Chemical Description:* Modified alkanolamide of higher fatty acids
*Category:* Detergent, thickener, viscosity builder for kerosene, etc.
*Applications:* Detergent–textiles, skins, papermaking, household, cleaners of street, buildings, hospitals
*Form:* Waxy solid; Color: Cream; Odor: Low
*Properties:* M.P.: 55 C
*Composition:* 90% active ingred.; other components: free diethanolamine (8% max.), free fatty acid (as oleic) (1% max.)
*Reaction:* Alkaline
*References:* Technical Data # 9

**Alkamide HTME.** Alkaril Chem.
*Chemical Description:* Monoethanolamide of higher fatty acids
*Category:* Detergent, opacifier, thickener for cosmetics; nonionic
*Form:* Waxy solid; Color: Cream; Odor: Low
*Properties:* M.P.: 85–90 C
*Composition:* 90% active ingred.; other components: free monoethanolamine (1.5 max.)
*Reaction:* Alkaline
*References:* Technical Data # 10

**Alkamide L7DE.** Alkaril Chem.
*Chemical Description:* Lauric-myristic diethanolamine
*Category:* Detergent, foam booster, viscosity builder, nonionic
*Applications:* Superfatting and thickening agent for toiletry preparations–shampoos, bubble bath; cleaning formulations
*Form:* Paste (liquid above 25 C); Color: White; Odor: Low
*Properties:* Solubility: In detergent systems, aromatic and chlorinated aliphatic solvents, petroleum solvents, mineral oils; Sp. gr.: 1.03–1.05; M.P.: 15 C
*Composition:* 92% active ingred.; other components: free diethanolamine (7% max.), free fatty acid (0.5% max.), amide ester (4.0% max.)
*Reaction:* Alkaline
*References:* Technical Data # 2

**Alkamide L7DE-BT.** Alkaril Chem.
*Chemical Description:* Liquefied

form of modified lauric-myristic diethanolamide
*Category:* Detergent, foam stabilizer, viscosity builder
*Applications:* Shampoo concentrates
*Form:* Liquid; Color: Straw; Odor: Bland
*Properties:* Sp. gr.: 0.99; Visc.: 275 cps @ 30 C; F.P.: 0 C; Cloud Pt.: 0 C max.
*Composition:* 84% active ingred.
*Reaction:* Alkaline
*References:* Technical Data sheet

**Alkamide L9DE.** Alkaril Chem.
*Chemical Description:* Lauric diethanolamine
*Category:* Detergent, foam stabilizer, viscosity builder, nonionic
*Applications:* Toiletry applications–shampoos, bubble bath (as active skin conditioner and foam stabilizer); cleaning formulations–rug shampoo, all-purpose cleaners; thickening agents
*Form:* Waxy solid (liquid > 25 C); Color: White; Odor: Low
*Properties:* Solubility: In detergent systems, aromatic and chlorinated aliphatic solvents, petroleum solvents, mineral oils; Sp. gr.: 1.04; F.P.: 25 C
*Composition:* 92% active ingred.; free diethanolamine (7% max.); free fatty acid (0.5% max.); amide ester (4.0% max.)
*Reaction:* Alkaline
*References:* Technical Data # 2

**Alkamide L9ME.** Alkaril Chem.
*Chemical Description:* Lauric monoethanolamide
*Category:* Detergent, foam stabilizer, viscosity builder
*Applications:* Detergent–laundering, household, cosmetics (active skin protector in technical rinses, barrier creams, hand cleaners, etc.), cleaners of street, buildings, hospital; pharmaceuticals
*Form:* Waxy solid; Color: White; Odor: Low
*Properties:* Solubility: In detergent systems, aromatic and chlorinated aliphatic solvents, petroleum solvents, mineral oils; M.P.: 60 C
*Composition:* 95% active ingred.; other components: free monoethanolamine (2% max.), free fatty acid (0.5% max.)
*Reaction:* Alkaline
*References:* Technical Data # 1

**Alkamide RDO.** Alkaril Chem.
*Chemical Description:* Ricinoleic diethanolamide
*Category:* Emulsifier, lubricant, nonionic
*Applications:* Making of cutting and soluble oils; textiles, skins, leather, household; street, buildings, hospital cleaners; antiblocking and mold release agent
*Form:* Clear liquid; Color: Brown; Odor: Low
*Composition:* 100% active ingred.
*Reaction:* Alkaline
*References:* Technical Data # 7

**Alkamide SDO.** Alkaril Chem.
*Chemical Description:* Modified oleic diethanolamide
*Category:* Foam stabilizer, viscosity builder
*Applications:* Textiles, skins, laun-

dering, leather, household, cosmetics (shampoo and bubble bath as an active skin conditioner), street, buildings, hospital cleaners; cleaning hangings; making of cutting and soluble oils; corrosion inhibitor
*Form:* Viscous liquid; Color: Amber; Odor: Low
*Properties:* Solubility: In detergent systems, aromatic and chlorinated aliphatic solvents, petroleum solvents, mineral oils; Sp. gr.: 1.04
*Composition:* 80% active ingred.; other components: free diethanolamine (12% max.), free fatty acid (0.5% max.)
*Reaction:* Alkaline
*References:* Technical Data # 6

**Alkamidox CME-2.** Alkaril Chem.
*Chemical Description:* Coconut monoethanolamide ethoxylate containing 2 moles ethylene oxide
*Category:* Detergent, foam stabilizer, nonionic
*Applications:* Dishwashing detergents, heavy duty liquid laundry detergents, cosmetic applications
*Form:* Solid; Color: Cream
*Properties:* Solubility: Soluble in ethanol, xylene; insol. kerosene, mineral oil; Stability: Unstable in aqueous systems above pH 12, stable to strong alkalies
*Composition:* 100% active ingred.
*References:* Technical Data sheet

**Alkamidox CME-5.** Alkaril Chem.
*Chemical Description:* Coconut monoethanolamide ethoxylate containing 5 moles ethylene oxide
*Category:* Detergent, foam stabilizer, nonionic
*Applications:* Dishwashing detergents, heavy duty liquid laundry detergents, cosmetic applications
*Form:* Liquid; Color: Straw
*Properties:* Solubility: In water, ethanol, xylene; insol. kerosene, mineral oil; Stability: Unstable in aqueous systems, stable to strong alkalies
*Composition:* 100% active ingred.
*References:* Technical Data sheet

**Alkaminox C2.** Alkaril Chem.
*Chemical Description:* Tertiary amine with one coconut derived alkyl group and two hydroxyethyl groups
*Category:* Antistatic agent
*Applications:* Antistatic agent and corrosion inhibitor in steam generating and circulating systems
*Form:* Clear liquid; Color: Amber
*Properties:* Solubility: Sol. in most organic solvents, mineral acid; insol. water; Sp. gr.: 0.87
*References:* Technical Data #18-1

**Alkaminox T2.** Alkaril Chem.
*Chemical Description:* Tertiary amine with one tallow fatty group and two oxyethylene groups
*Category:* Emulsifier, dispersant
*Applications:* Emulsifier, dispersant, internal antistat for polyethylene, antistat agent for synthetic fibers
*Form:* Paste; Color: 8 max. (Gardner)
*Properties:* Solubility: Sol. in most organic solvents, mineral acid; insol. water; Sp. gr.: 0.92
*References:* Technical Data #18-4

**Alkaminox T-5.** Alkaril Chem.
*Chemical Description:* Tertiary amine with one tallow fatty group and two polyoxyethylene groups
*Category:* Emulsifier, dispersant
*Applications:* Emulsifier and dispersant in textile industry; dispersing agent for pigments in coatings; corrosion inhibitor in steam generating and circulating systems
*Form:* Clear liquid; Color: 8 max. (Gardner)
*Properties:* Solubility: Sol. in most organic solvents, mineral acid; insol. water; Sp. gr.: 0.97
*References:* Technical Data #18-5

**Alkaminox T-15.** Alkaril Chem.
*Chemical Description:* Tertiary amine with one tallow fatty group and two polyoxyethylene groups
*Category:* Emulsifier, dispersant
*Applications:* Emulsifier and dispersing agent in textile industry; emulsifier for wax emulsions
*Form:* Clear liquid; Color: 8 max. (Gardner)
*Properties:* Solubility: Sol. in most organic solvents, mineral acid and water; Sp. gr.: 1.03
*References:* Technical Data #18-7

**Alkamox C20.** Alkaril Chem.
*Chemical Description:* Amine oxide of ethoxylated coco amine
*Category:* Detergent, wetting agent, foam booster
*Applications:* Foam booster and stabilizer for latex foam for carpets and carpet backing
*Form:* Liquid; Color: Yellow; Odor: Faint
*Properties:* F.P.: < 0 C
*Composition:* 33% active ingred.; other components: water
*Reaction:* Neutral

**Alkamox CAPO.** Alkaril Chem.
*Chemical Description:* Coco amido propyl dimethyl amine oxide
*Category:* Detergent, wetting agent, foaming agent and foam stabilizer
*Applications:* Foaming agent and foam stabilizer for shampoos, bath products, fine fabric cleaners
*Form:* Liquid; Color: Colorless; Odor: Faint
*Properties:* Solubility: Complete in water; F.P.: < 0 C
*Composition:* 30% active ingred.; other components: water
*Reaction:* Neutral

**Alkamox LO.** Alkaril Chem.
*Chemical Description:* Amine oxide of lauryl dimethyl amine
*Category:* Detergent, wetting agent, foaming agent and foam stabilizer, antistat agent, nonionic-cationic
*Applications:* Foaming agent and foam stabilizer for shampoos and detergents; foam rubber; electroplating; paper coating; as a pour point depressant for mineral oils
*Form:* Liquid; Color: Colorless; Odor: Faint
*Properties:* Solubility: Excellent in water; F.P.: < 0 C
*Composition:* 30% active ingred.; other components: water
*Reaction:* Neutral
*References:* Technical Data #19-2

**Alkamuls Series.** Alkaril Chem.
*Chemical Description:* Polyethylene glycol esters

*Category:* Emulsifier, dispersing agents, nonionic
*Applications:* General emulsifiers for O/W emulsions, dispersing agents for cosmetic and industrial use including: textiles; leather; papermaking; paints; varnishes; foods; household; dyeing; finishing; sizing; coating; waterproofing; insecticides, herbicides, fungicides; treatment of grains and forage; making of cutting and soluble oils; dispersing waxes, pigments, resins; printing; dispersing dry powders; preparation of emulsified paints
*Form, Properties, Composition:* see individual listings
*Reaction:* Neutral

**Alkamuls 400-DL.** Alkaril Chem.
*Chemical Description:* PEG 400 dilaurate
*Form:* Liquid
*Properties:* HLB: 10.0
*Composition:* 100% active ingred.

**Alkamuls 400-DO.** Alkaril Chem.
*Chemical Description:* Polyethylene glycol 400 dioleate
*Category:* Emulsifier, wetting agents, nonionic
*Applications:* Emulsifiers for defoamers, cosmetic creams, agricultural oils; thickeners in lotions; opacifiers for shampoos; finishing agents for textile fibers
*Form:* Clear Liquid; Color: Amber
*Properties:* Sp. gr.: 0.97–0.99; HLB: 7.2
*Composition:* 100% active ingred.
*References:* Technical Data 24-5

**Alkamuls 400-DS.** Alkaril Chem.
*Chemical Description:* Polyethylene glycol 400 distearate
*Category and Applications:* See Alkamuls 400-DO
*Form:* Waxy flakes; Color: White
*Properties:* HLB: 7.8; Solidification Pt.: 95 F typ
*Composition:* 100% active ingred.

**Alkamuls 400-MC.** Alkaril Chem.
*Chemical Description:* Polyethylene glycol 400 monococoate
*Category and Applications:* See Alkamuls 400-DO
*Form:* Liquid
*Properties:* HLB: 13.9
*Composition:* 100% active ingred.

**Alkamuls 400-ML.** Alkaril Chem.
*Chemical Description:* Polyethylene glycol 400 monolaurate
*Category and Applications:* See Alkamuls 400-DO
*Form:* Clear liquid; Color: Amber
*Properties:* Sp. gr.: 1.02–1.04; Cloud Pt.: 50 F max.; HLB: 12.8
*Composition:* 100% active ingred.
*References:* Technical Data 24-1

**Alkamuls 400-MO.** Alkaril Chem.
*Chemical Description:* Polyethylene glycol 400 monooleate
*Category and Applications:* See Alkamuls 400-DO; also used in dairy industry as an emulsifier to wet out and emulsify powder milk substitute for feeding of calves
*Form:* Clear liquid; Color: Amber
*Properties:* Sp. gr.: 1.01–1.03; HLB: 11.0; Solidification Pt.: $<$ 40 F
*Composition:* 100% active ingred.
*References:* Technical Data 24–2

**Alkamuls 400-MS.** Alkaril Chem.

*Chemical Description:* Polyethylene glycol 400 monostearate
*Category and Applications:* See Alkamuls 400-DO
*Form:* Paste
*Properties:* HLB: 11.2
*Composition:* 100% active ingred.

**Alkamuls 600-DO.** Alkaril Chem.
*Chemical Description:* Polyethylene glycol 600 dioleate
*Category and Applications:* See Alkamuls 400-DO
*Form:* Liquid; Color: Amber
*Properties:* Sp. gr.: 1.01-1.03; Solidification Pt.: 15 C typical
*Composition:* 100% active ingred.
*References:* Technical Data 24-11

**Alkamuls 600-ML.** Alkaril Chem.
*Chemical Description:* Polyethylene glycol 600 monolaurate
*Form:* Liquid
*Properties:* HLB: 14.6
*Composition:* 100% active

**Alkamuls 600-MO.** Alkaril Chem.
*Chemical Description:* Polyethylene glycol 600 monooleate
*Form:* Liquid
*Properties:* HLB: 13.0
*Composition:* 100% active

**Alkamuls 600-MS.** Alkaril Chem.
*Chemical Description:* Polyethylene glycol 600 monostearate
*Form:* Solid
*Properties:* HLB: 13.2
*Composition:* 100% active

**Alkamuls DEG Series.** Alkaril Chem.
*Chemical Description:* See individual listings
*Category:* Emulsifier, dispersing agents, nonionic
*Applications:* Detergent–skins, leather, papermaking, transport; Wetting agent–tanning, coating, insecticides, fungicides, herbicides; Emulsifiers–making of cutting and soluble oils; dispersing waxes, pigments, resins; printing; preparation of emulsified paints
*Composition:* 100% active ingred.
*Reaction:* Neutral

**Alkamuls DEG-DT.** Alkaril Chem.
*Chemical Description:* DEG Ditallate
*Category:* Dispersing agent, emulsifier
*Form:* Liquid; Color: Brown; Odor: Faint

**Alkamuls DEG-ML.** Alkaril Chem.
*Chemical Description:* DEG monolaurate
*Category:* Emulsifier
*Form:* Liquid; Color: Brown; Odor: Faint
*Properties:* HLB: 6.5

**Alkamuls DEG-MO.** Alkaril Chem.
*Chemical Description:* DEG monooleate
*Category:* Lubricant
*Form:* Liquid; Color: Brown; Odor: Faint
*Properties:* HLB: 4.7

**Alkamuls DEG-MS.** Alkaril Chem.
*Chemical Description:* DEG monostearate
*Category:* Opacifier, emulsifier, plasticizer
*Form:* Liquid; Color: Brown; Odor: Faint
*Properties:* HLB: 4.7

**Alkamuls GML.** Alkaril Chem.
*Chemical Description:* Glycerol monolaurate
*Category:* Emulsifier, nonionic
*Applications:* Textiles, leather, foods
*Form:* Paste
*Properties:* HLB: 3.0
*Composition:* 100% active
*Reaction:* Neutral

**Alkamuls GMO.** Alkaril Chem.
*Chemical Description:* Glycerol monooleate
*Category and Applications:* See Alkamuls 400-DO
*Form:* Liquid; Color: Light yellow
*Properties:* M.P.: 25 C max. HLB: 3.4
*Composition:* 100% active ingred.
*References:* Technical Data 24-33

**Alkamuls GMR.** Alkaril Chem.
*Chemical Description:* Glycerol monoricinoleate
*Category:* Emulsifier, stabilizer, nonionic
*Applications:* Textiles, leather, foods
*Form:* Liquid
*Composition:* 100% active
*Reaction:* Neutral

**Alkamuls GMS.** Alkaril Chem.
*Chemical Description:* Glycerol monostearate
*Category and Applications:* See Alkamuls 400-DO
*Form:* Solid
*Properties:* HLB: 3.4

**Alkamuls PEMO.** Alkaril Chem.
*Chemical Description:* Pentaerythritol monooleate
*Category:* Corrosion preventive in lubricants
*Applications:* Prevents corrosion in automotive and steel parts by depositing as a film on exposed parts
*Form:* Liquid; Color: Brown; Odor: Low
*Composition:* 100% active ingred.
*Reaction:* Neutral

**Alkamuls PSML-20.** Alkaril Chem.
*Chemical Description:* Ethoxylated sorbitan (20) monolaurate
*Category:* Emulsifier, solubilizer, nonionic surfactant
*Applications:* General purpose water-soluble surfactants, used as O/W emulsifiers
*Form:* Liquid; Color: Yellow
*Properties:* HLB: 16.7
*Composition:* 97% active ingred.; 3% max. water
*References:* Technical Data 25-11

**Alkamuls PSMO-5.** Alkaril Chem.
*Chemical Description:* Polyoxyethylene sorbitan (5) monooleate
*Category:* Emulsifier, nonionic surfactant
*Applications:* Preparation of emulsified paints; insecticides, herbicides, fungicides; foods; cosmetics
*Form:* Liquid; Color: Amber; Odor: Typical
*Properties:* Solubility: Water insol.; HLB: 10
*Composition:* 98.5% active ingred.
*Reaction:* Neutral
*References:* Technical Data 25-15

**Alkamuls PSMO-20.** Alkaril Chem.
*Chemical Description:* Polyoxyethylene sorbitan (20) monooleate

*Category:* Wetting agent, emulsifier, nonionic
*Applications:* Preparation of emulsified paints; insecticides; herbicides; fungicides; textiles; foods; cosmetics
*Form:* Liquid; Color: Yellow; Odor: Typical
*Properties:* Solubility: Water sol.; HLB: 15
*Composition:* 97% active ingred.
*Reaction:* Neutral
*References:* Technical Data 25-14

**Alkamuls PSMS-4.** Alkaril Chem.
*Chemical Description:* Polyoxyethylene sorbitan (4) monostearate
*Category:* Emulsifier, nonionic
*Applications:* Preparation of emulsified paints; insecticides, herbicides, fungicides; textiles; foods; cosmetics
*Form:* Solid; Color: Tan; Odor: Typical
*Properties:* Solubility: Water insol.; HLB: 9.6
*Composition:* 97% active
*Reaction:* Neutral
*References:* Technical Data 25-13

**Alkamuls PSMS-20.** Alkaril Chem.
*Chemical Description:* Polyoxyethylene sorbitan (20) monostearate
*Category:* Wetting agent, emulsifier, nonionic
*Applications:* Preparation of emulsified paints; insecticides, herbicides, fungicides; foods; cosmetics
*Form:* Viscous liquid, gel on standing; Color: Yellow; Odor: Typical
*Properties:* Solubility: Water sol.; HLB: 14.9
*Composition:* 97% active ingred.
*Reaction:* Neutral
*References:* Technical Data 25-12

**Alkamuls PSTO-20.** Alkaril Chem.
*Chemical Description:* Polyoxyethylene sorbitan (20) trioleate
*Category:* Emulsifier, nonionic
*Applications:* Preparation of emulsified paints; insecticides, herbicides, fungicides; foods; cosmetics
*Form:* Liquid; Color: Amber; Odor: Typical
*Properties:* Solubility: Water insol.; HLB: 11
*Composition:* 97% active ingred.
*Reaction:* Neutral

**Alkamuls PSTS-20.** Alkaril Chem.
*Chemical Description:* Polyoxyethylene sorbitan (20) tristearate
*Category:* Wetting agent, emulsifier, nonionic
*Applications:* Preparation of emulsified paints; insecticides, herbicides, fungicides; foods; cosmetics
*Form:* Solid; Color: Tan; Odor: Typical
*Properties:* Solubility: Water insol.; HLB: 10.5
*Composition:* 97% active ingred.
*Reaction:* Neutral
*References:* Technical Data 25-17

**Alkamuls SML.** Alkaril Chem.
*Chemical Description:* Sorbitan monolaurate
*Category:* Emulsifier, nonionic
*Applications:* Fat liquoring, dyeing, insecticides, herbicides, fungicides, textiles, foods, cosmetics
*Form:* Liquid (may form a sediment on standing which redissolves on warming); Color: Amber; Odor: Typical

*Properties:* Solubility: Water insol., moderately soluble in most alcohols, vegetable oils, mineral oils; Sp. gr.: 1.05 @ 60 F; HLB: 8.5
*Composition:* 97% active ingred.
*Reaction:* Neutral
*References:* Technical Data 25-1

**Alkamuls SMO.** Alkaril Chem.
*Chemical Description:* Sorbitan monooleate
*Category:* Wetting agent, emulsifier, nonionic
*Applications:* Making of cutting and soluble oils; dispersing waxes, pigments, resins; dyeing; insecticides, herbicides, fungicides; textiles; foods; cosmetics
*Form:* Liquid; Color: Amber; Odor: Typical
*Properties:* Solubility: Water insol., soluble in most vegetable and mineral oils, not soluble in glycol; Sp. gr.: 1.0; HLB: 4.3
*Composition:* 97% active ingred.
*Reaction:* Neutral
*References:* Technical Data 25-3

**Alkamuls SMS.** Alkaril Chem.
*Chemical Description:* Sorbitan monostearate
*Category:* Emulsifier, nonionic
*Applications:* Foods; cosmetics; dispersing waxes, pigments, resins
*Form:* Flakes; Color: Cream to tan; Odor: Typical
*Properties:* Solubility: Water insol.; HLB: 4.7
*Composition:* 98.5% active ingred.
*Reaction:* Neutral
*References:* Technical Data 25-2

**Alkamuls STO.** Alkaril Chem.
*Category:* Emulsifier, nonionic
*Applications:* Textiles; foods; cosmetics; insecticides, herbicides, fungicides
*Form:* Liquid; Color: Amber; Odor: Typical
*Properties:* Solubility: Water insol.; HLB: 1.8
*Composition:* 100% active ingred.
*Reaction:* Neutral

**Alkamuls STS.** Alkaril Chem.
*Category:* Emulsifier
*Applications:* Foods; cosmetics; dispersing waxes, pigments, resins
*Form:* Solid; Color: Tan; Odor: Typical
*Properties:* Solubility: Water insol.; HLB: 2.1
*Composition:* 100% active ingred.
*Reaction:* Neutral

**Alkanol 189-S.** DuPont Co.
*Chemical Description:* Sodium alkyl sulfonate
*Category:* Wetting agent, detergent, penetrant, anionic
*Applications:* Metal cleaning and pickling; bottle washing; general detergent applications
*Form:* Liquid; Color: Reddish-brown
*Properties:* Visc.: 30 cps @ 80 F; Stability: Good in acid, neutral and alkaline media; Surface Tension: 31.2 dynes/cm (1% sol'n. @ 90 F)
*Composition:* 31.5% active ingred.
*References:* Data sheet # 3

**Alkanol A-CN.** DuPont Co.
*Chemical Description:* Ethylene oxide condensate
*Category:* Antiprecipitant, dyeing

assistant, dye solubilizer, contrast agent, nonionic surfactant
*Applications:* Use in application of dyes to styled carpeting; aids dye solubility and exhaustion of two types of dye in same bath; aids level absorption of dyes
*Form:* Liquid; Color: Light amber
*Properties:* Solubility: > 50% in water; Density: 8.2 lb/gal @ 25 C; Stability: Indefinite in storage; Surface Tension: 38 dynes/cm (1% conc. @ 25 C)
*References:* Data sheet

**Alkanol CNR.** DuPont
*Chemical Description:* Amphoteric surfactant
*Category:* Dye assistant
*Applications:* Dye assistant for carpeting with both cationic- and anionic-dyeable nylon fibers; inhibits cross-staining; enhances level absorption of dyes
*Form:* Clear liquid; Color: Amber
*Properties:* Solubility: Miscible with water; Density: 8.61 lb/gal; Sp. gr.: 1.033
*References:* Bulletin

**Alkanol DOA.** DuPont Co.
*Chemical Description:* Amine polymer solution
*Category:* Dispersant
*Applications:* General dispersant for nonaqueous use; dispersant for drycleaning systems; paint manufacture; for use in reboilers in chemical manufacturing; to reduce sludge formation in storage of nonaqueous products; in quenching oils to prevent sludging
*Form:* Viscous liquid; Color: Light amber
*Properties:* Solubility: Complete in kerosene, hydrocarbons; insol. in acetone; Density: 7.4 lb/gal @ 25 C; Visc.: 540 cs @ 25 C; Flash Pt. (C.O.C.): 68 C; Pour Pt.: < 29 C; Stability: Excellent thermal stability
*Toxicity:* Acute oral toxicity, skin irritant
*References:* Data sheet

**Alkanol DW.** DuPont Co.
*Chemical Description:* Sodium alkylaryl sulfonate
*Category:* Emulsifier, wetting agent, detergent, anionic
*Applications:* Emulsifier for textile cleaning solvents; scouring, bleaching and dyeing textiles; defibering of waste paper; tar removal from wool; rewetting of textiles and paper mill felts; penetrant and spreader for shoe finishes; bottle washing; acid and alkaline cleaners; dairy and paint cleaners
*Form:* Liquid; Color: Yellow; Odor: Alcoholic
*Properties:* Solubility: Miscible in water; Density: 8.9 lb/gal; Cloud Pt.: 5 C (max.); Stability: Good in acid, alkali media, and storage
*Composition:* 28% active ingred.
*References:* Data sheet # 2

**Alkanol ND.** DuPont Co.
*Chemical Description:* Alkyl diaryl sulfonate
*Category:* Dyeing assistant, anionic surfactant
*Applications:* Dyeing assistant, aids leveling of acid dyes on nylon
*Form:* Liquid; Color: Yellow; Odor: Terpene

*Properties:* Solubility: Miscible with water; Density: 9.6 lb/gal; Visc.: Slightly viscous; Stability: Good in strong acids, alkalies, and in storage
*Composition:* 45% active ingred.
*References:* Data sheet

**Alkanol S.** DuPont Co.
*Chemical Description:* Sodium tetrahydronaphthalene sulfonate
*Category:* Dispersant, stabilizer, solubilizing agent, anionic
*Applications:* As dispersing, stabilizing and solubilizing agent in emulsions systems–textile printing compositions, cosmetic lotions and creams
*Form:* Small flakes; Color: Light cream
*Properties:* Solubility: Water sol. approx. 20%; Density: 2.6 lb/gal
*Composition:* 98% active ingred.
*References:* Data sheet # 3A

**Alkanol WXN.** DuPont Co.
*Chemical Description:* Sodium alkylaryl sulfonate
*Category:* Wetting, rewetting agent, dyeing assistant, anionic
*Applications:* Textile dyeing assistant and rewetting agent; leveling of acid dyes on corrugating medium and calender stains on boxboard; rewetting dried paper broke; rag cooking; dust control in coal piles
*Form:* Liquid; Color: Light yellow; Odor: Alcoholic
*Properties:* Solubility: (in water) miscible; Density: 8.6 lb/gal; Cloud Pt.: 0 C max.; Stability: Good in acid and alkali media, and storage; Surface Tension: 26 dynes/cm (1% sol'n. @ 25 C)
*Composition:* 30% active ingred.
*References:* Data sheet # 4

**Alkanol XC.** DuPont Co.
*Chemical Description:* Sodium alkylnaphthalene sulfonate
*Category:* Wetting agent, dispersant, penetrant, anionic
*Applications:* Rapid wetting and dispersing where excess foaming is undesirable and detergency is unnecessary; bleaching and dyeing of textiles, leather, paper; paper and paper mill felts; reduction of shrinkage in ceramics manufacture; dry colors manufacture
*Form:* Powder; Color: Light cream
*Properties:* Solubility: Soluble in ethyl alcohol, acetone, benzene, 8–10% in water @ 25 C; Sp. gr.: 0.55; Stability: Good in acid, alkali media and in storage
*References:* Data sheet # 1

**Alkaphos 3.** Alkaril Chem.
*Chemical Description:* Free acid of complex organic phosphate esters
*Category:* Detergent, wetting agent, emulsifier, solubilizer, antistatic agent, lubricants, anionic
*Applications:* Detergent–textiles, skins, leather, laundering esp. dry cleaning detergents, household and street, buildings, hospitals cleaners; esp. in high alkaline systems; Wetting agent–desizing and boiling off cottons; peroxide bleaching; finishing agent; insecticides, herbicides, fungicides; removing starches, wall paper; cleaning hangings; Emulsifer–Kier boil-

ing; making of cutting and soluble oils; printing
*Form:* Viscous liquid; Color: Pale yellow; Odor: Low
*Properties:* Solubility:Sol. in water and most aromatic and chlorinated solvents; more oil-sol. than Alkaphos-6 and -10; Sp. gr.: 1.025; Visc.: 863 cps @ 75 F; Stability: Excellent in acid and alkali media
*Composition:* 100% active ingred.
*Reaction:* Acid
*References:* Technical Data 45

**Alkaphos 6.** Alkaril Chem.
*Chemical Description:* Free acid of complex organic phosphate esters
*Category:* Detergent, wetting agent, emulsifier, solubilizer, antistatic agent, lubricant, anionic surfactant
*Applications:* See Alkaphos 3
*Form:* Viscous liquid; Color: Pale yellow; Odor: Low
*Properties:* Solubility: Sol. in water and most aromatic and chlorinated solvents; more oil-sol. than Alkaphos-10 and more hydrophilic than Alkaphos-3; Sp. gr.: 1.055; Visc.: 1045 @ 75 F; Stability: Excellent in acid and alkali media; Good rust inhibition properties
*Composition:* 100% active ingred.
*Reaction:* Acid
*References:* Technical Data 45

**Alkaphos 10.** Alkaril Chem.
*Chemical Description:* Free acid of complex organic phosphate esters
*Category:* Detergent, wetting agent, emulsifier, solubilizer, antistatic agent, lubricant, anionic
*Applications:* See Alkaphos 3; also water-repellent finish for cottons
*Form:* Viscous liquid; Color: Pale yellow; Odor: Low
*Properties:* Solubility: Sol. in water and most aromatic and chlorinated solvents; Sp. gr.: 1.085; Visc.: 2100 cps @ 75 F; Stability: Excellent in acid and alkali media; Good rust inhibition; Excellent cloud pt. depressant for nonionics
*Composition:* 100% active ingred.
*Reaction:* Acid
*References:* Technical Data 45

**Alkaphos K 380.** Alkaril Chem.
*Chemical Description:* Potassium salt of organic phosphate ester
*Category:* Solubilizer, hydrotropic surfactant, anionic
*Applications:* Solubilizer for built liquid detergents; liquid rinse aids to increase the solubility and raise cloud pt. of low foam nonionics
*Form:* Clear liquid
*Properties:* Cloud Pt.: $< 0$ C
*Composition:* 44% active ingred.
*References:* Technical Data 47

**Alkaphos QS.** Alkaril Chem.
*Chemical Description:* Free acid of complex organic phosphate ester
*Category:* Detergent, wetting agent, emulsifier, hydrotropic surfactant, solubilizer, anionic
*Applications:* Detergent–textiles, skins, laundering, household and street, building, hospital cleaners, esp. high alkaline and high electrolyte content cleaning preparations; Emulsifier–printing
*Form:* Liquid; Color: Colorless; Odor: Bland

*Properties:* Solubility: Excellent in water, caustic soda liquid and caustic potash liquid; Cloud Pt.: 68 C (1% sol'n. in 15% NaOH)
*Composition:* 90% active ingred.
*Reaction:* Alkaline
*References:* Technical Data 46

**Alkapol PEG 400.** Alkaril Chem.
*Chemical Description:* Polyoxyethylene glycol with average m.w. of 400
*Category:* Intermediate, plasticizer, solvent, lubricant
*Applications:* Intermediate for production of PEG-fatty acid esters; solvent or co-solvent for cosmetics, pharmaceuticals
*Form:* Liquid; Color: White
*Properties:* Solubility: Water sol.; Sp. gr.: 1.13 (20/15.5 C); F.P.: 4–8 C; Stability: Good physical chemical stability
*References:* Technical Data sheet

**Alkapol PEG 600.** Alkaril Chem.
*Chemical Description:* Polyoxyethylene glycol with average m.w. 600
*Category:* Intermediate, plasticizer, solvent, lubricant
*Applications:* Intermediate for production of PEG-fatty acid esters; solvent or co-solvent for cosmetics, pharmaceuticals
*Form:* Liquid; Color: White
*Properties:* Solubility: Water sol.; Sp. gr.: 1.13 (30/15.5 C); F.P.: 20–25 C; Stability: Good physical chemical stability
*References:* Technical Data sheet

**Alkaquat C.** Alkaril Chem.
*Chemical Description:* Dicoco imidazolinium quaternary
*Category:* Dyestuff retardant
*Applications:* Textiles processing
*Form:* Liquid; Color: Amber; Odor: Of solvent
*Properties:* Solubility: Dispersible in water; F.P.: $< 0$ C
*Composition:* 75% active ingred.; other components: isopropanol
*Reaction:* Neutral

**Alkaquat DMB-451.** Alkaril Chem.
*Chemical Description:* N-alkyl ($C_{12}$-$C_{14}$-$C_{16}$) methyl benzyl ammonium chloride
*Category:* Wetting agent, emulsifier, biocide, disinfectant, cationic
*Applications:* Bacteriocidal and cleaning formulations for beverage industry, dairy industry (plant equipment, milk tank trucks, milk pails cleaners), food processing; water treatment (algae control in swimming pools, water cooling systems); slimicides and mold control in paper industries; general disinfection and sanitization for public washrooms, hospitals, laundires; pest control; preservatives; antidandruff rinses
*Form:* Liquid; Color: Pale yellow (working solutions are odorless, colorless)
*Properties:* Solubility: Sol. in water, ethanol, acetone, many aliphatic solvents; Sp. gr.: 0.96; Stability: Good; Surface Tension: 33 dynes/cm @ 1% sol'n.
*Composition:* 50% min. active ingred.; other components: ethanol (10%), water (40% approx.)
*Toxicity:* Skin burns, severe eye

damage at concentration supplied
*References:* Technical Data sheet

**Alkaquat O.** Alkaril Chem.
*Chemical Description:* Oleic imidazoline quaternary
*Category:* Softening agent, cationic
*Applications:* Liquid form of Alkaquat T for use in high concentrations of fabric softeners; textiles; leather; laundering; papermaking; household; cosmetics
*Form:* Liquid; Color: Amber; Odor: Of solvent
*Properties:* Sp. gr.: 0.95; F.P.: < 0 C
*Composition:* 75% active ingred.; other components: isopropanol
*Reaction:* Neutral
*References:* Technical Data

**Alkaquat ST.** Alkaril Chem.
*Chemical Description:* Modified dialkyl imidazolinium quaternary based on stearyl derivative
*Category:* Fabric softener, antistat, cationic
*Applications:* Fabric softener and antistat for textile fibers, esp. for direct application to cloth; manufacture of sheet fabric softeners for dryer use
*Form:* Solid; Color: Cream; Odor: Isopropanol solvent
*Properties:* Solubility: Isopropanol solvent; M.P.: 45 C
*Composition:* 75% active ingred.; other components: isopropanol
*Reaction:* Neutral
*References:* Technical Data 71

**Alkaquat T.** Alkaril Chem.
*Chemical Description:* Ditallow imidazolinium quaternary
*Category:* Softening agent, cationic
*Applications:* Fabric softener for household and commercial laundry use and for textile finishing operations; papermaking; cosmetics
*Form:* Slightly viscous fluid; Color: Pale yellow; Odor: Typical
*Properties:* Solubility: Water insol.; Sp. gr.: 0.95
*Composition:* 74–76% active ingred.; other components: isopropanol
*Reaction:* Neutral
*References:* Technical Data 70

**Alkaquest EDTA.** Alkaril Chem.
*Chemical Description:* Tetrasodium ethylene diamine tetra acetate, aqueous solution
*Category:* Chelating agent
*Applications:* General purpose chelating agent; complexes calcium, magnesium, iron, etc., through wide pH range
*Form:* Liquid; Color: Straw
*Properties:* m.w.: 380 (anhydrous); Sp. gr.: 1.280; Chelation value: 10 (calcium, gms. of calcium carbonate sequestered by 100 gms. of product)
*Composition:* 37.5% active ingred.
*References:* Technical Data 60

**Alkasurf ALS.** Alkaril Chem.
*Chemical Description:* Ammonium lauryl sulfate
*Category:* Detergent, wetting agent, emulsifier, anionic
*Applications:* Hair shampoo and other cosmetics; household use
*Form:* Liquid; Color: Light; Odor: Typical

*Properties:* Solubility: Water sol.
*Composition:* 27% active ingred.; other components: water
*Reaction:* Neutral

**Alkasurf CA.** Alkaril Chem.
*Chemical Description:* Calcium alkaryl sulfonate
*Category:* Emulsifier
*Applications:* Emulsifier for agricultural oils; insecticides, herbicides, fungicides; also in combination with ethoxylated nonionics in self-dispersing liquids
*Form:* Liquid; Color: Amber; Odor: Typical
*Properties:* Solubility: Water insol.; HLB: 10.5
*Composition:* 60% active ingred.; other components: buthanol
*Reaction:* Neutral
*References:* Technical Data 44

**Alkasurf CBX16.** Alkaril Chem.
*Chemical Description:* Fatty alcohol polyglycol ether carboxylic salt
*Category:* Detergent, wetting agent, anionic
*Applications:* Detergent with improved resistance to alkali/electrolytes; textiles; laundering; household; street, buildings, hospitals cleaners; desizing and boiling off cottons
*Form:* Liquid; Color: Light yellow; Odor: None
*Properties:* Solubility: Good in alkali; F.P.: < 0 C
*Composition:* 70% active ingred.
*Reaction:* Neutral
*References:* Technical Data 40

**Alkasurf CBX16 Acid.** Alkaril Chem.
*Chemical Description:* Fatty alcohol polyglycol ether carboxylic acid
*Category:* Detergent, wetting agent, anionic
*Applications:* Detergent with improved resistance to alkali/electrolytes; textiles; laundering; household; street, buildings, hospitals cleaners; desizing and boiling off cottons
*Form:* Liquid; Color: Light yellow; Odor: None
*Properties:* Solubility: Good in alkali
*Composition:* 100% active ingred.
*Reaction:* Acid
*References:* Technical Data 40

**Alkasurf CO-25.** Alkaril Chem.
*Chemical Description:* Ethoxylated castor oil
*Category:* Detergent, wetting agent, emulsifier, nonionic
*Applications:* Textiles; cosmetics
*Form:* Liquid; Color: Yellow; Odor: Typical
*Composition:* 100% active ingred.
*Reaction:* Neutral

**Alkasurf DLS.** Alkaril Chem.
*Chemical Description:* Diethanolamine lauryl sulfate
*Category:* Detergent, wetting agent, emulsifier
*Applications:* Hair shampoos, cosmetics, household use
*Form:* Liquid; Color: Light; Odor: Typical
*Properties:* Solubility: Water sol.
*Composition:* 35% active ingred.; other components: water

*Reaction:* Neutral

**Alkasurf EA-60.** Alkaril Chem.
*Chemical Description:* Ammonium lauryl ether sulfate
*Category:* Detergent, wetting agent, emulsifier, anionic
*Applications:* General purpose surfactant, esp. good quality dishwashing detergent, shampoo, bubble bath; also laundering, dyeing, household, cosmetic, cleaners of street, buildings, hospitals; wetting agents—dyeing, spot removal
*Form:* Clear slightly viscous liquid; Color: Light amber; Odor: Typical
*Properties:* Solubility: Water sol.; Sp. gr.: 1.04 @ 20 C; Cloud Pt.: < 40 F
*Composition:* 58% active ingred.; other components: water, ethanol
*Reaction:* Neutral
*References:* Tech. Data

**Alkasurf ES-60.** Alkaril Chem.
*Chemical Description:* Sodium lauryl ether sulfate
*Category:* Detergent, emulsifier, wetting agent, anionic
*Applications:* General purpose surfactant esp. dishwashing detergent, shampoo, bubble bath; also textiles; laundering; dyeing; household; cosmetics; street, buildings, hospitals cleaners; dyeing and spot removal as wetting agent
*Composition:* 58% active ingred.; other components: water, ethanol
*Reaction:* Neutral
*References:* Tech. Data

**Alkasurf IPAM.** Alkaril Chem.
*Chemical Description:* Linear dodecyl benzene isopropyl amine sulfonate
*Category:* Emulsifier
*Applications:* Oil-soluble emulsifier, dry cleaning surfactant, emulsion degreasers, fuel oil additives, textiles, transport, insecticides, fungicides, herbicides
*Form:* Clear viscous liquid; Color: Pale yellow
*Properties:* Solubility: Water insol., sol. in most commonly used solvents; Sp. gr.: 1.06
*Composition:* 95% active ingred.
*Reaction:* Neutral
*References:* Tech. Data

**Alkasurf LA-3.** Alkaril Chem.
*Chemical Description:* Linear fatty alcohol ethoxylate
*Category:* Emulsifier, nonionic
*Applications:* Emulsifier and surfactant intermediate esp. in agricultural concentrates, dry cleaning soaps, degreasing compounds, wool lubricants, rubber based paints; making of cutting and sol. oils; dispersing waxes, pigments, resins
*Form:* Liquid; Color: Transluscent to opaque; Odor:
*Properties:* Solubility: Water insol.; Sp. gr.: 0.9 (25/25 C); Visc.: 19 (100 F); Cloud Pt.: 58–62 C; Stability: Good; HLB: 7.8
*Composition:* 100% active ingred.
*Reaction:* Neutral
*References:* Tech. Data sheet 21-1

**Alkasurf LA-7.** Alkaril Chem.
*Chemical Description:* Linear fatty alcohol ethoxylate
*Category:* Detergent, emulsifier,

wetting agent, nonionic
*Applications:* General purpose detergent–household, industrial cleaners; textiles; laundering; papermaking; paints, varnishes; transport; Wetting agent–caustic and mineral acid cleaners; dyeing; rewetting dried skins; pickling; electroplating; rotary washer; wetting pulp in beater; removing starches, wall paper; rewetting agent for paper towels, tissues; Emulsifier–making of cutting and soluble oils; dispersing waxes, pigments, resins; printing
*Form:* Liquid; Color: Light; Odor:
*Properties:* Solubility: Water sol.; Sp. gr.: 0.96 (50/25 C); Visc.: 32 cps @ 100 F; M.P.: 11–15 C; HLB: 12; Cloud Pt.: 44–48 C; Flash Pt.: 440 F (COC); Pour Pt.: 70 F; Stability: Good; Biodegradable
*Composition:* 100% active ingred.
*Reaction:* Neutral
*References:* Tech. Data sheet 21-2

**Alkasurf LA-9.** Alkaril Chem.
*Chemical Description:* Linear fatty alcohol ethoxylate
*Category:* Detergent, wetting agent, emulsifier, nonionic
*Applications:* General purpose surfactant–household and industrial cleaners; textiles; laundering; dyeing; papermaking; paints; varnishes; transport; cosmetics; Wetting agents–dyeing; rewetting dried skins; pickling; electroplating; rotary washer; wetting pulp in beater; caustic and mineral acid cleaners; Emulsifier–making of cutting and sol. oils; dispersing waxes, pigments, resins; printing
*Form:* Semisolid; Color: Light; Odor: Typical
*Properties:* Solubility: Water sol.; Sp. gr.: 1.01 (20 C); HLB: 13.1; Stability: Good; Biodegradable
*Composition:* 100% active ingred.
*Reaction:* Neutral
*References:* Tech. Data sheet 21-3

**Alkasurf LA-9 85%.** Alkaril Chem.
*Chemical Description:* See Alkasurf LA-9
*Category and Applications:* See Alkasurf LA-9
*Form:* Liquid form of Alkasurf LA-9
*Composition:* 85% active ingred.

**Alkasurf LA-12.** Alkaril Chem.
*Chemical Description:* Linear fatty alcohol ethoxylate
*Category:* Detergent, wetting agent, emulsifier, nonionic
*Applications:* See Alkasurf LA-9; also kier boiling; O/W emulsifier and solubilizer for topical cosmetic applications
*Form:* White solid; Odor: Low
*Properties:* Solubility: Water sol.; Sp. gr.: 1.0 (50/25 C); HLB: 14.4
*Composition:* 100% active ingred.
*Reaction:* Neutral

**Alkasurf LA-15.** Alkaril Chem.
*Chemical Description:* Linear fatty alcohol ethoxylate
*Category:* Emulsifier, detergent, nonionic
*Applications:* O/W emulsifier and solubilizer for topical cosmetic applications (has low sol. in nonpolar solvents); household and industrial cleaners (esp. high alka-

linity, high temperature cleaners); emulsion polymerization systems
*Form:* White solid
*Properties:* Biodegradable
*Composition:* 100% active ingred.
*References:* Tech. Data sheet 21-5

**Alkasurf LA-16.** Alkaril Chem.
*Chemical Description:* Straight chain polyoxyethylene fatty alcohol condensate
*Category:* Detergent, emulsifier, wetting agent, nonionic
*Applications:* Detergent–textiles, skins, leather, laundering, dyeing, papermaking, transport, household, cosmetics, industrial cleaners; Wetting agent–desizing and boiling off cottons; carbonizing woolens; dyeing; rewetting dried skins; tanning; pickling; electroplating; rotary washer; removing starches, wall paper; Emulsifer–making of cutting and sol. oils; dispersing waxes, pigments, resins; printing
*Form:* Clear colorless liquid; Odor: Low
*Properties:* Solubility: Water sol.; Sp. gr.: 1.0 (20/20 C); Visc.: 267 SUS (20 C); Biodegradable
*Composition:* 100% active ingred.
*Reaction:* Neutral
*References:* Tech. Data sheet 22-1

**Alkasurf LA Acid.** Alkaril Chem.
*Chemical Description:* High active sulfonic acid produced from the continuous $SO_3$ sulfonation of linear alkylate
*Category:* Detergent, emulsifier, penetrating and foaming agent
*Applications:* Dishwashing and heavy-duty liquid, powdered and paste detergents; car shampoos; bubble baths; degreasers and acid cleaners; dry cleaning detergents; metal cleaners; emulsion polymerization; agricultural emulsifiers
*Form:* Viscous liquid; Color: Brown
*Properties:* Stability: Good in hard and soft water, strong alkalies and strong acids; Biodegradable
*Composition:* 96% active ingred.
*References:* Tech. Data Sheet 41

**Alkasurf LAN-23.** Alkaril Chem.
*Chemical Description:* Linear fatty alcohol ethoxylate
*Category:* Emulsifier, nonionic surfactant
*Applications:* O/W emulsifier esp. topical cosmetic applications
*Form:* Solid; Color: Almost colorless; Odor: Characteristic
*Properties:* Biodegradable
*Composition:* 100% active ingred.
*References:* Tech. Data sheet 21-11

**Alkasurf MLS.** Alkaril Chem.
*Chemical Description:* Monoethanolamine lauryl sulfate
*Category:* Detergent, wetting agent
*Applications:* Surfactant for clear viscous shampoos; wax strippers where lower ammonia odor is required
*Form:* Liquid; Color: Pale yellow; Odor: Low
*Composition:* 32% active ingred.
*Reaction:* Neutral

**Alkasurf NP-4.** Alkaril Chem.

*Chemical Description:* Nonyl phenol ethoxylate
*Category:* Emulsifier, dispersant, intermediate, nonionic
*Applications:* Emulsifier for mineral oils, kerosene, chlorinated hydrocarbons; co-emulsifier with water soluble surfactants; oil soluble dispersant, surfactant intermediate; household, industrial, cosmetic use; fat liquoring; making of cutting and sol. oils; dispersing waxes, pigments, resins
*Form:* Liquid; Color: Light; Odor: Low
*Properties:* Solubility: Water insol.; Sp. gr.: 1.02; HLB: 9; Stability: Complete in acid or alkaline media
*Composition:* 100% active ingred.
*Reaction:* Neutral
*References:* Tech. Data sheet 23-1

**Alkasurf NP-5.** Alkaril Chem.
*Chemical Description:* Nonyl phenol ethoxylate
*Category:* Emulsifier, co-emulsifier, oil soluble dispersant, nonionic
*Applications:* See Alkasurf NP-4; also printing applications; preparation of emulsified paints
*Form:* Liquid; Color: Light; Odor: Low
*Properties:* Solubility: Water insol.; Sp. gr.: 1.03; HLB: 10; Stability: Complete in acid or alkaline media
*Composition:* 100% active ingred.
*Reaction:* Neutral
*References:* Tech. Data sheet 23-2

**Alkasurf NP-6.** Alkaril Chem.
*Chemical Description:* Nonyl phenol ethoxylate
*Category:* Emulsifier, co-emulsifier, oil soluble dispersant, nonionic
*Applications:* Household and industrial cleaners; intermediate; plasticizer and antistatic agent for plastics; emulsifier for mineral oils; insecticides, fungicides, herbicides; fat liquoring; making of cutting and sol. oils; dispersing waxes, pigments, resins; printing; preparation of emulsified paints
*Form:* Liquid; Color: Light; Odor: Characteristic
*Properties:* Solubility: Water insol.; Sp. gr.: 1.04; HLB: 11.0; Stability: Good
*Composition:* 100% active ingred.
*Reaction:* Neutral
*References:* Tech. Data sheet 23-3

**Alkasurf NP-8.** Alkaril Chem.
*Chemical Description:* Nonyl phenol ethoxylate
*Category:* Detergent, emulsifier, wetting agent
*Applications:* Low foaming general purpose detergent and wetting agent; household and industrial cleaners; textiles; leather; laundering; dyeing; papermaking; paints; cosmetics; dyeing; rewetting dried skins; electroplating; rotary washer; wetting pulp in beater; Kier boiling; fat liquoring; making of cutting and soluble oils; dispersing waxes, pigments, resins; printing

*Form:* Liquid; Color: Light; Odor: Low
*Properties:* Solubility: Water sol.; Sp. gr.: 1.05; HLB: 12.0; Stability:

*Composition:* 100% active ingred.
*Reaction:* Neutral
*References:* Tech. Data sheet 23-4

**Alkasurf NP-9.** Alkaril Chem.
*Chemical Description:* Nonyl phenol ethoxylate
*Category:* Detergent, wetting agent, emulsifier, nonionic
*Applications:* Detergent–household and industrial cleaners; textiles; skins; leather; laundering; dyeing; papermaking; paints; varnishes; cosmetics; Wetting agent–dyeing; rewetting dried skins; tanning; spot removal; pickling; electroplating; rotary washer; wetting pulp in beater; sizing; insecticides, fungicides, herbicides; Emulsifer–Kier boiling; fat liquoring; making of cutting and sol. oils; dispersing waxes, pigments, resins; printing; making of bitumen emulsions
*Form:* Liquid; Color: Light; Odor: Low
*Properties:* Solubility: Water sol.; Sp. gr.: 1.06; HLB: 13.4; Stability: Good
*Composition:* 100% active ingred.
*Reaction:* Neutral
*References:* Tech. Data sheet 23-5

**Alkasurf NP-10.** Alkaril Chem.
*Chemical Description:* Nonyl phenol ethoxylate
*Category:* Detergent, emulsifier, wetting agent, nonionic
*Applications:* Household and industrial cleaners; textiles; skins; leather; laundering; dyeing; papermaking; paints; varnishes; cosmetics; desizing and boiling off cottons; carbonizing woolens; rewetting dried skins; spot removal; pickling; electroplating; rotary washer; wetting pulp in beater; sizing; making of bitumen emulsions
*Form:* Liquid; Color: Light; Odor: Low
*Properties:* Solubility: Water sol.; Sp. gr.: 1.06; HLB: 13.5; Stability: Good
*Composition:* 100% active ingred.
*Reaction:* Neutral
*References:* Tech. Data sheet 23-6

**Alkasurf NP-11.** Alkaril Chem.
*Chemical Description:* Nonyl phenol ethoxylate
*Category:* Detergent, emulsifier, wetting agent, nonionic
*Applications:* Household and industrial cleaners; textiles; skins; leather; laundering; dyeing; paints; varnishes; cosmetics; dyeing; rewetting dried skins; spot removal; pickling; electroplating; rotary washer; wetting pulp in beater; removing starches, wall paper; cleaning hangings; dispersing waxes, pigments, resins; making of bitumen emulsions
*Form:* Liquid; Color: Light; Odor: Low
*Properties:* Solubility: Water sol.; Sp. gr.: 1.06; HLB: 13.8; Stability: Good
*Composition:* 100% active ingred.
*Reaction:* Neutral
*References:* Tech. Data sheet 23-7

**Alkasurf NP-15.** Alkaril Chem.
*Chemical Description:* Nonyl phenol ethoxylate

*Category:* Detergent, wetting agent, nonionic
*Applications:* High temperature general purpose detergent, compatible with higher conc. of electrolytes than lower NP series; textiles; skins; leather; laundering; dyeing; paints; varnishes; household; cosmetics; street, building, hospital cleaners; bleaching; pickling; electroplating; rotary washer; dispersing waxes, pigments, resins; making of bitumen emulsions
*Form:* Paste; Color: Light; Odor: Low
*Properties:* Solubility: Water sol.; Sp. gr.: 1.07; HLB: 15; Stability: Good
*Composition:* 100% active ingred.
*Reaction:* Neutral
*References:* Tech. Data sheet 23-8

**Alkasurf NP-40.** Alkaril Chem.
*Chemical Description:* Nonyl phenol ethoxylate with ≈ 40 moles of ethylene oxide
*Category:* Wetting agent, emulsifier, stabilizer, dispersant, penetrant, nonionic
*Applications:* Polymer emulsifier, high temperature applications; textiles; dyeing; paints; dispersing waxes, pigments, resins; printing; preparation of emulsified paints
*Form:* Liquid; Color: Light; Odor: Low
*Properties:* Solubility: Excellent water sol.; F.P.: 50–60 F
*Composition:* 70% active ingred.
*Reaction:* Neutral
*References:* Tech. Data sheet 23-11

**Alkasurf OP-5.** Alkaril Chem.
*Chemical Description:* Octyl phenoxy polyethoxy ethanol containing 5 moles of ethylene oxide
*Category:* Emulsifier, nonionic surfactant
*Applications:* Improves detergency, dispersability, wetting action of solvent based systems; emulsion type cleaners; co-emulsifier
*Form:* Liquid; Color: Light; Odor: Characteristic
*Properties:* Solubility: Oil sol.; Sp. gr.: 1.05 (15/15 C); HLB: 10.4
*Composition:* 100% active ingred.
*References:* Tech. Data sheet 20-2

**Alkasurf OP-8.** Alkaril Chem.
*Chemical Description:* Octyl phenoxy polyethoxy ethanol containing 7-8 moles of ethylene oxide
*Category:* Emulsifier, detergent, wetting agent, nonionic
*Applications:* Household and industrial cleaning and laundering, low foam level; agricultural pesticides; textiles; dyeing, cement, papermaking; paints; varnishes; transport; spot removal; pickling; electroplating; rotary washer
*Form:* Liquid; Color: Light; Odor: Low
*Properties:* Solubility: Water sol.; Sp. gr.: 1.06 (15/15 C); HLB: 12.4; Stability: Good
*Composition:* 100% active ingred.
*Reaction:* Neutral
*References:* Tech. Data sheet 20-3

**Alkasurf OP-10.** Alkaril Chem.
*Chemical Description:* Octyl phenoxy polyethoxy ethanol containing ≈ 10 moles of ethylene oxide
*Category:* Detergent, emulsifier, wetting agent, nonionic

*Applications:* Household and industrial cleaners; textiles; laundering; dyeing; paints; varnishes; transport; rewetting dried skins; spot removal; pickling; electroplating; rotary washer; caustic and mineral acid cleaners; dispersing waxes, pigments, resins; printing
*Form:* Liquid; Color: Light; Odor: Low
*Properties:* Solubility: Water sol.; Sp. gr.: 1.1 (15/15 C); HLB: 13.5; Stability: Good
*Composition:* 100% active ingred.
*Reaction:* Neutral
*References:* Tech. Data sheet 20-4

**Alkasurf OP-12.** Alkaril Chem.
*Chemical Description:* Octyl phenoxy polyethoxy ethanol containing 12-13 moles of ethylene oxide
*Category:* Detergent, emulsifier, wetting agent, nonionic
*Applications:* Household and industrial cleaners; textiles; laundering; dyeing; paints; varnishes; transport; spot removal; pickling; electroplating; rotary washer; removing starches, wall paper; cleaning hangings; wetting agent in caustic and mineral acid cleaners
*Form:* Liquid; Color: Light; Odor: Low
*Properties:* Solubility: Water sol.; Sp. gr.: 1.07 (25/25 C); HLB: 14.6; Stability: Good
*Composition:* 100% active ingred.
*Reaction:* Neutral
*References:* Tech. Data sheet 20-5

**Alkasurf OP-30 70%.** Alkaril Chem.
*Chemical Description:* Octyl phenoxy polyethoxy ethanol containing 30 moles of ethylene oxide
*Category:* Detergent, emulsifier, wetting agent, nonionic
*Applications:* Used where maximum water solubility is required at normal and high temperature and conc. electrolyte sol'ns.; stabilizer for synthetic latices
*Form:* Liquid; Color: Light; Odor: Low
*Properties:* Solubility: Excellent water sol.; Sp. gr.: 1.1; HLB: 17.3
*Composition:* 70% active ingred.
*References:* Tech. Data sheet 20-7

**Alkasurf OP-40.** Alkaril Chem.
*Chemical Description:* Octyl phenoxy polyethoxy ethanol containing 40 moles of ethylene oxide
*Category:* Detergent, emulsifier, wetting agent, nonionic
*Applications:* See Alkasurf OP-30; textiles; laundering; dyeing; household; industrial cleaners; pickling; electroplating; rotary washer
*Form:* Liquid; Color: Light; Odor: Low
*Properties:* Solubility: Excellent water sol.; Sp. gr.: 1.1; HLB: 18; Stability: Good
*Composition:* 70% active ingred.; other components: water
*Reaction:* Neutral
*References:* Tech. Data sheet 20-8

**Alkasurf SS-DA-6.** Alkaril Chem.
*Chemical Description:* Sodium sulfosuccinate ester of an ethoxylated $C_{10}$ to $C_{12}$ straight-chain alcohol mixture
*Category:* Detergent, foaming and wetting agent, anionic

*Applications:* Emulsion polymerization of vinyl acetate and acrylates
*Form:* Liquid; Color: Pale yellow; Odor: Bland
*Properties:* Solubility: Water sol.; Sp. gr.: 1.1; F.P. $<$ 0 C
*Composition:* 30% active ingred.; other components: water
*Reaction:* Neutral
*References:* Tech. Data sheet

**Alkasurf SS-L9DE.** Alkaril Chem.
*Chemical Description:* Sodium lauryl alkylolamide sulfosuccinate
*Category:* Detergent, anionic
*Applications:* Low irritation surfactant for toiletries; household use
*Form:* Liquid; Color: Light yellow; Odor: Bland
*Properties:* Solubility: Excellent water sol.; Sp. gr.: 1.17; Stability: Good; Biodegradable
*Composition:* 40% active ingred.; other components: water
*Reaction:* Neutral
*References:* Tech. Data sheet 32

**Alkasurf SS-L9ME.** Alkaril Chem.
*Chemical Description:* Sodium lauryl alkylolamide sulfosuccinate
*Category:* Detergent, anionic
*Applications:* High foaming surfactant for toiletry and carpet shampoos; laundering; cleaners of street, buildings, hospitals
*Form:* Viscous liquid; Color: Cream white; Odor: Bland
*Properties:* Sp. gr.: 1.00; Stability: Good; Biodegradable
*Composition:* 40% active ingred.
*Reaction:* Neutral
*References:* Tech. Data sheet 33

**Alkasurf SS-LA-3.** Alkaril Chem.
*Chemical Description:* Sodium polyglycol ether sulfosuccinate
*Category:* Detergent, emulsifier, wetting agent, anionic
*Applications:* Cosmetics esp. hair shampoos, bubble bath; emulsion polymerization; household and industrial cleaners
*Form:* Clear liquid; Color: Light; Odor: Typical
*Properties:* Solubility: Water sol.; Sp. gr.: 1.10
*Composition:* 30% active ingred.; other components: water
*Reaction:* Neutral
*References:* Tech. Data sheet 31

**Alkasurf SS-O-60.** Alkaril Chem.
*Chemical Description:* Sodium dioctyl sulfosuccinate
*Category:* Wetting agent, anionic
*Applications:* Textile industry—desizing, dyeing and printing, wetting and penetrating; dry cleaning; agriculture—insecticides, fungicides, treatment of grains and forage; activation of straw manures; cosmetics; paints; emulsion polymerization; rewetting dried skins; tanning; finishing; electroplating; abatement of dust; wetting pulp in beater; removing starches, wall paper; cleaning hangings
*Form:* Clear colorless liquid
*Properties:* Solubility: Water 1%; most organic liquids; Sp. gr.: 1.07; Surface Tension: 29 dynes/cm (0.1% active in water)
*Composition:* 60% active ingred.; other components: $\approx$ 13% ethanol, $\approx$ 27% water
*Reaction:* Neutral

*Toxicity:* Eye irritant
*References:* Tech. Data sheet 34

**Alkasurf SS-O-75.** Alkaril Chem.
*Chemical Description:* Sodium dioctyl sulfosuccinate
*Category and Applications:* Powerful wetting agent for aqueous and organic systems; see Alkasurf SS-O-60
*Form:* Clear colorless liquid; Odor: Typical
*Properties:* Solubility: 1% in water; soluble in most organic liquids; Sp. gr.: 1.09; F.P.: <–20 C; Surface Tension: 29 dynes/cm (0.1% active in dist. water)
*Composition:* 75% active ingred.; other components: ≈ 16% water, ≈ 9% ethanol
*Reaction:* Neutral
*Toxicity:* Eye irritant
*References:* Tech. Data sheet 35

**Alkasurf SS-OA.** Alkaril Chem.
*Chemical Description:* Disodium N-oleyl sulfosuccinamate
*Category:* Foaming agent
*Applications:* Foaming agent for rubber latex foam; emulsifying, dispersing, solubilizing, foam stabilizing action on soaps, etc.
*Form:* Liquid; Color: Amber; Odor: Low
*Properties:* Solubility: Water sol.; Visc.: 25–30 cps @ 70 F; F.P.: < 0 C
*Composition:* 31% active ingred.; other components: water
*Reaction:* Alkaline
*References:* Tech. Data sheet 37

**Alkasurf SS-O-ME.** Alkaril Chem.
*Chemical Description:* Fatty alkanolamide sulfosuccinate
*Category:* Emulsifier, anionic
*Applications:* Emulsifier for use in synthetic latices and other polymer emulsions
*Form:* Liquid
*Composition:* 40% conc.

**Alkasurf SS-TA.** Alkaril Chem.
*Chemical Description:* Disodium N-octadecyl sulfosuccinamate
*Category:* Detergent, foaming agent
*Applications:* Foaming agent for rubber latex foam; heavy duty alkaline cleaner; cement, brick, tile cleaners; emulsification of waxes for oil and wax polishes; emulsion polymerization of vinyl acetate
*Form:* Paste; Color: Cream; Odor: Low
*Properties:* Solubility: Poor in water
*Composition:* 35% active ingred.; other components: water
*Reaction:* Alkaline
*References:* Tech. Data sheet 36

**Alkasurf T.** Alkaril Chem.
*Chemical Description:* Cosmetic grade triethanolamine linear alkylate sulfonate
*Category:* Detergent, emulsifier, wetting agent
*Applications:* Hair shampoo, bubble bath, other cosmetic, household use
*Form:* Clear liquid; Color: Straw; Odor: Typical
*Properties:* Solubility: Complete in water; Visc.: 2500 cps max.
*Composition:* 60% active ingred.; other components: water
*Reaction:* Neutral

*References:* Tech. Data sheet 43

**Alkasurf TDA-7.5.** Alkaril Chem.
*Chemical Description:* Tridecyl alcohol ethoxylate
*Category:* Emulsifier, wetting agent, nonionic
*Applications:* Foam builder and solubilizer for alkyl aryl sulfates, etc.; emulsifier and co-emulsifier for O/W systems; wetting and rewetting agent in various industrial applications; chemical specialty manufacturing
*Form:* Opaque liquid; Color: Almost water white; Odor: Mild
*Properties:* Solubility: Water sol.; Sp. gr.: 1.00 (20/20 C)
*Composition:* 100% active ingred.
*References:* Tech. Data sheet

**Alkasurf TDA-12.** Alkaril Chem.
*Chemical Description:* Tridecyl alcohol ethoxylate
*Category:* Emulsifier, detergent, nonionic
*Applications:* Emulsifier and co-emulsifier for O/W systems; leveling agent; solubilizer; high-temperature detergent
*Form:* White solid; Odor: Mild
*Properties:* Solubility: Complete in water; M.P.: ≈ 27 C
*Composition:* 100% active ingred.
*References:* Tech. Data sheet

**Alkasurf TDA-15.** Alkaril Chem.
*Chemical Description:* Tridecyl alcohol ethoxylate
*Category:* Emulsifier, detergent, nonionic
*Applications:* Emulsifier and co-emulsifier for O/W systems; leveling agent, solubilizer, high-temperature detergent
*Form:* White solid; Odor: Mild
*Properties:* Solubility: Complete in water; M.P.: ≈ 30 C
*Composition:* 100% active ingred.
*References:* Tech. Data sheet

**Alkasurf TLS.** Alkaril Chem.
*Chemical Description:* Triethanolamine lauryl sulfate
*Category:* Detergent, emulsifier, wetting agent, anionic
*Applications:* Hair shampoos, bubble baths, and other toiletries where low cloud point properties are critical; household uses
*Form:* Liquid; Color: Light amber; Odor: Characteristic
*Properties:* Solubility: Water sol.; Sp. gr.: 1.02 (20/20 C); Cloud Pt.: 0 C
*Composition:* 40% active ingred.
*Reaction:* Neutral
*References:* Tech. Data sheet 26

**Alkasurf WADX.** Alkaril Chem.
*Chemical Description:* Modified diethanolamine lauryl sulfate
*Category:* Detergent, anionic
*Applications:* Cosmetic and pharmaceutical cleaning preparations
*Form:* Clear to slightly hazy viscous liquid; Color: Pale straw; Odor: Bland
*Properties:* Water sol.
*Composition:* 30% active ingred.
*References:* Tech. Data sheet 29

**Alkasurf WAQ.** Alkaril Chem.
*Chemical Description:* Sodium lauryl sulfate
*Category:* Detergent, emulsifier,

wetting agent, anionic
*Applications:* Hair shampoo, hand lotions, bubble bath, other cosmetics; institutional detergents, sanitizers, floor cleaners, rug and upholstery shampoos; laundry and dry cleaning uses
*Form:* Liquid; Color: White pearlescent; Odor: Typical
*Properties:* Solubility: Water sol.
*Composition:* 30% active ingred.
*Reaction:* Neutral
*References:* Tech. Data sheet 27

**Alkasurf WHC.** Alkaril Chem.
*Chemical Description:* Blend of emulsifiers suited to compounding waterless hand cleaner
*Category:* Wetting agent
*Applications:* Waterless hand cleaner base
*Form:* Clear viscous liquid; Color: Amber; Odor: Typical
*Composition:* 100% active ingred.
*Reaction:* Alkaline
*References:* Tech. Data sheet 65

**Alkaterge-E.** IMC Chem.
*Chemical Description:* Oxazoline-type compound
*Category:* Detergent, emulsifier, wetting agent
*Applications:* Salts and soaps; paper, textiles, metal cleaners; emulsion stabilizer; acid acceptor; pigment grinding and dispersion; antifoam agent; antioxidant
*Form:* Clear liquid; Color: 15 max. (Gardner)
*Properties:* Solubility: Sol. in most organic liquids, slight sol, in water; Density: 7.74 lb/gal; Sp. gr.: 0.9; Visc.: 155 cp @ 25 C; F.P.: ≈ – 31 C; Flash Pt.: > 212 F; Surface Tension: 40 dynes/cm (0.001% aqueous sol'n.)
*Composition:* 79% active
*Containers:* Can, drums
*References:* NP Data sheet 4B

**Alkaterge-T.** IMC Chem.
*Chemical Description:* Oxazoline-type compound
*Category:* Emulsifier, dispersant
*Applications:* Corrosion inhibitor; emulsifying agent; pigment dispersion
*Form:* Waxy solid; Color: Buff to brown
*Properties:* Solubility: Highly sol. in aromatic hydrocarbons; Flash Pt.: > 200 F (Tag closed cup); Surface Tension: 30.4 dynes/cm
*Toxicity:* Skin irritant
*Warnings and Cautions:* Avoid prolonged or repeated exposure to skin
*References:* NP Data sheet 4A

**Alkateric BC.** Alkaril Chem.
*Chemical Description:* Alkyl dimethyl betaine
*Category:* Detergent
*Applications:* Low irritation shampoo base and hair conditioning additive; coupling agent for quaternary based shampoos; amphoteric surfactant; dyeing assistant in textile industry; household, cosmetic use; cleaners of street, buildings, hospitals
*Form:* Liquid; Color: Pale straw; Odor: Low
*Properties:* Solubility: Water sol.; F.P.: <0 C; pH: 7.5–8.5

*Composition:* 31% active ingred.
*Reaction:* Neutral
*References:* Tech. Data sheet

**Alkateric CAB.** Alkaril Chem.
*Chemical Description:* Betaine of coconut alkylolamide
*Category:* Detergent, amphoteric foaming agent
*Applications:* Foaming/coupling agent for shampoos; household, cosmetic use; cleaners of street, buildings, hospitals
*Form:* Liquid; Color: Amber; Odor: Bland
*Properties:* Solubility: Water complete
*Composition:* 32% active ingred.; other component: water
*Reaction:* Neutral

**Alkateric CIB.** Alkaril Chem.
*Category:* Detergent
*Applications:* Amphoteric foaming agent and surfactant; baby shampoo; household and cosmetic use; cleaners of street, buildings, hospitals
*Form:* Liquid; Color: Amber; Odor: Bland
*Properties:* Sp. gr.: 1.1
*Composition:* 40% active ingred.; other components: water
*Reaction:* Alkaline
*Toxicity:* Low eye irritation

**Alkateric STB.** Alkaril Chem.
*Chemical Description:* Stearyl betaine
*Category:* Detergent, amphoteric surfactant/softener
*Applications:* Detergent, amphoteric surfactant/softener unaffected by hard water; household, cosmetic use; cleaners of street, buildings, hospitals
*Form:* Liquid; Color: Pale yellow
*Composition:* 50% active ingred.; other components: water/alcohol
*Reaction:* Neutral

**Alkateric TB.** Alkaril Chem.
*Chemical Description:* Tallow betaine
*Category:* Detergent; amphoteric surfactant/softener
*Applications:* Household, cosmetic use; cleaners of street, buildings and hospitals
*Form:* Liquid; Color: Amber
*Composition:* 70% active ingred.; other components: water/alcohol
*Reaction:* Neutral

**Alkatrope SX-40.** Alkaril Chem.
*Chemical Description:* Sodium xylene sulfonate
*Category:* Hydrotrope, solubilizer and coupling agent
*Applications:* Used to dissolve detergents in electrolyte sol'ns.; used to lower cloud point and viscosity of detergent systems
*Form:* Clear liquid; Color: Light
*Composition:* 41% active ingred.
*Containers:* 550-lb bung-type drums

**Alkazine C.** Alkaril Chem.
*Chemical Description:* Coconut hydroxyethyl imidazoline
*Category:* Emulsifier, antistat
*Applications:* Textiles, leather, household
*Form:* Paste; Color: Brown; Odor: Amine
*Properties:* Solubility: Sol. in aqueous acid; F.P.: 20 C; Stability:

Not stable to alkali
*Composition:* 85% active ingred.
*Reaction:* Alkaline
*References:* Tech. Data sheet 80

**Alkazine O.** Alkaril Chem.
*Chemical Description:* Oleic hydroxyethyl imidazoline
*Category:* Emulsifier, corrosion inhibitor, antistat agent
*Applications:* Emulsifier; asphalt emulsions; corrosion inhibitor; acid cleaners; solvent degreasers; textile antistat; leather and household uses
*Form:* Liquid; Color: Brown; Odor: Amine
*Properties:* Solubility: Sol. in aqueous acid and organics; F.P.: 5 C; Stability: Not stable in alkaline medium
*Composition:* 85% active ingred.
*Reaction:* Alkaline
*References:* Tech. Data sheet 81

**Alkazine ST.** Alkaril Chem.
*Chemical Description:* Stearic hydroxyethyl imidazoline
*Category:* Emulsifier, dispersant
*Applications:* Wax emulsions; corrosion inhibitor; household formulations; dispersing waxes, pigments, resins
*Form:* Waxy solid; Color: Tan; Odor: Amine
*Properties:* Solubility: Sol. in acid and organic systems
*Composition:* 85% active ingred.
*Reaction:* Alkaline
*References:* Tech. Data sheet 83

**Alkazine TO.** Alkaril Chem.
*Chemical Description:* Tall oil hydroxyethyl imidazoline
*Category:* Emulsifier, cationic
*Applications:* Asphalt emulsions; corrosion inhibitor—oil burning systems, pickling bath operations; dispersing waxes, pigments, resins; making of bitumen emulsions
*Form:* Liquid; Color: Brown; Odor: Amine
*Properties:* Solubility: Sol. in acid systems; Stability: Not stable to alkali
*Composition:* 80% active ingred.
*Reaction:* Alkaline
*References:* Tech. Data sheet 82

**Alkylate A215.** Monsanto Co.
*Chemical Description:* Alkyl benzene
*Category:* Intermediate, detergent
*Applications:* Intermediates for the manufacture of anionic liquid and solid detergents
*Form:* Liquid; Color: Colorless
*Properties:* m.w.: 231–241; Density: 7.2 lb/gal; Sp. gr.: 0.855–0.870; Biodegradable
*References:* Data sheet

**Alkylate A225.** Monsanto Co.
*Chemical Description:* Alkylbenzene
*Category:* Intermediate
*Applications:* See Alkylate A215
*Form:* See Alkylate A215
*Properties:* m.w.: 238–248; Density: 7.2 lb/gal; Sp. gr.: 0.855–0.870; Biodegradable

**Alkylate A230.** Monsanto Co.
*Chemical Description:* Alkylbenzene
*Category:* Intermediate

*Applications and Form:* See Alkylate A215
*Properties:* m.w.: 258–266; Density: 7.2 lb/gal; Sp. gr.: 0.855–0.870; Biodegradable

**Alkylate H230H.** Monsanto Co.
*Chemical Description, Category, Applications, Form:* See Alkylate H230L
*Properties:* m.w.: 360 (avg.); Ref. Index: 1.495–1.515; Sp. gr.: 0.890–0.910; Visc.: 140–220 SUS @ 100 F; I.B.P.: 290 C min.

**Alkylate H230L.** Monsanto Chem.
*Chemical Description:* By-product of alkylbenzene manufacture
*Category:* Intermediate
*Applications:* Anionic surfactant intermediates for formulating liquid and solid detergents
*Form:* Liquid; Color: Yellow
*Properties:* m.w.: 335 (avg.); Ref. Index: 1.495–1.515; Sp. gr.: 0.880–0.910; Visc.: 100–160 SUS @ 100 F; I.B.P.: 290 C min.

**Alkylene Series.** Hart Prod.
*Category:* Detergent–all except 8, 10, 11, 12; Wetting agent–17-24; Emulsifier–41-49
*Applications:* Detergents–Textiles; skins; leather; laundering; dyeing; metallurgy; mines and minerals; papermaking; transport; household; cosmetics; street, building, hospital cleaners; Wetting agents–Desizing and boiling off cottons; carbonizing woolens; bleaching; dyeing; rewetting dried skins; tanning; finishing; spot removal; Emulsifiers–Kier boiling; fat liquoring; making of cutting and soluble oils; dispersing waxes, pigments, resins; printing; dispersing dry powders; preparation of emulsified paints; making of bitumen emulsions
*Form:* Liquid; Color: Colorless; Odor: Practically odorless
*Properties:* Solubility: Water and organic solvents
*Composition:* Varies with grade; some in series are 100% active
*Reaction:* Neutral
*Containers:* Drums

**Alox 100.** Alox Corp.
*Applications:* Source of synthetic acids; manufacture of corrosion inhibitors
*Form:* Gel, paste; Color: Tan; Odor: Mild
*Properties:* Solubility: Nonpolar solvents; Density: 7.6–7.7 lb/gal; Sp. gr.: 0.92; M.P.: 100 F
*Composition:* 100% active ingred.
*Reaction:* Acid

**Alox 111.** Alox Corp.
*Category:* Mold release agent
*Applications:* Base for concrete mold release; rust preventive for metal forms
*Form:* Liquid; Color: Tan; Odor: Mild
*Properties;* Solubility: Nonpolar solvents; Density: 7.6–7.7 lb/gal; Sp. gr.: 0.92; Flash Pt.: 310 F min. (O.C.); Pour Pt.: 65 F max.
*Composition:* 100% active ingred.
*Reaction:* Slightly acid

**Alox 162.** Alox Corp.

*Category:* Lubricity agent, corrosion inhibitor
*Applications:* Lubricity agent in mineral oil sol'n.; railroad journal oils; making of cutting and sol. oils
*Form:* Paste; Color: Brown; Odor: Mild
*Properties:* Solubility: Nonpolar solvents; Density: 7.3 lb/gal; Sp. gr.: 0.88; M.P.: 86–89 F
*Composition:* 100% active ingred.
*Reaction:* Slightly acid
*References:* Bulletin

**Alox 164.** Alox Corp.
*Category:* Detergent, corrosion inhibitor
*Applications:* Base for crankcase oils for improved anti-wear, detergency, solvency, lubricity, antirust
*Form:* Liquid; Color: Brown; Odor: Mild
*Properties:* Solubility: Nonpolar solvents; Visc.: 35–45 SUS @ 100 F; Flash Pt.: 150 F min.; Pour Pt.: 60 F
*Composition:* 100% active ingred.
*Reaction:* Slightly acid

**Alox 318F.** Alox Corp.
*Category:* Corrosion inhibitor
*Applications:* Oil or solvent corrosion inhibitor, ferrous and nonferrous surfaces
*Form:* Gel, paste; Color: Brown; Odor: Mild
*Properties:* Nonpolar solvents; Density: 7.7 lb/gal @ 60 F; Sp. gr.: 0.91; M.P.: 100 F; Flash Pt.: 350 min. (O.C.)
*Composition:* 100% active ingred.
*Reaction:* Slightly acid
*Containers:* 15, 30, 55-gal containers, tankcar
*References:* Bulletin

**Alox 319F.** Alox Corp.
*Category:* Corrosion inhibitor
*Applications:* Oil or solvent corrosion inhibitor; making of cutting and sol. oils
*Form:* Paste; Color: Brown; Odor: Mild
*Properties:* Solubility: Nonpolar solvents; Density: 7.7 lb/gal @ 60 F; Sp. gr.: 0.91; M.P.: 100 F min.; Flash Pt.: 350 F min. (O.C.)
*Composition:* 100% active ingred.
*Reaction:* Acid
*Toxicity:* $LD_{50} > 5$ g/kg (Rats)
*References:* Bulletin

**Alox 350.** Alox Corp.
*Chemical Description:* Esters of an oxidized petroleum fraction
*Category:* Lubricity agent, corrosion inhibitor
*Applications:* Making of cutting and soluble oils
*Form:* Paste; Color: Brown; Odor: Mild
*Properties:* Solubility: Nonpolar solvents; Density: 7.7 lb/gal; Sp. gr.: 0.91; M.P.: 100 F
*Reaction:* Acid
*Toxicity:* $LD_{50} > 5$g/kg (Rat)
*Containers:* Drums, tank cars

**Alox 400L.** Alox Corp.
*Applications:* Source of synthetic acids; manufacture of corrosion inhibitors
*Form:* Liquid; Color: Dark; Odor: Mild
*Properties:* Solubility: Nonpolar solvents; Sp. gr.: 0.9; Pour Pt.: 10

F
*Composition:* 100% active ingred.
*Reaction:* Acid

**Alox 436.** Alox Corp.
*Category:* Mold release
*Applications:* Base for concrete mold release lubricants; corrosion inhibitor
*Form:* Dark liquid
*Properties:* Density: 7.7 lb/gal; Flash Pt.: 280 F; Stability: Good

**Alox 436A.** Alox Corp.
*Category and Applications:* See Alox 436–lower cost version of Alox 436
*Form:* Dark liquid; Odor: Mild
*Properties:* Solubility: Nonpolar solvents; Sp. gr.: 0.91
*Composition:* 100% active ingred.
*Reaction:* Slightly acid

**Alox 488.** Alox Corp.
*Chemical Description:* Mixture from oxidation of petroleum fractions
*Category:* Corrosion inhibitor, lubricant
*Applications:* Upper cylinder lubricant; additive for industrial burner fuels
*Form:* Liquid; Color: Dark; Odor: Mild
*Properties:* Solubility: Nonpolar solvents; Density: 7.3 lb/gal @ 60 F; Sp. gr.: 0.88; Visc.: 140-190 SUS @ 100 F; Flash Pt.: > 230 F
*Composition:* 100% active ingred.
*Reaction:* Slightly acid
*References:* Data sheet

**Alox 490.** Alox Corp.
*Applications:* Upper cylinder lubricant; engine scavenging additive
*Properties:* Density: 7.2 lb/gal; Sp. gr.: 0.86 @ 15.6 C; Flash Pt.: 200 F (O.C.)

**Alox 502.** Alox Corp.
*Chemical Description:* Oxygenated hydrocarbon partially neutralized with barium
*Category:* Corrosion inhibitor
*Applications:* Base for corrosion inhibitors; preservative lubricants
*Form:* Color: Dark; Odor: Mild
*Properties:* Solubility: Nonpolar solvents; Density: 8.1 lb/gal; Sp. gr.: 0.97; Flash Pt.: 390 F
*Composition:* 100% active
*Reaction:* Neutral
*References:* Data sheet

**Alox 575.** Alox Chem.
*Chemical Description:* Oxygenated hydrocarbon with barium and a sodium petroleum sulfonate
*Category:* Emulsifier, corrosion inhibitor
*Applications:* Oil soluble corrosion inhibitor (emulsifiable); making of cutting and soluble oils
*Form:* Liquid, gel; Color: Tan; Odor: Mild
*Properties:* Solubility: Oils; Density: 8.5 lb/gal; Sp. gr.: 1.02 @ 15.6 C; Flash Pt.: 400 F; Pour Pt. 80 F
*Composition:* 100% active ingred.
*Reaction:* Slightly acid
*References:* Bulletin

**Alox 600.** Alox Chem.
*Chemical Description:* Mixture of high m.w. organic acids, lactones and esters from petroleum
*Category:* Source of synthetic ac-

ids, manufacture of corrosion inhibitors; chemical intermediate
*Applications:* Rust preventives, metal-working compounds, automobile undercoating
*Form:* Solid; Color: Dark; Odor: Mild
*Properties:* Solubility: Nonpolar solvents; Density: 7.9 lb/gal @ 75 F; Sp. gr.: 0.95 @ 75 F; M.P.: 115 F; Flash Pt.: 345 F; Stability: Good
*Composition:* 100% active ingred.
*Reaction:* Acid
*Toxicity:* $LD_{50} > 5$ g/kg (Rat)
*Containers:* Drums, tankcars
*References:* Bulletin

**Alox 604.** Alox Corp.
*Chemical Description:* Oxygenated hydrocarbon partially neutralized with calcium
*Category:* Corrosion inhibitor
*Applications:* Base for soft film cut back rust preventives
*Form:* Paste; Color: Dark; Odor: Mild
*Properties:* Solubility: Nonpolar solvents; Density: 7.99 lb/gal; Sp. gr.: 0.96 @ 15.6 C; M.P.: 220 F; Flash Pt.: 400 F; Pour Pt.: 150 F
*Composition:* 100% active ingred.
*Containers:* Drums
*References:* Product bulletin

**Alox 606.** Alox Corp.
*Category:* Corrosion inhibitor
*Applications:* Outdoor rust preventive for protection of metal surfaces, with good salt resistance
*Form:* Solid; Color: Dark; Odor: Mild
*Properties:* Solubility: Nonpolar solvents; Density: 8.2 lb/gal; Sp. gr.: 0.98; M.P.: 250 F
*Composition:* 100% active ingred.
*Reaction:* Neutral
*References:* Tech. Data sheet

**Alox 607.** Alox Corp.
*Category and Applications:* See Alox 606; Alox 607 is a higher melting version of 606
*Properties:* Solubility: Nonpolar solvents; Density: 8.3 lb/gal; Sp. gr.: 0.98; M.P. 280 F
*Composition:* 100% active
*Reaction:* Neutral
*References:* Tech. Data sheet

**Alox 707.** Alox Corp.
*Chemical Description:* Mixture of organic acids, lactones and esters
*Category:* Corrosion inhibitor
*Applications:* Long term protection of steel during outdoor storage, with good salt resistance; undercoatings for automobiles
*Form:* Solid
*Properties:* Density: 8.8 lb/gal; M.P.: 185 F min.; Flash Pt.: 410 F (O.C.)
*Containers:* 100, 400 lb drums
*References:* Product bulletin

**Alox 800.** Alox Corp.
*Category:* Source of synthetic acids
*Applications:* Manufacture of corrosion inhibitors
*Form:* Liquid; Color: Dark; Odor: Mild
*Properties:* Solubility: Nonpolar solvents; Sp. gr.: 0.86; Pour Pt.: –25 F
*Composition:* 100% active ingred.
*Reaction:* Acid

**Alox 853.** Alox Corp.
*Category:* Corrosion inhibitor
*Applications:* Corrosion inhibitor esp. metals protection; fuel additive; anti-freeze additive
*Properties:* Solubility: Water insol.; soluble in ethanol, isopropanol; Density: 7.1 lb/gal; Sp. gr.: 0.85 @ 60 F; Flash Pt.: 155 F min. (O.C.); Pour Pt.: –25 F max.
*Containers:* Drums, tankcar

**Alox 901.** Alox Corp.
*Category:* Source of synthetic acids
*Applications:* Manufacture of corrosion inhibitors
*Form:* Liquid; Color: Dark; Odor: Mild
*Properties:* Solubility: Nonpolar solvents; Sp. gr.: 0.93; Pour Pt.: 30 F
*Composition:* 100% active ingred.
*Reaction:* Acid

**Alox 904.** Alox Corp.
*Category:* Corrosion inhibitor
*Applications:* Base for high flash fingerprint remover; corrosion inhibitor
*Form:* Gel, paste; Color: Dark
*Properties:* Solubility: Nonpolar solvents; Density: 8.16 lb/gal; Sp. gr.: 0.98 @ 77 F; Flash Pt.: 295 F (C.O.C.); Pour Pt.: 65 F
*Composition:* 100% active ingred.
*Reaction:* Slightly acid
*References:* Technical Data sheet

**Alox 940.** Alox Corp.
*Chemical Description:* Oxygenated hydrocarbon partially neutralized with calcium and a sodium petroleum sulfonate
*Category:* Corrosion inhibitor
*Applications:* Oil and solvent rust preventive base
*Form:* Viscous liquid; Color: Dark; Odor: Mild
*Properties:* Density: 8.0 lb/gal; Sp. gr.: 0.97; Flash Pt.: 300 F min. (O.C.)
*Composition:* 100% active ingred.
*Reaction:* Slightly acid
*Containers:* Drums, tankcars
*References:* Product bulletin

**Alox 1509H.** Alox Corp.
*Chemical Description:* Concentrated sol'n. of an inhibited resin in mineral spirits
*Applications:* Protective film for metal surfaces
*Properties:* Solubility: Gasoline, mineral spirits, aromatic and chlorinated solvents; Density: 8.2 lb/gal @ 60 F; Flash Pt.: 102 F; Pour Pt.: 40 F
*References:* Tech. data sheet

**Alox 1516.** Alox Corp.
*Category:* Corrosion inhibitor
*Applications:* Water soluble corrosion inhibitor for circulating cooling systems
*Form:* Viscous liquid; Color: Dark
*Properties:* Solubility: Water sol.; Density: 8.7 lb/gal; Sp. gr.: 1.04 @ 127 F; Flash Pt.: 220 F (O.C.); Pour Pt.: 30 F
*References:* Product bulletin

**Alox 1516B.** Alox Corp.
*Category:* Corrosion inhibitor
*Applications:* Water soluble corrosion inhibitor
*Form:* Viscous liquid; Color: Dark; Odor: Mild

*Properties:* Solubility: Water sol.; Sp. gr.: 1.04; Pour Pt.: 30 F
*Composition:* 100% active ingred.
*Reaction:* Neutral, amine salt

**Alox 1643.** Alox Corp.
*Applications:* Additive for industrial burner fuels promoting more efficient burning, better heat transfer, sludge suspension
*Form:* Liquid; Color: Brown
*Properties:* Density: 7.12 lb/gal
*References:* Tech. data sheet

**Alox 1680.** Alox Corp.
*Chemical Description:* Oxygenated hydrocarbon containing minor amount of a phosphatide
*Category:* Corrosion inhibitor
*Applications:* Penetrating oil additive
*Form:* Liquid; Color: Dark; Odor: Mild
*Properties:* Solubility: Nonpolar solvents; Density: 7.6 lb/gal; @ 60 F; Sp. gr.: 0.92 (60/60 F); Visc. 220 SUS @ 100 F; Flash Pt.: 350 F
*Composition:* 100% active ingred.
*Containers:* 15, 30, 55 gal drums and in bulk
*References:* Product bulletin

**Alox 1689.** Alox Corp.
*Chemical Description:* Ester of an oxygenated hydrocarbon
*Category:* Emulsifier, nonionic
*Applications:* Lubricity agent and nonionic emulsifier; making of cutting and soluble oils; rolling oil additive for stainless steel sheets and titanium sheets
*Form:* Solid; Color: Dark; Odor: Mild
*Properties:* Solubility: Nonpolar solvents; Density: 7.7 lb/gal; Sp. gr.: 0.93; M.P.: 115 F; Flash Pt.: 400 F (O.C.)
*Composition:* 100% active ingred.
*Reaction:* Slightly acid
*References:* Product bulletin

**Alox 1727.** Alox Corp.
*Category:* Emulsifier, corrosion inhibitor
*Applications:* Nonstaining slushing oil additive rust preventive for steel
*Form:* Viscous liquid; Color: Dark; Odor: Mild
*Properties:* Solubility: Nonpolar solvents; Density: 8.0 lb/gal; Sp. gr.: 0.98; Flash Pt.: 350 F
*Composition:* 100% active ingred.
*Reaction:* Slightly acid
*References:* Product bulletin

**Alox 1805.** Alox Corp.
*Category:* Corrosion inhibitor, anionic
*Applications:* Water soluble corrosion inhibitor; hydraulic and brake fluids; grinding coolants and metal cleaning compositions
*Form:* Viscous liquid; Color: Dark; Odor: Mild
*Properties:* Solubility: Water sol.; Density: 9.2 lb/gal; Sp. gr.: 1.1 @ 60 F; Flash Pt.: 212 F; Pour Pt.: 0 F; Surface Tension: 39 dynes/cm @ 0.1% conc.
*Composition:* 75% active ingred.; other components: 25% water
*Reaction:* Neutral; amine salt
*Toxicity:* Mild irritant to skin and eyes
*Warnings and Cautions:* Contains

para tertiary butyl benzoic acid
*References:* Tech. data sheet

**Alox 1832.** Alox Corp.
*Category:* Corrosion inhibitor
*Applications:* Oil soluble corrosion inhibitor for low viscosity, low temperature oils (not for use in internal combustion engines)
*Properties:* Density: 8.3 lb/gal; Sp. gr.: 0.99–1.0 @ 60 F; Flash Pt.: 440 F min.
*References:* Tech. data sheet

**Alox 1843.** Alox Corp.
*Chemical Description:* Amine salt of an organic acid
*Category:* Corrosion inhibitor
*Applications:* Water soluble corrosion inhibitor; low foaming surfactant
*Form:* Liquid; Color: Pale yellow; Odor: Mild
*Properties:* Solubility: Water, alcohols, glycols; Density: 9.3 lb/gal @ 60 F; Sp. gr.: 1.12 @ 60 F; Visc.: 138 cs @ 100 F; Pour Pt.: 15 F; Biodegradability: 80%
*Composition:* 75% active ingred.; other components: water (25%)
*Reaction:* Neutral; amine salt
*Toxicity:* Mild irritant to skin and eyes
*Warnings and Caution:* Contains para tertiary butyl benzoic acid
*References:* Product bulletin

**Alox 1846.** Alox Corp.
*Category:* Corrosion inhibitor
*Applications:* Corrosion inhibitor for jet fuels
*Form:* Liquid; Color: Tan; Odor: Mild
*Properties:* Solubility: Nonpolar solvents; Density: 7.39 lb/gal; Sp. gr.: 0.89; Flash Pt.: 175 F (C.O.C.); Pour Pt.: 20 F
*Composition:* 100% active ingred.
*Reaction:* Slightly acid
*References:* Tech. data sheet

**Alox 2000.** Alox Corp.
*Category:* Corrosion inhibitor
*Applications:* Base for corrosion inhibitor
*Form:* Viscous liquid; Color: Dark; Odor: Mild
*Properties:* Solubility: Nonpolar solvents; Density: 7.96 lb/gal; Sp. gr.: 0.95; Flash Pt.: 380 F; Pour Pt.: 40 F
*Composition:* 100% active ingred.
*Reaction:* Slightly acid
*References:* Product bulletin

**Alox 2019A.** Alox Corp.
*Category:* Corrosion inhibitor
*Applications:* Additive for fingerprint remover
*Form:* Liquid; Color: Dark; Odor: Mild
*Properties:* Solubility: Nonpolar solvents; Density: 7.9 lb/gal; Sp. gr.: 0.94; Visc.: 365 SUS @ 100 F; Flash Pt.: 137 F (C.C.)
*Composition:* 100% active ingred.
*Reaction:* Acid
*References:* Product bulletin

**Alox 2028.** Alox Corp.
*Chemical Description:* Blend of inhibitors including the barium soap of an oxygenated hydrocarbon
*Category:* Corrosion inhibitor
*Applications:* Base for nonemulsifiable, nonstaining, water displac-

ing rust preventive
*Form:* Paste; Color: Dark; Odor: Mild
*Properties:* Solubility: Nonpolar solvents; Density: 7.9 lb/gal; Sp. gr.: 0.95; M.P.: 95 F; Flash Pt.: 300 F (O.C.)
*Composition:* 100% active ingred.
*Reaction:* Neutral
*References:* Product bulletin

**Alox 2116A.** Alox Corp.
*Chemical Description:* Partial barium soap of an oxygenated hydrocarbon
*Category:* Corrosion inhibitor
*Applications:* Corrosion inhibitor for grease
*Form:* Paste; Color: Dark; Odor: Mild
*Properties:* Solubility: Nonpolar solvents; Sp. gr.: 0.97; M.P.: 86 F; Flash Pt.: 355 F (O.C.)
*Composition:* 100% active ingred.
*Reaction:* Acid
*References:* Tech. data sheet

**Alox 2138F.** Alox Corp.
*Category:* Corrosion inhibitor
*Applications:* Bullet lubricant used alone or with beeswax
*Form:* Solid; Color: Dark; Odor: Mild
*Properties:* Solubility: Nonpolar solvents; Density: 7.9 lb/gal @ 76 F; Sp. gr.: 0.95; M.P.: 160 F; Flash Pt.: 350 F
*Reaction:* Acid
*Containers:* Drums, bulk
*References:* Tech. data sheet

**Alox 2140S.** Alox Corp.
*Category:* Corrosion inhibitor
*Applications:* Additive for production of preservative lubricants and soft solvent cut-back rust preventives; automotive rustproofing with high resistance to salt spray
*Form:* Dark colored rubbery
*Properties:* Density: 7.9 lb/gal; Sp. gr.: 0.95; M.P.: 280 F min.; Flash Pt.: 490 F min.
*References:* Product bulletin

**Alox 2140S-70.** Alox Corp.
See Alox 2140S
*Composition:* 70% Alox 2140S, 30% mineral spirits

**Alox 2162.** Alox Corp.
*Chemical Description:* Barium soap of an oxygenated hydrocarbon
*Category:* Corrosion inhibitor
*Applications:* Base for thin film rust preventive for limited salt spray protection
*Form:* Paste; Color: Dark; Odor: Mild
*Properties:* Solubility: Nonpolar solvents; Density: 8.0 lb/gal @ 60 F; Sp. gr.: 0.96; M.P.: 95 F; Flash Pt.: 300 F
*Composition:* 100% active ingred.
*Reaction:* Slightly acid
*References:* Product bulletin

**Alox 2200.** Alox Corp.
*Category:* Corrosion inhibitor
*Applications:* Base for corrosion inhibitors
*Form:* Viscous liquid; Color: Brown; Odor: Mild
*Properties:* Solubility: Nonpolar solvents; Density: 8.39 lb/gal; Sp. gr.: 1.01; Flash Pt.: > 375 F

*Composition:* 100% active ingred.
*References:* Tech. data sheet

**Alox 2201.** Alox Corp.
*Category:* Corrosion inhibitor
*Applications:* Manufacture of water displacing, general purpose lubricating oil (low temperature)
*Form:* Liquid; Color: Dark; Odor: Mild
*Properties:* Solubility: Nonpolar solvents; Density: 8.0; Sp. gr.: 0.96; Flash Pt.: > 350 F (O.C.)
*Composition:* 100% active ingred.
*Reaction:* Slightly acid
*References:* Tech. data sheet

**Alox 2211K.** Alox Corp.
*Category:* Corrosion inhibitor
*Applications:* Thixotropic rust preventive depositing a thin film for long time protection from humidity and salt spray
*Form:* Film type–dry
*Properties:* Density: 7.9 lb/gal; Sp. gr.: 0.95; M.P.: ≈ 205 F; Flash Pt.: 350 F min.
*References:* Tech. data sheet

**Alox 2213.** Alox Corp.
*Category:* Corrosion inhibitor
*Properties:* Density: 7.7 lb/gal; Sp. gr.: 0.93; M.P.: 110–120 F; Flash Pt.: > 350 F
*References:* Tech. data sheet

**Alox 2213A.** Alox Corp.
*Category:* Corrosion inhibitor
*Applications:* Base for thin soft film water displacing rust preventive for household and industrial use; limited resistance to salt spray and acids; lubricant and penetrant
*Form:* Paste; Color: Dark; Odor: Mild
*Properties:* Solubility: Nonpolar solvents; Density: 7.75–8.08 lb/gal; Sp. gr.: 0.95; M.P.: 100 F; Flash Pt.: 300 F
*Composition:* 100% active ingred.
*Reaction:* Acid
*Containers:* Drums, bulk
*References:* Tech. data sheet

**Alox 2268.** Alox Corp.
*Category:* Corrosion inhibitor
*Applications:* Water emulsifiable corrosion inhibitor and lubricity agent; thin waxy film for protection of steel during extended indoor storage; excellent antiwear and lubricity characteristics
*Form:* Waxy soft solid; Color: Brown; Odor: Mild
*Properties:* Solubility: Water (emulsion); Density: 7.6 lb/gal; Sp. gr.: 0.92; M.P.: 135 F; Flash Pt.: > 275 F
*Composition:* 100% active ingred.
*Reaction:* Acid
*Containers:* Drums, bulk
*References:* Tech. data sheet

**Alrodyne 6104.** Ciba-Geigy
*Chemical Description:* Polyoxyalkylene fatty ester polymer
*Category:* Nonionic emulsifier
*Applications:* Nonionic emulsifier used with aromatic or aliphatic solvents and various toxicants
*Form:* Viscous liquid; Color: Dark amber

**Alromine RA.** Ciba-Geigy
*Chemical Description:* Fatty acid amide
*Applications:* Used in hair condi-

tioners and antistats
*Form:* Paste; Color: Off-white

**Alrosol B.** Ciba-Geigy
*Chemical Description:* Fatty acid amide
*Applications:* Used in liquid soaps and detergents for car washes, metal cleaners, dishwashing and hard surface cleaners
*Form:* Viscous liquid; Color: Amber

**Alrosol C.** Ciba-Geigy
*Chemical Description:* Fatty acid amide
*Category:* Foam booster, wetting agent
*Applications:* Used in hand soaps, hair and rug shampoos, metal cleaners
*Form:* Viscous liquid; Color: Light amber

**Alrosol Conc.** Ciba-Geigy
*Chemical Description:* Fatty acid amide
*Category:* Foam stabilizer
*Applications:* Foam stabilizer for shampoos, liquid soaps, shaving creams; alkaline hand and metal cleaners, degreasers, spot removers
*Form:* Clear viscous liquid; Color: Yellow

**Alrosol O.** Ciba-Geigy
*Chemical Description:* Fatty acid amide
*Category:* Emulsifier
*Applications:* Emulsifier for hydrocarbons, pigments, abrasives at pH 3–5; making of cutting oils; produces stable emulsions in presence of salt
*Form:* Viscous liquid; Color: Amber

**Alrosperse 11P Flake.** Ciba-Geigy
*Chemical Description:* Fatty acid amide
*Category:* Nonionic emulsifier, lubricant
*Applications:* Hair rinses, in hand modifiers for textiles, spreading agent in paste waxes and polishes
*Form:* Waxy flake; Color: Off white

**Alrosperse 100.** Ciba-Geigy
*Chemical Description:* Blend of nonionic and cationic surfactants
*Category:* Emulsifier, solubilizing agent
*Applications:* Solubilizing agent and emulsifier for nonaqueous media; in solvent-based metal cleaners; as dispersant in fuel oil and dry cleaning detergents
*Form:* Liquid; Color: Dark amber

**Alscoap SS-90.** Toho Chem.
*Category:* Detergent
*Applications:* Formulated hair shampoo; skins
*Form:* Paste; Color: Pale yellow
*Composition:* 90% active ingred.; other components: water

**Amacarrier CE.** AC&C
*Chemical Description:* Perc-biphenyl-benzoate
*Category:* Carrier, nonionic
*Applications:* Jet dyeing polyester with level dyeing, high-color yield, clean machines
*Form:* Clear liquid; Color: Pale yellow
*Properties:* Solubility: Self emul-

sifiable; Density: 10 lb/gal; Stability: Indefinite; Emulsifiability: 100%
*Composition:* 100% active ingred.
*Reaction:* Neutral, pH 8
*Containers:* 450 lb drums
*References:* Product bulletin

**Amacarrier DDQ.** AC&C
*Category:* Dye carrier, slightly anionic
*Applications:* Dye carrier for Quiana; atmospheric; mixed fiber dyeing
*Form:* Liquid; Color: Clear amber
*Properties:* Solubility: Dispersible in water; Stability: Indefinite; Emulsifiability: 100%
*Composition:* 100% active ingred.
*Reaction:* Neutral, pH 8
*References:* Product bulletin

**Amacarrier HB.** AC&C
*Chemical Description:* Self-emulsifying aromatic ester (benzoate)
*Category:* Dye carrier, nonionic-anionic
*Applications:* Dyeing polyester and triacetate; atmospheric conditions
*Form:* Liquid; Color: Clear amber
*Properties:* Solubility: Water sol.; Sp. gr.: 1.036; Stability: Indefinite; Emulsifiability: 100%; Biodegradable
*Composition:* 100% active ingred. (benzoate)
*Reaction:* Neutral, pH 8
*Containers:* 450 lb drums
*References:* Product bulletin

**Amacarrier L.** AC&C
*Chemical Description:* Blend TCB/biphenyl
*Category:* Dye carrier, nonionic
*Applications:* Polyester dyeing, atmospheric carrier
*Form:* Liquid; Color: Clear amber
*Properties:* Solubility: Self-emulsifiable at 100–120 F; Stability: Indefinite; Emulsifiability: 100%
*Composition:* 100% active ingred.
*Reaction:* Neutral, pH 8
*Containers:* 450 lb drums
*References:* Product bulletin

**Amacarrier LBP.** AC&C
*Chemical Description:* Liquid biphenyl
*Category:* Dye carrier, slightly anionic
*Applications:* Atmospheric carrier for polyester dyeing, mixed fiber dyeing
*Form:* Liquid; Color: Clear amber; Odor: Low
*Properties:* Solubility: Dispersible in warm water; Stability: Indefinite; Emulsifiability: 100%
*Composition:* 100% active ingred.
*Reaction:* Neutral, pH 8
*References:* Product bulletin

**Amafix PAM.** AC&C
*Category:* Dye fixative, anionic
*Applications:* Color fixation of acid dyes on nylon, dyed and printed fabrics
*Form:* Liquid; Color: Tan
*Properties:* Solubility: 100% water sol.; Density: 8.9 lb/gal; Sp. gr.: 1.07; Stability: Good; pH: 8
*Composition:* 25% active ingred.
*Reaction:* Neutral
*Containers:* 450 lb drums
*References:* Product bulletin

**Amaglide 53.** AC&C

*Category:* Dyebath lubricant, nonionic
*Applications:* Dyebath lubricant for synthetic and natural fabrics, esp. to control formation of crack marks on synthetic fabrics
*Form:* Liquid; Color: Milky white
*Properties:* Solubility: Water dispersible; Density: 8.3 lb/gal; Stability: Good; Emulsifiability: 100%
*Composition:* 30% active ingred.
*Reaction:* Neutral, pH 8
*Containers:* 450 lb drums
*References:* Product bulletin

**Amamigrant PA.** AC&C
*Chemical Description:* Polymeric
*Category:* Dyeing assistant, nonionic
*Applications:* Dye migration control in continuous dyeing
*Form:* Viscous liquid; Color: White
*Properties:* Solubility: 100% water sol.; Sp. gr.: 1.01; Visc.: 1000 cps (Brookfield); Stability: Good
*Reaction:* Neutral, pH 8
*Containers:* 450 lb drums
*References:* Product bulletin

**Amaquest.** AC&C
*Chemical Description:* EDTA–HEDTA
*Category:* Sequestering agent
*Applications:* Sequestering metals; Kier boiling; fat liquoring; bleaching; dyeing; finishing; insecticides, herbicides, fungicides; textiles; skins; leather; papermaking
*Form:* Liquid; Color: Clear
*Properties:* Solubility: 100%; Stability: Good
*Containers:* 450 lb drums
*References:* Product bulletin

**Amascour.** AC&C
*Category:* Detergent, wetting agent, anionic
*Applications:* Scouring agents for textiles; dyeing; desizing and boiling off cottons; bleaching; tanning; spot removal; Kier boiling; fat liquoring
*Form:* Liquid; Color: Clear white
*Properties:* Solubility: 100% water sol.; Stability: Indefinite
*Reaction:* Neutral, pH 8
*Containers:* 450 lb drums
*References:* Product bulletin

**Amasoft 16-7.** AC&C
*Category:* Fabric softener, cationic
*Applications:* Softener and antistat for synthetic fibers; softener for natural fibers; finishing
*Form:* Liquid; Color: Clear
*Properties:* Solubility: 100%; Density: 8.3 lb/gal
*Containers:* 450 lb drums
*References:* Product bulletin

**Amasperse N.** AC&C
*Category:* Anionic dispersing agent
*Applications:* Synthetic fiber dyeing; super dispersing agent for disperse color; tanning; sizing; textiles; skins; leather; metallurgy; mines and minerals; printing; dispersing dry powders; making of bitumen emulsions
*Form:* Liquid; Color: Clear
*Properties:* Solubility: 100% water sol.; Density: 9.7 lb/gal; Stability: Indefinite
*Composition:* 40% active ingred.
*Containers:* 450 lb drums
*References:* Product bulletin

**Amawet AMR.** AC&C
*Chemical Description:* Potassium salt of a long chain ester
*Category:* Wetting agent, anionic
*Applications:* Textile scouring, dyeing and finishing; specialty wetting agents; Kier boiling; continuous peroxide bleaching; vat dyeing; sulfur dyeing
*Form:* Clear liquid; Color: Light amber
*Properties:* Solubility: 100% water sol.; Stability: Good; pH: 8
*Containers:* 450 lb drums
*References:* Product bulletin

**Amawet DF.** AC&C
*Category:* Dye assistant
*Applications:* Nonfoaming auxiliary to increase flow of liquor in pump circulating machines; high temperature use
*Form:* Slightly viscous liquid; Color: Clear yellow
*Properties:* Density: 8.6 lb/gal; Sp. gr.: 1.03; Stability: Good
*References:* Product bulletin

**Amawet LF.** AC&C
*Chemical Description:* Exthoxylated linear alcohol
*Category:* Wetting, emulsifying, dispersing and penetrating agents; slightly anionic
*Form:* Water white liquid; Odor: Faint fatty alcohol
*Properties:* Biodegradable; low foaming
*Reaction:* Essentially neutral
*References:* Product bulletin

**Amawet P.** AC&C
*Chemical Description:* Sulfated fatty ester
*Category:* Dye assistant, anionic
*Applications:* Textile dyeing auxiliary aiding level dyeing; package, beam, cheese dyeing of cotton; preparation of dyeing of wool and synthetic knit, woven piece goods
*Form:* Liquid; Color: Dark amber
*Properties:* Solubility: Water sol; pH: 6.5
*References:* Product bulletin

**Ambidex.** Cincinnati Milacron Chem.
*Category:* Wetting agent, emulsifier, cationic
*Applications:* Wetting agent and emulsifier for oils and asphalts
*Form:* Viscous liquid; Color: Dark brown; Odor: Amine
*Properties:* Solubility: Oil sol.; Density: 8.14 lb/gal @ 60 F; Sp. gr.: 0.985 @ 60 F; Visc.: 6600 SUS @ 100 F; Flash Pt.: 450 F; Pour Pt.: 40 F
*Composition:* 100% active ingred.
*Containers:* 55 gal drums, tankcar
*References:* Tech. bulletin

**Ambidex-PB.** Cincinnati Milacron Chem.
*Category:* Wetting agent, emulsifier, cationic
*Applications:* Wetting agent for oils and asphalts
*Form:* Viscous liquid; Color: Dark brown; Odor: Slight amine
*Properties:* Solubility: Oil sol.; Density: 8.2 lb/gal; Sp. gr.: 0.985 @ 60 F; Visc.: 660 SUS @ 140 F; Flash Pt.: 370 F; Pour Pt.: 72 F
*Composition:* 100% active ingred.
*Containers:* 55-gal drums, tankcar
*References:* Tech. bulletin

**Amerchol Series.** Amerchol, a Unit of CPC International
*Chemical Description:* Sterol extracts
*Category:* Emulsifier, stabilizer, plasticizer, nonionic
*Applications:* Aerosols, baby toiletries, creams and lotions, dermatological vehicles, hair products, lotion detergents, pigmented makeup items
*References:* Products bulletin

**Amerchol 400.** Amerchol, a Unit of CPC International
*Category and Applications:* See Amerchol Series; Least effective emulsifier; adjust payoff of anhydrous makeups; suitable for medium to heavy creams and lotions
*Form:* Soft solid; Color: Yellow-ivory
*Properties:* M.P.: 48–52 C

**Amerchol BL.** Amerchol, a Unit of CPC International
*Category and Applications:* See Amerchol Series; Highest ester content of the series; ideal for heavy creams and lotions
*Form:* Semisolid; Color: Yellow-amber

**Amerchol C.** Amerchol, a Unit of CPC International
*Category and Applications:* See Amerchol Series; Moderate ester content; suitable for medium to heavy creams and lotions
*Form:* Soft solid; Color: Pale yellow-ivory
*Properties:* M.P.: 40–46 C

**Amerchol CAB.** Amerchol, a Unit of CPC International
*Category and Applications:* See Amerchol Series; Contains no esters or free fatty acids; hypoallergenic preparations; medium to heavy creams and lotions
*Form:* Soft solid; Color: Pale ivory
*Properties:* M.P.: 40–46 C

**Amerchol H-9.** Amerchol, a Unit of CPC International
*Category and Applications:* See Amerchol Series; Moderate ester content; spreads easily but retains consistency at body temp.; ideal for dermatologicals and heavy creams and lotions
*Form:* Soft solid; Color: Pale yellow
*Properties:* M.P.: 55–62 C

**Amerchol L-99.** Amerchol, a Unit of CPC International
*Category and Applications:* See Amerchol Series; Thin fluid form; economical alternative for less critical emulsions; light lotions to heavy creams
*Form:* Oily liquid; Color: Pale yellow; Odor: Faint, characteristic sterol
*Properties:* Sp. gr.: 0.856–0.866

**Amerchol L-101.** Amerchol, a Unit of CPC International
*Category and Applications:* See Amerchol Series; Most efficient stabilizer for emulsified makeups; plasticizer for hair sprays; hypoallergenic products; ideal for light lotions to heavy creams
*Form:* Oily liquid; Color: Pale yellow; Odor: Faint, characteristic sterol

*Properties:* Sp. gr.: 0.84–0.86

**Amerchol L-500.** Amerchol, a Unit of CPC International
*Category and Applications:* See Amerchol Series; Ideal for low oil/high wax content systems; economical conc. extract for light lotions to heavy creams
*Form:* Thick liquid to semisolid; Color: Amber; Odor: Characteristic sterol

**Amerchol RC.** Amerchol, a Unit of CPC International
*Category and Applications:* See Amerchol Series; High triterpene content; least effective emulsifier; shade developer for cosmetic pigments; ideal for anhydrous and emulsified makeups
*Form:* Soft solid; Color: Ivory
*Properties:* M.P.: 55–64 C

**Amerchol Polysorbate 80.** Amerchol, a Unit of CPC International
*Chemical Description:* Polyoxyethylene (20) sorbitan monooleate
*Category:* Emulsifier, solubilizer, nonionic
*Applications:* Used in cosmetic, toiletry, and pharmaceutical products
*Form:* Liquid
*Properties:* Solubility: Sol. in water and alcohol; HLB: 15.0
*Composition:* 100% active ingred.

**Ameroxol OE-2.** Amerchol, a Unit of CPC International
*Chemical Description:* Oleyl alcohol ethoxylate
*Category:* Nonionic emulsifier and solubilizer
*Applications:* W/O emulsifier; creams and lotions; ideal spreading and dispersing agent for bath oils
*Form:* Clear liquid; Color: Pale straw; Odor: Bland
*Properties:* Solubility: Sol. in mineral oil, isopropyl esters, anhydrous ethanol
*References:* Products bulletin

**Ameroxol OE-10.** Amerchol, a Unit of CPC International
*Chemical Description and Category:* See Amerchol OE-2
*Applications:* O/W emulsifier and solubilizer, esp. for fragrance products
*Form:* Semisolid; Color: White; Odor: Bland
*Properties:* Solubility: Sol. in water, ethanol, hydroalcoholics, glycols; Cloud Pt.: 47–55 C
*References:* Products bulletin

**Ameroxol OE-20.** Amerchol, a Unit of CPC International
*Chemical Description and Category:* See Amerchol OE-2
*Applications:* O/W emulsifier and solubilizer with soft afterfeel
*Form:* Waxy solid; Color: White; Odor: Bland
*Properties:* Solubility: Sol. in water, ethanol, hydroalcoholics, glycols; Cloud Pt.: 87–93 C

**Amidan.** Grindsted Products
*Chemical Description:* Distilled monoglycerides integrated with other food ingredients
*Category:* Emulsifier
*Applications:* Dough conditioner in bread, rolls, bread improvers

**Amidox C.** Stepan Chem.
*Chemical Description:* Ethoxylat-

ed coconut fatty acids monoethanol-amides
*Category:* Detergent, emulsifier
*Applications:* Shampoos, dishwashing detergents
*Form:* Liquid to paste
*Composition:* 100% active ingred.

**Amine C.** Ciba-Geigy
*Chemical Description:* Imidazoline derivative
*Category:* Emulsifier, dispersant, detergent, cationic
*Applications:* Acid cleaners, antistat agents for textiles, plastics; paints; corrosion inhibitor; making of cutting and rust preventive oils
*Form:* Yellow liquid
*Properties:* Solubility: Sol. in polar, organic solvents, hydrocarbons, relatively insol. water; Stability: Good
*Composition:* 100% active ingred.
*Toxicity:* Very irritating to skin and eyes
*References:* Tech. data

**Amine O.** Ciba-Geigy
*Chemical Description:* Oleyl imidazoline
*Category:* Emulsifier, dispersant, detergent, antistat, cationic
*Applications:* Antistat agent for textiles, plastics, paints; in acid cleaners; in corrosion inhibitor formulations; lubricating oils; asphalts
*Form:* Amber liquid
*Properties:* See Amine C
*Composition:* 100% active ingred.
*Toxicity:* Very irritating to skin and eyes
*References:* Tech. data

**Amine S.** Ciba-Geigy
*Chemical Description:* Stearyl imidazoline
*Category:* Emulsifier, dispersant, detergent, antistat, cationic
*Applications:* Antistat for textiles, paints; in acid cleaners; in corrosion inhibitors; emulsion waxes
*Form:* Tan wax
*Properties:* See Amine C
*Composition:* 100% active ingred.
*Toxicity:* Very irritating to skin and eyes
*References:* Tech. data

**Amine T.** Ciba-Geigy
*Chemical Description:* Tall oil fatty acid imidazoline
*Category and Applications:* See Amine O
*Form:* Dark amber liquid
*Composition:* 100% active ingred.
*Toxicity:* Very irritating to skin and eyes
*References:* Tech. data

**Aminogen DG Liquid.** . AC&C
*Category:* Foaming agent
*Applications:* Foam control for dyeing in combination with Aminogen K or SD; aids solubility of dyestuff
*Form:* Clear liquid
*Properties:* Density: 8.64 lb/gal; Stability: Stable and pourable at 25 F; High foam level; pH: 8.8
*Composition:* 100% active ingred.
*References:* Product bulletin

**Aminogen EB.** AC&C
*Category:* Dye assistant, anionic-nonionic
*Applications:* Energy saving dispersion, leveling antiprecipitant for

low temperature rapid dyeing of nylon stock, yarn, woven, knits, carpets
*Form:* Liquid; Color: Clear amber
*Properties:* Solubility: 100% water sol.; Stability: Indefinite
*Reaction:* Neutral, pH 7
*Containers:* 450 lb drums
*References:* Product bulletin

**Aminogen K-LF.** AC&C
*Category:* Dye assistant
*Applications:* Continuous dyeing of nylon yarn, carpet; maximize color value
*Form:* Light yellow liquid
*Properties:* Density: 8.14 lb/gal; Stability: Stable and pourable @ 25 F; Low foam level; pH: 9.0
*Composition:* 100% active ingred.
*References:* Product bulletin

**Aminogen P-LF.** AC&C
*Category:* Dye assistant
*Applications:* Printing of nylons; maximize color value
*Form:* Light yellow liquid
*Properties:* Density: 8.58 lb/gal; Stability: Stable and pourable @ 25 F; Moderate foam level
*Composition:* 100% active ingred.
*References:* Product bulletin

**Aminogen R.** AC&C
*Category:* Dye assistant
*Applications:* Energy saving low temperature dyeing of wool; aids rapid dyeing and dye penetration
*Form:* Liquid; Color: Clear
*Properties:* Solubility: 100% water sol.; Density: 8.7 lb/gal; Stability: Good ; pH 6.0–6.5
*Containers:* 450 lb drums
*References:* Product bulletin

**Aminogen SD-LF.** AC&C
*Category:* Dye assistant
*Applications:* Space dyeing; flat bed printing; yarn dyeing by pad application; nylon; carpet dyeing
*Form:* Light yellow liquid
*Properties:* Solubility: 100%; Density: 8.58 lb/gal; Stability: Stable and pourable @ 25 F; Low foam level; pH: 9.7
*Composition:* 100% active ingred.
*Containers:* 450 lb drums
*References:* Product bulletin

**Amisoft C-T-12.** Ajinomoto U.S.A.
*Chemical Description:* Monotriethanolamine *n*-cocoyl-*l*-glutamate
*Category:* Emulsifier
*Applications:* Cosmetics
*Form:* Liquid
*Composition:* 20% active ingred.

**Amisoft GS-11.** Ajinomoto U.S.A.
*Chemical Description:* Monosodium *n*-cocoyl-*l*-glutamate
*Category:* Emulsifier
*Applications:* Cosmetics
*Form:* Powder
*Composition:* 100% active ingred.

**Amisoft HS-11.** Ajinomoto U.S.A.
*Chemical Description:* Monosodium *n*-hydrogenated-tallowyl-*l*-glutamate
*Category:* Emulsifier
*Applications:* Cosmetics
*Form:* Powder
*Composition:* 100% active ingred.

**Amisoft HT 12.** Ajinomoto U.S.A.
*Chemical Description:* Monotriethanolamine *n*-hydrogenated-tallowyl-*l*-glutamate
*Category:* Emulsifier

*Application:* Cosmetics
*Form:* Liquid
*Composition:* 20% active ingred.

**Amisoft LS11.** Ajinomoto U.S.A.
*Chemical Description:* Monosodium *n*-lauroryl-*l*-glutamate
*Category:* Emulsifier
*Applications:* Cosmetics
*Form:* Powder
*Composition:* 100% active ingred.

**Amisoft LT-12.** Ajinomoto U.S.A.
*Chemical Description:* Monotriethanolamine *n*-lauroyl-*l*-glutamate
*Category:* Emulsifier
*Applications:* Cosmetics
*Form:* Liquid
*Composition:* 20% active ingred.

**Ammonyx 4.** Onyx Chem. Co.
*Chemical Description:* Stearyl dimethyl benzyl ammonium chloride
*Category:* Emulsifier, conditioner, softener, emollient, cationic
*Applications:* Hair rinses, skin creams and lotions
*Form:* Paste
*Properties:* Sp. gr.: 0.99 (25/20 C); Flash Pt.: > 200 F
*Composition:* 17–19% quaternary; 24–28% solids

**Ammonyx 4B.** Onyx Chem. Co.
*Chemical Description:* Stearyl dimethyl benzyl ammonium chloride
*Category:* Emulsifier, conditioner, softener, emollient, cationic
*Applications:* Hair rinses, skin creams and lotions
*Form:* Paste
*Properties:* Sp. gr.: 0.99 (25/20 C); Flash Pt.: > 200 F
*Composition:* 16–18% quaternary; 24–26% solids

**Ammonyx 485.** Onyx Chem. Co.
*Chemical Description:* Stearyl dimethyl benzyl ammonium chloride
*Category:* Emulsifier, conditioner, softener, emollient, cationic
*Applications:* Hair rinses, skin creams and lotions
*Form:* Powder
*Properties:* Sp. gr.: 0.45 (25/20 C); Flash Pt.: 122 F
*Composition:* 85% min. quaternary; 87% min. solids

**Ammonyx 490.** Onyx Chem. Co.
*Chemical Description:* Stearyl dimethyl benzyl ammonium chloride
*Category:* Emulsifier, conditioner, softener, emollient, cationic
*Applications:* Hair rinses, skin creams and lotions
*Form:* Powder
*Properties:* Sp. gr.: 0.47 (25/20 C); Flash Pt.: 130 F
*Composition:* 90% min. quaternary; 92 min. solids

**Ammonyx 856.** Onyx Chem. Co.
*Chemical Description:* Tallow dimethyl benzyl ammonium chloride
*Category:* Dispersant, suspending agent, deemulsifier, cationic
*Applications:* Industrial grade pigment dispersant and suspending agent; deemulsifier for solvent extraction processes

*Form:* Paste
*Properties:* Sp. gr. 0.90 (25/20 C); Flash Pt.: 64 F
*Composition:* 75% min. quaternary; 77% min. solids

**Ammonyx 4002.** Onyx Chem. Co.
*Chemical Description:* Stearyl dimethyl benzyl ammonium chloride
*Category:* Emulsifier, conditioner, softener, emollient, cationic
*Applications:* Hair rinses, skin creams and lotions
*Form:* Powder
*Properties:* Sp. gr.: 0.52 (25/20 C); Flash Pt.: 170 F
*Composition:* 94% min. quaternary; 96% min. solids

**Ammonyx 4080.** Onyx Chem. Co.
*Chemical Description:* 1-Methyl-1-tallow amido-ethyl-2-heptadecyl imidazolinium methosulfate
*Category:* Softener, cationic
*Applications:* Liquid fabric softener for home and industrial use; formulated over wide viscosity range
*Form:* Liquid
*Properties:* Sp. gr.: 0.97 (26/20 C); Flash Pt.: 68 F
*Composition:* 74–76% quaternary; 81% max. solids

**Ammonyx CDO.** Onyx Chem. Co.
*Chemical Description:* Coco amido propyl dimethyl amine oxide
*Category:* Wetting, foaming agent, foam stabilizer, conditioner, nonionic
*Applications:* Bubble baths, bath oils, dishwash concentrates, hair coloring systems, softeners and cleansers
*Form:* Liquid
*Properties:* Sp. gr.: 1.02 (25/20 C); Flash Pt.: > 200 F
*Composition:* 30% amine oxide

**Ammonyx CETAC.** Onyx Chem.
*Chemical Description:* Cetyl trimethyl ammonium chloride
*Category:* Emulsifier, conditioner, softener, emollient, cationic
*Applications:* Hair rinses, skin creams and lotions
*Form:* Liquid
*Properties:* Sp. gr.: 0.93 (25/20 C); Flash Pt.: > 200 F
*Composition:* 24–26% quaternary; 25% min. solids

**Ammonyx CO.** Onyx Chem. Co.
*Chemical Description:* Cetyl dimethyl amine oxide
*Category:* Conditioner, nonionic
*Applications:* Household, cosmetic and janitorial products
*Form:* Liquid
*Properties:* Sp. gr.: 0.96 (25/20 C); Flash Pt.: > 200 F; Stability: Stable in high electrolyte concentrations
*Composition:* 30% amine oxide

**Ammonyx DMCD-40.** Onyx Chem.
*Chemical Description:* Lauryl dimethyl amine oxide
*Category:* Wetting, foaming agent, foam stabilizer, nonionic
*Applications:* Cosmetic, household and janitorial products
*Form:* Liquid
*Properties:* Sp. gr.: 0.91 (25/20 C); Flash Pt.: 86 F
*Composition:* 40–42% amine oxide

**Ammonyx DME.** Onyx Chem. Co.
*Chemical Description:* Cetyl dimethyl benzyl ammonium chloride
*Category:* Emulsifier, wetting agent, cationic
*Applications:* Personal care products
*Form:* Paste
*Properties:* Sp. gr.: 0.96 (25/20 C); Flash Pt.: 52 F
*Composition:* 75% min. quaternary; 77% min. solids

**Ammonyx KP.** Onyx Chem. Co.
*Chemical Description:* Oleyl dimethyl benzyl ammonium chloride
*Category:* Conditioner, antistat, cationic
*Applications:* Clear hair rinses
*Form:* Liquid
*Properties:* Sp. gr.: 0.98 (25/20 C); Flash Pt.: > 200 F
*Composition:* 50% min. quaternary; 52% min. solids

**Ammonyx LO.** Onyx Chem. Co.
*Chemical Description:* Lauryl dimethyl amine oxide
*Category:* Wetting, foaming agent, foam stabilizer, grease emulsifier, nonionic
*Applications:* Shampoos, bath products, fine fabric cleaners, shaving creams, lotions
*Form:* Liquid
*Properties:* Sp. gr.: 0.96 (25/20 C); Flash Pt.: > 200 F
*Composition:* 30% amine oxide

**Ammonyx MCO.** Onyx Chem. Co.
*Chemical Description:* Myristyl/cetyl dimethyl amine oxide
*Category:* Wetting and foaming agent, foam stabilizer, nonionic
*Applications:* Cosmetic, household and janitorial products
*Form:* Liquid
*Properties:* Sp. gr.: 0.96 (25/20 C); Flash Pt.: > 200 F
*Composition:* 30% amine oxide

**Ammonyx MO.** Onyx Chem. Co.
*Chemical Description:* Myristyl dimethyl amine oxide
*Category:* Wetting and foaming agent, foam stabilizer, nonionic
*Applications:* Cosmetic, household and janitorial products
*Form:* Liquid
*Properties:* Sp. gr.: 0.96 (25/20 C); Flash Pt.: > 200 F
*Composition:* 30% amine oxide

**Ammonyx SO.** Onyx Chem. Co.
*Chemical Description:* Stearyl dimethyl amine oxide
*Category:* Conditioner, emulsifier, nonionic
*Applications:* Cream rinses, lotions, liquid bath products
*Form:* Paste
*Properties:* Sp. gr.: 0.99 (25/20 C); Flash Pt.: > 200 F
*Composition:* 25% amine oxide

**Ammonyx T.** Onyx Chem. Co.
*Chemical Description:* Cetyl dimethyl benzyl ammonium chloride
*Category:* Emulsifier, wetting agent, cationic
*Applications:* In specialty latex dispersions as emulsifier and wetting agent

*Form:* Liquid
*Properties:* Sp. gr.: 0.99 (25/20 C); Flash Pt.: > 200 F
*Composition:* 25% min. quaternary; 25% min. solids

**AMP.** IMC Chem.–NP Div.
*Chemical Description:* 2-amino-2-methyl-1-propanol
*Category:* Emulsifier, catalyst, anionic
*Applications:* As raw materials in polymer applications, e.g. oxazoline structures (as corrosion inhibitors, antifoam compounds, synthetic waxes); in coatings as pigment dispersants; in emulsions for cosmetics, cleaners, industrial insecticides, asphalt emulsions; in boiler-water treatment; as catalysts; in formaldehyde scavenging
*Form:* Solid; Color: 20 (APHA)
*Properties:* m.w.: 89.14; Density: 7.78 lb/gal; Sp. gr.: 0.928 (40/40C); Visc.: 102 cp @ 30 C; M.P.: 30 C; B.P.:165 C; Flash Pt.: 172F
*Composition:* 100% active ingred.
*Toxicity:* Skin irritant; causes eye burns
*Containers:* 55 gal drums or bulk; store in iron or steel
*Warnings and Caution:* Corrosive to copper, brass and aluminum; combustible; protective goggles and gloves should be worn
*References:* TDS No. 10

**AMP-95.** IMC Chem.–NP Div.
*Chemical Description:* 2-amino-2-methyl-1-propanol
*Category and Applications:* See AMP
*Form:* Liquid; Color: Colorless; Odor: Low
*Properties:* m.w.: 89.14; Density: 7.85 lb/gal; Sp. gr.: 0.942 (25/25C); Visc.: 147 cp; F.P.: –2 C; Flash Pt.: 182 F (Tag closed cup); Surface Tension: 36–38 dynes/cm
*Composition:* 95% active ingred.
*Toxicity:* Skin irritant, causes eye burns
*Containers:* See AMP
*Warnings and Caution:* See AMP
*References:* TDS No. 10A

**Ampho B11-34.** Armstrong Chem.
*Chemical Description:* Complex cocobetaine derived from dimethyl amine propyl amine
*Category:* High foaming amphoteric surfactant with outstanding foam stability; wetting agent
*Applications:* Cosmetic formulations
*Form:* Clear liquid; Color: Light yellow; Odor: Mild
*Properties:* Solubility: Solv. water; Density: 8 lb/gal; Sp. gr.: 0.96
*Composition:* 35% active ingred.
*Reaction:* Compatible with anionic, cationic and nonionic detergents; compatible with acid and alkaline systems

**Ampho T-35.** Armstrong Chem.
*Chemical Description:* Potassium salt of fatty amine polyethoxy carboxylate
*Category:* Detergent, amphoteric
*Applications:* Heavy duty surface cleaners; chelating agent; solubilizer for difficult materials; stabilizer; dispersant and penetrant; corrosion inhibitors; metal-to-metal lubricant; skin softening in dish-

washing compounds; laundry detergents
*Form:* Liquid; Color: Amber; Odor: Mild
*Properties:* Solubility: Water and polar solvents; Density: 8.7 lb/gal; Sp. gr.: 1.05 @ 80 F; Stability: Stable to acid or alkali; Biodegradable
*Composition:* 35% active ingred.
*Reaction:* Compatible with alkali or acid systems

**Ampholan U 203.** Diamond Shamrock
*Chemical Description:* Sodium salt of an amphoteric surfactant
*Category:* Detergent, wetting agent
*Applications:* Foaming agent for synthetic latex; effective in alkaline solution
*Form:* Liquid
*Composition:* 30% active ingred.
*Containers:* Drums

**Ampholan U203.** Lankro Chem. Ltd. Surfactants Div.
*Chemical Description:* Amphoteric surfactant with carboxyl and amino groups
*Category:* Detergent, foaming and stabilizing agent, dispersant, amphoteric
*Applications:* Industrial cleaners and detergents esp. in high electrolyte concentrations; auzilliary foaming and stabilizing agent in synthetic rubber latices in carpet backing industry; shampoo formulations; germicides; in textile industry as dispersant for pigments in manufacture of synthetic fibers
*Form:* Liquid; Color: Pale amber; Odor: Characteristic
*Properties:* Solubility: Water-sol. @ 20 C; Sp. gr.: 1.035 @ 20 C; Visc.: 132 cs @ 20 C; Flash Pt.: > 200 F (COC); Pour Pt.: < 0 C; pH: 6.0–8.0 (1% aq.)
*Composition:* 30% active ingred.
*Reaction:* Forms a cationic quaternary ammonium salt in acidic sol'ns.; forms anionic carboxylate salt in alkaline conditions
*Containers:* Store in SS, glass, resin or polythene lined equipment
*Warnings and Caution:* Avoid prolonged contact with concentrated product; should not be mixed with strong oxidizing or reducing agents (potentially explosive)

**Anionyx 12S.** Onyx Chem. Co.
*Chemical Description:* Disodium monooleamido diglycol sulfosuccinate
*Category:* Detergent, anionic
*Applications:* Low-irritation shampoos, bubble baths, dishwash detergents
*Form:* Liquid
*Properties:* Sp. gr.: 1.05 (25/20 C); Flash Pt.: 138 F
*Composition:* 20% active

**Antara LB-400.** GAF Chem. Prod.
*Chemical Description:* Fatty alcohol polyethyleneoxy phosphate ester acid
*Category:* Lubricant, anionic
*Applications:* Lubricant, extreme-pressure additive, rust inhibitor with good antiwear, antiweld, wetting, detergency, dispersancy, emulsification properties; chain

lubricants, circulating oils, cutting fluids, drawing compounds, diesel oils, greases, hydraulic fluids, rolling oils, rust preventives, steel-mill slushing oils, turbine oils, gear oils, glass cutting and polishing lubricants
*Form:* Opaque viscous liquid; Color: Amber
*Properties:* Solubility: Oil and water sol. and in aliphatic and aromatic solvents; Sp. gr.: 1.04
*Composition:* 99% active ingred.
*Toxicity:* Skin and eye irritant
*References:* Data sheet 450-65

**Antara LE-500.** GAF Chem. Prod.
*Chemical Description:* Free acid of a complex organic phosphate ester
*Category and Applications:* See Antara LB-400
*Form:* Clear viscous liquid; Color: Yellow
*Properties:* Solubility: Oil and water sol. and in aliphatic and aromatic solvents; Sp. gr.: 1.11

**Antara LE-700.** GAF Chem. Prod.
*Chemical Description:* Free acid of a complex organic phosphate ester
*Category and Applications:* See Antara LB-400
*Form:* Clear viscous liquid; Color: Pale yellow
*Properties:* Solubility: Water sol. and aromatic solvents; Sp. gr.: 1.12
*Composition:* 100% active

**Antara LF-200.** GAF Chem. Prod.
*Chemical Description:* Complex organic phosphate ester free acid
*Category and Applications:* See Antara LB-400
*Form:* Slight cloudy viscous liquid; Color: 8 (VCS)
*Composition:* 100% active ingred.

**Antara LK-500.** GAF Chem. Prod.
*Chemical Description:* Complex phosphate ester of an ethoxylated linear alcohol
*Category:* Low-foaming lubricant
*Applications:* Lubricant for either oil or synthetic water-based metalworking fluids, with excellent extreme-pressure, anti-wear and rust inhibition properties
*Form:* Slightly viscous, clear liquid Color: Pale yellow
*Properties:* Solubility: Sol. in xylene, ethanol, ethylene glycol, carbon tetrachloride, butyl cellosolve, water; Sp. gr.: 1.1; Visc.: 21 centistokes @ 100 C; Biodegradable
*Composition:* 100% active ingred.

**Antara LM-400.** GAF Chem. Prod.
*Chemical Description:* Free acid of a complex organic phosphate ester
*Category and Applications:* See Antara LB-400
*Form:* Clear liquid; Color: Yellow
*Properties:* Solubility: Oil sol. and aliphatic and aromatic solvents; Sp. gr.: 1.05
*Composition:* 100% active ingred.

**Antara LM-600.** GAF Chem. Prod.
*Chemical Description:* Free acid of a complex organic phosphate ester
*Category and Applications:* See Antara LB-400
*Form:* Clear viscous liquid; Color: Yellow

*Properties:* Solubility: Oil and water sol. and in aliphatic and aromatic solvents; Sp. gr.: 1.06
*Composition:* 100% active ingred.

**Antara LP-700.** GAF Chem. Prod.
*Chemical Description:* Free acid of a complex organic phosphate ester
*Category and Applications:* See Antara LB-400
*Form:* Clear viscous liquid; Color: Pale yellow
*Properties:* Solubility: Water, aromatic solvents; Sp. gr.: 1.20
*Composition:* 100% active ingred.

**Antara LS-500.** GAF Chem. Prod.
*Chemical Description:* Free acid of a complex organic phosphate ester
*Category and Applications:* See Antara LB-400
*Form:* Clear viscous liquid; Color: Pale yellow
*Properties:* Solubility: Oil and water sol. and in aliphatic and aromatic solvents; Sp. gr.: 1.00
*Composition:* 100% active ingred.

**Antaron FC-34.** GAF Chem. Prod.
*Chemical Description:* Complex fatty amido compound (fatty group derived from coconut oil)
*Category:* Detergent, wetting agent, emulsifier, dispersant, emollient, amphoteric surfactant
*Applications:* In textile processing, esp. fulling of woolens; leather degreasing; fat liquoring; household and industrial liquid-cleaners and shampoos (rugs, upholstery, floors, cars, white-wall tires); steel pickling; steam cleaning of bright aluminum; aerosol formulations; toiletries
*Form:* Clear semiviscous liquid; Color: Amber
*Properties:* Solubility: Water-sol. esp. high electrolyte content; Sp. gr.: 1.09; Stability: Good stability in acid, alkaline, or electrolyte systems; Surface Tension: 32.0 dynes/cm (0.0155% @ 20 C)
*Composition:* 37% active ingred.
*Reaction:* Compatible with wide range of anionic, cationic and nonionic surfactants
*Toxicity:* Skin irritant; causes eye burn
*Containers:* 10 lb pkg; 50, 500 lb drums; bulk avail. in tank cars
*References:* Bulletin 2303-004

**Antaron MC-44.** GAF Chem. Prod.
*Chemical Description:* Dicarboxylic coco imidazoline, sodium salt
*Category:* Emulsifier, solubilizer, coupling agent, amphoteric
*Applications:* Nonirritating shampoos, skin cleansers, cosmetics; industrial and household cleaners
*Form:* Viscous liquid; Color: Amber
*Composition:* 38% active ingred.

**Antaron PC-37.** GAF Chem. Prod.
*Chemical Description:* Sulfated fatty polyoxyethylene quaternary nitrogen compound
*Category:* Amphoteric shampoo surfactant
*Applications:* Cosmetics esp. shampoos due to skin and eye mildness, ease of formulation, hair manageability, broad surfactant and resin

compatibility
*Form:* Clear slightly viscous liquid; Color: Amber
*Properties:* Solubility (10% conc.): miscible—water, ethanol; Sp. gr.: 1.15; Visc.: 1100-1200 cps (Brookfield, 25C); Stability: Excellent > pH 5.5, in hard water and strong bases; Surface Tension: 39 dynes/cm (0.1%); Biodegradable
*Composition:* 75% active ingred.; 25% water
*Reaction:* Compatible with anionic, cationic, nonionics
*Toxicity:* Very mild to skin and eyes
*References:* Bulletin 9655-055

**Antarox BL-225.** GAF Chem. Prod.
*Chemical Description:* Modified linear aliphatic polyether
*Category:* Low-foaming nonionic surfactant with excellent detergency and wetting properties
*Applications:* Commercial dishwashing formulations; alkaline metal cleaners; machine floor-scrubbers; desizing baths; continuous bleaching, dyeing, finishing of textiles
*Form:* Clear liquid; Color: Slightly yellow; Odor: Bland
*Properties:* Solubility: Water and organic solvents except paraffins; Sp. gr.: 0.99; Visc.: 40 cps; Cloud Pt.: 25-29 C; Flash Pt.: 190 C (COC); Pour Pt.: < −18 C; Stability: Good to acids and alkalis; Surface Tension: 30 dynes/cm (0.1%); Biodegradable
*Composition:* 100% active ingred.
*Toxicity:* Eye irritant
*References:* Bulletin 9643-003

**Antarox BL-330.** GAF Chem. Prod.
*Chemical Description:* Aliphatic polyether
*Category:* Detergent, wetting agent, nonionic
*Applications:* Low foaming detergent and wetting agent; rinse aids, metal cleaning, mechanical dishwashing
*Form:* Liquid
*Composition:* 95% active ingred.

**Antarox BL-344.** GAF Chem. Prod.
*Chemical Description:* Aliphatic polyether
*Category:* Detergent, nonionic
*Applications:* High-temperature detergent; nondiscoloring dedusters for caustic powders
*Form:* Liquid
*Properties:* Stability: Good at high temperatures
*Composition:* 90% active ingred.

**Antarox BL-240.** GAF Chem. Prod.
*Chemical Description:* Modified linear aliphatic polyether
*Category:* See Antarox BL-225
*Applications:* High-temperature applications; metal spray cleaning; textile wet processing; rinse aid
*Form:* Liquid; Color: Slightly yellow; Odor: Bland
*Properties:* Solubility: More water sol. than Antarox BL-225; Sp. gr.: 0.99; Visc.: 35 cps; Cloud Pt.: 38-42 C; Flash Pt.: 124 C; Pour Pt.: < −21 C; Surface Tension: 28 dynes/cm (0.1%); Biodegradable
*Composition:* 100% active ingred.
*Toxicity:* Eye irritant

*References:* Bulletin 9643-003

**Antarox G-200.** GAF Chem. Prod.
*Chemical Description:* Amino amide from mixed fatty acids and mixed polyamines reacted with ethylene oxide
*Category:* Dispersant, nonionic
*Applications:* Dispersant substantive to ferrous metals; viscose spin bath additive (tire cords, textile filaments); deposit and sludge control in cellophane mfg., casting solutions, etc.; steelmaking process
*Form:* Opaque, viscous liquid; Color: Brown
*Properties:* Refractive Index: 1.48; Cloud Pt.: 64–68 C
*Composition:* 100% active ingred.

**Antarox LF-330.** GAF Chem. Prod.
*Chemical Description:* Aliphatic polyether
*Category:* Detergent, wetting agent, nonionic
*Applications:* Low-foaming surfactant for rinse aid concentrates; metal spray cleaners; machine dishwashing; textile wetting and scouring; aircraft, bottle, dairy, and food plant cleaning; laundry detergents; paper-machine-felt washing; pulp deresination; steam cleaning
*Form:* Clear liquid; Color: Pale yellow; Odor: Bland
*Properties:* Solubility: Water and most organic solvents except paraffins; Visc.: 29 centistokes; Cloud Pt.: 30 C; Pour Pt.: 3 C; Stability: Excellent to acid and strong alkali; Surface Tension: 29 dynes/cm (0.1%)
*Composition:* 95% active ingred.
*Reaction:* Slightly acid
*Toxicity:* Eye irritant
*References:* Bulletin 7543-097

**Antarox LF-344.** GAF Chem. Prod.
*Chemical Description:* Aliphatic polyether
*Category and Applications:* See Antarox LF-330. LF-344 has better solubility at high temperatures
*Form:* Clear liquid; Color: Pale yellow; Odor: Bland
*Properties:* Solubility: Water and most organic solvents except paraffins; Visc.: 61 centistokes; Cloud Pt.: 44 C; Pour Pt.: 6 C; Surface Tension: 33 dynes/cm (0.1%)
*Composition:* 90% active ingred.
*Reaction:* Slightly acid
*Toxicity:* Eye irritant
*References:* Bulletin 7543-097

**Antiprex 461.** Allied Colloids Inc.
*Chemical Description:* Polyacrylic acid
*Category:* Antiscalant
*Applications:* Antiscalant for use in industrial water systems, e.g., cooling towers, boilers, heat exchangers
*Form:* Liquid; Color: Light amber
*Properties:* m.w.: low; Sp. gr.: 1.26; Visc.: 400–700 cps
*Composition:* 50% active ingred.
*Containers:* 55 gal steel drums or bulk
*Warnings and Caution:* Keep away from skin and eyes
*References:* Technical data sheet

**Antiprex A.** Allied Colloids Inc.
*Chemical Description:* Acrylate polymer
*Category:* Scale inhibitor, dispersant, anionic

*Applications:* Scale inhibitor for use in aqueous systems, e.g., recirculating cooling systems, evaporators, boilers, in the chemical, mineral and oil industries
*Form:* Liquid; Color: Pale yellow; Odor: Slight
*Properties:* m.w.: 2–4000; Solubility: Water sol.; Sp. gr.: 1.3; Visc.: 1400 cp; Stability: Stable to high temperatures > 500 F
*Composition:* 45% active ingred.
*Containers:* 55 gal steel drums
*References:* Bulletin GID101

**Apex Level Dye #2123.** Apex
*Category:* Detergent
*Applications:* Solvent detergent blend; formulated for single bath scouring and dyeing of polyester fabrics
*Form:* Liquid
*Properties:* Solubility: Emulsifies in water
*Containers:* 55 gal drums

**Apex Mersapex #2251.** Apex
*Category:* Detergent
*Applications:* Scouring agent; removal of oily soils from synthetic fabrics and blends; dyeing applications
*Form:* Gel; Color: Water white
*Properties:* Solubility: In water
*Containers:* 55 gal drums

**Apex Mersynol #450.** Apex
*Category:* Detergent, wetting agent
*Applications:* Scouring and wetting agent; formulated for Kier boiling and bleaching; desizing and boiling
*Properties:* Solubility: Water
*Containers:* 55 gal drums

**Apex Mersynol #4150.** Apex
*Category:* Wetting agent, dispersant
*Applications:* Wetting and dispersing agent in Kier scouring and bleaching of cotton; dispersing and leveling assistant in dyeing; desizing and boiling off cottons
*Form:* Liquid; Color: Clear
*Containers:* 55 gal drums

**Apex Pentrapex #200.** Apex
*Category:* Detergent, wetting agent
*Applications:* Low foaming scouring agent for Kier processing; effective in baths containing high concentrations of salts and alkalis
*Form:* Liquid; Color: Amber
*Containers:* 55 gal drums

**Apex Pentrapex #1923.** Apex
*Category:* Wetting agent, penetrating agent
*Applications:* Wetting agent effective in solutions containing high concentrations of salts; wetting and penetrating agent effective and stable in Apex flame retardants
*Form:* Liquid; Color: Yellow
*Properties:* Solubility: Miscible with water
*Containers:* 55 gal drums
*Warnings and Caution:* Rinse well with water if spilled on skin or clothing

**Apex Tarem #20.** Apex
*Category:* Detergent, wetting a-

gent
*Applications:* Allows single bath scouring and dyeing of cellulosic/synthetic blends; applicable in pressure dyeing equipment
*Form:* Liquid
*Properties:* Solubility: Water dispersible
*Containers:* 55 gal drums

**Apex Tarem #1498.** Apex
*Category:* Detergent
*Applications:* Solvent based surfactant; solvent scour for cellulosics, synthetics and blends; bleaching and dyeing
*Form:* Liquid
*Properties:* Solubility: Forms emulsion in water
*Containers:* 55 gal drums

**Apexomine #1639.** Apex
*Category:* Detergent
*Applications:* Detergent; Kier scouring assistant; scouring agent for natural fibers; permits single bath scouring and dyeing of nylon or polyester fabrics; desizing and boiling off cottons; bleaching and dyeing
*Containers:* 55 gal drums

**Apexomine #1771.** Apex
*Category:* Detergent
*Applications:* Low foaming solvent scour; solvent emulsion for fabrics containing high concentrations of oils, grease, etc.
*Form:* Liquid; Color: Slightly yellow
*Properties:* Solubility: Emulsifies in water
*Containers:* 55 gal drums
*Warnings and Caution:* If spilled on skin or clothing, rinse well with water

**Apexscour #1962.** Apex
*Category:* Detergent
*Applications:* Continuous scouring agent
*Form:* Liquid
*Properties:* Solubility: Water
*Containers:* 55 gal drums

**Apexscour #2590.** Apex
*Category:* Detergent, wetting agent
*Applications:* Continuous scouring agent; economical anionic/nonionic surfactant for continuous scouring of printed fabrics
*Form:* Liquid
*Containers:* 55 gal drums

**Aremul AIS.** Arol Chem. Prod.
*Category:* Detergent, emulsifier
*Applications:* Industrial detergent base, emulsifier for aliphatic solvents; base for solvent degreasing compounds, dry cleaning soaps, waterless hand cleaners, emulsions of mineral oil and chlorinated solvents; Detergent—Textiles, skins, laundering, household, industrial; Wetting agent—Removing starches, wall paper; Emulsifier—Kier boiling, dispersing waxes, pigments, resins
*Form:* Liquid; Color: Amber; Odor: Mild
*Properties:* Solubility: Sol. in various aliphatic and chlorinated solvents; Sp. gr.: 1.03; Stability: Excellent
*Composition:* 90–95% active ingred.; other components: amine

salt of an alkyl aryl sulfonic acid
*Reaction:* Acid
*Toxicity:* Avoid ingestion; otherwise considered nontoxic
*Containers:* 55 gal drums
*Warnings and Cautions:* Avoid contact with eyes
*References:* Technical data sheet

**Aremul DCL.** Arol Chem. Prod.
*Chemical Description:* Blend of anionic sulfonates
*Category:* Detergent, emulsifier
*Applications:* Basic emulsifier for carrier solvents in polyester dyeing; dye leveler; wetting agent—dyeing and finishing; detergent—textiles, skins, laundering, cleaners of street, buildings, hospitals; dispersing waxes, pigments, resins
*Form:* Liquid; Color: Dark amber; Odor: Mild
*Properties:* Solubility: Miscible with water and ODCB, TCB, butyl benzoate and methyl salicylate; Sp. gr.: 1.04; Stability: Stable in presence of acids, anionic and nonionic dyeing assistants
*Composition:* 55–60% active ingred.; other components: blend of anionic sulfonates
*Reaction:* Acid
*Toxicity:* Avoid ingestion; otherwise considered nontoxic
*Containers:* 55 gal drums
*Warnings and Caution:* Avoid contact with eyes
*References:* Technical data sheet

**Aremul V-91.** Arol Chem. Prod.
*Chemical Description:* Blend of alkyl aryl sulfonates, modified nonionics and organic coupling agents
*Category:* Detergent, emulsifier
*Applications:* Concentrated emulsifier; synthetic detergent base; base for solvent type scouring, degreasing and cleaning agents; Detergent—textiles, household, street, buildings, hospitals cleaners; Emulsifier—Kier boiling; making of cutting and soluble oils; dispersing waxes, pigments, resins
*Form:* Liquid; Color: Dark amber; Odor: Mild
*Properties:* Solubility: Sol. in warm water, readily sol. in aliphatic solvents; Stability: Stable under mild acid and alkaline conditions up to boil and to hard water conditions
*Composition:* 90–95% active ingred.
*Reaction:* Acid
*Toxicity:* Avoid ingestion, otherwise considered nontoxic
*Containers:* 55 gal drums
*Warnings and Caution:* Avoid contact with eyes
*References:* Technical data sheet

**Aremul X-91.** Arol Chem. Prod.
*Chemical Description:* Blend of alkyl aryl sulfonates, modified nonionics and organic coupling agents
*Category:* Detergent, emulsifier
*Applications:* Concentrated emulsifier; synthetic detergent base; base for scouring agents, dye dispersant and scouring agent, degreaser, cleaner and emulsifier for fuel oil treatment; Detergent—textiles, skins, leather, laundering, dyeing, household, cosmetics,

cleaners of street, buildings, hospitals; Emulsifier–Kier boiling, dispersing waxes, pigments, resins
*Form:* Liquid; Color: Amber; Odor: Mild
*Properties:* Solubility: Sol. in warm water and aromatic solvents; Sp. gr.: 1.03; Stability: Stable to mild acids and alkalines up to boil, stable under hard water conditions
*Composition:* 90–95% active ingred.
*Reaction:* Acid
*Toxicity:* Avoid ingestion; otherwise considered nontoxic
*Containers:* 55 gal drums
*Warnings and Caution:* Avoid contact with eyes
*References:* Technical data sheet

**Arizona DR-22.** Arizona Chem.
*Category:* Emulsifier, detergent, wetting agent
*Applications:* Emulsifier for SBR rubber production; used to make Arizona disproportionated rosin soaps (DRS-40, 42, 43, 44)
*Properties:* Saponification No.: 165; Softening Pt.: 66 C

**Arizona DRS-40.** Arizona Chem.
*Chemical Description:* Potassium soap of disproportionated tall oil rosin
*Category:* Emulsifier, detergent, wetting agent, anionic
*Applications:* Emulsifier for SBR rubber production; Detergent–mines and minerals
*Form:* Paste; Color: Tan; Odor: Bland
*Properties:* Solubility: Water-dispersible; Density: 9.11 lb/gal @ 75 C; Sp. gr.: 1.1; Visc.: 500 cps @ 75 C; Stability: Good
*Composition:* 80% solids; DRS-40 contains an electrolyte to prevent stratification

**Arizona DRS-42.** Arizona Chem.
*Chemical Description:* Potassium soap of disproportionated tall oil rosin
*Category and Applications:* See Arizona DRS-40
*Form:* Paste; Color: Tan; Odor: Bland
*Properties:* Solubility: Water-dispersible; Density: 9.11 lb/gal @ 75 C; Sp. gr.: 1.1; Visc.: 800 cps @ 75 C; Stability: Good; Lower Acid No. and higher visc. than DRS-40
*Composition:* 80% solids

**Arizona DRS-43.** Arizona Chem.
*Chemical Description:* Sodium salt of disproportionated tall oil rosin
*Category and Applications:* See Arizona DRS-40
*Form:* Paste; Color: Tan; Odor: Bland
*Properties:* Solubility: Water-dispersible; Density: 8.94 lb/gal @ 75 C; Sp. gr.: 1.1; Visc.: 1700 cps @ 75 C; Stability: Good
*Composition:* 71% solids

**Arizona DRS-44.** Arizona Chem.
*Chemical Description:* Sodium salt of disproportionated tall oil rosin
*Category and Applications:* See Arizona DRS-40
*Form:* Paste; Color: Tan; Odor: Bland

*Properties:* Solubility: Water-dispersible; Density: 8.99 lb/gal @ 75 C; Sp. gr.: 1.1; Visc.: 750 cps @ 75 C; Stability: Good
*Composition:* 75% solids

**Arkomon SO.** American Hoechst Corp.
*Chemical Description:* Oleoyl sarcoside
*Category:* Intermediate, anionic
*Applications:* Intermediate in manufacture of anticorrosive oil; amine salts as emulsifiers for cutting oil emulsion
*Form:* Liquid

**Arkopal N040.** American Hoechst Corp.
*Chemical Description:* Nonyl phenoxy poly (ethyleneoxy) ethanol
*Category:* Detergent, wetting agent, emulsifier, nonionic
*Applications:* General industrial use
*Form:* Liquid
*Composition:* 100% conc.

**Arkopal N060.** American Hoechst Corp.
*Chemical Description:* Nonyl phenoxy poly (ethyleneoxy) ethanol
*Category and Applications:* See Arkopal N040
*Form:* Liquid
*Composition:* 100% conc.

**Arkopal N080.** American Hoechst Corp.
*Chemical Description:* Nonyl phenoxy poly (ethyleneoxy) ethanol
*Category and Applications:* See Arkopal N040
*Form:* Liquid
*Composition:* 100% conc.

**Arkopal N090.** American Hoechst Corp.
Chemical
*Chemical Description:* Nonyl phenoxy poly (ethyleneoxy) ethanol
*Category and Applications:* See Arkopal N040
*Form:* Liquid
*Composition:* 100% conc.

**Arkopal N100.** American Hoechst Corp.
*Chemical Description:* Nonyl phenoxy poly (ethyleneoxy) ethanol
*Category and Applications:* See Arkopal N040
*Form:* Liquid
*Composition:* 100% conc.

**Arkopal N110.** American Hoechst Corp.
*Chemical Description:* Nonyl phenoxy poly (ethyleneoxy) ethanol
*Category and Applications:* See Arkopal N040
*Form:* Liquid
*Composition:* 100% conc.

**Arkopal N130.** American Hoechst Corp.
*Chemical Description:* Nonyl phenoxy poly (ethyleneoxy) ethanol
*Category and Applications:* See

Arkopal N040
*Form:* Paste
*Composition:* 100% conc.

**Arkopal N150.** American Hoechst Corp.
*Chemical Description:* Nonyl phenoxy poly (ethyleneoxy) ethanol
*Category and Applications:* See Arkopal N040
*Form:* Paste
*Composition:* 100% conc.

**Arlacel 20.** I.C.I. United States
*Chemical Description:* Sorbitan monolaurate
*Category:* Emulsifier, nonionic
*Form:* Liquid; Color: Yellow
*Properties:* Solubility: Sol. in methanol, ethanol, mineral, cottonseed and corn oils, ethylene glycol; Sp. gr.: 1.0; Visc.: 4250 cps; HLB: 8.6; Flash Pt.: > 300 F; Fire Pt.: > 300 F
*Composition:* 100% active ingred.

**Arlacel 40.** I.C.I. United States
*Chemical Description:* Sorbitan monopalmitate
*Category:* Nonionic emulsifier
*Form:* Cream beads
*Properties:* Solubility: Sol. in isopropanol; Sp. gr.: ≈ 1; HLB: 6.7; Flash Pt.: > 300 F; Fire Pt.: > 300 F
*Composition:* 100% active ingred.

**Arlacel 60.** I.C.I. United States
*Chemical Description:* Sorbitan monostearate
*Category:* Nonionic surfactant
*Applications:* Cosmetic use
*Form:* Waxy beads; Color: Cream
*Properties:* Solubility: Sol. in isopropanol; HLB: 4.7; Flash Pt.: > 300 F; Fire Pt.: > 300 F; Pour Pt.: ≈ 53 C
*Composition:* 100% active ingred.
*Containers:* 51 gal fiber lever pkg.

**Arlacel 60.** I.C.I. United States
*Chemical Description:* Sorbitan monostearate
*Category and Applications:* See Arlacel C
*Form:* Solid
*Properties:* HLB: 4.7
*Composition:* 100% conc.

**Arlacel 80.** I.C.I. United States
*Chemical Description:* Sorbitan monooleate
*Category:* Nonionic surfactant
*Form:* Oil liquid; Color: Yellow
*Properties:* Solubility: Sol. in isopropanol, mineral and cottonseed oils; Sp. gr.: ≈ 1; Visc.: ≈ 1900 cps; HLB: 4.3; Flash Pt.: > 300 F; Fire Pt.: > 300 F
*Composition:* 100% active ingred.
*Containers:* 55 gal steel drum

**Arlacel 83.** I.C.I. United States
*Chemical Description:* Sorbitan sesquioleate
*Category:* and Applications: See Arlacel C. Arlacel 83 is cosmetic and pharmaceutical grade of Arlacel C
*Category:* Nonionic surfactant
*Form:* Oil liquid; Color: Yellow
*Properties:* Solubility: Sol. in

mineral and cottonseed oils, ethanol, isopropanol; Sp. gr.: ≈ 1; Visc.: ≈ 1500 cps; HLB: 3.7; Flash Pt.: > 300 F; Fire Pt.: > 300 F
*Composition:* 100% active ingred.

**Arlacel 85.** I.C.I. United States
*Chemical Description:* Sorbitan trioleate
*Category:* Nonionic surfactant
*Form:* Oily liquid; Color: Yellow
*Properties:* Solubility: Sol. in isopropyl alcohol, mineral oil, cottonseed oil, corn oil; Sp. gr.: ≈ 0.95; Visc.: ≈ 250 cps; HLB: 1.8; Flash Pt.: > 300 F; Fire Pt.: > 300 F
*Composition:* 100% active ingred.

**Arlacel 165.** I.C.I. United States
*Chemical Description:* Glycerol monostearate and polyoxethylene stearate blend
*Category:* Nonionic surfactant, emulsifier
*Applications:* Cosmetics field
*Form:* Beads; Color: White; Odor: Bland
*Properties:* Solubility: Dispersible in water; HLB: 11; Flash Pt.: > 300 F; Fire Pt.: > 300 F; Pour Pt.: ≈ 48 C; Stability: Acid stable; Self-emulsifying
*Composition:* 100% active ingred.

**Arlacel 186.** I.C.I. United States
*Chemical Description:* Mono- and diglycerides with 0.02% BHA and 0.01% citric acid in propylene glycol added as a preservative
*Category:* Nonionic surfactant, emulsifier
*Form:* Clear liquid; Color: Pale yellow
*Properties:* Solubility: Sol. in ethanol, isopropanol, cottonseed oil and mineral; Sp. gr.: ≈ 1.0; Visc.: ≈ 150 cps; HLB: 2.8; Flash Pt.: > 300 F; Fire Pt.: > 300 F
*Composition:* 100% active ingred.

**Arlacel 481.** I.C.I. United States
*Chemical Description:* Mixture of fatty acid esters
*Category:* Emulsifier, nonionic
*Applications:* Cosmetic creams and lotions
*Form:* Waxy
*Composition:* 100% conc.

**Arlacel 986.** I.C.I. United States
*Chemical Description:* Formulated product–mixture of fatty acid esters
*Category and Applications:* See Arlacel 481
*Form:* Solid
*Composition:* 100% conc.

**Arlacel 987.** I.C.I. United States
*Chemical Description:* Sorbitan mono fatty acid ester
*Category and Applications:* See Arlacel 481
*Form:* Liquid
*Composition:* 100% conc.

**Arlacel 988.** I.C.I. United States
*Chemical Description:* Polyoxyethylene glycerine sorbitan fatty acid ester
*Category and Applications:* See

Arlacel 481
*Form:* Solid
*Composition:* 100% conc.

**Arlacel 989.** I.C.I. United States
*Chemical Description:* Polyoxyethylene fatty glyceride
*Category and Applications:* See Arlacel 481
*Form:* Solid
*Composition:* 100% conc.

**Arlacel C.** I.C.I. United States
*Chemical Description:* Sorbitan sesquioleate
*Category:* Nonionic surfactant, W/O emulsifier
*Form:* Oily liquid; Color: Amber
*Properties:* Solubility: Sol. in mineral oil, cottonseed oil, ethanol, isopropanol; Sp. gr.: ≈ 1.0; Visc.: ≈ 1000 cps; HLB: 3.7; Flash Pt.: > 300 F; Fire Pt.: > 300 F
*Composition:* 100% active ingred.

**Arlamol E.** I.C.I. United States
*Chemical Description:* Polyoxypropylene (15) stearyl ether
*Category:* Emollient solvent
*Form:* Oily liquid
*Properties:* Solubility: Sol. in mineral oil, isopropyl esters, cottonseed oil, ethanol, isopropanol, hexadecyl alcohol; Sp. gr.: ≈ 0.95; Visc.: ≈ 90 cps; Pour Pt.: Typically < 0 C
*Composition:* 100% active ingred.

**Arlasolve 200.** I.C.I. United States
*Chemical Description:* Polyoxyethylene (20) isohexadecyl ether
*Category:* Nonionic surfactant, emulsifier, solubilizer
*Applications:* Cosmetics
*Form:* Soft white waxy solid
*Properties:* Solubility: Water, propylene glycol, ethanol; Sp. gr.: ≈ 1.0; HLB: 15.7; Flash Pt.: > 230 F; Fire Pt.: > 230 F
*Composition:* 100% active ingred.
*Toxicity:* Eye irritant

**Arlatone 285.** I.C.I. United States
*Chemical Description:* Polyoxyethylene fatty glyceride
*Category:* Coupling agent, solubilizer, nonionic
*Applications:* Cosmetics esp. for coupling and solubilizing perfumes and fragrances
*Form:* Semifluid
*Composition:* 100% conc.

**Arlatone 289.** I.C.I. United States
*Chemical Description:* Polyoxyethylene fatty glyceride
*Category and Applications:* See Arlatone 285
*Form:* Waxy
*Composition:* 100% conc.

**Arlatone 970.** I.C.I. United States
*Chemical Description:* Polyoxyethylene sorbitan fatty acid ester
*Category and Applications:* See Arlatone 285
*Form:* Liquid
*Composition:* 100% conc.

**Arlatone 983.** I.C.I. United States
*Chemical Description:* Polyoxyethylene fatty acid ester
*Category:* Emulsifier, nonionic
*Applications:* Emulsifier for oil/water creams and lotions
*Form:* Solid
*Composition:* 100% conc.

**Arlatone B.** I.C.I. United States
*Chemical Description:* Mixture of polyoxyethylene sorbitan fatty acid esters and alkyl aryl alcohols
*Category:* Emulsifier, nonionic
*Applications:* Blooming bath oils, waterless hand cleaners, lotion formulas, anhydrous aerosol foams based on mineral oil or deodorized kerosene; other cosmetic cleansing products
*Form:* Clear liquid; Color: Amber
*Properties:* Solubility: Miscible with mineral oil, deodorized kerosene, isopropyl myristate, isopropyl palmitate, hexadecyl alcohol, ethyl alcohol, isopropyl alcohol, cottonseed oil; Sp. gr.: ≈ 1; Visc.: ≈ 400 cp; HLB: 9; Flash Pt.: 300F; Pour Pt.: 0 F
*Composition:* 100% active ingred.

**Arlatone G.** I.C.I. United States
*Chemical Description:* Polyoxyethylene fatty glyceride
*Category:* Nonionic surfactant
*Applications:* For use in formulating clear gels
*Form:* Yellow viscous liquid to soft paste
*Properties:* Solubility: Sol. in water, ethanol and isopropanol; Sp. gr.: ≈ 1.0; HLB: 10.8; Flash Pt.: > 300 F; Fire Pt.: > 300 F
*Composition:* 100% active ingred.

**Arlatone T.** I.C.I. United States
*Chemical Description:* Polyoxyethylene (40) sorbitol septaoleate
*Category:* Emulsifier, solubilizer, antistat, lubricant, nonionic
*Applications:* General purpose water-soluble hydrophilic surfactants used as O/W emulsifiers, esp. with corresponding "Span" products; also in textile industry as antistat and fiber lubricant
*Form:* Liquid; Color: Yellow
*Properties:* Solubility: Sol. in vegetable oils, mineral oil, isopropyl myristate, isopropyl palmitate; dispersible in water; Sp. gr.: ≈ 1; Visc.: ≈ 175 cps; HLB: 9; Flash Pt.: > 300 F; Fire Pt.: > 300 F
*Composition:* 100% active ingred.

**Arlex.** I.C.I United States
*Chemical Description:* Aqueous solution of sorbitol and its anhydrides, and oxypropylene ether (83%)
*Category:* Polyhydric alcohol (polyol)
*Applications:* Industrial humectant
*Form:* Clear viscous liquid; Odor: Essentially odor free
*Properties:* Solubility: Sol. in water, alcohol, glycerin, propylene glycol; Sp. gr.: ≈ 1.32; Visc.: ≈ 1700 cps; Flash Pt.: > 300 F; Fire Pt.: > 300 F
*Composition:* 83% active ingred.
*Containers:* 55 gal closed-head steel drum

**Armac 18D.** Armak
*Chemical Description:* Octadecylamine acetate–distilled grade
*Category:* Wetting agent, emulsifier, cationic, corrosion inhibitor
*Applications:* Water-soluble cationic reagent for pigment flushing, froth flotation of minerals, flocculation, corrosion inhibition
*Form:* Flake; Color: White

*Properties:* Solubility: Water-sol.; Sp. gr.: 0.827 @ 80 C; M.P.: 85 C; Flash Pt.: 320 F (COC); Fire Pt.: 340 F (COC)
*Composition:* 100% active ingred.
*Toxicity:* Irritating to skin and eyes
*Containers:* 50 lb cardboard boxes, stored in dry place away from heat
*Warnings and Caution:* Protective goggles and gloves worn when handling
*References:* Bulletin 76-30

**Armac C.** Armak
*Chemical Description:* Cocoamine acetate
*Category:* See Armac 18D
*Applications:* See Armac 18D; also as stripping agent
*Form:* Solid; Color: Yellow
*Properties:* m.w.: 260; Solubility: Water-sol.; Sp. gr.: 0.852 @ 80 C; M.P.: 50 C; Flash Pt.: 285 F (COC); Fire Pt.: 300 F (COC)
*Containers:* 55 gal drum

**Armac HT.** Armak
*Chemical Description:* (hydrogenated-tallow)amine acetate
*Category and Applications:* See Armac 18D
*Form:* Solid; Color: White
*Properties:* Sp. gr.: 0.820 @ 80 C; M.P.: 60 C; Flash Pt.: 325 F (COC); Fire Pt.: 340 F (COC)
*Composition:* 100% active ingred.

**Armac OD.** Akzo Chemie
*Chemical Description:* Oleylamine acetate
*Category:* Corrosion inhibitor, flotation agent
*Applications:* Mineral flotation agent in mining industry; corrosion inhibitor
*Form:* Solid; Color: 13 max (Gardner)
*Properties:* M.P.: 55 C; Flash Pt.: 170 C (COC)
*Composition:* 95% min. active ingred.; 4% max. secondary amine
*Containers:* 172 kg net open head drums

**Armac T.** Akzo Chemie
*Chemical Description:* Tallowaamine acetate
*Category:* Flotation agent, corrosion inhibitor, lubricant
*Applications:* Mineral flotation agent in mining industry; corrosion inhibitor and lubricant in metal processing
*Form:* Solid; Color: 10 max. (Gardner)
*Properties:* Sp. gr.: 0.87 @ 60 C; M.P.: 45–61 C; Flash Pt.: 168 C (COC)
*Composition:* 94% min active ingred.; 5% max. secondary amine
*Toxicity:* Skin irritant, severe eye irritant
*Containers:* 172 kg net open head drums
*Warnings and Caution:* Avoid contact with strong oxidizing agents and strong mineral acids

**Armeen Series.** Armak

*Chemical Description:* Aliphatic amines (primary, secondary, or tertiary) with long-chain alkyl groups derived from fatty acids derived from coconut oil, soybean

oil and tallow
*Category:* Chemical intermediates, emulsifier, wetting agent
*Applications:* Preparation of amine salts of pesticide chemicals; synthesis of fuel oil and gasoline additives; amine oxides and quaternary ammonium salts for cosmetic formulations; amines and their salts as collectors in mineral flotation; salts and derivatives as asphalt emulsifiers, textile finishing agents, pigment dispersants, antistatic agents, corrosion inhibitors, metal polishes, antifoaming agents (textiles)
*Form and Properties:* See individual Armeen listings
*Reaction:* Armeen amines form salts with acids, amides with acylating agents, complexes with heavy metals, Schiff bases with aromatic aldehydes, ethoxylated tertiary amines with alkylene oxides and certain aldehydes, quaternary ammonium compounds with alkylating agents; ionic type: cationic
*Toxicity:* Severe burns to skin and eyes when in prolonged contact
*Containers:* 55-gal drums, tankcars
*References:* Product data bulletin 78-5

**Armeen 2C.** Armak
*Chemical Description:* Dicocoamine (secondary amine)
*Form:* Solid
*Properties:* Solubility: Sol. in chloroform, slightly sol. in isopropanol, toluene, carbon tetrachloride, kerosene; Sp. gr.: 0.793 @ 60/40 C; Visc.: 49.1 SSU @ 60 C; M.P.: 104–117 F; Flash Pt.: 350 F; Fire Pt.: 410 F; Pour Pt.: 80 F

**Armeen 2HT.** Armak
*Chemical Description:* Di(hydrogenated tallow)amine (secondary amine)
*Applications:* Corrosion inhibitor and pour point depressant for petroleum-refined products
*Form:* Solid
*Properties:* Solubility: Sol. in chloroform, slightly sol. in toluene, carbon tetrachloride, kerosene; Sp. gr.: 0.803 @ 60/4 C; Visc.: 63.9 SSU @ 70 C; M.P.: 140–149 F; Flash Pt.: 435 F; Fire Pt.: 480 F; Pour Pt.: 115 F

**Armeen 8D.** Armak
*Chemical Description:* Octylamine (primary amine)
*Form:* Liquid
*Properties:* Solubility: Sol. in methanol, ethanol, acetone, isopropanol, chloroform, toluene, carbon tetrachloride, kerosene; Sp. gr.: 0.793 @ 25/4 C; Visc.: 34.4 SSU @ 25 C; M.P.: 10–21 F; Flash Pt.: 145 F; Fire Pt.: 160 F; Pour Pt.: 15 F

**Armeen 12D.** Armak
*Chemical Description:* Dodecylamine (primary amine)
*Applications:* Lubricants (metal treatment industry)
*Form:* Liquid
*Properties:* Solubility: See Armeen 8D; Sp. gr.: 0.801 @ 25/4 C; Visc.: 42.2 SSU @ 25 C; Melting Range: 79–86 F; Flash Pt.: 235 F;

Fire Pt.: 250 F; Pour Pt.: 80 F

**Armeen 16D.** Armak
*Chemical Description:* Hexadecylamine (primary amine)
*Form:* Solid
*Properties:* Solubility: Sol. in methanol, ethanol, isopropanol, chloroform, toluene, carbon tetrachloride, slightly sol. in acetone, kerosene; Sp. gr.: 0.789 @ 60/4 C; Visc.: 37.5 SSU @ 55 C; Melting Range: 100–118 F; Flash Pt.: 285 F; Fire Pt.: 325 F; Pour Pt.: 100 F

**Armeen 18.** Armak
*Chemical Description:* Octadecylamine (primary amine)
*Applications:* Anticaking agent
*Form:* Solid
*Properties:* Solubility: Sol. in ethanol, isopropanol, chloroform, toluene, carbon tetrachloride, slightly sol. in acetone, kerosene; Sp. gr.: 0.792 @ 60/4 C; Visc.: 45.6 SSU @ 55 C; Melting Range: 122–133 F; Flash Pt.: 320 F; Fire Pt.: 365 F; Pour Pt.: 115 F

**Armeen 18D.** Armak
*Chemical Description:* Octadecylamine (primary amine)
*Form:* Solid
*Properties:* Solubility: Sol. in ethanol, isopropanol, chloroform, toluene, carbon tetrachloride, slightly sol. in methanol, kerosene; Sp. gr.: 0.791 @ 60/4 C; Visc.: 43.7 SSU @ 45 C; Melting Range: 122–133F; Flash Pt.: 300 F; Fire Pt.: 350 F; Pour Pt.: 110 F

**Armeen C.** Armak
*Chemical Description:* Cocoamine (primary amine)
*Applications:* Stripping agent for paints
*Form:* Liquid
*Properties:* Solubility: Sol. in methanol, ethanol, acetone, isopropanol, chloroform, toluene, carbon tetrachloride, kerosene; Sp. gr.: 0.805 @ 25/4 C; Visc.: 44.2 SSU @ 35 C; Melting Range: 54–59 F; Flash Pt.: 240 F; Fire Pt.: 270 F; Pour Pt.: 45 F

**Armeen CD.** Armak
*Chemical Description:* Cocoamine (primary amine)
*Form:* Liquid
*Properties:* Solubility: See Armeen C; Sp. gr.: 0.804 @ 25/4 C; Visc.: 43.0 SSU @ 35 C; Melting Range: 57–63 F; Flash Pt.: 230 F; Fire Pt.: 250 F; Pour Pt.: 55 F

**Armeen DM12D.** Akzo Chemie
*Chemical Description:* Dodecyldimethyl amine
*Category:* Chemical intermediate, cationic
*Applications:* Raw material for surfactants (e.g., amine oxides and quats)
*Form:* Liquid; Color: Yellow; Odor: Amine
*Properties:* Solubility: Water insol.; Sp. gr.: 0.78; Visc.: 3.30 cs; F.P.: −15 C; B.P.: 80–115 C @ 3 mm Hg.; Flash Pt.: 122 C (COC); Iodine Value: 3 max.
*Composition:* 98% active ingred.; 1% max. primary amine; 1% max. secondary amine

*Reaction:* Alkaline
*Toxicity:* Skin irritant, severe eye irritant
*Containers:* 200 L. bung type steel drums
*Warnings and Caution:* Avoid contact with strong oxidizing agents

**Armeen DM14D.** Akzo Chemie
*Chemical Description:* Tetradecyldimethyl amine
*Category:* Chemical intermediate, cationic
*Applications:* Raw material for manufacture of pharmaceutical quaternaries
*Form:* Liquid; Color: Yellow; Odor: Amine
*Properties:* Solubility: Insol. in water; Sp. gr.: 0.79; Visc.: 4.70 cs; F.P.: –8 C; B.P.: 100–125 C @ 3 mm Hg.; Flash Pt.: 28 C (COC); Iodine Value: 3 max.
*Composition:* 98% active ingred.; 1% max. prim. amine; 1% max. sec. amine
*Reaction:* Alkaline
*Toxicity:* Skin irritant, severe eye irritant
*Containers:* 200 L. bung type steel drums
*Warnings and Caution:* Avoid contact with strong oxidizing agents

**Armeen DM16D.** Akzo Chemie
*Chemical Description:* Hexadecyldimethyl amine
*Category:* Chemical intermediate, cationic
*Applications:* Raw material for surfactants (e.g., amine oxides and quaternaries)
*Form:* Liquid; Color: Yellow; Odor: Amine
*Properties:* Solubility: Insol. in water; Sp. gr.: 0.80; Visc.: 6.75 cs; F.P.: 8 C; B.P.: 100–136 C @ 3 mm Hg.; Flash Pt.: 150 C (COC); Iodine Value: 3 max.
*Composition:* 98% active ingred.; 1% max. prim. amine; 1% max. sec. amine
*Reaction:* Alkaline
*Toxicity:* Skin irritant, severe eye irritant

**Armeen DM18D.** Akzo Chemie
*Chemical Description:* Octadecyldimethyl amine
*Category:* Chemical intermediate, cationic
*Applications:* Raw material for surfactants (e.g., amine oxides and quaternaries)
*Form:* Liquid; Color: Yellow; Odor: Amine
*Properties:* Solubility: Insol. in water; Sp. gr.: 0.79; F.P.: 20 C; B.P.: 145–160 C @ 3 mm Hg.; Flash Pt.: 178 C (COC); Iodine Value: 3 max.
*Composition:* 98% active ingred.; 1% max. prim. amine; 1% max. sec. amine
*Reaction:* Alkaline
*Toxicity:* Skin irritant, severe eye irritant
*Containers:* 200 L. bung type steel drums
*Warnings and Caution:* Avoid contact with strong oxidizing agents and strong acids

**Armeen DMAD.** Akzo Chemie
*Chemical Description:* Alkyl ($C_{10}$–

$C_{18}$) dimethyl amine
*Category:* Chemical intermediate
*Applications:* Raw material for surfactants (e.g., quaternaries)
*Form:* Liquid; Color: Yellow; Odor: Amine
*Properties:* Solubility: Insol. in water; Sp. gr.: 0.79; Visc.: 4.44 cs; B.P.: 45-150 C @ 3 mm Hg.; Flash Pt.: 130 C (COC); Iodine Value: 3 max.
*Composition:* 98% active ingred.; 1% max. prim. amine; 1% max. sec. amine
*Reaction:* Alkaline
*Toxicity:* Skin irritant, severe eye irritant
*Containers:* 200 L. bung type steel drums
*Warnings and Caution:* Avoid contact with strong oxidizing agents

**Armeen DMCD**. Akzo Chemie
*Chemical Description:* Alkyl ($C_8$-$C_{18}$) dimethyl amine
*Category:* Chemical intermediate
*Applications:* Raw material for surfactants (e.g., amine oxides and quaternaries)
*Form:* Liquid; Color: Yellow; Odor: Amine
*Properties:* Solubility: Insol. in water; Sp. gr.: 0.79; Visc.: 4.20 cs; B.P.: 42-150 C @ 3 mm Hg.; Flash Pt.: 108 C (COC); Iodine Value: 3 max.
*Composition:* 98% active ingred.; 1% max. prim. amine; 1% max. sec. amine
*Reaction:* Alkaline
*Toxicity:* Skin irritant, severe eye irritant
*Containers:* 200 L. bung type steel drums
*Warnings and Caution:* Avoid contact with strong oxidizing agents

**Armeen DMHTD**. Akzo Chemie
*Chemical Description:* Hydrogenated tallow dimethyl amine
*Category:* Chemical intermediate
*Applications:* Raw material for surfactants (e.g., amine oxide, quaternaries)
*Form:* Liquid; Color: Yellow; Odor: Amine
*Properties:* Solubility: Insol. in water; Sp. gr.: 0.80; Visc.: 8.75 cs; F.P.: 18 C; B.P.: 100-155 C; Flash Pt.: 162 C (COC); Iodine Value: 3 max.
*Composition:* 98% active ingred.; 1% max. prim. amine; 1% max. sec. amine
*Reaction:* Alkaline
*Toxicity:* Skin irritant, severe eye irritant
*Containers:* 200 L. bung type steel drums
*Warnings and Caution:* Avoid contact with strong oxidizing agents

**Armeen DMMCD**. Akzo Chemie
*Chemical Description:* Alkyl ($C_{12}$-$C_{14}$) dimethyl amine
*Category:* Chemical intermediate
*Applications:* Raw material for surfactants (e.g., amine oxides, quaternaries)
*Form:* Liquid; Color: Yellow; Odor: Amine
*Properties:* Solubility: Insol. in water; Sp. gr.: 0.79; Visc.: 3.95 cs; F.P.: −19 C; B.P.: 90-125 C;

Flash Pt.: 125 C (COC); Iodine Value: 3 max.
*Composition:* 98% active ingred.; 1% max. prim. amine; 1% max. sec. amine
*Reaction:* Alkaline
*Toxicity:* Skin irritant, severe eye irritant
*Containers:* 200 L. bung type steel drums
*Warnings and Caution:* Avoid contact with strong oxidizing agents

**Armeen HT.** Armak
*Chemical Description:* (hydrogenated-tallow)amine (primary amine)
*Form:* Solid; Color: 9 (Gardner)
*Properties:* Solubility: Sol. in methanol, ethanol, isopropanol, chloroform, toluene, carbon tetrachloride, slightly sol. in kerosene; Sp. gr.: 0.795 @ 60/4 C; Visc.: 47.5 SSU @ 55 C; Melting Range: 79–136 F; Cloud Pt.: 115 F; Flash Pt.: 320 F; Fire Pt.: 365 F; Pour Pt.: 110 F

**Armeen HTD.** Armak
*Chemical Description:* (hydrogenated-tallow) amine (primary amine)
*Form:* Solid; Color: 2 (Gardner)
*Properties:* Solubility: See Armeen HT; Sp. gr.: 0.794 @ 60/4C; Visc.: 44.1 SSU @ 55 C; Melting Range: 70–120 F; Cloud Pt.: 110 F; Flash Pt.: 315 F; Fire Pt.: 360 F; Pour Pt.: 100 F

**Armeen L-10.** Armak
*Chemical Description:* Beta-amine (straight chain, saturated, petroleum derived chemicals with an amino functional group attached to an interior carbon atom)
*Category:* Cationic surfactant
*Form:* Clear liquid; Color: 1 (Gardner)
*Properties:* Solubility: Sol. in isopropyl alcohol, mineral oil, kerosene; Sp. gr.: 0.787; F.P.: –50 F; Cloud Pt.: – 50 F; Flash Pt.: 85 C; Fire Pt.: 94 C; Biodegradable
*References:* Bulletin 74-13

**Armeen L-11.** Armak
*Chemical Description:* See Armeen L-10
*Form:* Clear liquid; Color: 2 (Gardner)
*Properties:* Solubility: Sol. in isopropyl alcohols, mineral oil, kerosene; Sp. gr.: 0.802; M.P.: –20 F; Cloud Pt.: –20 F; Flash Pt.: 100C; Fire Pt.: 115 C; Biodegradable
*References:* Bulletin 74-13

**Armeen L-15.** Armak
*Chemical Description:* See Armeen L-10
*Form:* Clear liquid; Color: 5 (Gardner)
*Properties:* Solubility: Sol. in isopropyl alcohol, mineral oil, kerosene; Sp. gr.: 0.813; Cloud Pt.: 54 F; Flash Pt.: 157 C; Fire Pt.: 175 C; Pour Pt.: 50 F
*References:* Bulletin 74-13

**Armeen N-CMD.** Akzo Chemie
*Chemical Description:* *N*-coco morpholine
*Category:* Catalyst

*Applications:* Catalyst in polyurethane foams
*Form:* Liquid; Color: 3 (Gardner)
*Properties:* Solubility: Sol. in acetone, IPA; Sp. gr.: 0.87; Visc.: 50 SSU @ 55 C; M.P.: −29/−10 C; Flash Pt.: 155 C (COC); Pour Pt.: −10 C
*Composition:* 97% min. active ingred.
*Toxicity:* Skin irritant, severe eye irritant
*Warnings and Caution:* Avoid contact with strong oxidizing agents

**Armeen O.** Armak
*Chemical Description:* Oleylamine (primary amine)
*Form:* Paste; Color: 8 (Gardner)
*Properties:* Solubility: Sol. in acetone, methanol, ethanol, isopropanol, chloroform, toluene, carbon tetrachloride, kerosene, white mineral oil; Sp. gr.: 0.820 @ 38/4 C; Visc.: 57.0 SSU @ 25 C; Melting Range: 50–68 F; Flash Pt.: 320 F; Fire Pt.: 360 F

**Armeen OD.** Akzo Chemie
*Chemical Description:* Oleylamine
*Category:* Wetting agent
*Applications:* Lube oil additive (anticorrosive)
*Form:* Paste; Color: 2 (Gardner)
*Properties:* Solubility: Insol. in water; Sp. gr.: 0.79 @ 60 C; Visc.: 56.6 SSU; M.P.: 8–18 C; Flash Pt.: 154 C (COC); Iodine Value: 78 min.
*Composition:* 98% active ingred.; 3% max. sec. amine
*Reaction:* Alkaline
*Toxicity:* Skin irritant, severe eye irritant
*Containers:* 200 L. bung type steel drums
*Warnings and Caution:* Avoid contact with strong mineral acids, strong oxidizing agents

**Armeen SD.** Armak
*Chemical Description:* Soyaamine (primary amine)
*Form:* Paste; Color: 3 (Gardner)
*Properties:* Solubility: Sol. in methanol, ethanol, acetone, isopropanol, chloroform, toluene, carbon tetrachloride; Sp. gr.: 0.81 @ 38/4 C; Visc.: 46.2 SSU @ 35 C; Melting Range: 81–86 F; Cloud Pt.: 85 F; Flash Pt.: 305 F; Fire Pt.: 345 F; Pour Pt.: 70 F

**Armeen SZ.** Akzo Chemie
*Chemical Description:* Sodium salt of *N*-coco β-amino butyric acid
*Category:* Detergent, wetting agent, stabilizer, corrosion inhibitor
*Applications:* Paints; dispersion and softening aid for pigments; dyeing and levelling agents; fungicides (paint films); stabilizer for styrene-butadiene based paints and other latex based compositions; corrosion inhibitor
*Form:* Liquid; Color: Yellow
*Properties:* Solubility: Sol. in water; Sp. gr.: 0.96; Visc.: 19 cps @ 55 C; Flash Pt.: 40 C (Abel-Pensky closed cup); pH: 9.9–10.8
*Composition:* 39–41% active in-

gred.; 18% isopropylalcohol; 6% max. free amine; 40% water
*Reaction:* Alkaline
*Toxicity:* Moderate skin irritant, severe eye irritant
*Containers:* 200 L. bung type steel drums
*Warnings and Caution:* Avoid contact with strong oxidizing agents/ strong acids

**Armeen T.** Armak
*Chemical Description:* Tallowamine (primary amine)
*Form:* Paste; Color: 11 (Gardner)
*Properties:* Solubility: Sol. in isopropanol, methanol, ethanol, chloroform, toluene, carbon tetrachloride; Sp. gr.: 0.813 @ 38/4 C; Visc.: 47.0 SSU @ 35 C; Melting Range: 64–117 F; Cloud Pt.: 100 F; Flash Pt.: 315 F; Fire Pt.: 355 F; Pour Pt.: 70 F
*Composition:* 95% active ingred.
*Toxicity:* Corrosive to skin and eyes
*Containers:* 55-gal open-head drums; tank cars, tank wagons
*Warnings and Caution:* Protective goggles and gloves
*References:* Bulletin 77–7

**Armeen TD.** Armak
*Chemical Description:* Tallowamine (primary amine)
*Form:* Paste; Color: 2 (Gardner)
*Properties:* Solubility: Same as Armeen T; Sp. gr.: 0.812 @ 38/ 4C; Visc.: 45.2 SSU @ 35 C; Melting Range: 64–118 F; Cloud Pt.: 102 F; Flash Pt.: 305 F; Fire Pt.: 345 F; Pour Pt.: 70 F
*Composition:* 98% active ingred.

**Armeen Z.** Armak
*Chemical Description:* N-coco-beta-aminobutyric acid
*Applications:* Cosmetic; pigment softening and dispersing agent; freeze-thaw stabilizer for latex paints; lubricating aid in printing inks; dispersing agent in latex rubber reclamation; cooling tower corrosion inhibitor
*Form:* Pumpable slurry; Color: 8 (Gardner)
*Properties:* Solubility: Sol. in water, isopropanol; Sp. gr.: 0.98 @ 25/4 C; Visc.: 247 SSU @ 25 C; Flash Pt.: Foamed to 210 F; Pour Pt.: 65 F
*Reaction:* Amphoteric; compatible with cationic, anionic and nonionic surfactants

**Armeen Z-9.** Akzo Chemie
*Chemical Description:* *N*-coco β-amine butyric acid
*Category:* Emulsifier
*Applications:* Pigment dispersion aid
*Form:* Liquid; Color: Yellow
*Properties:* Solubility: Sol. in water; Sp. gr.: 0.90; Flash Pt.: 33 C (Abel-Pensky closed cup)
*Composition:* 46–50% active ingred.; 10% isopropylalcohol; 6% max. free amine; 40% water
*Reaction:* Neutral
*Toxicity:* Skin irritant, severe eye irritant
*Containers:* 200 L. bung type steel drums
*Warnings and Caution:* Avoid contact with strong oxidizing agents

**Armid 18.** Armak
*Chemical Description:* Octadecanamide
*Category:* Antiblock agent, stabilizers, dispersants
*Applications:* Antiblock, slip agents in manufacturing operations, in coatings and films; in manufacture of food packaging materials; as additive for printing inks, dyes, carbon paper coatings, fusible coatings for glassware and ceramics due to good viscosity control and improvement in dye solubility; foam viscosity stabilizers and boosters for detergents; raw materials for synthetic waxes; water repellents for textiles; pigment dispersing aids; thickening agents for paints and varnish removers
*Form:* Flake; Color: 7 (Gardner)
*Properties:* M.P.: 99–109 C; Flash Pt.: 225 C; Fire Pt.: 250 C
*Composition:* 90% active ingred.
*Containers:* Cardboard cartons, 55-gal steel drums
*References:* Bulletin 73–5

**Armid C.** Armak
*Chemical Description:* Cocoamide
*Category and Applications:* See Armid 18
*Form:* Flake; Color: 10 (Gardner)
*Properties:* M.P.: 85 C; Flash Pt.: 174 C; Fire Pt.: 185 C
*Composition:* 90% active ingred.
*Containers:* Cardboard cartons, 55-gal steel drums
*References:* Bulletin 73–5

**Armid E.** Akzo Chemie
*Chemical Description:* Erucamide
*Category:* Plastic additive
*Applications:* Slip and antiblocking agent
*Form:* Solid, flakes; Color: Yellow
*Properties:* Solubility: Insol. in water; M.P.: 79 C; Flash Pt.: 230 C (COC); Iodine Value: 72–80
*Composition:* 90% active ingred.; 3.5% free fatty acid
*Containers:* 50 kg. fiberboard keg
*Warnings and Caution:* Avoid contact with strong oxidizing agents

**Armid HT.** Armak
*Chemical Description:* (hydrogenated-tallow)amide
*Category and Applications:* See Armid 18
*Form:* Flake, powder; Color: 7 (Gardner)
*Properties:* M.P.: 98–103 C; Flash Pt.: 225 C; Fire Pt.: 250 C
*Composition:* 90% active ingred.
*Containers:* Cardboard cartons, 55-gal steel drums
*References:* Bulletin 73–5

**Armid O.** Armak
*Chemical Description:* Oleamide
*Category and Applications:* See Armid 18; also as release agent in cosmetics; penetrant in paper manufacture
*Form:* Flake, solid pellet; Color: 7 (Gardner); Odor: Bland
*Properties:* M.P.: 68 C; Flash Pt.: 207 C; Fire Pt.: 228 C
*Composition:* 90% active
*Containers:* Cardboard cartons, 55-gal steel drums
*References:* Bulletin 73–5

**Armix 146.** Arjay, Inc.
*Chemical Description:* Blended

adjuvant
*Category:* Anionic surfactant
*Applications:* Specialty surfactant to enhance the effectiveness of MSMA formulations; used for DSMA and high caustic industrial detergent cleaners
*Form:* Liquid
*Composition:* 100% conc.

**Armoflo 48.** Armak
*Category:* Conditioning agent
*Applications:* Conditioning agent to control caking and/or dusting, esp. effective for ammonium sulfate, ammonium nitrate, diammonium phosphate, urea; also in treatment of thermosetting plastics, detergents, insecticides
*Form:* Liquid; Color: 10 (Gardner)
*Properties:* Sp. gr.: 0.854 @ 60 C; Visc.: 69.1 SSU @ 60 C; M.P.: –4 C

**Armoflo 65.** Armak
*Category and Applications:* See Armoflo 48
*Form:* Liquid; Color: 12 (Gardner)
*Properties:* Sp. gr.: 0.824; Visc.: 52.2 SSU @ 60 C; M.P.: 5 C

**Armoflo 66.** Armak
*Category and Applications:* See Armoflo 48
*Form:* Semiliquid
*Properties:* Sp. gr.: 0.826 @ 60 C; Visc.: 59.0 SSU @ 60 C; M.P.: 38 C

**Armogloss 100.** Armak
*Applications:* Concentrated liquid polishes for use in auto car washers, with excellent rinse beading to decrease drying time
*Form:* Liquid; Color: Clear yellow
*Properties:* Sp. gr.: 0.88; Visc.: 500 cp; Pour Pt.: 35 F

**Armogloss 123.** Armak
*Applications:* See Armogloss 100
*Form:* Liquid; Clear golden
*Properties:* Sp. gr.: 0.85; Visc.: 100 cp; Flash Pt.: > 200 F; Pour Pt.: 22 F

**Armohib 25.** Armak
*Chemical Description:* Amine-based
*Category:* Corrosion inhibitor, cationic
*Applications:* Metal cleaning and pickling; plant cleaning; corrosion inhibitor in sulfuric, phosphoric, sulfamic and citric acids; oil well acidizing
*Form:* Liquid; Color: Amber
*Properties:* Solubility: Readily sol. in acids; Density: 7.4 lb/gal; Sp. gr.: 0.89; Cloud Pt.: 15 F; Flash Pt.: 70 F (COC); Pour Pt.: < 0 F
*Composition:* 50% active ingred.
*Toxicity:* Irritant to skin and eyes
*Containers:* 55 gal steel drums with bungs; 5 and 1 gal cans
*Warnings and Caution:* Goggles and gloves worn
*References:* Bulletin 73-9

**Armohib 28.** Armak
*Chemical Description:* Amine-based
*Category:* Corrosion inhibitor, cationic
*Applications:* Metal cleaning and pickling; corrosion inhibitor in hydrochloric acid and HCl/HF

mixtures
*Form:* Liquid; Color: Amber
*Properties:* Solubility: Readily sol. in acids; Density: 7.69 lb/gal; Sp. gr.: 0.95; Flash Pt.: 90 F (COC); Pour Pt.: 52 F
*Composition:* 75% active ingred.
*Toxicity:* Irritant to skin and eyes
*Containers:* 55 gal steel drums with bungs, 5 and 1 gal cans
*Warnings and Caution:* Goggles and gloves
*References:* Bulletin 73-9

**Armohib 31.** Armak
*Category:* Corrosion inhibitor, cationic
*Applications:* Corrosion inhibitor in sulfuric, phosphoric, sulfamic and citric acids; metal cleaning and pickling
*Form:* Liquid; Color: Amber
*Properties:* Solubility: Readily sol. in acids; Density: 8.66 lb/gal; Sp. gr.: 1.042; Cloud Pt.: 28 F; Flash Pt.: 340 F (COC); Pour Pt.: 24 F
*Composition:* 100% active ingred.
*Toxicity:* Irritant to skin and eyes
*Containers:* 55 gal steel drums with bungs; 5 and 1 gal cans
*Warnings and Caution:* Goggles and gloves
*References:* Bulletin 73-9

**Armosoft L.** Armak
*Chemical Description:* Ditallow dimethyl ammonium chloride
*Category:* Fabric softener, cationic
*Applications:* Household and commercial fabric softener with antistatic, lubricating, bacteriostatic properties
*Form:* Hazy to clear liquid; Color: 6 (Gardner)
*Properties:* Density: 7.09 lb/gal; Sp. gr.: 0.85 @ 100 F; Visc.: 45.5 cps @ 120 F; Pour Pt.: 70–75 F
*Composition:* 75% active ingred.
*Toxicity:* Skin and eye irritant
*Containers:* 55 gal steel drums, tank trucks and tank cars
*Warnings and Caution:* Flammable; gloves and goggles should be worn
*References:* Bulletin 75-21

**Armotan MO.** Akzo Chemie Italia
*Chemical Description:* Sorbitan monooleate
*Category:* Emulsifier
*Applications:* Cosmetic and pharmaceutical preparations; making of cutting and soluble oils
*Form:* Liquid; Color: 8 max. (Gardner)
*Properties:* Sp. gr.: 1.01 @ 20 C; Visc.: 9.5–11 poise; Pour Pt.: $\leq$ −12 C
*Composition:* > 99% active ingred.
*Reaction:* Neutral

**Armotan MS.** Akzo Chemie Italia
*Chemical Description:* Sorbitan monostearate
*Category:* Emulsifier
*Applications:* Cosmetic and pharmaceutical preparations
*Form:* Needle-like; Color: Cream
*Properties:* M.P.: 51–58 C
*Composition:* > 99% active ingred.
*Containers:* 25 kg bags

**Armotan PMD20.** Akzo Chemie Italia
*Chemical Description:* Polysor-

bate 80
*Category:* Wetting agent, emulsifier
*Applications:* Cosmetic and pharmaceutical preparations; rewetting dried skins
*Form:* Liquid; Color: 7 max (Gardner)
*Properties:* Sp. gr.: 1.02 @ 20 C; Visc.: 345–445 cs; Pour Pt.: −16 C
*Reaction:* Neutral

**Arneel 18D.** Armak
*Chemical Description:* Stearyl nitrile
*Category:* Detergent, wetting agent, rust inhibitor
*Applications:* Metal-wetting with mineral oils; plasticizer for synthetic rubbers and plastics
*Form:* Color: 6 (Gardner)
*Properties:* F.P.: 39 C; B.P.: 330–360 C

**Arneel C.** Armak
*Chemical Description:* Coco nitrile
*Category and Applications:* See Arneel 18D

**Arneel OD.** Armak
*Chemical Description:* Oleyl nitrile
*Category and Applications:* See Arneel 18D
*Form:* Color: 3 (Gardner)
*Properties:* Sp. gr.: 0.83; F.P.: 5 C; B.P.: > 325 C

**Arneel T.** Armak
*Chemical Description:* Tallow nitrile
*Category and Applications:* See Arneel 18D
*Properties:* F.P.: 4 C; B.P.: 330–360 C

**Arodet 25 S.** Arol Chem. Prod.
*Chemical Description:* Blend of amine condensate and alkyl aryl sulfonates
*Category:* Detergent, wetting agent, emulsifier
*Applications:* Textile fulling agent, industrial detergent base; base for hair shampoos, car washes, glass cleaners, textile detergents, all purpose cleaners; detergent–textiles, skins, laundering, foods, household, cleaners of street, buildings, hospitals
*Form:* Liquid; Color: Amber; Odor: Mild
*Properties:* Solubility: Complete in water; Sp. gr.: 0.98; Stability: Excellent
*Composition:* 100% active ingred.
*Reaction:* Alkaline
*Toxicity:* Avoid ingestion and eye contact
*Containers:* 55 gal drums
*References:* Technical data sheet

**Arodet 60 T Soft.** Arol Chem.
*Chemical Description:* Triethanolamine salt of linear alkyl sulfonate
*Category:* Detergent, wetting agent, emulsifier
*Applications:* High foaming, liquid detergent concentrate and coupling agent; shampoos, liquid dishwashing detergents, bubble baths, auto shampoos; textiles; skins; laundering; foods; household; cosmetics; street, building, hospital cleaners

*Form:* Liquid; Color: Pale yellow; Odor: Mild
*Properties:* Solubility: Complete in water; Sp. gr.: 1.09; Stability: Excellent
*Composition:* 60% active ingred.
*Reaction:* Alkaline
*Toxicity:* Avoid ingestion and contact with eyes
*Containers:* 55 gal drums
*References:* Technical data sheet

**Arodet AA-350.** Arol Chem. Prod.
*Chemical Description:* Alkyl aryl sulfonate with glycol coupling agents
*Category:* Detergent, wetting agent, dyeing assistant
*Applications:* Scouring and wetting agent; pre- and after-scouring agent, leveler and retarder, dye dispersant; finishing agent; emulsifier–dispersing waxes, pigments, resins
*Form:* Liquid; Color: Pale yellow; Odor: Mild
*Properties:* Solubility: Readily miscible in water at ambient temperature; Sp. gr.: 1.04; Stability: Excellent
*Composition:* 35% active ingred.
*Reaction:* Acid
*Toxicity:* Avoid ingestion and contact with eyes
*Containers:* 55 gal drums
*References:* Technical data sheet

**Arodet AN-100.** Arol Chem. Prod.
*Chemical Description:* Modified nonionic derivative
*Category:* Detergent, wetting agent, emulsifier, penetrant, leveling agent, scouring agent, dye dispersant
*Applications:* Detergent–textiles, dyeing; Wetting agent–desizing and boiling off cottons, dyeing, finishing; Emulsifier–Kier boiling
*Form:* Liquid; Color: Water white to pale yellow; Odor: Mildly alcoholic
*Properties:* Solubility: Complete in water; Sp. gr.: 1.11; Stability: Excellent
*Composition:* 68–70% active ingred.
*Reaction:* Acid
*Toxicity:* Avoid ingestion and contact with eyes
*Containers:* 55 gal drums
*References:* Technical data sheet

**Arodet BN-100.** Arol Chem. Prod.
*Chemical Description:* Ethoxylated alcohol
*Category:* Detergent, wetting agent, emulsifier, dyeing assistant
*Applications:* Detergent–textiles, dyeing; Wetting agent–dyeing, finishing; Emulsifier–for solvents, oils, waxes; dispersing waxes, pigments, resins
*Form:* Liquid; Color: Clear; Odor: Mild
*Properties:* Solubility: Sol. in lukewarm water; Stability: Excellent
*Composition:* 100% active ingred.
*Reaction:* Acid
*Toxicity:* Avoid ingestion and contact with eyes
*Containers:* 55 gal drums
*References:* Technical data sheet

**Arodet F.** Arol Chem. Prod.
*Chemical Description:* Fatty amide derivative

*Category:* Detergent, wetting agent
*Applications:* Scouring agent, Kier boiling, bleaching and dyeing assistant; all purpose textile surfactant; desizing and boiling off cottons
Form:
*Form:* Gel; Color: Pale straw; Odor: Mildly perfumed
*Properties:* Solubility: Readily sol. in warm water; Sp. gr.: 1.07; Stability: Excellent
*Composition:* 14–16% active ingred.
*Reaction:* Acid
*Toxicity:* Avoid ingestion and contact with eyes
*Containers:* 55 gal drums
*References:* Technical data sheet

**Arodet L-70 M.** Arol Chem. Prod.
*Chemical Description:* Ethylene oxide derivative
*Category:* Detergent, wetting agent, emulsifier, dye assistant, nonionic
*Applications:* All purpose detergent with high cloud point; Detergent–textiles, skins, leather, laundering, dyeing, housekhold, cleaners of street, buildings, hospitals; Wetting agent–desizing and boiling off cottons, bleaching, dyeing, finishing, spot removal; Emulsifier–Kier boiling, dispersing waxes, pigments, resins
*Form:* Liquid; Color: Water white; Odor: Mildly alcoholic
*Properties:* Solubility: Water sol.; Sp. gr.: 1.03; Stability: Stable to acids and alkalies due to nonionic character, stable in hard water conditions
*Toxicity:* Avoid ingestion and contact with eyes
*Containers:* 55 gal drums
*References:* Technical data sheet

**Arodet RA.** Arol Chem. Prod. Co.
*Chemical Description:* Sulfated modified alcohol derivative
*Category:* Detergent, scouring agent, penetrant, dyeing assistant
*Applications:* Textiles, laundering
*Form:* Liquid; Color: Pale amber; Odor: Mild
*Properties:* Solubility: Readily sol. in warm water; Stability: Stable to hard water conditions, hypochlorite and peroxide bleaching solutions, alkalis and mild acid conditions
*Composition:* 32–35% active ingred.
*Reaction:* Acid
*Toxicity:* Avoid ingestion and contact with eyes
*Containers:* 55 gal drums
*References:* Technical data sheet

**Arodet SC Special.** Arol Chem.
*Chemical Description:* Mixture of alcohol sulfates, alkyl aryl sulfonates, amine condensates and sequesterants
*Category:* Detergent, emulsifier, wetting agent
*Applications:* General purpose cleaning agent; rug shampoo base; hair shampoos; liquid car washes; bar glass cleaners; textiles; laundering; foods; household; cosmetics; cleaners of street, buildings, hospitals; wetting agent–removal of starches, wall paper

*Form:* Liquid; Color: Clear; Odor: Mild
*Properties:* Solubility: Readily sol. in water over wide temperature range; Sp. gr.: 1.09; Stability: Excellent
*Reaction:* Alkaline
*Toxicity:* Avoid ingestion and contact with eyes
*Containers:* 55 gal drums
*References:* Technical data sheet

**Arodet TX-7.** Arol Chem. Prod.
*Chemical Description:* Modified nonionic blend
*Category:* Detergent, wetting agent, emulsifier, nonionic
*Applications:* Detergent–skins, leather, textiles, household and commercial uses; Wetting agent–bleaching, dyeing, finishing, sizing; Emulsifier– for Varsol and after scour to improve crock resistance of polyester fibers
*Form:* Liquid; Color: Amber; Odor: Mildly alcoholic
*Properties:* Solubility: Readily sol. in water; Sp. gr.: 1.01; Stability: Excellent
*Composition:* 50–55% active
*Reaction:* Acid
*Toxicity:* Avoid ingestion and contact with eyes
*Containers:* 55 gal drums
*References:* Technical data sheet

**Arofos 200 Concentrate.** Arol
*Chemical Description:* Phosphate ester blend
*Category:* Detergent, wetting agent, emulsifier, penetrant, dye leveling agent
*Applications:* Kier boiling; dispersing waxes, pigments, resins
*Form:* Liquid; Color: Water white to pale yellow; Odor: Mild
*Properties:* Solubility: Readily sol. in water; Sp. gr.: 1.08; Stability: Excellent
*Composition:* 85% active ingred.
*Reaction:* Mild alkaline
*Toxicity:* Avoid ingestion and contact with eyes
*Containers:* 55 gal drums
*References:* Technical data sheet

**Arofos 326.** Arol Chem. Prod. Co.
*Chemical Description:* Poly phosphorylated surfactant
*Category:* Detergent, emulsifier, wetting agent
*Applications:* Detergent base for alkaline cleaners, textiles, dyeing, household, cleaners of street, buildings, hospitals; Excellent emulsifier for chlorinated and aromatic solvents in steam jenny cleaners, all purpose cleaners, alkaline cleaners
*Form:* Liquid; Color: Pale yellow; Odor: Mild
*Properties:* Solubility: Complete in water; Sp. gr.: 1.09; Stability: Excellent
*Composition:* 99+% active ingred.
*Reaction:* Acid
*Toxicity:* Irritant to skin, fingers, eyes
*Containers:* 55 gal drums
*Warnings and Caution:* Avoid contact with eyes and skin; product is an acid
*References:* Technical data sheet

**Arofos TD-19 Special.** Arol Chem.
*Chemical Description:* Phosphory-

lated ethylene oxide adduct in acid form
*Category:* Detergent, emulsifier, wetting agent
*Applications:* Phosphoric acid thickener; used in hair preparations, cosmetics, heavy-duty cleansers, floor/wax strippers, metal processing, textile wetting, scouring and emulsifying agent; household and commercial detergents; wetting agent–bleaching and dyeing; emulsifier–dispersing waxes, pigments, resins and printing applications
*Form:* Gel; Color: Light amber; Odor: Mild
*Properties:* Solubility: Excellent in aromatic and chlorinated solvents, completely sol. in water; Sp. gr.: 1.10; Stability: Excellent
*Composition:* 98% active ingred.
*Reaction:* Alkaline
*Toxicity:* Avoid ingestion and contact with eyes
*Containers:* 55 gal drums
*References:* Technical data sheet

**Arol Defoamer JT.** Arol Chem.
*Chemical Description:* Complex alkyl ethoxylate blend
*Category:* Defoamer
*Applications:* Nonsilicone defoamer; dye dispersant; leveling agent; defoamer for jet machine dyeing; dispersing waxes, pigments, resins; preparation of emulsified paints
*Form:* Liquid; Color: Water white; Odor: Mild
*Properties:* Solubility: Readily dispersible at ambient temperatues in water; Stability: Excellent
*Composition:* 100% active ingred.
*Reaction:* Acid
*Toxicity:* Avoid ingestion and contact with eyes
*Containers:* 55 gal drums
*References:* Technical data sheet

**Arolterge 100 M.** Arol Chem. Prod.
*Chemical Description:* Blend of fatty acid alkanolamides and phosphate esters
*Category:* Detergent, emulsifier, wetting agent
*Applications:* Industrial detergent base or used by itself; floor stripper cleaner; auto engine cleaners, tanks, garage floors; textiles; removal of starches, wall paper
*Form:* Liquid; Color: Dark amber; Odor: Mild
*Properties:* Solubility: Complete in water; Sp. gr.: 1.0; Stability: Excellent
*Composition:* 100% active ingred.
*Reaction:* Alkaline
*Toxicity:* Avoid ingestion and contact with eyes
*Containers:* 55 gal drums
*References:* Technical data sheet

**Aromox 18/12.** Armak
*Category and Applications:* See Aromox C/12
*Form:* Soft paste; Color: 2 (Gardner)
*Properties:* Solubility: Solv. aqueous isopropanol; Sp. gr.: 0.927 @ 100 F; Visc.: 54 cps @ 30 C; Flash Pt.: 130 F; Pour Pt.: > 120 F
*Composition:* 50% active ingred.
*Containers:* See Aromox C/12
*Warnings and Caution:* Protective clothing, goggles, gloves should be worn

*References:* Bulletin 74-21

**Aromox C/12.** Armak
*Chemical Description:* bis(2-hydroxyethyl) cocoamine oxide
*Category:* Wetting agent, emulsifier, stabilizer, foaming agent, antistat, cationic
*Applications:* Detergent additive, shampoos, cosmetics, textiles, metal plating, petroleum additives, paper, plastics, rubber, agricultural chemicals, paints and dyes, lubricant compounding, chemical specialties
*Form:* Clear liquid; Color: 2 (Gardner)
*Properties:* Solubility: Solv. aqueous isopropanol; Sp. gr.: 0.949; Visc.: 52 cp; Cloud Pt.: 18 F; Flash Pt.: 82 F; Fire Pt.: 175 F; Pour Pt.: 0 F; Surface Tension: 33 dynes/cm (0.1%); Biodegradable
*Composition:* 50% active ingred.
*Containers:* SS, glass or fiberglass reinforced polyester tanks
*Warnings and Caution:* Protective clothing, gloves, goggles should be worn
*References:* Bulletin 74-21

**Aromox C/12-W.** Armak
*Chemical Description:* bis(2-hydroxyethyl) cocoamine oxide (water-based)
*Category and Applications:* See Aromox C/12
*Form:* Clear liquid; Color: 2 (Gardner)
*Properties:* Solubility: Solv. water; Sp. gr.: 0.997; Visc.: 2097 cp (Brookfield #2 spindle); Flash Pt.: > 212 F; Pour Pt.: 35 F; Surface Tension: 30.8 dynes/cm (0.1%); Biodegradable
*Composition:* 40% active ingred.
*Containers:* See Aromox C/12
*Warnings and Caution:* Protective clothing, goggles, gloves should be worn
*References:* Bulletin 74-21

**Aromox CD/12.** Armak
*Chemical Description:* bis(2-hydroxyethyl) cocoamine oxide
*Category and Applications:* See Aromox C/12
*Form:* Clear liquid; Color: 1 (Gardner)
*Properties:* Solubility: Solv. aqueous isopropanol; Sp. gr.: 0.949; Visc.: 52 cp; Cloud Pt.: 18 F; Flash Pt.: < 80 F; Fire Pt.: 175 F; Pour Pt.: 0 F; Biodegradable
*Composition:* 50% active ingred.
*Containers:* See Aromox C/12
*Warnings and Caution:* Protective clothing, goggles, gloves should be worn
*References:* Armak catalogue

**Aromox DM14D-W.** Akzo Chemie
*Chemical Description:* Tetradecyldimethylamine oxide
*Category:* Detergent
*Applications:* Thickener of sodium hypochlorite; household and cosmetic preparations
*Form:* Liquid; Color: 1 max. (Gardner)
*Properties:* Sp. gr.: 0.96; M.P.: –3 C; Flash Pt.: Nonflammable; pH: 7–8
*Composition:* 30% min. active ingred.; 70% water; 0.5% max. free amine; 0.1% max. free peroxide

*Reaction:* Neutral
*Containers:* Bung-type polythene drum; net 200 kg
*Warnings and Caution:* Avoid contact with heavy metals, salts of Fe, Cu, Ni, etc.

**Aromox DM16.** Armak
*Chemical Description:* Dimethylhexadecylamine oxide
*Category and Applications:* See Aromox C/12
*Form:* Clear liquid; Color: 1 (Gardner)
*Properties:* Solubility: Solv. aqueous isopropanol; Sp. gr.: 0.885; Visc.: 19 cp; Cloud Pt.: 44 F; Flash Pt.: 80 F; Fire Pt.: 183 F; Pour Pt.: 0 F; Surface Tension: 31.6 dynes/cm (0.1%); Biodegradable
*Composition:* 40% active ingred.
*Containers:* See Aromox C/12
*Warnings and Caution:* Protective clothing, goggles, gloves should be worn
*References:* Bulletin 74-21

**Aromox DM16D-W.** Akzo Chemie
*Chemical Description:* Hexadecyldimethylamine oxide
*Category:* Detergent, emollient
*Applications:* Emollient in cosmetics
*Form:* Liquid/paste; Color: 1 max. (Gardner)
*Properties:* Sp. gr.: 0.94 @ 60 C; M.P.: 33 C; Flash Pt.: Nonflammable; pH: 7–8
*Composition:* 25% min. active ingred.; 75% water; 0.5% max. free amine; 0.1% max. free peroxide
*Reaction:* Neutral
*Containers:* 100 kg net open-head polythene drums
*Warnings and Caution:* Avoid contact with heavy metals, salts of Fe, Cu, Ni, etc.

**Aromox DM18D-W.** Akzo Chemie
*Chemical Description:* Octadecyldimethylamine oxide
*Category:* Detergent, emollient
*Applications:* Emollient in cosmetics
*Form:* Paste
*Properties:* Solubility: Water-sol.; pH: 7–8
*Composition:* 25% min. active ingred.; 75% water; 0.5% max. free amine; 0.1% max. free peroxide
*Reaction:* Neutral
*Containers:* 100 kg net open-head polythene drums
*Warnings and Caution:* Avoid contact with heavy metals, salts, Fe, Cu, Ni, etc.

**Aromox DMC.** Armak
*Chemical Description:* Dimethylcocoamine oxide
*Category and Applications:* See Aromox C/12
*Form:* Clear liquid; Color: 1 (Gardner)
*Properties:* Solubility: Solv. aqueous isopropanol; Sp. gr.: 0.89; Visc.: 18 cp; Cloud Pt.: 6 F; Flash Pt.: < 80 F; Fire Pt.: 100 F; Pour Pt.: < –25 F; Surface Tension: 33.1 dynes/cm (0.1%); Biodegrad.
*Composition:* 40% active ingred.
*Containers:* See Aromox C/12
*Warnings and Caution:* Protective clothing, goggles, gloves should be worn

*References:* Bulletin 74-21

**Aromox DMCD.** Akzo Chemie
*Chemical Description:* Alkyl ($C_8$–$C_{18}$)dimethylamine oxide
*Category:* Detergent
*Applications:* Thickener of NaOCl; household and cosmetic preparations
*Form:* Liquid; Color: 1 max. (Gardner)
*Properties:* Sp. gr.: 0.89; Flash Pt.: 21 C (Abel-Pensky closed cup); pH: 6–9
*Composition:* 39% min. active ingred.; 30% water; 30% isopropylalcohol; 1.5% max. free amine; 0.35% max. free oxide
*Containers:* 100 kg net open-head polythene drums
*Warnings and Caution:* Avoid contact with heavy metals, salts of Fe, Cu, Ni, etc.

**Aromox DMC-W.** Armak
*Chemical Description:* Dimethylcocoamine oxide (water-based)
*Category and Applications:* See Aromox C/12
*Form:* Clear liquid; Color: 1 (Gardner)
*Properties:* Solubility: Solv. water; Sp. gr.: 0.971; Visc.: 17 cp; Cloud Pt.: 34 F; Flash Pt.: > 212 F; Pour Pt.: 35 F; Surface Tension: 32.5 dynes/cm (0.1%); Biodegradable
*Composition:* 30% active ingred.
*Containers:* See Aromox C/12
*Warnings and Caution:* Protective clothing, goggles, gloves should be worn
*References:* Bulletin 74-21

**Aromox DMHT.** Armak
*Chemical Description:* Dimethyl (hydrogenated-tallow)amine oxide
*Category and Applications:* See Aromox C/12
*Form:* Clear liquid; Color: 1 (Gardner)
*Properties:* Solubility: Aqueous isopropanol; Sp. gr.: 0.886; Visc.: 32 cp; Cloud Pt.: 64 F; Flash Pt.: < 85 F; Fire Pt.: 105 F; Pour Pt.: 35 F; Surface Tension: 36.2 dynes/cm (0.1%)
*Composition:* 40% active ingred.
*Containers:* See Aromox C/12
*Warnings and Caution:* Protective clothing, goggles, gloves should be worn
*References:* Bulletin 74-21

**Aromox DMHTD.** Akzo Chemie
*Chemical Description:* Hydrogenated tallow dimethylamine oxide
*Category:* Detergent
*Applications:* Cosmetics (shampoos); household preparations
*Form:* Color: 1 max. (Gardner)
*Properties:* Sp. gr.: 0.89; Flash Pt.: 29 C (PMCC)
*Composition:* 39% min. active ingred.; 1.5% max. free amine; 30% water; 30% isopropylalcohol; 0.34% free peroxide
*Toxicity:* Severe skin irritation; minimal eye irritation
*Containers:* 100 kg net open-head polythene drums
*Warnings and Caution:* Avoid contact with heavy metals, salts of Fe, Cu, Ni, etc.

**Aromox DMMC-W.** Armak
*Chemical Description:* Dimethyl-

cocoamine oxide (middle-cut coco fraction)–water-based
*Category and Applications:* See Aromox C/12
*Form:* Clear liquid; Color: 1 (Gardner)
*Properties:* Solubility: Solv. water; Sp. gr.: 0.96; Visc.: 90 cp; Cloud Pt.: 22 F; Flash Pt.: >212F; Pour Pt.: 36 F; Surface Tension: 31.2 (0.1%)
*Composition:* 30% active ingred.
*Containers:* See Aromox C/12
*Warnings and Caution:* Protective clothing, goggles, gloves should be worn
*References:* Bulletin 74-21

**Aromox T/12.** Armak
*Chemical Description:* bis(2-hydroxyethyl)tallow amine oxide
*Category and Applications:* See Aromox C/12
*Form:* Clear liquid; Color: 4 (Gardner)
*Properties:* Solubility: Solv. aqueous isopropanol; Sp. gr.: 0.94; Visc.: 77 cp; Cloud Pt.: 60 F; Flash Pt.: 90 F; Fire Pt.: 180 F; Pour Pt.: 55 F; Surface Tension: 33.0 dynes/cm (0.1%)
*Composition:* 50% active ingred.
*Containers:* See Aromox C/12
*Warnings and Caution:* Protective clothing, goggles, gloves should be worn
*References:* Bulletin 74-21

**Arosurf AA-22.** Sherex
*Category:* Emulsifier, cationic
*Applications:* Asphalt mixing-grade emulsions

**Arosurf AA-23.** Sherex
*Chemical Description:* Diamine type
*Category:* Emulsifier, cationic
*Applications:* Asphalt rapid-set and mixing-grade emulsions

**Arosurf AA-54.** Sherex
*Chemical Description:* Modified diamine type
*Category:* Emulsifier, cationic
*Applications:* Asphalt rapid-set and mixing-grade emulsions
*Form:* Liquid

**Arosurf AA-55.** Sherex
*Chemical Description:* Modified diamine type
*Category:* Emulsifier, cationic
*Applications:* Asphalt rapid-set and mixing-grade emulsions
*Form:* Liquid

**Arosurf AA-57.** Sherex
*Category:* Emulsifier, cationic
*Applications:* Asphalt slow-set emulsions

**Arosurf AA-60.** Sherex
*Category:* Emulsifier, cationic
*Applications:* Asphalt rapid-set and mixing-grade emulsions

**Arosurf AA-76.** Sherex
*Category:* Emulsifier, cationic
*Applications:* Asphalt slurry-seal applications

**Arosurf TA-100.** Sherex
*Chemical Description:* Dimethyl distearyl ammonium chloride

*Category:* Softener, cationic
*Applications:* Creme rinse concentrate for hair care; fabric softener for home and commercial launderies, textile processing
*Form:* Powder
*Composition:* 97% solids content

**Arowet NF.** Arol Chem. Prod.
*Category:* Wetting agent, anionic
*Applications:* Nonfoaming type wetting agent with good dye leveling properties; dyeing and bleaching applications
*Form:* Liquid; Color: Reddish brown
*Properties:* Solubility: Readily sol. in water; Sp. gr.: 1.04; Stability: Compatible under mild acid and alkaline conditions to boil; stable to hard water conditions
*Reaction:* Neutral
*Toxicity:* Avoid ingestion and contact with eyes
*Containers:* 55 gal drums
*References:* Technical data sheet

**Arowet S.** Arol Chem. Prod.
*Chemical Description:* Blend of sulfonated esters
*Category:* Wetting agent
*Applications:* Wetting and rewetting agent; excellent for pasting-out dyestuffs, good leveling properties; desizing and boiling off cottons; bleaching and dyeing
*Form:* Liquid; Color: Amber; Odor: Mild
*Properties:* Solubility: Readily sol. in water; Sp. gr.: 1.03; Stability: Excellent
*Toxicity:* Avoid ingestion and contact with eyes
*Containers:* 55 gal drums
*References:* Technical data sheet

**Arquad 2C-75.** Armak
*Chemical Description:* Dimethyl dicocoammonium chloride
*Category and Applications:* See Arquad 12-33
*Form:* Semiliquid; Color: 7 (Gardner)
*Properties:* m.w.: 447; Solubility: Solv. aqueous isopropanol; Sp. gr.: 0.89; HLB: 11.4; Flash Pt.: < 80 F; Pour Pt.: 10 F; Surface Tension: 30 dynes/cm (0.1%)
*Composition:* 75% active ingred.
*Toxicity:* Skin irritant, severe eye irritant
*Containers:* See Arquad 12-33
*Warnings and Caution:* Flammable; protective clothing, goggles, gloves should be worn
*References:* Bulletins 76-15–17

**Arquad 2HT-75.** Armak
*Chemical Description:* Dimethyldi-(hydrogenated-tallow)ammonium chloride
*Category and Applications:* See Arquad 12-33; also as bacteriostat
*Form:* Soft paste; Color: White
*Properties:* m.w. (of active): 573; Solubility: Solv. aqueous isopropanol; Density: 7.22 lb/gal; Sp. gr.: 0.87; Visc.: 47.5 cps @ 120 F; F.P.: 95 F; HLB: 9.7; Flash Pt.: 112 F; Pour Pt.: 90–100 F; Surface Tension: 37 dynes/cm (0.1%)
*Composition:* 75% active ingred.
*Toxicity:* Skin and eye irritant
*Containers:* See Arquad 12-33

*Warnings and Caution:* Combustible; protective gloves and glasses should be worn
*References:* Bulletins 76-15-17

**Arquad 12-33.** Armak
*Chemical Description:* Trimethyldodecylammonium chloride
*Category:* Emulsifier, foaming, wetting, dispersion agents, corrosion inhibitor, antistat, cationic
*Applications:* Antistatic agents, corrosion inhibitors, fabric softeners, hair conditioners, combing aids, latex foaming agents, leather softeners, metal cleaning and polishing emulsions, mineral flotation agents, liquid starch stabilizers, water and sewage treatment flocculants
*Form:* Liquid; Color: 7 (Gardner)
*Properties:* m.w. (of active): 263; Solubility: Solv. aqueous isopropanol; Sp. gr.: 0.98; F.P.: 5 F; HLB: 17.1; Flash Pt.: 140 F; Stability: Good; Biodegradable
*Composition:* 33% active ingred.
*Toxicity:* Skin irritant, severe eye irritant
*Containers:* 55-gal nonreturnable epoxy-phenolic lined drums; bulk available SS tank trucks
*Warnings and Cautions:* Flammable; protective clothing, goggles, gloves should be worn
*References:* Bulletins 76-15-17

**Arquad 12-50.** Armak
*Chemical Description:* Trimethyldodecylammonium chloride
*Category and Applications:* See Arquad 12-33
*Form:* Liquid; Color: 1 (Gardner)
*Properties:* m.w. (of active): 263; Solubility: Solv. aqueous isopropanol; Sp. gr.: 0.89; F.P.: 13 F; HLB: 17.1; Flash Pt.: < 80 F; Stability: Good; Surface Tension: 33 dynes/cm (0.1% conc.); Biodegradable
*Composition:* 50% active ingred.
*Toxicity:* Skin irritant, severe eye irritant
*Containers:* See Arquad 12-33
*Warnings and Caution:* Flammable; protective clothing, goggles, gloves should be worn
*References:* Bulletins 76-15-17

**Arquad 16-29.** Armak
*Chemical Description:* Trimethylhexadecylammonium chloride
*Category and Applications:* See Arquad 12-33
*Form:* Liquid; Color: 6 (Gardner)
*Properties:* m.w.: 319; Solubility: Solv. aqueous isopropanol; Sp. gr.: 0.96; F.P.: 61 F; HLB: 15.8; Flash Pt.: > 212 F; Stability: Good; Biodegradable
*Composition:* 29% active aqueous
*Toxicity:* Skin irritant, severe eye irritant
*Containers:* See Arquad 12-33
*Warnings and Caution:* Corrosive; protective goggles, gloves should be worn
*References:* Bulletins 76-15-17

**Arquad 16-50.** Armak
*Chemical Description:* Trimethylhexadecylammonium chloride
*Category and Applications:* See Arquad 12-33
*Form:* Liquid; Color: 6 (Gardner)

*Properties:* m.w. (of active): 319; Solubility: Solv. aqueous isopropanol; Sp. gr.: 0.88; HLB: 15.8; Flash Pt.: < 80 F; Stability: Good; Surface Tension: 34 dynes/cm (0.1%); Biodegradable
*Composition:* 50% active ingred.
*Toxicity:* Skin irritant, severe eye irritant
*Containers:* See Arquad 12-33
*Warnings and Caution:* Flammable; protective clothing, goggles, gloves should be worn
*References:* Bulletins 76-15–17

**Arquad 18-50.** Armak
*Chemical Description:* Trimethyloctadecylammonium chloride
*Category and Applications:* See Arquad 12-33; also as dye-leveling agents, viscosity stabilizers, fuel oil additives, in lubricant compounding
*Form:* Liquid; Color: 7 (Gardner)
*Properties:* m.w. (of active): 347; Solubility: Solv. aqueous isopropanol; Sp. gr.: 0.88; HLB: 15.7; Flash Pt.: <80 F; Stability: Good; Surface Tension: 34 dynes/cm (0.1%); Biodegradable
*Composition:* 50% active ingred.
*Toxicity:* Skin irritant, severe eye irritant
*Containers:* See Arquad 12-33
*Warnings and Cautions:* Flammable; protective clothing, goggles, gloves, should be worn
*References:* Bulletins 76-15–17

**Arquad B-50 USP.** Akzo Chemie
*Chemical Description:* Alkyl ($C_8$–$C_{18}$) dimethylbenzyl ammonium chloride USP
*Category:* Disinfectant, germicide
*Form:* Liquid; Color: 2 max. (Gardner)
*Properties:* Sp. gr.: 0.98
*Composition:* 50% active ingred.
*Reaction:* Neutral
*Toxicity:* Skin irritant, severe eye irritant
*Containers:* 100 kg net open-head polythene drums
*Warnings and Caution:* Avoid contact with strong oxidizing agents, anionics; kills bacteria in biological disposal systems

**Arquad B-90 USP.** Akzo Chemie
*Chemical Description:* Alkyl ($C_8$–$C_{18}$) dimethylbenzyl ammonium chloride USP
*Category:* Disinfectant
*Applications:* Disinfectant, germicide, fungicide
*Form:* Viscous mass @25 C; Color: 6 max. (Gardner)
*Composition:* 90% active ingred.
*Toxicity:* Skin irritant, severe eye irritant
*Containers:* 100 kg net open-head polythene drums
*Warnings and Caution:* Avoid contact with strong oxidizing agents, anionics; kills bacteria in biological disposal systems

**Arquad B-100.** Armak
*Chemical Description:* Dimethylalkylbenzyl ammonium chloride
*Category:* Bacteriocide
*Applications:* EPA approved bacteriocide, sanitizer for schools, hospitals, food processing plants

*Form:* Liquid; Color: 2 (Gardner)
*Properties:* m.w.: 380; Solubility: Solv. aqueous isopropanol; Flash Pt.: < 80 F
*Composition:* 50% active ingred.
*Toxicity:* Irritating to skin and eyes
*Warnings and Caution:* Flammable, protective gloves and goggles should be worn
*References:* Bulletins 76-1, 76-15–17

**Arquad C-33.** Armak
*Chemical Description:* Trimethylcocoammonium chloride
*Category and Applications:* See Arquad 12-33
*Form:* Liquid; Color: 7 (Gardner)
*Properties:* m.w. (of active): 278; Solubility: Solv. aqueous isopropanol; Sp. gr.: 0.98; F.P.: 25 F; HLB: 16.5; Stability: Good; Biodegradable
*Composition:* 33% active ingred.
*Toxicity:* Skin irritant, severe eye irritant
*Containers:* See Arquad 12-33
*Warnings and Caution:* Corrosive; protective clothing, goggles, gloves should be worn
*References:* Bulletins 76-15–17

**Arquad C-33-W.** Akzo Chemie
*Chemical Description:* Cocotrimethyl ammonium chloride
*Category:* Emulsion-break retardant in cosmetics
*Form:* Liquid
*Properties:* Sp. gr.: 0.96; F.P.: –3 C; Flash Pt.: Nonflammable; pH: 5–8
*Composition:* 32–25% active

**Arquad C-50.** Armak
*Chemical Description:* Trimethylcocoammonium chloride
*Category and Applications:* See Arquad 12-33
*Form:* Liquid; Color: 7 (Gardner)
*Properties:* m.w. (of active): 278; Solubility: Solv. aqueous isopropanol; Sp. gr.: 0.89; F.P.: 5 F; HLB: 16.5; Flash Pt.: < 80 F; Stability: Good; Surface Tension: 31 dynes/cm (0.1%); Biodegradable
*Composition:* 50% active ingred.
*Toxicity:* Skin irritant, severe eye irritant
*Containers:* See Arquad 12-33
*Warnings and Caution:* Flammable; protective clothing, goggles, gloves should be worn
*References:* Bulletins 76-15–17

**Arquad DM14B-90.** Akzo Chemie
*Chemical Description:* Tetradecyldimethylbenzyl ammonium chloride
*Category:* Bactericide, fungicide, germicide, disinfectant
*Form:* Powder; Color: White
*Composition:* 90% active ingred.
*Toxicity:* Skin irritant, severe eye irritant
*Containers:* 50 kg net fiberboard kegs
*Warnings and Caution:* Avoid contact with strong oxidizing agents, anionics; product will kill bacteria in biological disposal systems

**Arquad DM18B-90.** Akzo Chemie

*Chemical Description:* Octadecyldimethylbenzyl ammonium chloride
*Category:* Conditioner, emulsifier, softener
*Applications:* Cosmetics and pharmaceuticals; conditioner/softener in after shampoo creams
*Form:* Powder; Color: White
*Properties:* Solubility: 100% water sol.; pH: 5–8 (1% sol'n.)
*Composition:* 90% min. active
*Toxicity:* Skin irritant, severe eye irritant
*Containers:* 50 kg fiberboard kegs
*Warnings and Caution:* Avoid contact with strong oxidizing agents, anionics; product will kill bacteria in biological disposal systems

**Arquad DMHTB-75.** Akzo Chemie
*Chemical Description:* Hydrogenated tallow dimethylbenzyl ammonium chloride
*Category:* Bactericide, disinfectant, softening agent
*Applications:* Disinfection; textile softening agent
*Form:* Soft paste; Color: 4 max. (Gardner)
*Properties:* F.P.: 60 C
*Composition:* 75% active ingred.
*Reaction:* Neutral
*Toxicity:* Skin irritant, severe eye irritant
*Containers:* 50 kg net fiberboard kegs
*Warnings and Caution:* Avoid contact with strong oxidizing agents, anionics; product will kill bacteria in biological disposal systems

**Arquad DMMCB-50.** Akzo Chemie
*Chemical Description:* Alkyl ($C_{12}$–$C_{14}$, $C_{16}$) dimethyl benzyl ammonium chloride
*Category:* Antistat, flocculant, emulsifier, softener, corrosion inhibitor
*Applications:* Cosmetics industry; soil stabilization; antistatic agent in textile industry; flocculant; fabric softener; corrosion inhibitor; fungicides, bactericides
*Form:* Liquid
*Composition:* 50% active

**Arquad DMMCB-75.** Akzo Chemie
*Chemical Description:* Alkyl ($C_{12}$–$C_{14}$) dimethylbenzyl ammonium chloride
*Category:* Disinfectant, bacteri-
Cloud Pt.: −36 F; Flash Pt.: $<$ 80 F; Biodegradable
*Toxicity:* Skin and eye irritant
*Warnings and Caution:* Flammable, protective gloves and goggles should be worn

**Arquad L-11.** Armak
*Chemical Description:* Quarternized beta-amines
*Category:* Emulsifier, cationic
*Properties:* Solubility: Sol. in water; Sp. gr.: 0.885; M.P.: −90F;
cide, germicide, fungicide
*Form:* Liquid; Color: 2 max. (Gardner)
*Properties:* Sp. gr.: 0.98; Flash Pt.: 37 C
*Composition:* 50% active ingred.
*Toxicity:* Skin irritant, severe eye irritant

*Containers:* 100 kg net open-head polythene drums
*Warnings and Caution:* Avoid contact with strong oxidizing agents, anionics; product will kill bacteria in biological disposal systems

**Arquad L-15.** Armak
*Chemical Description:* Quarternized beta-amines
*Category:* Emulsifier, cationic
*Form:* Clear liquid; Color: 9 (Gardner)
*Properties:* Solubility: Water-sol.; Sp. gr.: 0.877; M.P.: 46 F; Cloud Pt.: 42 F; Flash Pt.: 100 F
*Composition:* 50% active ingred.

**Arquad S-50.** Armak
*Chemical Description:* Trimethylsoyaammonium chloride
*Category and Applications:* See Arquad 12-33
*Applications:* Used in butumen emulsions; to control growth of organism in spray water for air conditioning; algae and slime control agents in water systems
*Form:* Liquid; Color: 8 max. (Gardner)
*Properties:* Sp. gr.: 0.89; F.P.: 20 C; pH: 5–8
*Composition:* 49–52% active
*Toxicity:* Skin irritant, severe eye irritant
*Containers:* 180 kg net bung type steel drums
*Warnings and Caution:* Avoid contact with strong oxidizng agents, anioncs; avoid temperatures > 70 C product will kill bacteria in biological disposal systems

*References:* Bulletins 76-15–17

**Arquad S-2C-50.** Akzo Chemie
*Chemical Description:* 1:1 mixture of oleyltrimethyl ammonium chloride and dicocodimethyl ammonium chloride
*Category:* Bactericide, wetting agent, corrosion inhibitor
*Applications:* Bactericide; oil recovery as wetting agent
*Form:* Liquid; Color: 10 max. (Gardner)
*Properties:* Sp. gr.: 9.87; F.P.: –15 C; pH: 5–8
*Composition:* 50–55% active
*Containers:* 172 kg net bung type drums

**Arquad T-50.** Armak
*Chemical Description:* Trimethyltallowammonium chloride
*Category and Applications:* See Arquad 12-33
*Applications:* In manufacture of tetramycin (oxytetracycline) antibiotics
*Form:* Liquid; Color: 8 max. (Gardner)
*Form:* Liquid; Color: 8 (Gardner)
*Properties:* m.w.: 340; Solubility: Solv. aqueous isopropanol; Sp. gr.: 0.88; HLB: 14.2; Flash Pt.: < 80 F; Stability: Good; Biodegradable
*Composition:* 50% active ingred.
*Toxicity:* Skin irritant, severe eye irritant
*Containers:* 180 kg net bung type steel drums
*Warnings and Caution:* Avoid contact with strong oxidizing agents,

anionics; avoid temperatures > 70 C; will kill bacteria in biological disposal systems Flammable; protective clothing, goggles, gloves should be worn
*References:* Bulletins 76-15-17

**Arquad T-2C-50.** Armak
*Chemical Description:* 1:1 mixture of trimethyltallowammonium chloride and dimethyldicocoammonium chloride
*Category and Applications:* See Arquad 12-33
*Form:* Liquid; Color: 8 (Gardner)
*Properties:* m.w. (of active): 394; Solubility: Solv. aqueous isopropanol; Sp. gr.: 0.87; HLB: 13.0; Flash Pt.: < 80 F
*Composition:* 50% active ingred.
*Toxicity:* Skin irritant, severe eye irritant
*Containers:* See Arquad 12-33
*Warnings and Caution:* Flammable, protective gloves and goggles should be worn
*References:* Bulletins 76-15-17

**Arquest 665-S.** Arjay, Inc.
*Chemical Description:* Phosphate ester
*Category:* Scale inhibitor, anionic
*Applications:* Scale inhibitor for calcium carbonate and calcium sulfate
*Form:* Liquid

**Arquest 669-S.** Arjay, Inc.
*Chemical Description:* Acrylic polymer
*Category:* Scale inhibitor, anionic
*Applications:* Controls barium sulfate, calcium sulfate, and calcium carbonate scale
*Form:* Liquid

**Arquest 674-S.** Arjay, Inc.
*Chemical Description:* Phosphonate
*Category:* Scale inhibitor, sequestering and deflocculating agent, anionic
*Form:* Liquid

**Arquest 675-S.** Arjay, Inc.
*Chemical Description:* Phosphonate
*Category:* Scale inhibitor, anionic
*Applications:* Scale inhibitor for calcium carbonate, calcium sulfate, and barium sulfonate
*Form:* Liquid

**Arsul 458-S.** Arjay, Inc.
*Chemical Description:* Alcohol ether sulfate
*Category:* Detergent, foaming agent, anionic
*Applications:* Used in light duty dishwash detergents, all purpose cleaners, bubble baths, upholstery and rug cleaners; emulsion polymerization; foaming agent for wall board manufacturing, air drilling, foam markers
*Form:* Liquid
*Composition:* 60% conc.

**Arsul DDB.** Arjay, Inc.
*Chemical Description:* Branched chain dodecyl benzene sulfonic acid
*Category:* Chemical intermediate,

anionic
*Applications:* Intermediate for reaction with different alkali bases to form the metal or amine salts; sulfonate derivatives used as emulsifiers, dispersants, cleaning compounds, emulsion breakers, and corrosion inhibitors
*Form:* Liquid
*Composition:* 90% conc.

**Arsul LAS.** Arjay, Inc.
*Chemical Description:* Biodegradable linear dodecyl benzene sulfonic acid
*Category and Applications:* See Arsul DDB
*Form:* Liquid
*Composition:* 98% conc.

**Arsul WAQ.** Arjay, Inc.
*Chemical Description:* Sodium lauryl sulfate
*Category:* Anionic surfactant
*Applications:* Used in applications requiring good foam stability; used in emulsion polymerization and cosmetic formulations
*Form:* Liquid
*Composition:* 30% conc.

**Artodan.** Grinsted Prod.
*Chemical Description:* Sodium stearyl-2-lactylate
*Category:* Emulsifier, anionic
*Applications:* Food emulsifier, dough conditioner, starch complexing agents for antistaling, structure improvement, increased volume in bread
*Form:* Powder
*Properties:* HLB: 9
*Composition:* 100% active ingred.

**Artrads 6524, 6923, 7522.** Arjay, Inc.
*Chemical Description:* Dimer trimer polybasic acids
*Category:* Intermediate
*Applications:* Intermediate for reaction with polyfunctional materials, e.g., diols, diamines polyols, and polyamines, to produce good corrosion inhibitors
*Form:* Liquid
*Composition:* 100% conc.

**Arylan CA.** Lankro Chem. Ltd.
*Chemical Description:* Straighter chain dodecyl benzene sulfonate, calcium salt
*Category:* Emulsifier, anionic
*Applications:* Emulsifier for certain pesticidal and herbicidal preparations; hydrocarbon solvents and waxes; oils and aromatic solvents in degreaser formulations
*Form:* Viscous liquid; Odor: Alcoholic
*Properties:* Solubility: Water-insol.; Sp. gr.: 1.012 @ 20 C; Visc.: 3025 cs @ 20 C; Flash Pt.: 94 F (Abel CC); Pour Pt.: $<$ 0 C; pH: 5–8 (1% aq.)
*Composition:* 70% active ingred.
*Toxicity:* Avoid prolonged contact with concentrated form
*Containers:* Store in SS, glass, resin, or polythene-lined equipment
*Warnings and Caution:* Contains flammable solvent; do not mix with strong oxidizing or reducing agents (potentially explosive)

**Arylan DA 36.** Lankro Chem. Ltd.

*Chemical Description:* Blended liquid detergent with nonionic foam stabilizer
*Category:* Detergent, anionic/nonionic
*Applications:* Concentrated blend of liquid detergents. Fully biodegradable.
*Form:* Liquid;
*Properties:* Biodegradable
*Composition:* 36% Conc.

**Arylan DT.** Lankro Chem. Ltd.
*Chemical Description:* Specially blended liquid detergent concentrate based on alkyl aryl sulfonates with foam stabilizers
*Category:* Detergent
*Applications:* High active detergents, hard surface cleaners, upholstery cleaners, industrial cleaners
*Form:* Clear liquid; Color: Light amber; Odor: Mild
*Properties:* Sp. gr.: 1.05 @ 20 C; Visc.: 330 cs @ 20 C; Flash Pt.: > 200 F (COC); Pour Pt.: < 0 C; Stability: Good; pH: 6–7.5 (1% aq.)
*Composition:* 30% active ingred.
*Containers:* Store in SS, glass, resin or polythene-lined equipment
*Warnings and Caution:* Avoid prolonged contact with concentrated form; do not mix with strong oxidizing or reducing agents (potentially explosive)

**Arylan HAL.** Lankro Chem. Ltd.
*Chemical Description:* Specially blended liquid detergent concentrate based on alkyl aryl sulfonates with foam stabilizers
*Category:* Detergent
*Applications:* High active detergents, hard surface cleaners, upholstery cleaners
*Form:* Clear liquid; Color: Amber; Odor: Mild, alcoholic
*Properties:* Sp. gr.: 1.04 @ 20 C; Visc.: 300 cs @ 20 C; Flash Pt.: 90 F (Abel CC); Pour Pt.: < 0 C; Stability: Good; pH: 6.5–7.5 (1% aq.)
*Composition:* 87% active ingred.
*Containers:* Store in SS, glass, resin or polythene-lined equipment
*Warnings and Caution:* Avoid prolonged contact with concentrated form; do not mix with strong oxidizing or reducing agents (potentially explosive)

**Arylan LQ.** Lankro Chem. Ltd.
*Chemical Description:* Specially blended liquid detergent concentrate based on alkyl aryl sulfonates with foam stabilizers
*Category:* Detergent
*Applications:* High active detergents, hard surface cleaners, upholstery cleaners, industrial cleaners
*Form:* Clear liquid; Color: Light amber; Odor: Mild
*Properties:* Sp. gr.: 1.053 @ 20 C; Visc.: 350 cs @ 20 C; Flash Pt.: > 200 F (COC); Pour Pt.: < 0 C; Stability: Good; pH: 7–9 (1% aq.)
*Composition:* 40% active ingred.
*Containers:* SS, glass, resin or polythene lined equipment
*Warnings and Caution:* Avoid prolonged contact with concentrated form; do not mix with strong oxidizing or reducing agents (potentially explosive)

**Arylan PWS.** Lankro Chem. Ltd.
*Chemical Description:* Amine alkylaryl sulfonate
*Category:* Emulsifier
*Applications:* Emulsifier for mineral oils, kerosene, waxes and chlorinated solvents
*Form:* Liquid
*Properties:* Visc.: 4200 cs @ 20 C
*Composition:* 90% active ingred.
*Reaction:* Acid

**Arylan S Acid.** Diamond Shamrock Europe/Process Chemicals Div.
*Chemical Description:* Dodecyl benzene sulfonic acid (branched chain).
*Category:* Intermediary, anionic
*Applications:* Intermediary for preparation of anionic surfactants
*Form:* Liquid
*Composition:* 96% Conc.

**Arylan S90 Flake.** Lankro Chem. Ltd.
*Chemical Description:* Sodium dodecyl benzene sulfonate flake (branched chain)
*Category:* Wetting agent, emulsifier, anionic
*Applications:* Primary emulsifier in the emulsion polymerization of PVA: wetting agent for wettable powders
*Form:* Liquid
*Composition:* 95% Conc.

**Arylan S Flake.** Lankro Chem. Ltd.
*Chemical Description:* Sodium dodecylbenzene sulfoante
*Category:* Wetting agent, emulsifier
*Applications:* Emulsion polymerization; wettable powders
*Form:* Flake; Color: Cream
*Composition:* 82% active ingred.
*Reaction:* Neutral

**Arylan SBC 25.** Lankro Chem. Ltd.
*Chemical Description:* Straighter chain dodecyl benzene sulfonate, sodium salt
*Category:* Detergent, emulsifier, anionic
*Applications:* Liquid detergent base and primary emulsifier for emulsion polymerization
*Form:* Liquid
*Composition:* 25% Conc.

**Arylan SBC Acid.** Lankro Chem. Ltd.
*Chemical Description:* Straighter chain dodecyl benzene sulfonic acid
*Category:* Detergent, anionic
*Applications:* Biodegradable intermediate for detergents
*Form:* Liquid
*Properties:* Biodegradable
*Composition:* 97% Conc.

**Arylan SC 30.** Lankro Chem. Ltd.
*Chemical Description:* Sodium dodecylbenzene sulfonate
*Category:* Detergent, emulsifier
*Applications:* Base for liquid detergents; primary emulsifier for emulsion polymerization
*Form:* Liquid

*Composition:* 30% active ingred.
*Reaction:* Neutral

**Arylan SC Acid.** Lankro Chem. Ltd.
*Chemical Description:* Dodecylbenzene sulfonic acid
*Category:* Intermediate
*Applications:* Intermediate for detergents, esp. liquid detergents
*Form:* Liquid; Color: Brown
*Properties:* Visc.: 1800 cs @ 20 C
*Composition:* 97% active ingred.
*Reaction:* Acid
*Toxicity:* Strong organic acid, hence irritating to eyes and skin
*Containers:* Resin-lined drums

**Arylan SNS.** Diamond Shamrock
*Chemical Description:* Sodium naphthalene sulfonic acid formaldehyde condensate
*Category:* Dispersing agent
*Applications:* Dispersing solid particles in aqueous suspensions
*Form:* Powder; Color: Buff
*Composition:* 90% active ingred.
*Reaction:* Neutral

**Arylan SO60 Acid.** Lankro Chem. Ltd.
*Chemical Description:* Branched chain tridecyl benzene sulfonic acid
*Category:* Detergent, emulsifier, wetting agent, emulsion polymerization aids
*Applications:* Base surfactant for heavy duty powder detergent formulations
*Form:* Viscous liquid; Color: Brown; Odor: Characteristic
*Properties:* Solubility: Water-sol. @ 20 C; Sp. gr.: 1.045 @ 20 C; Visc.: > 25,000 cs @ 20 C; Flash Pt.: > 250 F (COC); Pour Pt.: < 0 C; pH: 2.0 (1% aq.)
*Composition:* > 96% active ingred.
*Toxicity:* Corrosive organic acid; extreme eye and skin irritant
*Containers:* Store in SS, resin, polythene or glass-lined equipment
*Warnings and Caution:* Protective goggles, gloves and clothing should be worn; use with adequate ventilation (may evolve sulfur dioxide odor); in case of fire, toxic sulfur dioxide fumes may be produced

**Arylan SX Flake.** Lankro Chem. Ltd.
*Chemical Description:* Straighter chain dodecyl benzene sulfonate flake, sodium salt
*Category:* Detergent, wetting agent, anionic
*Applications:* Base for detergents; wetting agent for powders
*Form:* Liquid
*Composition:* 75% Conc.

**Arylan TE/C.** Lankro Chem. Ltd.
*Chemical Description:* Straighter chain dodecyl benzene sulfonate (TEA salt)
*Category:* Detergent, emulsifier, wetting agent, emulsion polymerization aids, anionic
*Applications:* Specialty wax emulsifier; co-emulsifier with an Ethylan nonionic for the emulsification of phenolics and chlorinated

solvents
*Form:* Clear liquid; Odor: Mild
*Properties:* Solubility: Water-sol. @ 20 C; Sp. gr.: 1.060 @ 20 C; Visc.: 900 cs @ 20 C; Flash Pt.: > 200 F (COC); Pour Pt.: < 0 C; Stability: Good in alkaline, mild acid, low electrolyte concentration and in hard or soft waters; pH: 7.0–8.5 (1% aq.); Biodegradable
*Composition:* 40% active ingred.
*Containers:* Store in SS, glass, resin, or polythene-lined equipment
*Warnings and Caution:* Avoid prolonged contact with concentrated form; in case of fire, toxic fumes of sulfur dioxide may be produced

**Arylene M40.** Hart Chem.
*Chemical Description:* Dioctyl sulfosuccinate
*Category:* Wetting agent, anionic
*Applications:* Scouring of cellulosics and in desizing process
*Form:* Liquid; Color: Clear
*Properties:* Refractive Index: 1.3932; Density: 1.03; Visc.: 50 cps
*Composition:* 40% active ingred.
*Reaction:* Neutral
*Toxicity:* Skin and eye irritant
*Containers:* T/L/Drums
*Warnings and Caution:* Avoid contact with skin and eyes

**Arylene M60.** Hart Chem.
*Chemical Description:* Dioctyl sulfosuccinate
*Category:* Wetting agent
*Applications:* Scouring of cellulosics and in desizing processes
*Form:* Liquid; Color: Clear
*Properties:* Refractive Index: 1.3932; Density: 1.06; Visc.: 50 cps
*Composition:* 60% active ingred.
*Toxicity:* Skin and eye irritant
*Containers:* T/L/Drums
*Warnings and Caution:* Avoid contact with eyes and skin

**Arzoline 215.** Arjay, Inc.
*Chemical Description:* Alkyl cyclic amido amine
*Category:* Corrosion inhibitor base, cationic
*Applications:* Corrosion inhibitor base; may be further modified to produce oil soluble, water dispersible, or water soluble inhibitors
*Form:* Liquid
*Composition:* 100% conc.

**Asco-66.** Berol Kemi AB
*Chemical Description:* Polyamine derivative
*Category:* Antistripping agent
*Applications:* Antistripping agent in road construction, jointsealing compounds, undersealing compounds for cars; promotes passive and active adhesion between binder and aggregate
*Form:* Liquid; Color: Dark; Odor: Ammoniacal
*Properties:* Solubility: Sol. in asphalt, oils and many solvents; Flash Pt.: > 150 C (PM); Stability: Good heat stability
*Reaction:* Alkaline
*Toxicity:* May cause skin and eye injuries

*Containers:* 200 kg net drums
*Warnings and Caution:* Protective face shields, goggles should be worn

**Asol.** Lucas Meyer
*Chemical Description:* Lecithin fraction
*Category:* Antispatter agent, release agent, emulsifier, nonionic
*Applications:* Antispatter agent, release agent for margarine; emulsifier for cosmetics, pharmaceuticals
*Form:* Liquid
*Composition:* 40–100% conc.

**Aston AP Concentrate.** Onyx Chem. Co.
*Chemical Description:* Cationic polyamine
*Category:* Antistatic agent, softener, cationic
*Applications:* Water and solvent soluble antistatic agent and softener for industrial and textile use
*Form:* Liquid
*Properties:* Solubility: Water and solvent sol.; Sp. gr.: 1.04 (25/20 C); Flash Pt.: > 200 F
*Composition:* 100% solids

**Aston BO.** Onyx Chem. Co.
*Chemical Description:* Special cationic
*Category:* Antistat, cationic
*Applications:* Antistatic agent for rug shampoos which reduces resoiling after shampooing
*Form:* Liquid
*Properties:* Sp. gr.: 0.99 (25/20 C); Flash Pt.: > 200 F
*Composition:* 25% min. solids

**Atcopen SD.** Bostik South
*Chemical Description:* Alkyl phosphate
*Category:* Wetting agent, anionic
*Applications:* Low foaming, surfactant wetting agent, stable in acid and alkaline solvents
*Form:* Liquid
*Composition:* 78% conc.

**Atcowet W.** Bostik South
*Chemical Description:* Dioctyl sodium sulfosuccinate
*Category:* Wetting and rewetting agent, anionic
*Applications:* Textile wetting and rewetting applications
*Form:* Liquid
*Composition:* 75% conc.

**Atlas G-263.** I.C.I. United States
*Chemical Description:* Cetyl ethyl morpholinium ethosulfate (35% aqueous sol'n.)
*Category:* Antistatic agent, cationic
*Applications:* Textile specialties
*Form:* Liquid
*Properties:* HLB: 30.0
*Composition:* 35% active ingred.

**Atlas G-263 Anhydrous.** ICI United States
*Chemical Description:* Cetyl ethyl morpholinium ethosulfate
*Category and Applications:* See Atlas G-263
*Form:* Waxy solid; Color: Tan to brown; Odor: Fruity
*Properties:* Solubility: Sol. in water and ethanol forming cloudy solutions
*Composition:* 99% active ingred.

**Atlas G-271.** I.C.I. United States
*Chemical Description:* Soya ethyl morpholinium ethosulfate (35% aqueous solution)
*Category:* Deodorant, cosmetic specialty, cationic
*Form:* Liquid
*Properties:* HLB: 30.0
*Composition:* 35% active ingred.

**Atlas G-695.** I.C.I. United States
*Chemical Description:* Mono- and diglycerides of edible fats or oils
*Category:* Food emulsifier
*Form:* Paste; Odor and Taste: Bland
*Properties:* Solubility: Sol. in lower alcohols, vegetable and mineral oils; Sp. gr.: 0.96; Visc.: 528 cps; HLB: ≈ 3; Clear Pt.: 40 C; Flash Pt.: > 300 F; Fire Pt.: > 300 F

**Atlas G-711.** I.C.I. Unites States
*Chemical Description:* Alkyl aryl sulfonate
*Category:* Detergent, anionic
*Applications:* Dry cleaning detergent
*Form:* Liquid; Color: Light yellow
*Properties:* Solubility: Sol. in chlorinated solvents, ethylene glycol, lower alcohols, aniline and petroleum solvents; Sp. gr.: ≈ 1; Visc.: ≈ 5500 cps; HLB: 11.7; Flash Pt.: > 300 F; Fire Pt.: > 300 F
*Composition:* 100% active ingred.

**Atlas G-750.** I.C.I. United States
*Chemical Description:* 80% sodium sorbitol borate
*Category:* Water-sol. resin
*Form:* Viscous liquid; Color: Light yellow
*Properties:* Refractive Index: 1.45–1.456; Solubility: Sol. in water; Sp. gr.: 1.375–1.390; Visc.: ≈ 4500 cp

**Atlas G-950.** I.C.I. United States
*Chemical Description:* Sorbide dioleate
*Category:* Nonionic surfactant
*Form:* Liquid; Color: Orange-red
*Properties:* Solubility: Dispersible in water; sol. in aromatic and aliphatic hydrocarbons, vegetable oils; Sp. gr.: ≈ 1; Visc.: ≈ 250 cps; HLB: 2.7; Flash Pt.: > 300 F; Fire Pt.: > 300 F
*Containers:* 55 gal closed head steel drum

**Atlas G-991.** I.C.I. United States
*Chemical Description:* Mono- and diglycerides with 0.02% BHA and 0.01% citric acid added as preservatives
*Category:* Food emulsifier
*Form:* Ivory white, plastic solid
*Properties:* Solubility: Sol. above its melting point in vegetable oils and mineral oil; M.P.: ≈ 118 F; HLB: 2.8; Flash Pt.: > 300 F; Fire Pt.: > 300 F

**Atlas G-1045.** I.C.I. United States
*Chemical Description:* Polyoxyethylene sorbitol laurate
*Category:* Emulsifier, nonionic
*Applications:* Fiber lubricant; general purpose emulsifier, esp. for insecticides
*Form:* Oily liquid; Color: Light yellow

*Properties:* Solubility: Dispersible with haze in water; sol. in toluene, mineral spirits, acetone, diethyl ether, dioxane, cellsolve, ethyl acetate, aniline; Sp. gr.: ≈ 1; Visc.: ≈ 175 cp; HLB: 11.5; Flash Pt.: > 300 F; Fire Pt.: > 300 F
*Composition:* 100% active ingred.

**Atlas G-1052.** I.C.I. United States
*Chemical Description:* Polyoxyethylene sorbitol distearate
*Category:* Nonionic surfactant
*Form:* Wax; Color: Cream
*Properties:* Solubility: Sol. in isopropanol; HLB: 8.5; Flash Pt.: > 300 F

**Atlas G-1069.** I.C.I. United States
*Chemical Description:* Polyoxyethylene sorbitol ester of mixed fatty acids
*Category:* Nonionic surfactant
*Form:* Liquid; Color: Amber
*Properties:* Solubility: Dispersible in water, sol. in isopropanol, most vegetable oils, hydrocarbon solvent; Sp. gr.: ≈ 1.0; Visc.: ≈ 200 cps; HLB: 9.4; Pour Pt.: ≈ 45 F

**Atlas G-1086.** I.C.I. United States
*Chemical Description:* Polyoxyethylene sorbitol hexaoleate
*Category:* Nonionic emulsifier
*Form:* Oily liquid; Color: Yellow
*Properties:* Solubility: Sol. in toluene, mineral spirits, many vegetable oils, acetone, carbon tetrachloride, cellosolve, methanol, lower alcohols, aniline; Sp. gr.: ≈ 1; Visc.: ≈ 200 cps; HLB: 10.2; Flash Pt.: > 300 F; Fire Pt.: > 300 F
*Composition:* 100% active ingred.

**Atlas G-1096.** I.C.I. United States
*Chemical Description:* Polyoxyethylene (50) sorbitol hexaoleate
*Category:* Nonionic surfactant
*Form:* Liquid; Color: Amber
*Properties:* Solubility: Sol. in vegetable oils, acetone, cellosolve, lower alcohols, some aromatic solvents, carbon tetrachloride; Sp. gr.: ≈ 1.0; Visc.: ≈ 220 cps; HLB: 11.4; Flash Pt.: > 300 F; Fire Pt.: > 300 F
*Composition:* 100% active ingred.

**Atlas G-1186.** I.C.I. United States
*Chemical Description:* Polyoxyethylene sorbitol oleate
*Category:* Nonionic surfactant
*Applications:* Fiber lubricant; general purpose emulsifier, esp. for insecticides
*Form:* Oily liquid; Color: Amber
*Properties:* Solubility: Sol. in lower alcohols, vegetable and mineral oils; Sp. gr.: ≈ 1; Visc.: ≈ 230 cps; HLB: 10.2; Flash Pt.: >300 F; Fire Pt.: > 300 F
*Composition:* 100% active ingred.

**Atlas G-1223.** I.C.I. United States
*Chemical Description:* Polyoxyethylene esters of mixed fatty and resin acids with 0.025% Antifoam added
*Category:* Detergent, nonionic
*Form:* Oily liquid; Color: Light amber
*Properties:* Solubility: Sol. in water, diethyl ether, dioxane, cellosolve, ethyl acetate, lower alcohols, toluene, acetone; Sp. gr.: ≈ 1.1; Visc.: ≈ 350 cps; HLB: 13.8;

Flash Pt.: > 300 F; Fire Pt.: > 300 F

**Atlas G-1236.** I.C.I. United States
*Category:* Blend of nonionic lubricants, emulsifiers, and rewetters
*Form:* Liquid; Color: Clear amber
*Properties:* Solubility: Dispersible in water; sol. in alcohol, chloroform; Visc.: ≈ 600 cps; Flash Pt.: > 300 F; Fire Pt.: 300 F; Pour Pt.: ≈ 0 C

**Atlas G-1292.** I.C.I. United States
*Chemical Description:* Polyoxyethylene fatty glyceride
*Category:* Nonionic surfactant
*Form:* Pale yellow liquid to soft paste
*Properties:* Solubility: Sol. in water, ethanol, isopropanol; Sp. gr.: ≈ 1.0; HLB: 10.8; Flash Pt.: > 300 F; Fire Pt.: > 300 F
*Containers:* 55 gal closed head steel drums

**Atlas G-1295.** I.C.I. United States
*Chemical Description:* Polyoxyethylene fatty glyceride
*Category:* Nonionic surfactant
*Form:* White to cream colored, hard, waxy solid
*Properties:* Solubility: Sol. in water, lower alcohols, acetone, and chloroform; Sp. gr.: ≈ 1.05 @ 60 C; HLB: 17.5; Flash Pt.: > 300 F; Fire Pt.: > 300 F

**Atlas G-1300.** I.C.I. United States
*Chemical Description:* Polyoxyethylene glyceride ester
*Category:* Emulsifier, nonionic
*Applications:* Emulsifiers for use in textile chemical formulations
*Form:* Hard waxy solid; Color: Tan
*Properties:* Solubility: Sol. in water, lower alcohols, acetone, chloroform; HLB: ≈ 18.1; Flash Pt.: > 300 F; Fire Pt.: > 300 F

**Atlas G-1441.** I.C.I. United States
*Chemical Description:* Polyoxyethylene (40) sorbitol lanolin derivative
*Category:* Emulsifier, nonionic
*Applications:* Cosmetic use
*Form:* Solid; Color: Red brown
*Properties:* Solubility: Sol. in water (with haze), isopropanol; Sp. gr.: ≈ 1; HLB: 14; Flash Pt.: > 300 F; Fire Pt.: > 300 F; Pour Pt.: ≈ 33 C

**Atlas G-1471.** I.C.I. United States
*Chemical Description:* Polyoxyethylene (75) lanolin derivative
*Category:* Nonionic surfactant, emulsifiers
*Applications:* Cosmetic use
*Form:* Soft paste; Color: Amber
*Properties:* Solubility: Sol. in isopropyl alcohol, ethylene glycol; HLB: 16; Flash Pt.: > 300 F; Fire Pt.: > 300 F

**Atlas G1556.** I.C.I. United States
*Chemical Description:* Blend of nonionic waxy esters of polyhydric alcohols
*Form:* Waxy solid; Color: Light tan
*Properties:* Solubility: Dispersible in hot water, sol. in carbon tetrachloride, chloroform, anhydrous methanol; Sp. gr.: ≈ 1.05; Flash

Pt.: > 300 F; Fire Pt.: > 300 F; Pour Pt.: > 43 C

**Atlas G-1564.** I.C.I. United States
*Chemical Description:* Blend of nonionic waxy esters of polyhydric alcohols
*Category:* Textile lubricant and softener
*Form:* Waxy granules; Color: Light ivory
*Properties:* Solubility: Sol. in hot carbon tetrachloride, chloroform, anhydrous methanol; dispersible in warm water; Sp. gr.: ≈ 1.04; Flash Pt.: > 300 F; Pour Pt.: ≈ 50 C
*Composition:* 100% active ingred.

**Atlas G-1565.** I.C.I. United States
*Chemical Description:* Blend of nonionic waxy esters of polyhydric alcohols
*Category:* Textile lubricant and softener
*Form:* Waxy solid; Color: Light ivory
*Properties:* Solubility: Dispersible in warm water; sol. in hot carbon tetrachloride, chloroform, anhydrous methanol; Sp. gr.: ≈ 1.05; HLB: 10.1; Flash Pt.: > 300 F; Fire Pt.: > 300 F; Pour Pt.: ≈ 50 C
*Composition:* 100% active ingred.

**Atlas G-1649.** I.C.I. United States
*Chemical Description:* Polyoxyethylene alkyl aryl ether
*Category:* Nonionic surfactant
*Form:* Waxy solid; Color: Cream
*Properties:* Solubility: Sol. in water, toluol, methanol, isopropanol, carbon tetrachloride; Sp. gr.: ≈ 1.1; HLB: 16.0; Flash Pt.: > 300 F; Fire Pt.: > 300 F

**Atlas G-1702.** I.C.I. United States
*Chemical Description:* Polyoxyethylene (6) sorbitol beeswax derivative
*Category:* Nonionic surfactant
*Form:* Waxy solid; Color: Tan
*Properties:* Solubility: Slightly sol. in vegetable oils; HLB: 5; Flash Pt.: > 300 F; Fire Pt.: > 300 F

**Atlas G-1726.** I.C.I. United States
*Chemical Description:* Polyoxyethylene (20) sorbitol beeswax derivative
*Category:* Nonionc surfactant
*Form:* Waxy solid; Color: Tan
*Properties:* Solubility: Slightly sol. in vegetable oils; HLB: 5; Flash Pt.: > 300 F; Fire Pt.: > 300 F

**Atlas G-1790.** I.C.I. United States
*Chemical Description:* Polyoxyethylene (20) lanolin derivative
*Category:* Nonionic, emulsifier
*Applications:* Cosmetic use
*Form:* Solid; Color: Brown
*Properties:* Solubility: Sol. in water; HLB: 11; Flash Pt.: > 300 F; Fire Pt.: > 300 F

**Atlas G-1795.** I.C.I. United States
*Chemical Description:* Polyoxyethylene (50) lanolin derivative
*Category:* Nonionic emulsifier
*Form:* Waxy solid; Color: Tan
*Properties:* Solubility: Sol. in warm water and ethanol; Sp. gr.: ≈ 1.15; HLB: 17.0; Flash Pt.: > 300 F; Fire Pt.: > 300 F

**Atlas G-2079.** I.C.I. United States
*Chemical Description:* Polyoxyethylene (20) palmitate
*Category:* Nonionic surfactant
*Form:* Waxy solid; Color: White
*Properties:* Solubility: Sol. in water and ethanol; HLB: 15.5; Flash Pt.: > 300 F; Fire Pt.: > 300 F

**Atlas G-2090.** I.C.I. United States
*Chemical Description:* Blend of polyoxyethylene fatty amine and polyoxyethylene sorbitol oleate
*Category:* Emulsifier blend
*Applications:* Detergent, emulsifier, detergent specialty
*Form:* Mobile liquid; Color: Red-brown
*Properties:* Solubility: Dispersible in water; sol. in isopropanol, xylene, mineral spirits, kerosene; Sp. gr.: ≈ 0.95; Visc.: ≈ 140 cps; HLB: 12.5; Flash Pt.: > 300 F; Fire Pt.: > 300 F
*Composition:* 100% active ingred.

**Atlas G-2109.** I.C.I. United States
*Chemical Description:* Polyoxyethylene coconut fatty acid
*Category:* Lubricant, nonionic
*Applications:* Heat resistant textile lubricant
*Form:* Liquid
*Composition:* 100% active ingred.

**Atlas G2127.** I.C.I. United States
*Chemical Description:* Polyoxyethylene laurate
*Category:* Nonionic surfactant
*Form:* Liquid; Color: Light yellow
*Properties:* Solubility: Sol. in isopropyl alcohol; dispersible in water and many vegetable oils; Sp. gr.: ≈ 1.0; HLB: 12.8; Flash Pt.: > 300 F

**Atlas G2143.** I.C.I. United States
*Chemical Description:* Polyoxyethylene (10) monoleate
*Category:* Nonionic emulsifier
*Form:* Liquid; Color: Light yellow
*Properties:* Solubility: Dispersible in water; sol. in most mineral oils, lower alcohols; Flash Pt.: > 300 F; Pour Pt.: ≈ −10 C

**Atlas G-2151.** I.C.I. United States
*Chemical Description:* Polyoxyethylene (30) stearate
*Category:* Nonionic surfactant
*Form:* Waxy solid; Color: Ivory
*Properties:* Solubility: Sol. in propylene glycol, ethylene glycol, water; HLB: 16; Flash Pt.: > 300 F; Fire Pt.: > 300 F

**Atlas G-2160.** I.C.I. United States
*Chemical Description:* Polyoxyethylene (25) propylene glycol monostearate
*Category:* Nonionic surfactant
*Form:* Semisolid; Color: Light tan
*Properties:* Solubility: Sol. in water, methanol and ethanol; HLB: 16; Flash Pt.: > 300 F; Fire Pt.: > 300 F

**Atlas G-2162.** I.C.I. United States
*Chemical Description:* Polyoxyethylene (25) propylene glycol stearate
*Category:* Nonionic surfactant
*Form:* Semisolid; Color: Cream

*Properties:* Solubility: Sol. in water, methanol, ethanol; HLB: 16; Flash Pt.: > 300 F; Fire Pt.: > 300 F
*Composition:* 100% active ingred.

**Atlas G-2183.** I.C.I. United States
*Chemical Description:* Propylene glycol monopalmitate
*Category:* Food emulsifier
*Form:* Waxy solid; Color: Ivory
*Properties:* Solubility: Sol. in vegetable oils, common organic solvents; M.P.: ≈ 87 F; HLB: 3.2

**Atlas G-2185.** I.C.I. United States
*Chemical Description:* Propylene glycol monooleate
*Category:* Food emulsifier
*Form:* Liquid; Color: Yellow
*Properties:* Solubility: Sol. in vegetable oils, common organic solvents; HLB: 3.2; Congealing Pt.: 40 F or less

**Atlas G-2198.** I.C.I. United States
*Chemical Description:* Polyoxyethylene (40) stearate
*Category:* Nonionic surfactant
*Form:* Waxy solid; Color: Light tan
*Properties:* Solubility: Sol. in water, acetone, ether, dioxane, cellosolve, carbon tetrachloride, lower alcohols, ethyl acetate, aniline and many organic solvents; HLB: 16.9; Flash Pt.: > 300 F; Fire Pt.: > 300 F; Pour Pt.: ≈ 38 C

**Atlas G-2240.** I.C.I. United States
*Chemical Description:* Polyoxyethylene (6) sorbitol
*Category:* Polyhydric alcohol (polyol)
*Form:* Syrupy pale amber transparent liquid; Odor: Faint bland
*Properties:* Solubility: Sol. in water, sorbitol sol'ns., ethanol, methanol, cellosolve, dioxane, aniline; Sp. gr.: ≈ 1.2; Visc.: ≈ 1100 cp; Flash Pt.: > 300 F; Fire Pt.: > 300 F

**Atlas G-2320.** I.C.I. United States
*Chemical Description:* Polyoxyethylene (20) sorbitol
*Category:* Polyhydric alcohol
*Form:* Liquid; Color: Light
*Properties:* Solubility: Sol. in water, lower alcohols, ethylene glycol, cellosolve, dioxane, aniline; Sp. gr.: ≈ 1.15; Visc.: ≈ 400 cps; Flash Pt.: > 300 F; Fire Pt.: > 300 F

**Atlas G-2401.** I.C.I. United States
*Chemical Description:* Aqueous solution of hydroxypropyl sorbitol
*Category:* Polyhydric alcohol
*Form:* Liquid; Color: Straw; Taste: Slightly bitter
*Properties:* Refractive Index: ≈ 1.475; Solubility: Sol. in water, lower alcohols, propylene glycol, ethylene glycol, cellosolve; Sp. gr.: ≈ 1.25; Visc.: ≈ 2000 cps

**Atlas G-2402.** I.C.I. United States
*Chemical Description:* Hydroxy propyl sorbitol
*Category:* Polyhydric alcohol
*Form:* Liquid; Color: Straw
*Properties:* Solubility: Sol. in water, lower alcohols, propylene glycol, ethylene glycol, cellosolve; Sp. gr.: ≈ 1.3; Visc.: ≈ 12,000 cps;

Flash Pt.: > 300 F; Fire Pt.: > 300 F

**Atlas G-2684.** I.C.I. United States
*Chemical Description:* Blend of sorbitan monooleate and polyoxyethylene esters of mixed fatty and rosin acids
*Category:* Detergent, nonionic
*Form:* Liquid; Color: Amber
*Properties:* Solubility: Dispersible in water; sol. in acetone, toluene, dioxane, carbon tetrachloride, cyclohexanol, aniline, mineral oils; Sp. gr.: ≈ 1; Visc.: ≈ 500 cps; HLB: 7.8; Flash Pt.: > 300 F; Fire Pt.: > 300 F; Pour Pt.: ≈ 15 F

**Atlas G-3300.** I.C.I. United States
*Chemical Description:* Alkyl aryl sulfonate
*Category:* Anionic surfactant
*Applications:* Dispersant, pigments
*Form:* Liquid; Color: Amber
*Properties:* Solubility: Sol. at low conc. in water and hard water; sol. in acetone, carbon tetrachloride, cellosolve, diethyl ether, dioxane, ethylene glycol, methanol, aniline, mineral spirits, kerosene, toluene; Sp. gr.: ≈ 1.0; Visc.: ≈ 6000 cps; HLB: 11.7; Flash Pt.: > 300 F; Fire Pt.: > 300 F
*Composition:* 90% active ingred.

**Atlas G-3634A.** I.C.I. United States
*Chemical Description:* Quaternary ammonium derivative
*Category:* Antistat
*Applications:* Liquid textile antistatic agent; textile softener
*Form:* Liquid; Color: Light amber
*Properties:* Solubility: Sol. in water, lower alcohols, acetone, propylene glycol, carbon tetrachloride, perchlorethylene; Sp. gr.: ≈ 1.05; Visc.: ≈ 2500 cps; HLB: 18.5; Flash Pt.: > 300 F
*Containers:* 55 gal closed head steel drums

**Atlas G-3684.** I.C.I. United States
*Chemical Description:* Polyoxyethylene fatty amine
*Category:* Polyoxyethylene alkyl amine surfactant
*Form:* Liquid; Color: Amber
*Properties:* Solubility: Sol. in acetone, lower alcohols, 8% sulfuric acid, 6% sodium hydroxide, water; Sp. gr.: ≈ 1.03; Visc.: ≈ 150 cps; Flash Pt.: > 300 F; Fire Pt.: > 300 F; Pour Pt.: ≈ 20 F

**Atlas G-3780A.** I.C.I. United States
*Chemical Description:* Polyoxyethylene alkyl amine
*Category:* Surfactant
*Form:* Liquid (may become hazy); Color: Amber
*Properties:* Solubility: Sol. in water, lower alcohols, acetone, ethyl acetate, ethylene glycol; Sp. gr.: ≈ 1.04; Visc.: ≈ 250 cps; HLB: ≈ 15.5; Flash Pt.: > 300 F

**Atlas G-3801.** I.C.I. United States
*Chemical Description:* Polyoxyethylene (2) alkyl alcohol (antioxidant added)
*Category:* Nonionic emulsifier
*Form:* Pourable liquid; Color: White
*Properties:* Sp. gr.: ≈ 0.89; Visc.: ≈ 60 cps; HLB: 5.3; Stability:

Alkalai stable

**Atlas G-3811.** I.C.I. United States
*Chemical Description:* Polyoxyethylene (10) alkyl alcohol (antioxidant added)
*Category:* Nonionic emulsifier
*Form:* Pourable liquid; Color: White
*Properties:* Sp. gr.: ≈0.992; Visc.: ≈ 415 cps; HLB: 12.9; Stability: Alkalai stable

**Atlas G-3821.** I.C.I. United States
*Chemical Description:* Polyoxyethylene (20) alkyl alcohol (antioxidant added)
*Category:* Nonionic emulsifier
*Form:* Soft wax
*Properties:* Sp. gr.: ≈ 1.04; HLB: 15.7; Pour Pt.: ≈ 32 C; Stability: Alkali stable

**Atlas G-7166P.** I.C.I. United States
*Chemical Description:* Polyoxyethylene (16) sorbitan tristearate
*Category:* Nonionic surfactant
*Form:* Waxy solid; Color: Tan
*Properties:* Solubility: Dispersible in warm water, tending to gel at higher concentrations; dispersible in most organic solvents, in vegetable and mineral oils; Sp. gr.: ≈ 1.05; HLB: ≈ 10

**Atlas G-7596H.** I.C.I. United States
*Chemical Description:* Polyoxyethylene sorbitan laurate
*Category:* Nonionic emulsifier
*Form:* Oily liquid; Color: Orange
*Properties:* Solubility: Sol. in water; Sp. gr.: ≈ 1.1; Visc.: ≈ 400 cps; HLB: 14.3; Flash Pt.: > 300 F; Fire Pt.: > 300 F

**Atlas G-7596-J.** I.C.I. United States
*Chemical Description:* Polyoxyethylene (10) sorbitan monolaurate
*Category:* Detergent, nonionic
*Form:* Liquid; Color: Yellow
*Properties:* Solubility: Sol. in water, acetone, dioxane, cellosolve, lower alcohols, ethyl acetate, aniline; Sp. gr.: ≈ 1.1; Visc.: ≈ 400 cps; HLB: 14.9; Flash Pt.: > 300 F; Fire Pt.: > 300 F

**Atlas G-9046T.** I.C.I. United States
*Chemical Description:* Polyoxyethylene (20) mannitan monolaurate
*Category:* Nonionic surfactant
*Form:* Oily liquid; Color: Yellow
*Properties:* Solubility: Sol. in water, ethanol, methanol, propylene glycol; Sp. gr.: ≈ 1.1; HLB: 16.7; Flash Pt.: > 300 F; Fire Pt.: > 300 F

**Atlox 775.** I.C.I. United States
*Category:* Nonionic emulsifier
*Applications:* Emulsifier for agricultural spray oils
*Form:* Liquid
*Composition:* 100% active ingred.

**Atlox 1045A.** I.C.I. United States
*Chemical Description:* Polyoxyethylene sorbitol oleate-laurate
*Category:* Nonionic agricultural surfactant
*Form:* Liquid; Color: Yellow
*Properties:* Solubility: Sol. in most organic solvents and pesticides; dispersible in water; Sp. gr.:

≈ 1.0; Visc.: 200 cps; HLB: 200 cps; HLB: 11.4; Flash Pt.: > 300F

**Atlox 1087**. I.C.I. United States
*Chemical Description:* Polyoxyethylene sorbitol septaoleate
*Category:* Nonionic emulsifier
*Form:* Liquid; Color: Yellow
*Properties:* Solubility: Sol. in most vegetable and mineral oil; dispersible in water; Sp. gr.: ≈ 1; Visc.: 175 cps; HLB: 9.2; Flash Pt.: > 300 F; Fire Pt.: > 300 F

**Atlox 1196**. I.C.I. United States
*Chemical Description:* Polyoxyethylene sorbitol oleate
*Category:* Nonionic emulsifier
*Form:* Oily liquid; Color: Amber
*Properties:* Solubility: Dispersible in water; sol. in xylol, methanol, acetone, most agricultural concentrates; Sp. gr.: ≈ 1; Visc.: ≈ 450 cp; HLB: 11.4; Flash Pt.: ≈ 475 F; Fire Pt.: ≈ 600 F

**Atlox 1256**. I.C.I. United States
*Chemical Description:* Polyoxyethylene sorbitol esters of mixed fatty and resin acids
*Category:* Nonionic surfactant
*Form:* Liquid; Color: Dark amber
*Properties:* Solubility: Sol. in most organic solvents and pesticide formulations; dispersible in water; Sp. gr.: ≈ 1.0; Visc.: ≈ 1000 cps; HLB: 9.7; Flash Pt.: > 300 F; Fire Pt.: > 300 F

**Atlox 2081 B**. Atlas Chemical Industries
*Chemical Description:* Polyoxyethylene sorbitan esters of fatty and resin acids–biodegradable alkylaryl sulfonate blend
*Category:* Emulsifier
*Applications:* Surfactants used with all types of agricultural emulsifiable concentrates
*Form:* Liquid
*Properties:* Biodegradable

**Atlox 3300 B**. Atlas Chemical Industries
*Chemical Description:* Polyoxyethylene sorbitan esters of fatty and resin acids; alkyl aryl sulfonate blend
*Category:* Anionic surfactants
*Applications:* Surfactants used with all types of agricultural emulsifiable concentrates
*Form:* Liquid
*Properties:* Biodegradable

**Atlox 3335**. I.C.I. United States
*Chemical Description:* Alkyl aryl sulfonate blended with polyoxyethylene sorbitan ester of tall oil
*Category:* Agricultural surfactant
*Form:* Liquid; Color: Amber
*Properties:* Solubility: Dispersible in water; sol. in most aromatic solvents and pesticide formulations; Sp. gr.: ≈ 1.1 @ 20 C; Visc.: ≈ 2200 cps; HLB: 13; Flash Pt.: ≈ 160 F

**Atlox 3335 B**. Atlas Chemical Industries
*Chemical Description:* Polyoxyethylene sorbitan esters of fatty and resin acids; alkyl aryl sulfonate blend
*Category:* Anionic surfactants
*Applications:* Surfactants used

with all types of agricultural emulsifiable concentrates
*Form:* Liquid
*Properties:* Biodegradable

**Atlox 3386 B.** Atlas Chemical Industries
*Chemical Description:* Polyoxyethylene sorbitol fatty esters; alkyl aryl sulfonate blend
*Category:* Anionic surfactants
*Applications:* Surfactants used with all types of agricultural emulsifiable concentrates
*Form:* Liquid
*Properties:* Biodegradable

**Atlox 3387.** I.C.I. United States
*Chemical Description:* Anionic-nonionic surfactant blend
*Category:* Agricultural surfactant
*Form:* Viscous liquid; Color: Amber
*Properties:* Solubility: Sol. in most aromatic solvents and pesticide formulations; Sp. gr.: ≈ 1 @ 20 C; Visc.: ≈ 4000 cps

**Atlox 3387 B.** Atlas Chemical Industries
*Chemical Description:* Polyoxyethylene glyceride ester; alkyl aryl sulfonate blend
*Category:* Anionic surfactants
*Applications:* Surfactants used with all types of agricultural emulsifiable concentrates
*Form:* Liquid
*Properties:* Biodegradable

**Atlox 3400 B.** Atlas Chemical Industries
*Chemical Description:* Polyoxyethylene alkyl phenol; alkyl aryl sulfonate blend
*Category:* Anionic surfactants
*Applications:* Surfactants used with all types of agricultural emulsifiable concentrates
*Form:* Liquid
*Properties:* Biodegradable

**Atlox 3403F.** I.C.I. United States
*Chemical Description:* Anionic/nonionic blend
*Category:* Emulsifier
*Applications:* Pesticide emulsifier
*Form:* Liquid; Color: Amber
*Properties:* Solubility: Sol. in most aromatic solvents used in pesticide formulations; dispersible in water; Density: ≈ 8.4 lb/gal @ 20 C; Sp. gr.: ≈ 1.01 @ 20 C; Visc.: ≈ 100 cp @ 77 F; Flash Pt.: ≈ 58 F; Pour Pt.: ≈ 25 F
*Toxicity:* Eye irritant; poison
*Warnings and Caution:* Flammable liquid

**Atlox 3404F.** I.C.I. United States
*Chemical Description:* Anionic/nonionic blend
*Category:* Emulsifier
*Applications:* Pesticide emulsifier
*Form:* Liquid; Color: Amber
*Properties:* Solubility: Sol. in kerosene and most aromatic solvents used in pesticide formulations; dispersible in water; Density: ≈ 8.7 lb/gal; Sp. gr.: ≈ 1.04; Visc.: ≈ 1400 cp @ 77 F; Flash Pt.: ≈ 64 F; Pour Pt.: ≈ –15 F
*Toxicity:* Poison; skin and eye irritant
*Warnings and Caution:* Flammable liquid

**Atlox 3406F.** I.C.I. United States
*Chemical Description:* Anionic/nonionic blend
*Category:* Emulsifier
*Applications:* Pesticide emulsifier
*Form:* Liquid; Color: Amber
*Properties:* Solubility: Sol. in most aromatic solvents used in pesticide formulations; dispersible in water; Density: ≈ 8.6 lb/gal @ 20 C; Sp. gr.: ≈ 1.03 @ 20 C; Visc.: ≈ 200 cp @ 77 F; Flash Pt.: ≈ 61 F; Pour Pt.: ≈ 20 F
*Toxicity:* Poison; eye irritant
*Warnings and Caution:* Flammable

**Atlox 3407F.** I.C.I. United States
*Chemical Description:* Anionic/nonionic blend
*Category:* Pesticide emulsifier
*Form:* Liquid; Color: Amber
*Properties:* Solubility: Sol. in most aromatic solvents used in pesticide formulations; dispersible in water; Sp. gr.: ≈ 1.03 @ 20/20 C; Visc.: ≈ 250 cp @ 70 F; Flash Pt.: ≈ 59 F; Pour Pt.: ≈ 10 F
*Toxicity:* Poisonous if ingested; eye contact and repeated skin exposure can cause chemical burns
*Warnings and Caution:* Flammable liquid; goggles and gloves should be worn

**Atlox 3408F.** I.C.I. United States
*Chemical Description:* Anionic/nonionic blend
*Category:* Pesticide emulsifier
*Form:* Liquid; Color: Amber
*Properties:* Solubility: Sol. in most aromatic solvents used in pesticide formulations; dispersible in water; Density: ≈ 8.5 lb/gal @ 20 C; Sp. gr.: ≈ 1.02 @ 20/20C; Visc.: ≈ 150 cp @ 77 F; Flash Pt.: ≈ 57 F; Pour Pt.: ≈ 20 F
*Toxicity:* Poisonous if ingested; eye contact and repeated skin exposure can cause chemical burns
*Warnings and Caution:* Goggles and gloves should be worn; avoid inhalation. Flammable

**Atlox 3409F.** I.C.I. United States
*Chemical Description:* Anionic/nonionic blend
*Category:* Pesticide emulsifier
*Form:* Liquid; Color: Amber
*Properties:* Solubility: Sol. in most aromatic solvents used in pesticide formulations; dispersible in water; Density: ≈ 8.4 lb/gal @ 20C; Sp. gr.: ≈ 1.01 @ 20 C; Visc.: ≈ 150 cp @ 77 F; Flash Pt.: ≈ 54 F; Pour Pt.: ≈ 40 F
*Toxicity:* Poisonous if ingested; eye irritant
*Warnings and Caution:* Flammable liquid;

**Atlox 3414.** I.C.I. United States
*Chemical Description:* Anionic/nonionic blend
*Category:* Emulsifier
*Form:* Viscous liquid
*Properties:* Solubility: Sol. in kerosene, fuel oil, xylene, heavy aromatic naptha; Sp. gr.: ≈ 1.03; Flash Pt.: > 240 F

**Atlox 3421F.** I.C.I. United States
*Chemical Description:* Anionic/nonionic surfactant blend
*Category:* Emulsifier
*Applications:* Agricultural emulsifiers for improved concentrate

ageing; esp. effective with organophosphate pesticides
*Form:* Hazy liquid
*Properties:* Solubility: Sol. in most armoatic solvents and pesticide formulations; dispersible in water; Sp. gr.: 1.04; Visc.: 250 cp; Flash Pt.: 48 F; Pour Pt.: 55 F
*Warnings and Caution:* Flammable

**Atlox 3422F.** I.C.I. United States
*Chemical Description:* Anionic/nonionic surfactant blend
*Category:* Emulsifier
*Applications:* See Atlox 3421F
*Form:* Hazy liquid
*Properties:* Solubility: Sol. in most aromatic solvents and pesticide formulations; dispersible in water; Sp. gr.: 1.04; Visc.: 900 cp; Flash Pt.: 58 F; Pour Pt.: 50 F
*Warnings and Caution:* Flammable

**Atlox 3450F.** I.C.I. United States
*Chemical Description:* Anionic/nonionic blend
*Category:* Emulsifier
*Applications:* Agricultural surfactant; emulsifier for aerial spray formulations that are applied at dilution rates as low as 1:1; in more dilute sprays for ground application;emulsifies high-poundage formulations
*Form:* Hazy liquid; Color: Dark brown
*Properties:* Solubility: Sol. in most aromatic solvents and pesticide formulations; dispersible in water; Density: 8.7 lb/gal; Sp. gr.: ≈ 1.04; Visc.: ≈ 700 cp; Flash Pt.: ≈ 120 F; Pour Pt.: ≈ 60 F; Stability: Good storage stability
*Warnings and Caution:* Combustible

**Atlox 3453F.** I.C.I. United States
*Chemical Description:* Nonionic/anionic blend
*Category:* Emulsifier
*Applications:* General purpose emulsifiers for pesticide concentrates where high flash point is required
*Form:* Liquid
*Composition:* 100% active ingred.

**Atlox 3454F.** I.C.I. United States
See Atlox 3453F

**Atlox 3455F.** I.C.I. United States
See Atlox 3453F

**Atlox 3459F.** I.C.I. United States
*Chemical Description:* Anionic/nonionic surfactant blend
*Category:* Emulsifier
*Applications:* Agricultural emulsifiers for improved concentrate ageing; sometimes required for optimum emulsification of extremely high poundage phosphate formulas
*Form:* Clear liquid
*Properties:* Solubility: Sol. in most aromatic solvents and pesticide formulations; sol. in water; Sp. gr.: 1.05; Visc.: 500 cp; Flash Pt.: 64 F; Pour Pt.: 15 F
*Warnings and Caution:* Flammable

**Atlox 4851 B.** Atlas Chemical Industries
*Chemical Description:* Polyoxyethylene triglyceride alkyl aryl sulfonate blend
*Category:* Anionic surfactants

*Applications:* Surfactants used with all types of agricultural emulsifiable concentrates
*Form:* Liquid
*Properties:* Biodegradable

**Atlox 4853 B.** Atlas Chemical Industries
*Chemical Description:* Blend of ionic and nonionic surfactants
*Category:* Anionic, nonionic
*Applications:* Surfactants used with all types of agricultural emulsifiable concentrates
*Form:* Liquid
*Properties:* Biodegradable

**Atlox 4855 B.** Atlas Chemical Industries
*Chemical Description:* Polyoxyethylene triglyceride; alkyl aryl sulfonate blend
*Category:* Anionic surfactants
*Applications:* Surfactants used with all types of agricultural emulsifiable concentrates
*Form:* Liquid
*Properties:* Biodegradable

**Atlox 4856 B.** Atlas Chemical Industries
*Chemical Description:* Polyoxyethylene alcohol alkyl aryl sulfonate blend
*Category:* Anionic surfactants
*Applications:* Surfactants used with all types of agricultural emulsifiable concentrates
*Form:* Liquid
*Properties:* Biodegradable

**Atlox 4857 B.** Atlas Chemical Industries
*Chemical Description:* Polyoxyethylene alkyl phenol; alkyl aryl sulfonate blend
*Category:* Anionic surfactants
*Applications:* Surfactants used with all types of agricultural emulsifiable concentrates
*Form:* Liquid
*Properties:* Biodegradable

**Atlox 4858 B.** Atlas Chemical Industries
*Chemical Description:* Polyoxyethylene sorbitan esters of fatty and resin acids; alkyl aryl sulfonate blend
*Category:* Anionic surfactants
*Applications:* Surfactants used with all types of agricultural emulsifiable concentrates
*Form:* Liquid
*Properties:* Biodegradable
*Composition:* 100% conc.

**Atlox 4860.** Atlas Chemical Industries
*Chemical Description:* Polyoxyethylene alkyl ether and fatty acids; alkyl aryl sulfonate blend
*Category:* Anionic surfactants
*Applications:* Surfactants used with all types of agricultural emulsifiable concentrates
*Form:* Powder
*Properties:* Biodegradable
*Composition:* 100% conc.

**Atlox 4862.** Atlas Chemical Industries
*Chemical Description:* Anionic suspending agent
*Category:* Wetting/suspending a-

gent, anionic
*Applications:* Agricultural wettable powders
*Form:* Powder
*Composition:* 100% conc.

**Atlox 4868B.** Atlas Chemical Industries
*Chemical Description:* Polyoxyethylene triglyceride/alkyl aryl sulfonate blend
*Category:* Wetting/suspending agent
*Applications:* Agricultural wettable powders
*Form:* Liquid
*Properties:* Biodegradable

**Atlox 4870 B.** Atlas Chemical Industries
*Chemical Description:* Polyethylene sorbitan esters of fatty and resin acids; biodegradable alkyl aryl sulfonate blend
*Category:* Wetting/suspending agent
*Applications:* Agricultural wettable powders
*Form:* Liquid
*Properties:* Biodegradable

**Atlox 4991.** Atlas Chemical Industries
*Chemical Description:* Polyoxyethylene alkyl aryl ether
*Category:* Nonionic surfactant
*Form:* Liquid
*Composition:* 100% conc.

**Atlox 4995.** Atlas Chemical Industries
*Category:* Wetting/suspending agent, nonionic
*Applications:* Agricultural wettable powders
*Form:* Powder
*Composition:* 100% conc.

**Atlox 4996 B.** Atlas Chemical Industries
*Category:* Suspending/wetting agent, anionic
*Applications:* Agricultural wettable powders
*Form:* Powder

**Atlox 8916 P.** Atlas Chemical Industries
*Chemical Description:* Polyoxyethylene esters of mixed fatty and resin acids
*Category:* Nonionic surfactant
*Form:* Liquid
*Composition:* 100% conc.

**Atlox 8916P.** I.C.I. United States
*Chemical Description:* Polyoxyethylene (16) sorbitan ester of tall oil
*Category:* Nonionic surfactant
*Applications:* Agricultural surfactant
*Form:* Liquid; Color: Amber
*Properties:* Solubility: Sol. in water and most aromatic solvents and pesticide formulations; Sp. gr.: ≈ 1.1; Visc.: ≈ 1320 cps; HLB: 14.6; Flash Pt.: > 300 F

**Atlox 8916PF.** I.C.I. United States
*Chemical Description:* Alkoxylated polyol fatty acid ester
*Category:* Emulsifier
*Applications:* Agricultural surfactant; pesticides

*Form:* Liquid; Color: Amber
*Properties:* Solubility: Sol. with haze in most aromatic solvents used in pesticide formulations; gels upon addition to water, sol. in water after heating; Density: ≈ 9.0 lb/gal; Sp. gr.: ≈ 1.08; Visc.: ≈ 500 cp; Flash Pt.: > 200 F; Pour Pt.: ≈ –20 F
*Warnings and Caution:* Combustible liquid

**Atlox 8916TF.** I.C.I. United States
*Chemical Description:* Alkoxylated polyol fatty acid ester
*Category:* Emulsifier
*Applications:* Agricultural surfactant; pesticides
*Form:* Liquid; Color: Amber
*Properties:* Solubility: Sol. in most aromatic solvents used in pesticide formulations; sol. with haze in high aromatic naphtha, sol. without haze in xylene; sol. in water after heating; Density: ≈ 8.3 lb/gal; Sp. gr.: ≈ 1.0; Visc.: ≈ 500 cp; Flash Pt.: > 200 F; Pour Pt.: ≈ –20 F
*Warnings and Caution:* Combustible liquid

**Atmer 507.** I.C.I. United States
*Chemical Description:* Aliphatic polyether
*Category:* Nonionic surfactant
*Form:* Liquid; Color: Amber
*Properties:* Solubility: Dispersible in water; sol. in acetone, ethylene acetate, dioxane, cellosolve, lower alcohols; Sp. gr.: ≈ 1.1; Visc.: ≈ 300 cps; Flash Pt.: > 300 F; Fire Pt.: > 300 F

**Atmer 508.** I.C.I. United States
*Chemical Description:* Aromatic polyether
*Category:* Nonionic surfactant
*Form:* Liquid; Color: Light
*Properties:* Solubility: Sol. in lower alcohols, cellosolve, many petroleum solvents; Sp. gr.: ≈ 1.0; Visc.: ≈ 220 cps; Flash Pt.: > 300 F; Fire Pt.: > 300 F

**Atmer 665.** I.C.I. United States
*Chemical Description:* Nonionic ester
*Category:* Nonionic surfactant
*Form:* Waxy beads; Color: Tan
*Properties:* Solubility: Sol. in isopropanol; HLB: 2.1; Flash Pt.: > 300 F; Fire Pt.: > 300 F; Pour Pt.: ≈ 53 C

**Atmos 150.** I.C.I. United States
*Chemical Description:* Mono- and diglycerides
*Category:* Food emulsifier
*Form:* Ivory white beads; Odor and Taste: Bland
*Properties:* Solubility: Sol. above its melting point in vegetable oils, mineral oil and isopropyl alcohol; M.P.: ≈ 140 F; HLB: 3.2; Flash Pt.: > 300 F; Fire Pt.: > 300 F
*Containers:* 47 gal fiber lever pak

**Atmos 300.** I.C.I. United States
*Chemical Description:* Mono- and diglycerides with 0.02% BHA and 0.01% citric acid in propylene glycol added as preservatives
*Category:* Food emulsifier, nonionic
*Form:* Clear liquid; Color: Yellow

*Properties:* Solubility: Sol. in cottonseed oil, mineral oil, isopropyl alcohol; Sp. gr.: ≈ 0.96; Visc.: ≈ 130 cps; HLB: 2.8; Flash Pt.: > 300 F; Fire Pt.: > 300 F
*Composition:* 100% active ingred.

**Atmos 729.** I.C.I. United States
*Chemical Description:* Hydrated mono- and diglycerides, polysorbate 60 (11.3%), and sorbitan monostearate (11.3%); with phosphoric acid, sodium propionate and sodium benzoate as preservatives
*Category:* Food emulsifier
*Applications:* Used in various cake types for improved batter aeration
*Form:* Creamy white soft plastic
*Properties:* Solubility: Disperses in water; Flash Pt.: > 300 F; Fire Pt.: > 300 F
*Composition:* 45% active ingred.

**Atmos 758.** I.C.I. United States
*Chemical Description:* Hydrated sorbitan, monostearate (22.5%), polysorbate 60 (13.5%), and mono- and diglycerides, with phosphoric acid, sodium propionate and sodium benzoate added as preservatives
*Category:* Food emulsifier
*Applications:* Cake doughnut emulsifier for improved batter aeration
*Form:* Creamy white soft plastic
*Properties:* Solubility: Disperses in water; Flash Pt.: > 300 F; Fire Pt.: > 300 F
*Composition:* 45% active ingred.

**Atmos 2462.** I.C.I. United States
*Chemical Description:* Hydrated mono- and diglycerides, polysorbate 60 (13%), sorbitan monostearate (13%) and propylene glycol; with phosphoric acid, sodium propionate and sodium benzoate added as preservatives
*Category:* Food emulsifier
*Applications:* Cake emulsifier for improved batter aeration in various cake types
*Form:* Creamy white soft plastic
*Properties:* Solubility: Disperses in water; HLB: 7.4; Flash Pt.: > 300 F; Fire Pt.: > 300 F
*Composition:* 39% active ingred.

**Atmul 80.** I.C.I. United States
*Chemical Description:* Mono- and diglycerides with 0.02% BHA and 0.01% citric acid added as preservatives
*Category:* Food emulsifier, nonionic
*Applications:* Food shortenings applications
*Form:* Ivory white plastic solid
*Properties:* Solubility: Sol. above its melting point in isopropyl alcohol, vegetable oils, mineral oil; M.P.: ≈ 115 F; HLB: 2.8; Flash Pt.: > 300 F; Fire Pt.: > 300 F
*Composition:* 100% active ingred.

**Atmul 84.** I.C.I. United States
*Chemical Description:* Mono- and diglycerides
*Category:* Food emulsifier
*Form:* Ivory white beads or flakes; Odor and Taste: Bland
*Properties:* Solubility: Sol. above

its melting point in alcohol, vegetable oils, mineral oil; M.P.: ≈ 140 F; HLB: 2.8; Flash Pt.: > 300 F; Fire Pt.: > 300 F

**Atmul 84K.** I.C.I. United States
*Chemical Description:* Mono- and diglycerides
*Category:* Food emulsifier
*Applications:* Certified as Kosher
*Form:* Ivory-white flakes; Odor and Taste: Bland
*Properties:* Solubility: Sol. above its melting point in alcohol, vegetable oils, mineral oil; M.P.: ≈ 144 F; HLB: 2.8; Flash Pt.: > 300 F; Fire Pt.: > 300 F

**Atmul 86K.** I.C.I. United States
*Chemical Description:* Mono- and diglycerides with 0.02% BHA and 0.01% citric acid added as preservatives
*Category:* Food emulsifier
*Applications:* For Kosher use
*Form:* Ivory white plastic solid
*Properties:* Solubility: Sol. above its melting point in isopropyl alcohol, vegetable oils, mineral oil; M.P.: ≈ 117 F; HLB: 2.8; Flash Pt.: > 300 F; Fire Pt.: > 300 F

**Atmul 122.** I.C.I. United States
*Chemical Description:* Mono- and diglycerides with 0.02% BHA and 0.01% citric acid added as preservatives
*Category:* Food emulsifier
*Form:* Ivory white plastic solid
*Properties:* Solubility: Sol. above its melting point in isopropyl alcohol, vegetable oil, mineral oil; M.P.: ≈ 126 F; HLB: 3.5; Flash Pt.: > 300 F; Fire Pt.: > 300 F

**Atmul 124.** I.C.I. United States
*Chemical Description:* Mono- and diglycerides
*Category:* Food emulsifier
*Form:* Ivory white flakes; Odor and Taste: Bland
*Properties:* Solubility: Sol. above its melting point in isopropyl alcohol, vegetable oils, mineral oil; M.P.: ≈ 140 F; HLB: 3.5; Flash Pt.: > 300 F; Fire Pt.: > 300 F
*Containers:* 50 lb bags

**Atmul 500.** I.C.I. United States
*Chemical Description:* Mono- and diglycerides with 0.02% BHA and 0.01% citric acid added as preservatives
*Category:* Food emulsifier
*Form:* Ivory white plastic solid
*Properties:* Solubility: Sol. above its melting point in isopropyl alcohol, vegetable oils, mineral oils; M.P.: ≈ 132 F; HLB: 3.5; Flash Pt.: > 300 F; Fire Pt.: > 300 F

**Atmul 651K.** I.C.I. United States
*Chemical Description:* Mono- and diglycerides with 0.02% BHA and 0.01% citric acid added as preservatives
*Category:* Food emulsifier
*Applications:* For Kosher use
*Form:* Ivory white plastic solid
*Properties:* Solubility: Sol. above its melting point in isopropyl alcohol, vegetable oils, mineral oil; M.P.: ≈ 133 F; HLB: 3.5; Flash Pt.: > 300 F; Fire Pt.: > 300 F

**Atmul 700H.** I.C.I. United States

*Chemical Description:* Hydrated mono- and diglycerides with polysorbate 60 (1%), and sodium propionate and lactic acid added as preservatives
*Category:* Food emulsifier
*Applications:* Softener for yeast-raised bakery products
*Form:* Soft plastic; Color: Creamy white
*Properties:* Solubility: Dispersible in water; HLB: 4.0; Flash Pt.: > 300 F; Fire Pt.: > 300 F
*Composition:* Total monoglycerides: ≈ 25%

**Atolex AST/2.** Standard Chem.
*Category:* Antistat, nonionic
*Applications:* Antistatic liquid and fiber lubricant for textile industry

**Atolex AST/3.** Standard Chem.
*Category:* Antistat, cationic
*Applications:* Textile antistatic agent
*Form:* Paste

**Atolex BT.** Standard Chem.
*Category:* Dyebath auxiliary
*Applications:* Backtanning agent particularly for acid dyes on nylon

**Atolex DA/25.** Standard Chem.
*Category:* Dyebath auxiliary
*Applications:* Anionic levelling and dispersing agent

**Atolex DCP/2.** Standard Chem.
*Category:* Dyebath auxiliary
*Applications:* Dye carrier for high temperature pressure dyeing of polyester

**Atolex LDA/40.** Standard Chem.
*Category:* Dyebath auxiliary, nonionic
*Applications:* Levelling and dispersing agent

**Atolex LS/3.** Standard Chem.
*Chemical Description:* Sulphated fatty alcohol
*Category:* Detergent and wetting agent, nonionic/anionic
*Form:* Liquid

**Atolex NI/100B.** Standard Chem.
*Category:* Detergent, wetting agent, nonionic
*Applications:* Concentrated detergent and wetting agent for textiles
*Form:* Liquid

**Atolex PFC.** Standard Chem.
*Category:* Lubricant
*Applications:* Cationic polyethylene emulsion substantive lubricant and softener for textile industry

**Atolex PFN.** Standard Chem.
*Category:* Lubricant
*Applications:* Nonionic polyethylene emulsion lubricant and softener

**Atolex PMC.** Standard Chem.
*Category:* Lubricant, cationic
*Applications:* Modified cationic polyethylene emulsion substantive lubricant and softener also applicable in dyebath

**Atpet 80.** I.C.I. United States
*Chemical Description:* Sorbitan

monooleate
*Category:* Nonionic rust inhibitor
*Form:* Liquid; Color: Amber
*Properties:* Solubility: Dispersible in water; sol. in most petroleum oils, most vegetable oils, lower alcohols and chlorinated solvents; Sp. gr.: ≈ 1; Visc.: ≈ 1200 cps; HLB: 4.3; Flash Pt.: > 300 F; Fire Pt.: > 300 F; Pour Pt.: ≈ 0 F
*Composition:* 100% active ingred.

**Atpet 100.** I.C.I. United States
*Chemical Description:* Sorbitan fatty acid ester
*Category:* Nonionic rust inhibitor
*Form:* Viscous liquid (may gel on standing); Color: Red brown
*Properties:* Solubility: Sol. in most mineral oils and vegetable oils, lower alcohols, chlorinated solvents; dispersible in water; Sp. gr.: ≈ 1.0; Visc.: ≈ 900 cps; HLB: 4.3; Flash Pt.: > 300 F; Fire Pt.: > 300 F

**Atpet 200.** I.C.I. United States
*Chemical Description:* Sorbitan fatty acid ester
*Category:* Nonionic rust inhibitor
*Form:* Liquid; Color: Reddish brown
*Properties:* Solubility: Sol. in most mineral oils, vegetable oils, lower alcohols, chlorinated solvents; dispersible in water; Sp. gr.: ≈ 1.0; Visc.: ≈ 1200 cps; HLB: 4.3; Flash Pt.: > 300 F; Fire Pt.: > 300 F; Pour Pt.: ≈ 0 F
*Containers:* 55 gal closed head steel drums

**Atplus 300F.** I.C.I. United States
*Chemical Description:* Nonionic surfactants with coupling agents
*Category:* Agricultural surfactant, nonionic
*Applications:* Surfactant portion of crop oil concentrate adjuvant for pesticides
*Form:* Liquid; Color: Amber
*Properties:* Solubility: Sol. in most aromatic solvents used in pesticide formulations; dispersible in water; sol. in agricultural grade paraffin oil at 3 to 50%; Density: ≈ 8.6 lb/gal @ 20 C; Sp. gr.: ≈ 1.03 @ 20 C; Visc.: ≈ 400 cp; Flash Pt.: > 200 F; Pour Pt.: ≈ 15 F
*Composition:* 92% solids
*Warnings and Caution:* Combustible

**Atplus 400.** I.C.I. United States
*Category:* Agricultural surfactant, anionic
*Form:* Liquid; Color: Pale yellow
*Properties:* Solubility: Sol. in water; Density: ≈ 9 lb/gal @ 20 C; Sp. gr.: ≈ 1.08; Visc.: < 100 cps @ 20 C

**Atplus 401.** I.C.I. United States
*Chemical Description:* Anionic surfactant with coupling agents
*Category:* Agricultural surfactant, wetting agent and adjuvant
*Applications:* For use with monosodium acid methane arsonate solutions and other herbicides containing MSMA; also with high salt concentration formulations; low phytotoxicity to crops
*Form:* Liquid; Color: Colorless or light yellow
*Properties:* Solubility: Disperses

readily without heat; Sp. gr.: ≈ 1.21; Visc.: ≈ 31 centistokes; Flash Pt.: ≈ 118 F; Pour Pt.: ≈ −20 F
*Toxicity:* May cause skin and eye irritation on contact
*Containers:* 55 gal closed head steel drums; tank wagons; tank cars available

**Atplus 411F.** I.C.I. United States
*Chemical Description:* Oil/surfactant concentrate
*Category:* Agricultural surfactant
*Applications:* Widely used with many herbicides and harvest aid chemicals; shows activity with pre-emergent herbicides

**Atplus 522.** I.C.I. United States
*Chemical Description:* Formulated product
*Category:* Agricultural surfactant
*Applications:* For use in sodium chlorate solutions; improves wetting properties by reduction of surface tension
*Form:* Liquid; Color: Amber
*Properties:* Solubility: Disperses readily, sol. in water; Density: ≈ 10.0 lb/gal @ 20 C; Sp. gr.: ≈ 1.2 @ 20 C; Visc.: ≈ 35 centistokes; Flash Pt.: ≈ 115 F; Pour Pt.: < 32 F
*Toxicity:* May cause eye and skin irritation on contact
*Containers:* 55 gal closed head steel drums; tank wagons and tank cars available

**Atplus 526.** I.C.I. United States
*Chemical Description:* Concentrated blend of fatty acids, fatty acid esters and alkoxylated polyhydric alcohol fatty acid esters
*Category:* Agricultural surfactant, nonionic
*Applications:* Spray tank adjuvant for improving the wetting, spreading and sticking properties of agricultural sprays; specifically developed for repackaging by formulators
*Form:* Liquid

**Atplus 535.** I.C.I. United States
*Category:* Agricultural surfactant; anionic
*Applications:* Stabilizer for liquid fertilizer-pesticide mixtures; for repackaging/resale
*Form:* Light amber liquid
*Properties:* Solubility: Complete in liquid fertilizers; Density: 10.5 lb/gal; Sp. gr.: ≈ 1.25; Visc.: ≈ 40 cps; Flash Pt.: > 200 F

**Atplus 540.** I.C.I. United States
*Chemical Description:* Alkoxylated alcohol ether
*Category:* Agricultural foaming agent
*Applications:* Spray tank additive for short life foam with a stability of 15 to 30 min.
*Form:* Liquid; Color: Amber
*Properties:* Density: ≈ 8.1 lb/gal; Sp. gr.: ≈ 0.97; Visc.: ≈ 30 cps; Flash Pt.: ≈ 65 F; Pour Pt.: ≈ 20 F
*Composition:* 75% alkoxylated alcohol ether; 25% constituents ineffective as foaming agents

**Atplus 555.** I.C.I. United States
*Chemical Description:* Nonionic

surfactant and coupling agents
*Category:* Agricultural surfactant, nonionic
*Applications:* Very effective with post-emergent herbicides; powerful wetting agent

**Atplus 989.** I.C.I. United States
*Chemical Description:* Nonionic surfactant and coupling agents
*Category:* Agricultural surfactant
*Applications:* Shows activity with pre-emergent Caporal®

**Atplus 1034.** I.C.I. United States
*Chemical Description:* Aromatic sulfates and polyhydroxy compounds
*Category:* Agricultural foaming agent
*Applications:* Spray tank additive for long life foam with stability of 2 to 24 hours
*Form:* Liquid; Color: Amber
*Properties:* Density: ≈ 10 lb/gal; Sp. gr.: ≈ 1.2; Visc.: ≈ 140 cps; Flash Pt.: > 200 F; Pour Pt.: ≈ −35 F

**Atplus 1364.** I.C.I. United States
*Chemical Description:* Nonionic surfactant with sticker and coupler
*Category:* Agricultural surfactant
*Applications:* Used with post-emergent WP pesticides

**Atplus 1399.** I.C.I. United States
*Chemical Description:* Nonionic surfactant and coupling agents
*Category:* Agricultural surfactant
*Applications:* Active with some pre-emergent herbicides; medium HLB

**Atplus 1400.** I.C.I. United States
*Chemical Description:* Nonionic surfactant with coupling agents
*Category:* Agricultural surfactant
*Applications:* Medium-high HLB; for activation, reduction of crop injury, pre-emergent herbicides

**Atplus 1401.** I.C.I. United States
*Chemical Description:* Nonionic surfactant and coupling agents
*Category:* Agricultural surfactant
*Applications:* High HLB; used with growth regulators; increases pesticide soil movement

**Atplus 1402.** I.C.I. United States
*Chemical Description:* Nonionic surfactant with coupling agents
*Category:* Agricultural surfactant
*Applications:* High HLB; similar to Atplus 1401 but more effective in some cases

**Atplus 1403.** I.C.I. United States
*Chemical Description:* Anionic surfactant and coupling agents
*Category:* Agricultural surfactant
*Applications:* Wetting agent, activation; used with post-emergent herbicides

**Atsurf 2801.** I.C.I. United States
*Chemical Description:* Polyol oleate
*Category:* Industrial grade surfactant, emulsifier, nonionic
*Applications:* Polyethylene, carnauba, and paraffin wax emulsions for paper coatings, floor polishes, furniture polishes, fiberboard and particle board production; silicone oil emulsions for mold release a-

gents, polishes, antifoam agents; perchloroethylene, mineral spirits, kerosene emulsions for metal cleaning; mineral oil emulsions for metalworking
*Form:* Liquid; Color: Amber
*Properties:* Solubility: Sol. in isopropanol, perchloroethylene, xylene, cottonseed oil, paraffinic oil; Density: 1.00 g/ml; Visc.: 379 cs; HLB: 5; Pour Pt.: –39 F
*Composition:* 100% active ingred.

**Atsurf 2802.** I.C.I. United States
*Chemical Description:* Polyol trioleate
*Category and Applications:* See Atsurf 2801
*Form:* Liquid; Color: Amber
*Properties:* Solubility: Sol. in perchloroethylene, xylene, cottonseed and paraffinic oils; Density: 0.94 g/ml; Visc.: 204 cs; HLB: 2.5; Pour Pt.: $< -35$ F
*Composition:* 100% active ingred.

**Atsurf 2803.** I.C.I. United States
*Chemical Description:* Polyol stearate
*Category and Applications:* See Atsurf 2801
*Form:* Solid; Color: Light tan
*Properties:* Solubility: Sol. in isopropanol (1%), perchloroethylene, xylene; Density: 1.06 g/ml; Visc.: 165 cs @ 160 F; HLB: 5; Pour Pt.: 125 F
*Composition:* 100% active ingred.

**Atsurf 2821.** I.C.I. United States
*Chemical Description:* Ethoxylated polyol oleate
*Category and Applications:* See Atsurf 2801
*Form:* Liquid; Color: Amber
*Properties:* Solubility: Sol. in water, isopropanol, perchloroethylene (10%), xylene (1%), cottonseed oil (1%); Density: 1.06 g/ml; Visc.: 846 cs; HLB: 16; Pour Pt.: 55 F
*Composition:* 100% active ingred.

**Atsurf 2822.** I.C.I. United States
*Chemical Description:* Ethoxylated polyol trioleate
*Category and Applications:* See Atsurf 2801
*Form:* Liquid; Color: Amber
*Properties:* Solubility: Sol. in isopropanol, perchloroethylene, xylene, cottonseed and paraffinic oils; Density: 1.02 g/ml; Visc.: 349 cs; HLB: 11; Pour Pt.: 15 F
*Composition:* 100% active ingred.

**Atsurf 2823.** I.C.I. United States
*Chemical Description:* Ethoxylated polyol stearate
*Category and Applications:* See Atsurf 2801
*Form:* Semisolid; Color: Amber
*Properties:* Solubility: Sol. in water, isopropanol, xylene (1%), cottonseed oil (1%); Density: 1.08 g/ml; Visc.: 50 cs @ 160 F; HLB: 16; Pour Pt.: 90 F
*Composition:* 100% active ingred.

**Avirol 200.** Henkel Inc.
*Chemical Description:* Special ammonium lauryl sulfate
*Category:* Foaming agent, dispersant
*Applications:* Foaming and dispersing agent for dry wall manu-

facture; foaming agent and suspension aid in rock drilling; foaming agent for fire fighting; dispersing agent for dyes and pigments
*Form:* Liquid; Color: Light amber; Odor: Typical
*Properties:* Visc.: 1000 cps; Cloud Pt.: 10 C max.
*Composition:* 28% fatty alcohol sulfate; 0.5% ammonium chloride; 1.5% ammonium sulfate

**Avirol 300.** Henkel Inc.
*Chemical Description:* TEA lauryl sulfate
*Applications:* Base for shampoos, bubble baths, and other cosmetic preparations for cleansing purposes
*Form:* Liquid; Color: Light amber; Odor: Pleasant, mild
*Properties:* Visc.: 500 cps max.
*Composition:* 39–41% fatty alcohol sulfate

**AXS.** Conoco Chemical
*Chemical Description:* Ammonium xylene sulfonate
*Applications:* Hydrotrope for detergent formulations
*Form:* Liquid; Color: Near colorless; Odor: Slight ammonia
*Properties:* Density: 9.5; Sp. gr.: 1.141; Visc.: 4.8; M.P.: 14; B.P.: 212
*Composition:* 42% active ingred.; other components: water

# B

**Barium Petronate 50S Neutral.** Witco Chem. Sonneborn Div.
*Applications:* Additive for lube and industrial oils, fuels, specialty oils and greases
*Form:* Color: Dark brown
*Properties:* Sp. gr.: 1.0 @ 60 F
*Composition:* 45% active ingred.; 54% mineral oil; 1% water
*Toxicity:* May cause eye burns
*Containers:* Drums, tankwagons, tankcars

**Barium Petronate 50-S Neutral.** Witco Chem. Sonneborn Div.
*Chemical Description:* Synthetic sulfonate
*Category:* Lubricant
*Applications:* Additives for lube and industrial oils, fuel oil, undercoating and greases
*Properties:* Visc.: 100 SSU
*Composition:* 50–52% barium sulfonate

**Barium Petronate B-70.** Witco Chem. Sonneborn Div.
*Applications:* Lube and industrial oil and fuel additive
*Form:* Viscous liquid; Color: Dark brown
*Properties:* Sp. gr.: 1.0 @ 60 F
*Composition:* 46% active ingred.; 53.5% mineral oil; 0.5% water
*Reaction:* Acid
*Toxicity:* May cause eye burn
*Containers:* Drums, tankwagons, tank cars

**Barium Petronate Basic.** Witco Chem. Sonneborn Div.
*Chemical Description:* Barium carbonate salt of hydrocarbon sulfonic acid
*Category:* Lubricant
*Applications:* Additives for lube and industrial oils and fuels, specialty oils, greases
*Properties:* Visc.: 250 SSU
*Composition:* 40–45% barium sulfonate

**Base 11-T.** Cincinnati Milacron Chem. Inc.
*Category:* Emulsifier
*Applications:* Preparation of o/w emulsions for metal cutting, rust proofing and quenching

*Form:* Liquid; Color: Brown
*Properties:* Solubility: Oil sol.; Density: 8.7 lb/gal @ 60 F; Sp. gr.: 1.05 @ 60 F; Visc.: 190 SUS @ 210 F
*Composition:* 100% active ingred.
*Containers:* 55 gal drums

**Berol 02.** Berol Kemi AB
*Chemical Description:* Alkyl phenol ethylene oxide adduct
*Category:* Emulsifier, nonionic
*Applications:* Emulsifier for waxes, oils, solvents; emulsion polymerization
*Form:* Clear liquid; Color: < 200 (Hazen)
*Properties:* Solubility: Sol. in ethanol, xylene, trichloroethylene, light fuel oil; disp. in water, paraffin oil, white spirit; Density: 1.04 $g/cm^3$ @ 20 C; Visc.: ≈ 350 cps @ 20 C; HLB: 10.9; Cloud Pt.: 60–70 C; Clear Pt.: < 0 C; Flash Pt.: ≫ 100 C; Pour Pt.: < 0 C; Stability: Stable to weak acids, bases, oxidizing and reducing agents; Surface Tension: 31 dynes/cm; Biodegradable in normal biological sewage purification conditions; pH: 6–7 (1% aq.)
*Composition:* 100% active ingred.
*Toxicity:* Degreasing effect on skin in concentrated form
*Containers:* 200 kg net drums

**Berol 07.** Berol Kemi AB
*Chemical Description:* Fatty alcohol ethoxylate
*Category:* Detergent, equalizing and dispersing agent, nonionic
*Applications:* Low-foam washing machine detergents in powder form; textile industry for high-temperature washing; equalizing and dispersing agent for dyes
*Form:* Wax; Color: White
*Properties:* Solubility: Sol. in water, ethanol, xylene, trichloroethylene; disp. in paraffin oil, white spirit, light fuel oil; Density: 1.09 $g/cm^3$; Visc.: ≈ 50 cps @ 50 C; M.P.: 36–38 C; HLB: 15.0; Cloud Pt.: 74–75 C; Flash Pt.: ≫ 100 C; Stability: Stable to weak acids, bases, oxidizing and reducing agents; Surface Tension: 40 dynes/cm; Biodegradable; pH: 6–7 (1% aq.)
*Composition:* 100% active ingred.
*Toxicity:* Degreasing effect on skin in concentrated form
*Containers:* 100 kg net drums

**Berol 08 Powder.** Berol Kemi AB
*Chemical Description:* Fatty alcohol ethoxylate
*Category:* Detergent, dispersant, nonionic
*Applications:* Low foaming powder detergents; in its melted form as dispersant for other powders with very small particle size (to form prills)
*Form:* Powder; Color: White
*Properties:* Solubility: Sol. in water, ethanol, trichloroethylene; disp. in xylene; Density: 0.55–0.65 $g/cm^3$; Visc.: ≈ 200 cps (@ 70 C); M.P.: 52–54 C; HLB: 18.7; Cloud Pt.: 76–77 C; Flash Pt.: ≫ 100 C; Stability: Stable to weak acids, bases, oxidizing and reducing agents; Surface Tension: 48 dynes/cm; Biodegradable; pH: 6–7 (1% aq.)

*Composition:* 100% active ingred.
*Toxicity:* Degreasing effect on skin in concentrated form
*Containers:* 25 kg net sacks

**Berol 09.** Berol Kemi AB
*Chemical Description:* Alkyl phenol ethylene oxide adduct
*Category:* Detergent, wetting agent, emulsifier, dispersant, nonionic
*Applications:* Detergent; textile wetting and cleaning agent for natural and synthetic fibers, esp. for washing virgin wool; emulsifier for waxes, oil, solvents; emulsion polymerization; dispersant for pigments and tinting pastes in paints and varnishes; emulsifier for paint resins
*Form:* Clear liquid; Color: < 150 (Hazen)
*Properties:* Solubility: Sol. in water, ethanol, xylene, trichloroethylene; disp. in paraffin oil, white spirit, light fuel oil; Density: 1.05 g/cm$^3$ @ 20 C; Visc.: ≈ 350 cps @ 20 C; HLB: 13.3; Cloud Pt.: 52–58 C; Clear Pt.: ≈ 7 C; Flash Pt.: ≫ 100 C; Pour Pt.: ≈ 5 C; Stability: Stable to weak acids, bases, oxidizing and reducing agents; Surface Tension: 33 dynes/cm; Biodegradable under normal biological sewage purification conditions; pH: 6–7 (1% aq.)
*Composition:* 100% active ingred.
*Toxicity:* Degreasing effect on skin in concentrated form
*Containers:* 200 kg net drums

**Berol 17.** Berol Kemi AB
*Chemical Description:* Resin acid ethylene oxide adduct
*Category:* Lubricant, emulsifier, nonionic
*Applications:* Textile industry as a spinning oil for natural and synthetic fibers; with alkyl aryl sulphonates as emulsifier for biocides and aromatic solvents
*Form:* Clear liquid; Color: 15 (Gardner)
*Properties:* Solubility: Sol. in water, ethanol, trichloroethylene, xylene; disp. in acetone, propylene glycol; Density: 1.11 g/cm$^3$ @ 20 C; Visc.: 50 cps @ 20 C; HLB: 14; Cloud Pt.: 45 C (1% in 10% NaCl sol'n.); Clear Pt.: ≈ 17 C; Flash Pt.: ≫ 100 C; Pour Pt.: ≈ 15 C; Stability: Stable to weak acids and bases; Surface Tension: 38 dynes/cm; pH: 7.5–8.5 (1% aq.)
*Composition:* 100% active ingred.
*Toxicity:* Degreasing effect on skin in concentrated form

**Berol 26.** Berol Kemi AB
*Chemical Description:* Alkyl phenol ethylene oxide adduct
*Category:* Emulsifier, detergent additive, base for sulfation, nonionic
*Applications:* Emulsifier for waxes, oils, solvents; detergent additive to improve degreasing properties; base for sulfation to produce ether sulfate for dishwashing liquids
*Form:* Clear liquid; Color: < 250 (Hazen)
*Properties:* Solubility: Sol. in ethanol, xylene, trichloroethylene, paraffin oil, white spirit, light fuel oil; disp. in water; Density: 1.02

g/cm$^3$ @ 20 C; Visc.: ≈ 350 cps @ 20 C; HLB: 8.9; Cloud Pt.: 42–46 C; Clear Pt.: < 0 C; Flash Pt.: ≫ 100 C; Pour Pt.: < 0 C; Stability: Stable to weak acids, bases, oxidizing and reducing agents; Biodegradable in normal conditions in biological sewage purification; pH: 6–7 (1% aq. disp.)
*Composition:* 100% active ingred.
*Toxicity:* Degreasing effect on skin in concentrated form
*Containers:* 200 kg net drums

**Berol 048.** Berol Kemi AB
*Chemical Description:* Alcohol ethylene oxide adduct
*Category:* Wetting agent, detergent, nonionic
*Applications:* Wetting and cleaning agent in industrial cleaners containing strong acids and alkalis; pickling and phosphating of metals; rust removal; degreaser for cleaning bath for solution reuse; textile wetting agent (adhesive removal, bleaching); wetting and dispersing talc and magnesium stearate
*Form:* Clear liquid; Color: < 100 (Hazen)
*Properties:* Solubility: Sol. in water and ethanol; disp. in propylene glycol, benzene, trichloroethylene, paraffin oil, white spirit; Density: 1.02 g/cm$^3$ @ 20 C; Visc.: 170 cps @ 20 C; HLB: 13.5; Cloud Pt.: ≈ 70 C; Clear Pt.: ≈ 4 C; Flash Pt.: ≫ 100 C; Pour Pt.: ≈ 0 C; Stability: Stable to strong acids and bases; Surface Tension: 27 dynes/cm; pH: 6–7 (1% aq.)
*Composition:* 85% active ingred.
*Toxicity:* Degreasing effect on skin in concentrated form

**Berol 049.** Berol Kemi AB
*Chemical Description:* n-primary alcohol ethylene oxide adduct
*Category:* Detergent, nonionic
*Applications:* Liquid cleaners, dishwashing liquids, general purpose high foam cleaners
*Form:* Clear liquid; Color: < 100 (Hazen)
*Properties:* Solubility: Sol. in water, ethanol, xylene, trichloroethylene, light fuel oil, white spirit (hazy); Density: 0.99 g/cm$^3$ @ 20 C; Visc.: ≈ 40 cps @ 20 C; HLB: 12.0; Cloud Pt.: 55–60 C; Clear Pt.: ≈ 15 C; Flash Pt.: ≫ 100 C; Pour Pt.: ≈ 7 C; Stability: Stable to weak acids, bases, oxidizing and reducing agents; Surface Tension: ≈ 30 dynes/cm; Biodegradable; pH: 6–7 (1% aq.)
*Composition:* 100% active ingred.
*Toxicity:* Degreasing effect on skin in concentrated form

**Berol 050.** Berol Kemi AB
*Chemical Description:* Fatty alcohol ethoxylate
*Category:* Sulfonation raw material, emulsifier, nonionic
*Applications:* Sulfonation raw material to produce alkyl ether sulfates for hair shampoos, bubble baths, washing-up liquids, etc.; emulsifier for waxes, oils, solvents
*Form:* Clear liquid; Color: < 100 (Hazen)
*Properties:* Solubility: Sol. in ethanol, xylene, trichloroethylene, paraffin oil, white spirit, light fuel

oil; disp. in water; Density: 0.93 g/cm$^3$ @ 20 C; Visc.: 30 cps @ 20 C; HLB: 8.4; Cloud Pt.: 48–52 C; Clear Pt.: ≈ 20 C; Flash Pt.: ≫ 100 C; Pour Pt.: ≈ 0 C; Stability: Stable to weak acids, bases, oxidizing and reducing agents; Biodegradable; pH: 6–7 (1% aq. disp.)
*Composition:* 100% active ingred.
*Toxicity:* Degreasing effect on skin in concentrated form
*Containers:* 185 kg net drums

**Berol 055.** Berol Kemi AB
*Chemical Description:* n-primary alcohol ethylene oxide adduct
*Category:* Sulfonation raw material,emulsifier, nonionic
*Applications:* Sulfonation raw material to produce alkyl ether sulfates for hair shampoos, bubble baths, etc.; emulsifier for waxes, oils, solvents
*Form:* Clear liquid; Color: < 100 (Hazen)
*Properties:* Solubility: Sol. in ethanol, xylene, trichloroethylene, paraffin oil, white spirit, light fuel oil; disp. in water; Density: 0.90 g/cm$^3$ @ 20 C; Visc.: ≈ 30 cps @ 20 C; HLB: 6.5; Cloud Pt.: 35–40 C; Clear Pt.: ≈ 15 C; Flash Pt.: ≫ 100 C; Pour Pt.: ≈ 0 C; Stability: Stable to weak acids, bases, oxidizing and reducing agents; Biodegradable; pH: 6–7 (1% aq. disp.)
*Composition:* 100% active ingred.
*Toxicity:* Degreasing effect on skin in concentrated form
*Containers:* 185 kg net drums

**Berol 057.** Berol Kemi AB
*Chemical Description:* Fatty alcohol ethoxylate
*Category:* Emulsifier, nonionic
*Applications:* Emulsion polymerization, alone or in combination with less hydrophilic nonionics or with anionic surfactants to obtain a very small particle size
*Form:* Soft wax; Color: < 300 (Hazen)
*Properties:* Solubility: Sol. in water, ethanol, xylene, trichloroethylene; disp. in paraffin oil, white spirit; Density: 1.00 g/cm$^3$ @ 60 C; Visc.: ≈ 60 cps @ 40 C; HLB: 15.5; Cloud Pt.: 74–76 C (1% in 10% NaCl sol'n.); Flash Pt.: ≫ 100 C; Pour Pt.: ≈ 30 C; Stability: Stable to weak acids, bases, oxidizing and reducing agents; Surface Tension: 39 dynes/cm; Biodegradable; pH: 6–7 (1% aq.)
*Composition:* 100% active ingred.
*Toxicity:* Degreasing effect on skin; otherwise nontoxic
*Containers:* 200 kg net drums

**Berol 059.** Berol Kemi AB
*Chemical Description:* n-primary alcohol ethylene oxide adduct
*Category:* Detergent, nonionic
*Applications:* Liquid cleaners, dishwashing liquids, etc.
*Form:* Clear liquid; Color: < 100 (Hazen)
*Properties:* Solubility: Sol. in water, ethanol, xylene, trichloroethylene, light fuel oil, white spirit (hazy); Density: 0.97 g/cm$^3$ @ 20 C; Visc.: ≈ 35 cps; HLB: 11.3; Cloud Pt.: 35–40 C (1% aq.) @ 20 C; Clear Pt.: ≈ 13 C; Flash Pt.: ≫ 100 C; Pour Pt.: ≈ 5 C; Stability: Stable to weak acid, bases, oxidiz-

ing and reducing agents; Surface Tension: ≈ 29 dynes/cm; Biodegradable; pH: 6–7 (1% aq.)
*Composition:* 100% active ingred.
*Toxicity:* Degreasing effect on skin in concentrated form

**Berol 065.** Berol Kemi AB
*Chemical Description:* Fatty alcohol ethoxylate
*Category:* Detergent, nonionic
*Applications:* Low-foam washing machine detergents, esp. excellent grease and pigment removal; detergents for spray drying
*Form:* Soft wax; Color: White
*Properties:* Solubility: Sol. in water, ethanol, xylene, trichloroethylene; disp. in paraffin oil, white spirit, light fuel oil; Density: 0.98 g/cm$^3$ @ 40 C; Visc.: ≈ 30 cps @ 50 C; HLB: 12.6; Cloud Pt.: 78–82 C (1% aq.); Clear Pt.: ≈ 44 C; Flash Pt.: ≫ 100 C; Pour Pt.: ≈ 33 C; Stability: Stable to weak acids, bases, oxidizing and reducing agents; Surface Tension: 34 dynes/cm; Biodegradable; pH: 6–7 (1% aq.)
*Composition:* 100% active ingred.
*Toxicity:* Degreasing effect on skin in concentrated form
*Containers:* 200 kg net wt drums

**Berol 79.** Berol Kemi AB
*Category:* Emulsifier, nonionic
*Applications:* In emulsifiable solvents for cleaning, degreasing auto parts, workshop floors, car bodies, etc.; antistatic and emulsifiable spinning oil for textile industry
*Form:* Clear liquid; Color: Yellow
*Properties:* Solubility: Sol. in ethanol, benzene, trichloroethylene, paraffin oil, light mineral oil; disp. in water; Density: 0.97 g/cm$^3$ @ 20 C; Pour Pt.: 0 C
*Composition:* 100% active ingred.
*Containers:* 200 kg net drums

**Berol 081 Flakes.** Berol Kemi AB
*Chemical Description:* Fatty alcohol ethoxylate
*Category:* Detergent, emulsifier, dispersant, stabilizer, nonionic
*Applications:* Low foam detergents in powder form; emulsifier for waxes; dispersant for pigments; latice stabilizer
*Form:* Flakes; Color: White
*Properties:* Solubility: Sol. in water, ethanol, trichloroethylene; disp. in xylene; Density: 0.55–0.65 g/cm$^3$; M.P.: 46–50 C; HLB: 16.7; Cloud Pt.: 75–76 C (1% in 10% NaCl sol'n.); Flash Pt.: ≫ 100 C; Stability: Stable to weak acids, bases, oxidizing and reducing agents; Surface Tension: 43 dynes/cm; Biodegradable; pH: 6–7 (1% aq.)
*Composition:* 100% active ingred.
*Toxicity:* Degreasing effect on skin in concentrated form
*Containers:* 25 kg net sacks

**Berol 081 Powder.** Berol Kemi AB
*Chemical Description:* Fatty alcohol ethoxylate
*Category:* Detergent, emulsifier, dispersant, stabilizer, nonionic
*Applications:* Low foam powder detergents; emulsifier for waxes; dispersing agent for pigments; latice stabilizer
*Form:* Powder; Color: White

*Properties:* Solubility: Sol. in water, ethanol, trichloroethylene; disp. in xylene; Density: 0.55–0.65 g/cm$^3$; M.P.: 46–50 C; HLB: 16.7; Cloud Pt.: 75–76 C; Flash Pt.: ⩾ 100 C; Stability: Stable to weak acids, bases, oxidizing and reducing agents; Surface Tension: 43 dynes/cm; Biodegradable; pH: 6–7 (1% aq.)
*Composition:* 100% active ingred.
*Toxicity:* Degreasing effect on skin in concentrated form
*Containers:* 25 kg net sacks

**Berol 087.** Berol Kemi AB
*Chemical Description:* n-primary alcohol surfactant
*Category:* Detergent, nonionic
*Applications:* Low foam detergent for machine dishwashing, high pressure spraying and brushing; combined with a phosphate ester to increase the solubility at high temperatures and strong electrolyte concentrations
*Form:* Clear liquid; Color: 75 (Hazen)
*Properties:* Solubility: Sol. in water, ethanol, trichloroethylene; disp. in xylene; Density: 0.997 g/cm$^3$ @ 20 C; Visc.: 85 cps @ 20 C; HLB: ≈ 11.5; Cloud Pt.: 42 C; Clear Pt.: 8 C; Flash Pt.: ⩾ 100 C; Pour Pt.: −1 C; Stability: Stable to weak acids, bases, oxidizing and reducing agents; Surface Tension: 31.9 dynes/cm; Biodegradable under normal sewage treatment conditions; pH: 5–7 (1% aq.)
*Composition:* 100% active ingred.
*Toxicity:* Degreasing effect on skin in concentrated form

**Berol 106.** Berol Kemi AB
*Chemical Description:* Alkyl glyceride ethylene oxide adduct
*Category:* Emulsifier, nonionic
*Applications:* O/W emulsifier; in combination with more hydrophobic surfactants for biocides and pesticides
*Form:* Cloudy liquid; Color: < 4 (Gardner)
*Properties:* Solubility: Sol. in water, ethanol, propylene glycol, xylene, trichloroethylene; Density: 1.055 g/cm$^3$; Visc.: ≈ 650 cps; HLB: ≈ 15; Cloud Pt.: 54–58 C; Clear Pt.: ≈ 25 C; Flash Pt.: ⩾ 100 C; Pour Pt.: ≈ 15 C; Stability: Stable to weak acids and bases; Surface Tension: 44 dynes/cm; Biodegradable; pH: 5–7 (1% aq.)
*Composition:* 100% active ingred.
*Toxicity:* Degreasing effect on skin in concentrated form

**Berol 121.** Berol Kemi AB
*Chemical Description:* Fatty alcohol ethoxylate
*Category:* Detergent, nonionic
*Applications:* Liquid detergents, dishwashing liquids, general purpose high foam cleaners (in combination with anionics, e.g., alkyl aryl sulfonate or alkyl ether sulfate)
*Form:* Clear/opal liquid; Color: < 100 (Hazen)
*Properties:* Solubility: Sol. in ethanol, white spirit, light fuel oil; opaquely sol. in water and xylene; disp. in trichloroethylene, paraffin oil; Density: 0.97 g/cm$^3$ @ 20 C; HLB: 11.0; Cloud Pt.: 75–80 C (in 15% n-propanol); Clear Pt.: ≈

26 C; Flash Pt.: $\gg$ 100 C; Pour Pt.: $\approx$ 20 C; Stability: Stable to weak acids, bases, oxidizing and reducing agents; Surface Tension: 29 dynes/cm; Biodegradable; pH: 6–7 (1% aq.)
*Composition:* 97.5% active ingred.
*Toxicity:* Degreasing effect on skin in concentrated form
*Containers:* 200 kg net drums

**Berol 151.** Berol Kemi AB
*Chemical Description:* Fatty acid alkanolamide type based on lauric/myristic acid diethanolamide
*Category:* Foam stabilizer and modifier, dispersant, nonionic
*Applications:* Foam stabilizer and booster in hair shampoos, bubble baths, cleaners, etc. in combination with alkyl sulfates and alkyl ether sulfates; viscosity booster; dispersant for calcium soaps, liquid hand soap
*Form:* Clear liquid; Color: $\leqslant$ 200 (Hazen)
*Properties:* Solubility: Sol. in ethanol, propylene glycol, xylene; disp. in water, white spirit, paraffin oil; Density: 0.98 g/cm$^3$ @ 20 C; Visc.: 1100 cps @ 20 C; HLB: $\approx$ 12; Clear Pt.: $\approx$ 3 C; Flash Pt.: $\gg$ 100 C; Pour Pt.: $\approx$ –5 C; Stability: Stable to weak acids and bases; Surface Tension: 29.5 dynes/cm; Biodegradable; pH: $\approx$ 9.5 (1% aq.)
*Composition:* 100% active ingred.
*Toxicity:* Degreasing effect on skin in concentrated form

**Berol 160.** Berol Kemi AB
*Chemical Description:* Fatty alcohol ethoxylate
*Category:* Detergent, emulsifier, nonionic
*Applications:* Liquid detergents; emulsifier for waxes, oils, solvents
*Form:* Clear liquid; Color: < 100 (Hazen)
*Properties:* Solubility: Sol. in water, ethanol, xylene, trichloroethylene, light fuel oil; disp. in white spirit; Density: 0.98 g/cm$^3$ @ 20 C; Visc.: $\approx$ 50 cps @ 20 C; HLB: 11.8; Cloud Pt.: 40–50 C; Clear Pt.: $\approx$ 19 C; Flash Pt.: $\gg$ 100 C; Pour Pt.: $\approx$ 15 C; Stability: Stable to weak acids, bases, oxidizing and reducing agents; Surface Tension: 29 dynes/cm; Biodegradable; pH: 6–7 (1% aq.)
*Composition:* 100% active ingred.
*Toxicity:* Degreasing effect on skin in concentrated form
*Containers:* 200 kg net drums

**Berol 173.** Berol Kemi AB
*Chemical Description:* Fatty alcohol ethoxylate
*Category:* Detergent, nonionic
*Applications:* Low foaming machine washing detergents; pigment and fatty soil removal over wide temperature range for natural and synthetic fabrics; spray dried and mechanically mixed detergent powders
*Form:* Paste; Color: White
*Properties:* Solubility: Sol. in xylene, trichloroethylene, light fuel oil; opaquely sol. in water and ethanol; Density: 0.99 g/cm$^3$ @ 20 C; Visc.: $\approx$ 30 cps @ 50 C; HLB: 12.5; Cloud Pt.: 65–70 C (1% aq.); Clear Pt.: $\approx$ 44 C; Flash Pt.: $\gg$ 100 C; Pour Pt.: $\approx$ 25 C;

Stability: Stable to weak acids, bases, oxidizing and reducing agents; Surface Tension: 33 dynes/cm (0.1% sol'n.); pH: 6–7 (1% aq.); Easily biodegradable
*Composition:* 100% active ingred.
*Toxicity:* Degreasing effect on skin in concentrated form
*Containers:* 200 kg net drums

**Berol 185.** Berol Kemi AB
*Chemical Description:* Modified fatty alcohol ethoxylate
*Category:* Detergent, nonionic
*Applications:* Concentrated liquid detergents; dishwashing liquids and general purpose cleaners with high foaming and cleaning; strong pigment and fatty soil removal; textile washing processes
*Form:* Clear liquid; Color:Colorless; Odor: Less tendency than similar products to unpleasant odor > 30 C
*Properties:* Solubility: Sol. in water, ethanol; disp. in xylene, trichloroethylene, white spirit, light fuel oil; Density: 1.02 g/cm$^3$ @ 20 C; Visc.: ≈ 100 cps @ 20 C; HLB: 13.5; Cloud Pt.: 65–70 C; Clear Pt.: < 5 C; Flash Pt.: ≫ 100 C; Pour Pt.: < 0 C; Stability: Stable to weak acids, bases, oxidizing and reducing agents; Surface Tension: 31.5 dynes/cm (0.1% sol'n.); pH: 6–7 (1% aq.); Easily biodegradable
*Composition:* 90% active ingred.
*Toxicity:* Degreasing effect on skin in concentrated form
*Containers:* 200 kg net drums

**Berol 186.** Berol Kemi AB
*Chemical Description:* Modified fatty alcohol ethoxylate
*Category:* Detergent, nonionic
*Applications:* Concentrated liquid detergents; dishwashing liquids and general purpose cleaners with high foaming and cleaning; strong pigment and fatty soil removal; textile washing processes
*Form:* Clear/opal liquid; Color: < 100 (Hazen)
*Properties:* Solubility: Sol. in water, ethanol, xylene, trichloroethylene, white spirit, light fuel oil: Density: 1.01 g/cm$^3$ @ 20 C; Visc: ≈ 100 cps @ 20 C; HLB: 13.5; Cloud Pt.: 65–70 C (1% aq.); Clear Pt.: ≈ 18 C; Flash Pt.: ≫ 100 C; Pour Pt.: ≈ 10 C; Stability: Stable to weak acids, bases, oxidizing and reducing agents; Surface Tension: 31.5 dynes/cm (0.1% sol'n.); pH: 6–7 (1% aq.); Biodegradable
*Composition:* 100% active ingred.
*Toxicity:* Degreasing effect on skin in concentrated form
*Containers:* 200 kg net drums

**Berol 196.** Berol Kemi AB
*Chemical Description:* Alkyl glyceride ethylene oxide adduct type
*Category:* Emulsifier, nonionic
*Applications:* Emulsifier in connection with foodstuffs; superfatting agent in bath oils, hand cleaners
*Form:* Clear liquid; Color: ⩽5 Gardner;
*Properties:* Solubility: Sol. in ethanol, propylene glycol, trichloroethylene, white spirity; opaquely sol. in water· Density: 1.020 g/cm$^3$ @ 20 C; Visc.: ≈ 500 cps @

20 C; HLB: 9; Cloud Pt.: ≈ 58 C; Clear Pt.: < 0 C; Flash Pt.: ≫ 100 C; Pour Pt.: - 15 C; Stability: Stable to weak acids and bases; pH: 5–7 (1% aq.); Biodegradable
*Composition:* 100% active ingred.
*Toxicity:* Degreasing effect on skin in concentrated form

**Berol 252.** Berol Kemi AB
*Chemical Description:* Alkyl phenol ethylene oxide adduct type
*Category:* Emulsifier, nonionic
*Applications:* Emulsifier in emulsion polymerization; combined with anionic surfactants such as alkyl aryl sulfonates, alkyl and alkyl ether sulfates
*Form:* Clear liquid; Color: Max. 200 (Hazen)
*Properties:* Solubility: Sol. in water, ethanol; Density: 1.09 g/cm$^3$ @ 20 C; Visc.: 1000 cps @ 20 C; Cloud Pt.: 77 C; Clear Pt.: ≈ 17 C; Flash Pt.: ≫ 100 C; Pour Pt.: ≈ 12 C; Stability: Stable to weak acids, bases, oxidizing and reducing agents; Surface Tension: 41 dynes/cm; pH: 6–7 (1% aq.); Biodegradable
*Composition:* 70% active ingred.
*Toxicity:* Degreasing effect on skin in concentrated form

**Berol 259.** Berol Kemi AB
*Chemical Description:* Alkyl phenol ethylene oxide adduct type
*Category:* Stabilizer, nonionic
*Applications:* Foaming in industrial processes; intermediary product of lubricating oil additives and as a stabilizer for rubber
*Form:* Clear liquid; Color: ≤ 500 (Hazen)
*Properties:* Solubility: Sol. in ethanol, xylene, trichloroethylene, paraffin oil, white spirit, light fuel oil; dispersible in water; Density: 1.01 g/cm$^3$ @ 20 C; Visc.: ≈ 550 cps @ 20 C; HLB: 5.7; Cloud Pt.: 50–54 C; Clear Pt.: < 0 C; Flash Pt.: ≫ 100 C; Pour Pt.: < 0 C; Stability: Stable to weak acids, bases, oxidizing and reducing agents; pH: 6–7; Biodegradable
*Composition:* 100% active ingred.
*Toxicity:* Degreasing effect on skin in concentrated form
*Containers:* 200 kg net drums

**Berol 263.** Berol Kemi AB
*Chemical Description:* Alkyl phenol ethylene oxide adduct type
*Category:* Detergent, nonionic
*Applications:* Detergents and cleaners; dispersing agent for pigments
*Form:* Soft wax; Color: < 200 (Hazen)
*Properties:* Solubility: Sol. in water, ethanol, xylene, trichloroethylene; dispersible in paraffin oil, white spirit; Density: 1.06 g/cm$^3$ @ 30 C; Visc.: ≈ 325 cps @ 25 C; HLB: 14.7; Cloud Pt.: 92–94 C (1% aq. sol'n.); Clear Pt.: ≈ 24 C; Flash Pt.: ≫ 100 C; Pour Pt.: ≈ 20 C; Stability: Stable to weak acids, bases, oxidizing and reducing agents; Surface Tension: 36 dynes/cm (0.1% sol'n.); pH: 6–7 (1% aq.); Biodegradable
*Composition:* 100% active ingred.
*Toxicity:* Degreasing effect on skin in concentrated form

**Berol 267.** Berol Kemi AB
*Chemical Description:* Alkyl phenol ethylene oxide adduct type
*Category:* Emulsifier, detergent, nonionic
*Applications:* Used as an emulsifier for waxes, oils and solvents; dishwashing compounds; washing of virgin wool
*Form:* Clear liquid; Color: $<$ 200 (Hazen)
*Properties:* Solubility: Sol. in ethanol, xylene, trichloroethylene; disp. in water, paraffin oil, white spirit, light fuel oil; Density: 1.04 g/cm$^3$ @ 20 C; Visc.: $\approx$ 350 cps @ 20 C; HLB: 12.3; Cloud Pt.: $\approx$ 10 C (1% aq.); Clear Pt.: $<$ 0 C; Flash Pt.: $\gg$ 100 C; Pour Pt.: $<$ 0 C; Stability: Stable to weak acids, bases, oxidizing and reducing agents; Surface Tension: 31 dynes/cm; pH: 6–7; (1% aq.); Biodegradable
*Composition:* 100% active ingred.
*Toxicity:* Degreasing effect on skin in concentrated form
*Containers:* 200 kg net drums

**Berol 269.** Berol Kemi AB
*Chemical Description:* Dialkyl phenol ethylene oxide adduct type
*Category:* Emulsifier, nonionic
*Applications:* Emulsifier for heavy paraffin and aromatic oils
*Form:* Opal liquid; Color: $<$ 500 (Hazen)
*Properties:* Solubility: Sol. in ethanol, xylene, trichloroethylene, light fuel oil; disp. in water and white spirit; Density: 1.02 g/cm$^3$ @ 20 C; Visc.: $\approx$ 550 cps @ 20 C; HLB: 10.5; Cloud Pt.: $\approx$ 60 C (in 15% *n*-propanol); Flash Pt.: $\gg$ 100 C; Pour Pt.: $<$ 0 C; Stability: Stable to weak acids, bases, oxidizing and reducing agents; pH: 6–7 (1% aq.)
*Composition:* 100% active ingred.
*Toxicity:* Degreasing effect on skin in concentrated form
*Containers:* 200 kg net drums

**Berol 272.** Berol Kemi AB
*Chemical Description:* Dialkyl phenol ethylene oxide adduct type
*Category:* Detergent, nonionic
*Applications:* Low-foaming washing machine detergents in powder form; good stability for spray drying
*Form:* Paste; Color: Light yellow
*Properties:* Solubility: Sol. in water, ethanol, xylene, trichloroethylene; disp. in paraffin oil; Density: 1.04 g/cm$^3$ @ 20 C; Visc.: $\approx$ 300 cps @ 35 C; M.P.: $\approx$ 34 C; HLB: 13.4; Cloud Pt.: $\approx$ 60 C (1% aq.); Flash Pt.: $\gg$ 100 C; Pour Pt.: $\approx$ 23 C; Stability: Stable to weak acids, bases, oxidizing and reducing agents; Surface Tension: 30 dynes/cm (0.1% sol'n.); pH: 6–7 (1% aq.)
*Composition:* 100% active ingred.
*Toxicity:* Degreasing effect on skin in concentrated form
*Containers:* 200 kg net drums

**Berol 274.** Berol Kemi AB
*Chemical Description:* Alkyl phenol ethylene oxide adduct type
*Category:* Emulsifier, nonionic
*Applications:* Emulsifier in emulsion polymerization; increases sta-

bility of the latice
*Form:* Soft wax; Color: < 200 (Hazen)
*Properties:* Solubility: Sol. in water, ethanol, xylene, trichloroethylene; Density: 1.07 g/cm$^3$ @ 50 C; Visc.: ≈ 200 cps @ 50 C; HLB: ≈ 17; Cloud Pt.: 77 C (1% in 10% NaCl sol'n.); Clear Pt.: ≈ 44 C; Flash Pt.: ≫ 100 C; Pour Pt.: ≈ 41 C; Stability: Stable to weak acids, bases, oxidizing and reducing agents; Surface Tension: 41 dynes/cm (0.1% sol'n.); pH: 6–7 (1% aq.); Easily biodegradable under normal biological sewage purification conditions
*Composition:* 100% active ingred.
*Toxicity:* Degreasing effect on skin in concentrated form

**Berol 277.** Berol Kemi AB
*Chemical Description:* Alkyl phenol ethylene oxide adduct type
*Category:* Emulsifier, nonionic
*Applications:* Emulsion polymerization; increases stability of latice
*Form:* Clear liquid; Color: < 200 (Hazen)
*Properties:* Solubility: Sol. in water and ethanol; Density: 1.09 g/cm$^3$ @ 20 C; Visc.: 1000 cps @ 20 C; HLB: 17; Cloud Pt.: 77 C (1% in 10/ NaCl sol'n.); Clear Pt.: ≈ 17 C; Flash Pt.: ≫ 100 C; Pour Pt.: ≈ 12 C; Stability: Stable to weak acids, bases, oxidizing and reducing agents; Surface Tension: 41 dynes/cm (0.1% sol'n.); pH: 6–7 (1% aq.); Easily biodegradable under normal biological sewage purification conditions
*Composition:* 70% active ingred.
*Toxicity:* Degreasing effect on skin in concentrated form

**Berol 278.** Berol Kemi AB
*Chemical Description:* Alkyl phenol ethylene oxide adduct type
*Category:* Emulsifier, nonionic
*Applications:* Emulsion polymerization; combined with anionics to obtain a very small particle size
*Form:* Clear liquid; Color: < 200 (Hazen)
*Properties:* Solubility: Sol. in water and ethanol; Density: 1.08 g/cm$^3$ @ 20 C; Visc.: ≈ 625 cps @ 20 C; HLB: 15.2; Cloud Pt.: 97–99 C (1% aq.); Clear Pt.: ≈ 1 C; Flash Pt.: ≫ 100 C; Pour Pt.: ≈ 0 C; Stability: Stable to weak acids, bases, oxidizing and reducing agents; Surface Tension: 37 dynes/cm (0.1% sol'n.); pH: 6–7 (1% aq.); Easily biodegradable under normal biological sewage purification conditions
*Composition:* 80% active ingred.
*Toxicity:* Degreasing effect on skin in concentrated form
*Containers:* 200 kg net drums

**Berol 281.** Berol Kemi AB
*Chemical Description:* Alkyl phenol ethylene oxide type
*Category:* Emulsifier, nonionic
*Applications:* Emulsion polymerization; combined with anionics to obtain a very small particle size
*Form:* Clear viscous liquid; Color: < 200 (Hazen)
*Properties:* Solubility: Sol. in water and ethanol; Density: 1.09 g/cm$^3$ @ 20 C; Visc.: ≈ 625 cps @ 20 C; HLB: 16.0; Cloud Pt.: 73 C

(1% in 10% NaCl sol'n.); Clear Pt.: ≈ 14 C; Flash Pt.: ≫ 100 C; Pour Pt.: ≈ 7 C; Stability: Stable to weak acids, bases, oxidizing and reducing agents; Surface Tension: 39 dynes/cm (0.1% sol'n.); pH: 6–7 (1% aq.); Easily biodegradable under normal biological sewage purification conditions
*Composition:* 80% active ingred.
*Toxicity:* Degreasing effect on skin in concentrated form
*Containers:* 200 kg net drums

**Berol 282.** Berol Kemi AB
*Chemical Description:* Alkyl phenol ethylene oxide adduct type
*Category:* Emulsifier, nonionic
*Applications:* Emulsion polymerization; with anionics to obtain a very small particle size; increases stability of the latice
*Form:* Soft paste; Color: < 200 (Hazen)
*Properties:* Solubility: Sol. in water and ethanol; Density: 1.09 g/cm$^3$ @ 30 C; Visc.: 625 cps @ 20 C; HLB: 17; Cloud Pt.: 77 C (1% in 10% NaCl sol'n.); Clear Pt.: ≈ 24 C; Flash Pt.: ≫ 100 C; Pour Pt.: ≈ 21 C; Stability: Stable to weak acids, bases, oxidizing and reducing agents; Surface Tension: 41 dynes/cm (0.1% sol'n.); pH: 6–7 (1% aq.); Easily biodegradable under normal biological sewage purification conditions
*Composition:* 80% active ingred.
*Toxicity:* Degreasing effect on skin in concentrated form

**Berol 291.** Berol Kemi AB
*Chemical Description:* Alkyl phenol ethylene oxide adduct type
*Category:* Emulsifier, nonionic
*Applications:* Emulsion polymerization; with anionics to obtain a very small particle size; increase stability of the latice
*Form:* Hard wax; Color: < 200 (Hazen)
*Properties:* Solubility: Sol. in water, ethanol, xylene, trichloroethylene; Density: ≈ 1.07 g/cm$^3$ @ 60 C; Visc.: ≈ 300 cps @ 60 C; HLB: 18.2; Cloud Pt.: 77 C (1% in 10% NaCl sol'n.); Clear Pt.: ≈ 53 C; Flash Pt.: ≫ 100 C; Pour Pt.: ≈ 51 C; Stability: Stable to weak acids, bases, oxidizing and reducing agents; Surface Tension: 45 cm; pH: 6-7 (1% aq.)
*Composition:* 100% active ingred.
*Toxicity:* Degreasing effect on skin in concentrated form
*Containers:* 100 kg net drums

**Berol 292.** Berol Kemi AB
*Chemical Description:* Alkyl phenol ethylene oxide adduct type
*Category:* Emulsifier nonionic
*Applications:* Emulsion polymerization; with anionics to obtain a very small particle size
*Form:* Soft wax; Color: < 200 (Hazen)
*Properties:* Solubility: Sol. in water, ethanol, xylene, trichloroethylene, disp. in white spirit; Density: 1.07 g/cm$^3$ @ 40 C; Visc.: ≈ 200 cps @ 40 C; HLB: 16.0; Cloud Pt.: 73 C (1% in 10% NaCl sol'n'); Clear Pt.: ≈ 35 C; Flash Pt.: ≫ 100 C; Pour Pt.: ≈ 33 C; Stability: Stable to weak acids, bases, oxidizing and reducing a-

gents; Surface Tension: 39 dynes/cm; pH: 6–7 (1% aq.); Biodegradable
*Composition:* 100% active ingred.
*Toxicity:* Degreasing effect on skin in concentrated form
*Containers:* 100 kg net drums

**Berol 295.** Berol Kemi AB
*Chemical Description:* Alkyl phenol ethylene oxide adduct type
*Category:* Emulsifier, nonionic
*Applications:* Emulsion polymerization; with anionics to obtain a very small particle size; increases stability of the latice
*Form:* Clear liquid; Color: < 300 (Hazen)
*Properties:* Solubility: Sol. in water, ethanol; Density: 1.10 g/cm$^3$ @ 20 C Visc.: 1500 cps @ 20 C; HLB: 17.8; Cloud Pt.: 77 C (1% in 10% NaCl sol'n.); Clear Pt.: ≈ 19 C; Flash Pt.: ≫ 100 C; Pour Pt.: ≈ 14 C; Stability: Stable to weak acids, bases, oxidizing and reducing agents; Surface tension: 43 dynes/cm; pH: 5–7 (1% aq.)
*Composition:* 70% active ingred.
*Toxicity:* Degreasing effect on skin in concentrated form
*Containers:* 200 kg net drums

**Berol 296.** Berol Kemi AB
*Chemical Description:* Alkyl phenol ethylene oxide adduct type
*Category:* Emulsifier, nonionic
*Applications:* Emulsion polymerization; with anionics to obtain a very small particle size; biocides and petroleum solvents
*Form:* Soft paste; Color: < 200 (Hazen)
*Properties:* Solubility: Sol. in water, ethanol, xylene, trichloroethylene; disp. in light fuel oil; Density: 1.07 g/cm$^3$ @ 30 C; Visc.: ≈ 375 cps @ 30 C; HLB: 15.2; Cloud Pt.: 97–99 C (1% aq.); Clear Pt.: ≈ 27 C; Flash Pt.: ≫ 100 C; Pour Pt.: ≈ 26 C; Stability: Stable to weak acids, bases, oxidizing and reducing agents; Surface tension: 37 dynes/cm; pH: 6–7 (1% aq.); Biodegradable
*Composition:* 100% active ingred.
*Toxicity:* Degreasing effect on skin in concentrated form
*Containers:* 200 kg net drums

**Berol 297.** Berol Kemi AB
*Chemical Description:* Alkyl phenol ethylene oxide adduct type
*Category:* Emulsifier, nonionic
*Applications:* Used for various types of waxes, oils and solvents
*Form:* Clear liquid; Color: < 250 (Hazen)
*Properties:* Solubility: Sol. in ethanol, xylene, trichloroethylene, paraffin oil, white spirit, light fuel oil; disp. in water; Density: 1.03 g/cm$^3$ @ 20 C; Visc.: ≈ 350 cps @ 20 C; HLB: 9.5; Cloud Pt.: 48–54 C; Clear Pt.: < 0 C; Flash Pt.: ≫ 100 C; Pour Pt.: < 0 C; Stability: Stable to weak acids, bases, oxidizing and reducing agents; pH: 6–7 (1% aq.); Biodegradable
*Composition:* 100% active ingred.
*Toxicity:* Degreasing effect on skin in concentrated form

**Berol 305.** Berol Kemi AB
*Chemical Description:* Amine oxide

*Category:* Foam controller, nonionic
*Form:* Clear liquid; Color: Pale yellow
*Properties:* Solubility: Sol. in water, ethanol, acetone; Density: 0.96 @ 20 C; Cloud Pt.: 95 C (9% NaOH sol'n.); Clear Pt.: - 1 C; Pour Pt.: 0 C; Stability: Stable to acid and alkaline solutions; Surface Tension: 31 dynes/cm
*Composition:* 28% active ingred.
*Containers:* 200 kg net drums

**Berol 370.** Berol Kemi AB
*Chemical Description:* Ethylene oxide/propylene oxide block polymer
*Category:* Detergent, rinse aid additive, nonionic
*Applications:* Low foaming detergent for use with strong mechanical action, spraying under pressure, e.g., machine dishwashing detergents, alkaline degreasers; rinse aid additive for dishwashers, car cleaners; windscreen washer fluids
*Form:* Clear liquid; Color: $<$ 150 (Hazen)
*Properties:* m.w.: $\approx$ 1400; Solubility: Sol. in ethanol, xylene, trichloroethylene, white spirit; opaquely sol. in water; Density: 1.05 g/cm$^3$ @ 20 C; Visc.: $\approx$ 230 cps @ 20 C; Cloud Pt.: 30–34 C (1% aq.); Clear Pt.: $<$ –20 C; Flash Pt.: $\gg$ 100 C; Pour Pt.: $<$ –10 C; Stability: Stable to weak acids, bases, oxidizing, and reducing agents; Surface Tension: 41 dynes/cm (0.1% sol'n.); pH: 5–7 (1% aq.)
*Composition:* 100% active ingred.
*Toxicity:* Degreasing effect on skin in concentrated form
*Containers:* 200 kg net drums

**Berol 371.** Berol Kemi AB
*Chemical Description:* Ethylene oxide/propylene oxide block polymer
*Category:* Detergent, antistat, emollient, solvent, nonionic
*Applications:* Low foaming detergent for use with strong mechanical action or spraying under pressure; textile industry as water-sol. spinning oil, antistat agent (esp. for synthetic fibers); cosmetics as emollient and solvent for deodorants and antiperspirants
*Form:* Soft paste; Color: $<$ 150 (Hazen)
*Properties:* m.w.: $\approx$ 2900; Solubility: Sol. in water, ethanol, xylene, trichloroethylene; Density: 1.05 g/cm$^3$ @ 20 C; Visc.: $\approx$ 700 cps @ 30 C; Cloud Pt.: 62–66 C (1% aq.); Clear Pt.: $\approx$ 30 C; Flash Pt.: $\gg$ 100 C; Pour Pt.: $\approx$ 28 C; Stability: Stable to weak acids, bases, oxidizing and reducing agents; Surface Tension: 38 dynes/cm (0.1% sol'n.); pH: 5–7 (1% aq.)
*Composition:* 100% active ingred.
*Toxicity:* Degreasing effect on skin in concentrated form
*Containers:* 200 kg net drums

**Berol 372 Flakes.** Berol Kemi AB
*Chemical Description:* Ethylene oxide/propylene oxide block polymer
*Category:* Emulsifier, emollient, nonionic
*Applications:* Emulsifier in emul-

sion polymerization; improves storage and freeze/thaw stabilities in finished latice; emollient for deodorants and antiperspirants; produces material in block form, e.g., perfume, insect repellents, etc.
*Form:* Flakes; Color: White
*Properties:* m.w.: ≈ 9000; Solubility: Sol. in water, ethanol, trichloroethylene; Density: 0.55-0.65 g/cm$^3$; M.P.: ≈ 50 C; Cloud Pt.: ≈ 70 C (1% in 10% NaCl sol'n.); Flash Pt.: ≫ 100 C; Stability: Stable to weak acids, bases, oxidizing and reducing agents; Surface Tension: 42 dynes/cm (0.1% sol'n.); pH: 5-7 (1% aq.)
*Composition:* 100% active ingred.
*Toxicity:* Degreasing effect on skin in concentrated form
*Containers:* 25 kg net sacks

**Berol 374.** Berol Kemi AB
*Chemical Description:* Ethylene oxide/propylene oxide block polymer
*Category:* Emulsifier, foam depressant/detergent, nonionic
*Applications:* Low foaming emulsifier in emulsion polymerization esp. suited to manufacture of paint, etc.; foam depressant/detergent in foodstuffs industry
*Form:* Clear liquid; Color: < 100 (Hazen)
*Properties:* m.w.: ≈ 2200; Solubility: Sol. in ethanol, xylene, trichloroethylene, white spirit; disp. in water; Density: 1.05 g/cm$^3$ @ 20 C; Visc.: ≈ 450 cps @ 20 C; Cloud Pt.: 24-26 C (1% aq.); Clear Pt.: < −20 C; Flash Pt.: ≫ 100 C; Pour Pt.: < −10 C; Stability: Stable to weak acids, bases, oxidizing and reducing agents; Surface Tension: 40 dynes/cm (0.1% sol'n.); pH: 5-7 (1% aq.)
*Composition:* 100% active ingred.
*Toxicity:* Degreasing effect on skin in concentrated form
*Containers:* 200 kg net drums

**Berol 452.** Berol Kemi AB
*Chemical Description:* Alkyl ether sulfate type consisting of the sodium salt of sulfated technical lauryl alcohol polyglycol ether
*Category:* Wetting and foaming agent, emulsifier, anionic
*Applications:* Good foaming surfactant with strong degreasing action; in combination with less strongly degreasing surfactants to produce hair shampoos, foam baths, manual dishwashing compounds, general detergents; emulsion polymerization of vinyl chloride, vinyl acetate, acrylic acids, etc.
*Form:* Clear liquid; Color: < 100 (Hazen)
*Properties:* Solubility: Sol. in water, ethanol, propylene glycol; disp. in acetone; Density: 1.04 g/cm$^3$ @ 20 C; Visc.: ≈ 200 cps @ 20 C; Clear Pt.: < 12 C; Flash Pt.: ≫ 100 C; Stability: Stable to weak bases and very weak acids; Surface Tension: 37 dynes/cm (0.1% sol'n.); pH: 6.5-8.0 (1% aq.); Biodegradable
*Composition:* 27-29% active
*Toxicity:* Degreasing effect on skin in concentrated form

**Berol 463.** Berol Kemi AB

*Chemical Description:* Alkyl sulfate type consisting of the diethanolamine salt of sulfated technical lauryl alcohol
*Category:* Foaming agent, anionic
*Applications:* Hair shampoos, bath foams, hand cleaners
*Form:* Clear liquid; Color: < 100 (Hazen)
*Properties:* Solubility: Sol. in water, ethanol, propylene glycol, acetone; Density: 1.050 g/cm$^3$ @ 20 C; Visc.: ≈ 1000 cps @ 20 C; Clear Pt.: < 15 C; Flash Pt.: ≫ 100 C; Stability: Stable to weak bases and very weak acids; Surface Tension: ≈ 30 dynes/cm (0.1% sol'n.); pH: 6.5–7.5 (1% aq.); Biodegradable
*Composition:* 33–35% active
*Toxicity:* Degreasing effect on skin in concentrated form

**Berol 472.** Berol Kemi AB
*Chemical Description:* Alkyl sulfate type consisting of sodium cetyl sulfate
*Category:* Emulsifier, dispersant, anionic
*Applications:* Emulsion polymerization of vinyl chloride, vinyl acetate, styrene/butadiene, acrylics, etc.; dispersing agent for pigments in manufacture of latex based paints and adhesives
*Form:* Cloudy liquid; Color: < 150 (Hazen)
*Properties:* Solubility: Sol. in propylene glycol; disp. in water, ethanol; Density: 1.055 g/cm$^3$ @ 20 C; Visc.: ≈ 20 cps @ 20 C; Flash Pt.: ≫ 100 C; Stability: Stable to weak bases and very weak acids; Surface Tension: 37 dynes/cm (0.1% sol'n.); pH: 7–8 (1% aq.); Biodegradable
*Composition:* 32 ± 1% active
*Toxicity:* Degreasing effect on skin in concentrated form

**Berol 473.** Berol Kemi AB
*Chemical Description:* Nonylphenol ether sulfate type
*Category:* Detergent, anionic
*Applications:* Detergent with good fatty soil removal, foaming properties for industrial detergents, dishwashing liquids
*Form:* Clear liquid; Color: < 100 (Hazen)
*Properties:* Density: 1.075 g/cm$^3$ @ 20 C; Visc.: 205 cps @ 20 C; Clear Pt.: ≈ 2 C; Flash Pt.: ≫ 100 C; Pour Pt.: ≈ 2 C; Stability: Stable to weak bases and very weak acids; pH: ≈ 3.9; Biodegradable
*Composition:* ≈ 30% active ingred.
*Toxicity:* Degreasing effect on skin in concentrated form

**Berol 474.** Berol Kemi AB
*Chemical Description:* Alkyl sulfate type consisting of the sodium salt of sulfated technical lauryl alcohol
*Category:* Emulsifier, anionic
*Applications:* Emulsion polymerization of vinyl chloride and styrene/butadiene; in combination with other surfactants in polymerization of acrylic acids and vinyl acetate
*Form:* Soft paste; Color: < 50 (Hazen)
*Properties:* Solubility: Sol. in water, ethanol, propylene glycol;

Density: 1.030 g/cm$^3$ @ 20 C; Clear Pt.: ≈ 25 C; Flash Pt.: ≫ 100 C; Stability: Stable to weak bases and very weak acids; Surface Tension: 25 dynes/cm (0.1% sol'n.); pH: 7-8 (1% aq.); Biodegradable
*Composition:* 39-41% active
*Toxicity:* Degreasing effect on skin in concentrated form

**Berol 475.** Berok Kemi AB
*Chemical Description:* Alkyl ether sulfate type consisting of the sodium salt of sulfated *n*-primary alcohol polyglycol ether
*Category:* Degreaser, dispersant, anionic
*Applications:* Cleaner with good pigment removal and balanced degreasing; esp. for washing fur coats, leathers, sensitive woollen materials; spray-dried household detergents; cream hair shampoos, hand cleaners; dispersant for water-based pigment pastes
*Form:* Soft paste; Color: < 300 (Hazen)
*Properties:* Solubility: Sol. in propylene glycol and water (slightly opal sol'n.); disp. in ethanol, trichloroethylene, xylene; Density: 0.97 g/cm$^3$ @ 20 C; Flash Pt.: ≫ 100 C; Stability: Stable to weak bases and very weak acids; Surface Tension: 37 dynes/cm (0.1% sol'n.); pH: 6.5-8.0 (1% aq.); Biodegradable
*Composition:* 39-41% active
*Toxicity:* Degreasing effect on skin in concentrated form

**Berol 480.** Berol Kemi AB
*Chemical Description:* Alkyl sulfate type consisting of the triethanol amine salt of sulfated technical lauryl alcohol
*Category:* Detergent, foaming agent, anionic
*Applications:* Hair shampoos, bath foams, hand cleaners
*Form:* Clear liquid; Color: < 100 (Hazen)
*Properties:* Solubility: Sol. in water, ethanol, propylene glycol; Density: 1.04 g/cm$^3$ @ 20 C; Visc.: 50 cps @ 20 C; Clear Pt.: 12 C; Flash Pt.: ≫ 100 C; Stability: Stable to weak bases and very weak acids; Surface Tension: 26 dynes/cm (0.1% sol'n.); pH: 6.5-7.5 (1% aq.); Biodegradable
*Composition:* 39-41% active
*Toxicity:* Degreasing effect on skin in concentrated form

**Berol 484.** Berol Kemi AB
*Chemical Description:* Alkyl sulfate type consisting of mono ethanol amine salt of sulfated technical lauryl alcohol
*Category:* Foaming, anionic
*Applications:* Hair shampoos, bath foams, hand cleaners
*Form:* Clear liquid; Color: ⩽ 200 (Hazen)
*Properties:* Solubility: Sol. in water, propylene glycol, acetone; disp. in ethanol; Density: 1.020 g/cm$^3$ @ 20 C; Visc.: 150-350 cps @ 20 C; Clear Pt.: ⩽ 15 C; Flash Pt.: ≫ 100 C; Stability: Stable to weak and strong bases, very weak acids, oxidizing and reducing agents; Surface Tension: 33 dynes/cm; pH: 6.5-7.5 (1% aq.); Bio-

degradable
*Composition:* 32–34% active ingred.
*Toxicity:* Degreasing effect on skin in concentrated form

**Berol 490.** Berol Kemi AB
*Chemical Description:* Sodium alkyl aryl sulfonate
*Category:* Wetting agent, foaming and emulsifier, anionic
*Applications:* Because of its good wetting, foaming and emulsifying properties, improved emulsifying power can be achieved when combined with other surfactants
*Form:* Clear liquid; Color: < 300 (Hazen)
*Properties:* Solubility: Sol. in water, ethanol, propylene glycol, acetone; disp. in light mineral oil, white spirit; Density: 1.037 g/cm$^3$ @ 20 C; Visc.: ≈ 250 cps @ 20 C; Clear Pt.: < 18 C; Flash Pt.: ≫ 100 C; Pour Pt.: ≈ 5 C; Stability: Stable to high concentrations of acids and bases; Surface Tension: 35 dynes/cm; pH: 6–8 (1% aq.); Biodegradable
*Composition:* 24–26% active ingred.
*Toxicity:* Skin and eye irritant

**Berol 493.** Berol Kemi AB
*Chemical Description:* Sodium alkyl aryl sulfonate
*Category:* Detergent, foaming, wetting agent, anionic
*Applications:* High foaming and effective washing and cleaning agents, used in textile industry as a general wetting and penetration agent
*Form:* Clear liquid; Color: < 150 (Hazen)
*Properties:* Solubility: Sol. in water, ethanol, propylene glycol; disp. in acetone, trichloroethylene; Density: 1.100 g/cm$^3$ @ 20 C; Visc.: ≈ 5000 cps @ 20 C; Clear Pt.: < 15 C; Flash Pt.: ≫ 100 C; Stability: Stable to strong acids and bases; Surface tension: 35 dynes/cm; pH: 6–8 (1% aq.); Biodegradable
*Composition:* 34–36% active

**Berol 496.** Berol Kemi AB
*Chemical Description:* Linear Sodium alkyl aryl sulfonate
*Category:* Detergent, foaming, wetting, penetrant, emulsifier, anionic
*Applications:* Good cleaning effect, foaming and wetting power; used in textile industry as a general wetting and penetration agent; emulsifier for polymerization
*Form:* Soft paste; Color: 150 (Hazen)
*Properties:* Solubility: Sol. in water, ethanol, propylene glycol, acetone; disp. in trichloroethylene, spindle oil, white spirit; Density: 1.060 g/cm$^3$ @ 20 C; Clear Pt.: ≈ 0 C; Flash Pt.: ≫ 100 C; Stability: Stable to strong acids and bases; Surface tension: 35 dynes/cm; pH: 6–8; Biodegradable
*Composition:* 80% active ingred.
*Toxicity:* Degreasing effect on skin in concentrated form

**Berol 513.** Berol Kemi AB
*Chemical Description:* Alkyl ether phosphate ester and an *n*-primary

alcohol ethylene adduct type
*Category:* Detergent, anionic/nonionic
*Applications:* Good cleaning effect with corrosion-inhibiting qualities;
*Form:* Paste; Color: < 5 Gardner
*Properties:* Solubility: Sol. in ethanol, acetone, xylene, propylene glycol, white spirit; disp. in water, trichloroethylene; Density: 1.050 g/cm$^3$ @ 30 C; Visc.: ≈ 330 cps @ 30 C; Flash Pt.: ≫ 100 C; Stability: Stable at pH 8–12; Surface tension: 34.4 dynes/cm; Biodegradable
*Composition:* ≈ 100% active ingred.
*Warnings and Caution:* Avoid contact with skin

**Berol 516.** Berol Kemi AB
*Chemical Description:* Mixture of an alkyl ether phosphate ester and an *n*-primary alcohol ethylene oxide adduct
*Category:* Detergent, corrosion inhibitor, anionic/nonionic
*Applications:* Detergents and cleaning agents with builders which may cause corrosion to soft metals; washing machines
*Form:* Paste; Color: < 5 (Gardner)
*Properties:* Solubility: Sol. in ethanol, acetone, xylene, propylene glycol, white spirit; disp. in water; Density: 1.050 g/cm$^3$ @ 30 C; Visc.: ≈ 330 cps @ 30 C; Clear Pt.: ≈ 40 C; Flash Pt.: ≫ 100 C; Pour Pt.: ≈ 30 C; Stability: Stable at pH 8–12 in its neutralized form; limited stability in presence of water; Surface Tension: 34.4 dynes/cm (0.1% sol'n.); pH: 2.3 (1% aq.); Biodegradable
*Composition:* 100% active ingred.
*Toxicity:* Acid–avoid contact with skin

**Berol 518.** Berol Kemi AB
*Chemical Description:* Mixture of an alkyl ether phosphate ester and an *n*-primary alcohol ethylene oxide adduct
*Category:* Detergent, foam controller, anionic/nonionic
*Applications:* Machine detergents, spray-dried detergents
*Form:* Soft wax; Color: < 5 (Gardner)
*Properties:* Solubility: Sol. in ethanol, acetone, xylene, trichloroethylene, white spirit, light fuel oil, propylene glycol (slightly opal sol'n); disp. in water; Density: 1.010 g/cm$^3$ @ 50 C; Visc.: 1500 cps @ 35 C, 125 cps @ 50 C; Clear Pt.: ≈ 35 C; Flash Pt.: ≫ 100 C; Pour Pt.: ≈ 35 C; Stability: Stable at pH 8–12 in neutralized form; limited stability in presence of water; Surface Tension: 38 dynes/cm (0.1% sol'n.); pH: 2.3 (1% aq.); Biodegradable
*Composition:* 100% active ingred.
*Toxicity:* Acid–avoid contact with skin

**Berol 521.** Berol Kemi AB
*Chemical Description:* Water sol'n. of potassium alkyl phosphate esters
*Category:* Corrosion inhibitor, solubilizer, anionic
*Applications:* Corrosion inhibitor; solubilizes nonionic surfactants in

presence of high electrolyte conc.; detergents and cleaning agents with builders which can corrode soft metals; liquid and powder products
*Form:* Clear liquid; Color: $<$ 300 (Hazen)
*Properties:* Solubility: Sol. in water, propylene glycol: Density: 1.145 g/cm$^3$ @ 20 C; Visc.: 315 cps @ 20 C; Clear Pt.: 25 C; Flash Pt.: $\gg$ 100 C; Pour Pt.: −10 C; Stability: Stable at pH 8–12; pH: 7–8 (1% aq.); Biodegradable
*Composition:* $\approx$ 40% active
*Toxicity:* Causes burning and eye irritation; avoid skin contact

**Berol 522.** Berol Kemi AB
*Chemical Description:* Water sol'n. of a potassium alkyl phosphate ester
*Category:* Solubilizer, anionic
*Applications:* Solubilizes nonionic surfactants in high electrolyte conc. (esp. for nonyl phenol ethylene oxide condensates); household and commercial liquid laundry detergents; industrial cleaning
*Form:* Clear liquid; Color: $<$ 300 (Hazen)
*Properties:* Solubility: Sol. in water, propylene glycol; Density: $<$ 1.245 g/cm$^3$ @20 C; Visc.: $<$ 125 cps @ 20 C; Clear Pt.: $<$ 0 C; Flash Pt.: $\gg$ 100 C; Pour Pt.: $<$ −10 C; Stability: Stable at pH 8–12; Surface Tension: 31 dynes/cm (0.1% conc.); pH: $<$ 8-9 (1% aq.); Biodegradable
*Composition:* 40–42% active
*Toxicity:* Causes burning and eye irritation; avoid contact with skin

**Berol 525.** Berol Kemi AB
*Chemical Description:* Mixture of an alkyl ether phosphate ester and a straight-chain primary alcohol ethylene oxide adduct
*Category:* Detergent, solubilizer, anionic/nonionic
*Applications:* Solubilizes nonionic surfactants in presence of inorganic salts; liquid alkaline industrial cleaners; household and institutional slightly alkaline detergents; spray-dried detergents
*Form:* Paste; Color: $<$ 150 (Hazen)
*Properties:* Solubility: Sol. in water, ethanol, acetone, propylene glycol (slightly opal), trichloroethylene, xylene, white spirit; Density: 1.060 g/cm$^3$; Visc.: 1500–2000 cps; Clear Pt.: $\approx$ 30 C; Flash Pt.: $\gg$ 100 C; Pour Pt.: $\approx$ 23 C; Stability: Stable at pH 8–12 in neutralized form; limited stability in presence of water; Surface Tension: 33 dynes/cm (0.1% conc.); pH: 2.3 (1% aq.); Biodegradable
*Composition:* 100% active ingred.
*Toxicity:* Acid—avoid skin contact

**Berol 548.** Berol Kemi AB
*Chemical Description:* Amphoteric surfactant
*Category:* Detergent, wetting agent, amphoteric
*Applications:* Industrial detergents
*Form:* Clear liquid; Color: 50 (Hazen)
*Properties:* Solubility: Sol. in 30% caustic sol'ns.; Visc.: 70 cps @ 20 C; Clear Pt.: −5 C; Flash Pt.: $>$ 100 C; Surface Tension: 28 dynes/cm (0.1% conc.); pH: 9.3 (1% aq.)
*Composition:* 35% active ingred.

*Toxicity:* Slightly corrosive; degreasing effect on skin in concentrated form
*Containers:* 200 liter trauser drum

**Berol 563.** Berol Kemi AB
*Chemical Description:* Alkyl polyglycol ether ammonium methyl sulfate
*Category:* Detergent, cationic
*Applications:* Alkaline degreasing and cleaning agents with good solubility and stability in alkaline conditions
*Form:* Clear liquid; Color: ⩽ 7 (Gardner)
*Properties:* Solubility: Sol. in water, ethanol, propylene glycol, xylene; Density: 1.110 g/cm$^3$ @ 20 C; Visc.: 500 cps @ 20 C; Cloud Pt.: ≈ 85 C (1% in 9% NaOH); Clear Pt.: 18 C; Flash Pt.: ⩾ 100 C; Pour Pt.: 10 C; Stability: Stable to weak acids, bases; Surface Tension: 43 dynes/cm (0.1% conc.); pH: 7–9 (1% aq.)
*Composition:* 100% active ingred.
*Toxicity:* Degreasing effect on skin in concentrated form

**Berol 594.** Berol Kemi AB
*Chemical Description:* Hydroxyethyl-2-alkyl imidazolin type
*Category:* Adhesion aid, dispersing agent, wetting and foaming agent, corrosion inhibitor
*Applications:* Strong cationic activity in acid sol'ns. and oils and solvents; temporary corrosion inhibitor; aids adhesion in solvent-based paints and varnishes, emulsions of oils, waxes and asphalt; dispersant for pigments; in acid cleaning agents
*Form:* Clear liquid; Color: ⩽ 12 (Gardner)
*Properties:* m.w.: ≈ 360; Solubility: Sol. in ethanol, xylene, trichloroethylene, white spirit, paraffin oil and dilute sulfuric, hydrochloric, phosphoric and acetic acids; disp. in water; Density: 0.950 g/cm$^3$ @ 20 C; Visc.: 300 cps @ 25 C; Clear Pt.: ⩽ –10 C; Flash Pt.: ⩾ 100 C; Pour Pt.: < 0 C; Stability: Stable to dilute acids; Surface Tension: 32 dynes/cm (0.1% sol'n.); pH: 10–11 (1% aq.)
*Composition:* 100% active ingred.
*Toxicity:* Can burn and irritate skin
*Warnings and Caution:* Protective clothing and goggles should be worn

**Berol 716.** Berol Kemi AB
*Chemical Description:* Alkyl phenol ethylene oxide adduct
*Category:* Detergent, nonionic
*Applications:* Low foaming machine washing detergents in powder form; good pigment and grease removal; spray drying
*Form:* Cloudy liquid; Color: < 500 (Hazen)
*Properties:* Solubility: Sol. in ethanol, xylene, trichloroethylene, paraffin oil (opaque); disp. in water, white spirit, light fuel oil; Density: 1.04 g/cm$^3$ @ 20 C; Visc.: ≈ 800 cps @ 20 C; HLB: 13; Cloud Pt.: 80–85 C (in 10% *n*-propanol); Clear Pt.: ≈ 35 C; Flash Pt.: ⩾ 100 C; Pour Pt.: ≈ 20 C; Stability: Stable to weak acids, bases, oxi-

dizing and reducing agents; pH: 6–7 (1% aq.)
*Composition:* 100% active ingred.
*Toxicity:* Degreasing effect on skin in concentrated form
*Containers:* 200 kg net drums

**Berol 730.** Berol Kemi AB
*Chemical Description:* Polypropylene glycols
*Category:* Antifoaming agent, lubricant, release agent, hydraulic fluid lubricant
*Applications:* Substitute for different oils; dust adhesive fluid in air filters
*Form:* Clear liquid; Color: $< 100$ (Hazen)
*Properties:* m.w.: 1200; Solubility: Sol. in mineral spirit, acetone, xylene, methanol, carbontetrachloride @ 25 C (5% sol'n); sol. in water @ 25 C ($<$ 2% sol'n.); Density: 1.0 g/cm$^3$ @ 25 C; Visc.: 160 cps @ 25 C; Flash Pt.: $>$ 200 C; Pour Pt.: – 40 C
*Composition:* 100% active ingred.
*Containers:* 200 kg net drums

**Berol 733.** Berol Kemi AB
*Chemical Description:* Potassium salt of a phosphated alkyl phenol ethylene oxide adduct
*Category:* Detergent, anionic
*Applications:* Solubilizes nonionic surfactants in presence of inorganic salts; liquid alkaline industrial cleaners; household and institutional slightly alkaline detergents; spray-dried detergents
*Form:* Clear liquid; Color: $< 100$ (Hazen)
*Properties:* Solubility: Sol. in water, 6% sodium hydroxide sol'n., 35% tetra-potassium pyrophosphate sol'n., 50% sodium metasilicate sol'n., propylene glycol; disp. in acetone; Density: 1.050 g/cm$^3$ @ 20 C; Visc.: $\approx$ 400 cps @ 20 C; Clear Pt.: – 5 C; Flash Pt.: $\gg$ 100 C; Pour Pt.: –7 C; Stability: Stable at pH 8–12; Surface Tension: 41 dynes/cm; pH: 9–10 (1% aq.); Biodegradable
*Composition:* 39–41% active
*Toxicity:* Degreasing effect on skin in concentrated form

**Berol 740.** Berol Kemi AB
*Chemical Description:* Polypropylene glycols
*Category:* Antifoaming agent, lubricant, release agent, hydraulic fluid lubricant
*Applications:* Substitute for different oils; dust adhesive fluid in air filters
*Form:* Clear liquid; Color: $< 100$ (Hazen)
*Properties:* m.w.: 1800; Solubility: Sol. in mineral spirit, acetone, xylene, methanol, carbontetrachloride @ 25 C (5% sol'n.); sol. in water @ 25 C ($<$ 0.1% sol'n.); Density: 1.0 g/cm$^3$ @ 25 C; Visc.: 200 cps @ 25 C; Flash Pt.: $>$ 200 C; Pour Pt.: – 35 C
*Composition:* 100% active ingred.
*Containers:* 200 kg net drums

**Berol 752.** Berol Kemi AB
*Chemical Description:* Surfactant
*Category:* Foam stabilizer, anionic
*Applications:* Stabilization of mechanically whipped PVC foam; uniform and even distribution of

the size of the air pores
*Form:* Viscous liquid; Color: < 7 Gardner
*Properties:* Solubility: Disp. in water, propylene glycol, dioctylphthalate, tributylphosphate; Density: 1.070 g/cm$^3$; Visc.: ≈ 10,000 cps @ 25 C; Flash Pt.: ≫ 100 C; Pour Pt.: ≈ 25 C; Stability: Stable to weak acids and alkalies; pH: 6–8 (1% sol'n. in 1:1 water/ethanol)
*Composition:* 95–100% active
*Toxicity:* Degreasing effect on skin in concentrated form

**Berol 755.** Berol Kemi AB
*Chemical Description:* Linear alkyl-aryl-sulfonate and of a *n*-primary alcohol ethylene oxide adduct
*Category:* Detergent, anionic/nonionic
*Applications:* Combination giving excellent detergency and relatively strong and stable foaming
*Form:* Clear liquid; Color: ⩽ 5 Gardner
*Properties:* Solubility: Sol. in water, ethanol, isopropanol, xylene; disp. in white spirit, light fuel oil; Density: 1.050 g/cm$^3$ @ 20 C; Visc.: 150 cps @ 25 C; Clear Pt.: ≈ 0 C; Flash Pt.: ≫ 100 C; Pour Pt.: ≈ -10 C; Stability: Stable to weak acids and bases; Surface Tension: 36 dynes/cm; pH: 6–7; Biodegradable
*Composition:* 60% active ingred.
*Toxicity:* Degreasing effect on skin in concentrated form

**Berol 767.** Berol Kemi AB
*Chemical Description:* Sodium salt of an alkyl ether sulfate and an alkyl phenol ethoxylate
*Category:* Emulsifier, anionic/nonionic
*Applications:* Emulsion polymerization of vinyl acetate and vinyl acetate/acrylate
*Form:* Clear liquid; Color: Max. 100 (Hazen)
*Properties:* Solubility: Sol. in water, ethanol, acetone, propylene glycol; disp. in light fuel oil; Density: 1.060 g/cm$^3$ @ 20 C; Visc.: 230 cps @ 20 C; Clear Pt.: below 0 C; Flash Pt.: ≫ 100 C; Pour Pt.: below 0 C; Stability: Stable to weak bases and very weak acids; Surface tension: 45 dynes/cm; pH: 6–7 (1% aq.)
*Composition:* 34–36% active
*Toxicity:* Degreasing effect on skin in concentrated form

**Berol 777.** Berol Kemi AB
*Chemical Description:* Linear alkyl aryl sulfonate and alkyl phenol ethylene oxide adduct
*Category:* Detergent, anionic/nonionic
*Applications:* Excellent detergency and relatively strong foaming; used in textile industry as a general wetting and cleaning agent; excellent pigment dispersing properties;
*Form:* Clear liquid; Color: ⩽ 250 (Hazen)
*Properties:* Solubility: Sol. in water, propylene glycol; disp. in ethanol, acetone, trichloroethylene, xylene; Density: 1.055 g/cm$^3$ @ 20 C; Visc.: 425 cps @ 20 C; Clear Pt.: ≈ 1 C; Flash Pt.: ≫ 100 C; Pour Pt.: ≈ 0 C; Stability: Stable

to weak acids and bases; Surface tension: ≈ 38 dynes/cm; pH: 6–7 (1% aq.); Biodegradable
*Composition:* 34% active ingred.
*Toxicity:* Degreasing effect on skin in concentrated form

**Berol 779.** Berol Kemi AB
*Category:* Anionic surfactant
*Applications:* Solvent degreasing
*Form:* Clear liquid; Color: < 5 (Gardner)
*Properties:* Solubility: Sol. in water (opal sol'n.), ethanol, xylene, trichloroethylene, white spirit, light fuel oil; Density: 0.98 g/cm$^3$ @ 20 C; Visc.: ≈ 150 cps @ 20 C; Clear Pt.: < –20 C; Flash Pt.: 70 C; Pour Pt.: < –20 C; Stability: Stable to moderate concentrations of weak acids and bases; pH: 5.5–6.5 (1% aq.); Biodegradable
*Composition:* 100% active ingred.
*Toxicity:* Degreasing effect on skin in concentrated form

**Berol 782.** Berol Kemi AB
*Chemical Description:* Blended emulsifier
*Category:* Emulsifer
*Applications:* Dissolve large quantities of grease, oil, asphalt
*Form:* Clear liquid; Color: 12 Gardner
*Properties:* Solubility: Sol. in ethyl alcohol, xylene, trichloroethylene, white spirit, light fuel oil; Density: 0.943 g/cm$^3$; Visc.: 60 cps @ 25 C; Clear Pt.: -30 C; Flash Pt.: 72 C; Pour Pt.: –30 C; pH: 6 ± 0.1; Biodegradable
*Composition:* 100% active ingred.
*Containers:* 195 kg net drums
*Warnings and Caution:* Skin and eye irritant

**Berol 786.** Berol Kemi AB
*Chemical Description:* Oil-in-water emulsion
*Category:* Emulsifier
*Applications:* Corrosion inhibitor
*Form:* Liquid; Color: Yellow-white
*Properties:* Density: 0.994 g/cm$^3$ @ 20 C; Visc.: 100 cps @ 20 C; Flash Pt.: > 100 C; Pour Pt.: 0 C;
*Containers:* 200 kg net drums

**Berol 788.** Berol Kemi AB
*Chemical Description:* Linear alkyl aryl sulfonate and an alkyl phenol ethylene oxide adduct
*Category:* Detergent, anionic/nonionic
*Applications:* Grease and pigment removing properties and relatively strong foaming
*Form:* Clear liquid; Color: 500 (Hazen)
*Properties:* Solubility: Sol. in water, ethanol, propylene glycol; disp. in acetone, trichloroethylene, white spirit; Density: 1.050 g/cm$^3$ @ 20 C; Visc.: 450 cps @ 20 C; Clear Pt.: ≈ 5 C; Flash Pt.: ≫ 100 C; Pour Pt.: ≈ 0 C; Stability: Stable to weak acids and bases; Surface tension: 37 dynes/cm; pH: 6–7; Biodegradable
*Composition:* 30% active ingred.
*Toxicity:* Degreasing effect on skin in concentrated form

**Berol 792.** Berol Kemi AB
*Category:* Emulsifier, nonionic
*Applications:* Emulsifier for oils

and solvents; high penetrating and excellent degreasing and cleaning properties for engine parts, workshop floors, car bodies, etc.
*Form:* Clear liquid; Color: Yellow
*Properties:* Solubility: Sol. in ethanol, paraffin oil, light mineral oil; disp. in water, benzene, trichloroethylene; Density: 1.02 g/cm$^3$ @ 20 C; Pour Pt.: < –10 C
*Containers:* 200 kg net drums

**Berol 808.** Berol Kemi AB
*Category:* Detergent, softener, antistat, anionic/cationic/nonionic
*Applications:* Textile softening and antistatic agent; liquid light duty detergents; post additive to powder formulations
*Form:* Clear liquid; Color: Yellow
*Properties:* Solubility: Sol. in ethanol, propylene glycol, white spirit, light fuel oil; opaquely sol. in water; Density: 1.013 g/cm$^3$; Visc.: 105 cps; Cloud Pt.: > 90 C (1% aq.); Clear Pt.: ≈ 2 C; Flash Pt.: 60 C; Flow Pt.: –4 C; Stability: Stable to weak acids and bases; Surface Tension: 32 dynes/cm (0.1%); pH: 5.5 (1% aq.); Biodegradable under normal sewage treatment conditions
*Composition:* 80% active ingred.
*Toxicity:* Causes burning and eye irritation; degreasing effect on skin in concentrated form

**Berol 809.** Berol Kemi AB
*Chemical Description:* Combined anionic/cationic/nonionic surfactant
*Category:* Softener, antistat, detergent, anionic/cationic/nonionic
*Applications:* Textile softening and antistatic agent; light duty detergents
*Form:* Clear liquid; Color: Light yellow
*Properties:* Solubility: Sol. in water, ethanol, white spirit; Density: 1.04 g/cm$^3$; Visc.: 170 cps; Cloud Pt.: > 55 C (opal, 1% aq.); Clear Pt.: 1 C; Flash Pt.: ≫ 100 C; Pour Pt.: –4 C; Stability: Stable to weak acids and bases; Surface Tension: 31.5 dynes/cm (0.1% sol'n.); pH: 6.3 (1% aq.); Biodegradable under normal sewage treatment conditions
*Composition:* 80% active ingred.
*Toxicity:* Causes burning and eye irritation; degreasing effect on skin in concentrated form

**Berol 822.** Berol Kemi AB
*Chemical Description:* Alkyl aryl sulfonate type
*Category:* Emulsifier, anionic
*Applications:* Emulsifier of biocide (pesticide) concentrates, and some highly aromatic and chlorinated solvents
*Form:* Clear viscous liquid; Color: < 10 (Gardner)
*Properties:* Solubility: Sol. in ethanol, trichloroethylene, paraffin oil, xylene, white spirit, light fuel oil; disp. in water; Density: 1.03 g/cm$^3$ @ 20 C; Visc.: 1200 cps @ 20 C; Flash Pt.: ≈ 35 C; Pour Pt.: ≈ 5 C; Stability: Stable to high concentrations of strong acids and bases; pH: 5–8 (1% aq.); Biodegradable
*Composition:* 60–62% active; solvent: butanol

*Toxicity:* Degreasing effect on skin in concentrated form
*Containers:* 200 kg net drums

**Berol Dinonylphenol.** Berol Kemi AB
*Chemical Description:* Dinonylphenol
*Category:* Chemical intermediate
*Applications:* Reaction products include salts with mono- and polyvalent metal ions, ethers, resins with good oil solubility, polyoxyethylene derivatives (used as emulsifiers, detergents and dispersants); as component in lubricating oils, corrosion inhibitors, stabilizer for rubber, plasticizer for polymers and resins; hydrophobe for nonionic surfactants
*Form:* Viscous liquid; Color: Yellow; Odor: Phenol-like
*Properties:* Ref. Index: 1.4985; Solubility: Sol. in ethanol, acetone, trichloroethylene, butyl glycol, white spirit, mineral oil; Density: 0.911 g/cm$^3$ @ 20 C; Visc.: $\approx$ 400 cps @ 40 C; B.P.: 326–334 C (5%); Flash Pt.: > 500 C; Pour Pt.: $\approx$ 0 C; Hydroxyl No.: 165
*Toxicity:* Severe skin and eye irritant; respiratory irritant on prolonged exposure to vapor
*Containers:* Store in stainless steel or aluminum
*Warnings and Caution:* Protective clothing, goggles and gloves should be worn

**Berol EGA 07.** Berol Kemi AB
*Chemical Description:* Fatty alcohol ethoxylate
*Category:* Dye levelling and dispersing agent, nonionic
*Applications:* Textile dye levelling and dispersing agent; protective colloid; stripping of vat dyes
*Form:* Clear liquid; Color: < 150 (Hazen)
*Properties:* Solubility: Sol. in water, ethanol, xylene, trichloroethylene; disp. in paraffin oil, white spirit, light fuel oil; Density: 1.01 g/cm$^3$ @ 20 C; Visc.: $\approx$ 10 cps @ 20 C; HLB: 15.0; Cloud Pt.: 74–76 C (5% in 10% NaCl sol'n.); Clear Pt.: < 0 C; Flash Pt.: $\gg$ 100 C; Pour Pt.: < 0 C; Stability: Stable to weak acids, bases, oxidizing and reducing agents; Surface Tension: 40 dynes/cm (0.1%); pH: 6–7 (5% aq.); Biodegradable
*Composition:* 20% active ingred.
*Toxicity:* Degreasing effect on skin in concentrated form
*Containers:* 200 kg net drums

**Berol EGA 192.** Berol Kemi AB
*Chemical Description:* Alkyl glyceride ethylene oxide adduct type
*Category:* Equalizing and dispersing agent, nonionic
*Applications:* Textile dye equalizing and dispersing agent
*Form:* Clear liquid; Color: < 3 (Gardner)
*Properties:* Solubility: Sol. in water, ethanol, acetone; disp. in propylene glycol, xylene; Density: 1.080 g/cm$^3$ @ 20 C; Visc.: 120 cps @ 20 C; HLB: $\approx$ 18; Cloud Pt.: 62 C (1% in 10% NaCl sol'n.); Clear Pt.: $\approx$ 14 C; Flash Pt.: $\gg$ 100 C; Pour Pt.: $\approx$ 13 C; Stability: Stable to weak acids and bases; Surface Tension: 42 dynes/cm;

pH: 6–7 (1% aq.); Biodegradable
*Composition:* 85% active ingred.
*Toxicity:* Degreasing effect on skin in concentrated form

**Berol Fintex 10.** Berol Kemi AB
*Chemical Description:* Fatty acid ethylene oxide adduct type
*Category:* Softener, antistat, equalizing agent, emulsifier, nonionic
*Applications:* Textile softener and antistat esp. on cotton and rayon; adhesive bath additive; emulsifier for greases and oils (under certain conditions with foodstuffs)
*Form:* Wax; Color: White
*Properties:* Solubility: Sol. in ethanol, trichloroethylene, xylene; disp. in water, propylene glycol, acetone, white spirit; Density: 1.010 g/cm$^3$ @ 36 C; Visc.: ≈ 80 cps @ 36 C; HLB: ≈ 12; Cloud Pt.: 75–80 C; Clear Pt.: ≈ 38 C; Flash Pt.: ≫ 100 C; Pour Pt.: ≈ 35 C; Stability: Stable to weak acids and bases; Surface Tension: 42 dynes/cm (0.1%); pH: 6–7 (1% aq.); Biodegradable
*Composition:* 100% active ingred.
*Toxicity:* Degreasing effect on skin in concentrated form

**Berol Fintex 41.** Berol Kemi AB
*Category:* Softener, cationic
*Applications:* Textile softener esp. for cotton and rayon; in anticrease treatment with resins; adhesive bath additive
*Form:* Paste; Color: Yellowish
*Properties:* Density: 0.900–1.000 g/cm$^3$; Flash Pt.: ≫ 100 C; Stability: Stable to weak acids and bases; pH: ≈ 9 (1% aq.)
*Composition:* 50% active ingred.
*Toxicity:* Degreasing effect on skin in concentrated form

**Berol Fintex 42.** Berol Kemi AB
*Chemical Description:* Weakly cationic surfactant
*Category:* Softener, antistat, cationic
*Applications:* Textile softener for natural and synthetic fibers, esp. for cotton and rayon; in anticrease treatment with resins; adhesive bath additive
*Form:* Liquid dispersion; Color: Light yellow
*Properties:* Density: 1.000 g/cm$^3$ @ 20 C; Visc.: ≈ 500 cps @ 20 C; Flash Pt.: ≫ 100 C; Pour Pt.: ≈ 0 C; Stability: Stable to weak acids and bases; pH: ≈ 9 (1% aq.)
*Composition:* 25% active ingred.
*Toxicity:* Degreasing effect on skin in concentrated form

**Berol Fintex 573.** Berol Kemi AB
*Chemical Description:* Cationic surfactant consisting of alkyl polyglycol ether ammonium ethyl sulfate and a nonionic wetting agent
*Category:* Antistat, bactericide, cationic
*Applications:* Antistatic agent for synthetic textiles and plastics; bactericide for gram-postive bacteria
*Form:* Clear liquid; Color: ≤ 7 (Gardner)
*Properties:* Solubility: Sol. in water, ethanol, propylene glycol,

acetone, trichloroethylene, xylene, white spirit; Density: 1.050 g/cm$^3$ @ 20 C; Visc.: 750 cps @ 20 C; Cloud Pt.: ≈ 70 C (1% in 10% NaCl); Clear Pt.: 10 C; Flash Pt.: ≫ 100 C; Pour Pt.: ≈ 2 C; Stability: Stable to weak acids and bases; Surface Tension: 35 dynes/cm (0.1% sol'n.); pH: 7.0 ± 0.5 (1% aq.)
*Composition:* 100% active ingred.
*Toxicity:* Degreasing effect on skin in concentrated form

**Berol Fintex 577.** Berol Kemi AB
*Chemical Description:* Quaternary ammonium compound
*Category:* Softener, rewetting agent, antistat, cationic
*Applications:* Softening, antistatic and rewetting agent for natural and synthetic textiles; hair cream rinses; antibacterial properties
*Form:* Clear liquid; Color: < 300 (Hazen)
*Properties:* Solubility: Disp. in water (hot or cold); sol. in ethanol, xylene, propylene glycol, white spirit; Density: 0.94 g/cm$^3$ @ 20 C; Visc.: 100 cps @ 20 C; Clear Pt.: < 15 C; Flash Pt.: 25 C; Pour Pt.: ≈ 5 C; Stability: Stable to weak acids and bases; moderately stable to strong acids and bases; Surface Tension: 32 dynes/cm; pH: 6–7 (1% aq.)
*Composition:* ≈ 70% active
*Toxicity:* Degreasing effect on skin in concentrated form

**Berol Fintex 713.** Berol Kemi AB
*Chemical Description:* Highly concentrated emulsion of polyethylene wax
*Category:* Finishing agent, nonionic
*Applications:* Finishing agent for cotton and rayon to improve tear-, wear- and crease-resistance; softening agent; textile filler
*Form:* Milky-white emulsion
*Properties:* Density: 0.980 g/cm$^3$ @ 20 C; Visc.: ≈ 100 cps @ 20 C; Flash Pt.: ≫ 100 C; Stability: Stable to acids and bases; pH: ≈ 9 (2% emulsion)
*Composition:* 45% active ingred.
*Toxicity:* Degreasing effect on skin in concentrated form

**Berol Fintex 755.** Berol Kemi AB
*Chemical Description:* Mixture of anionic (linear alkyl aryl sulfonate) and nonionic (*n*-primary alcohol) surfactants
*Category:* Wetting agent, dispersing agent, detergent, anionic/nonionic
*Applications:* Textile wetting, cleaning, dispersing agents; levelling agent in dyeing
*Form:* Clear liquid; Color: ≤ 5 (Gardner)
*Properties:* Solubility: Sol. in water, ethanol, isopropanol, trichloroethylene (opal), xylene; Density: 1.050 g/cm$^3$ @ 20 C; Visc.: 150 cps @ 25 C; Clear Pt.: 0 C; Flash Pt.: ≫ 100 C; Pour Pt.: −10 C; Stability: Stable to weak acids and bases; Surface Tension: 36 dynes/cm (0.1% sol'n.); pH: 6–7 (1% aq.); Biodegradable
*Composition:* 70% active ingred.
*Toxicity:* Degreasing effect on skin in concentrated form

**Berol Fintex 776.** Berol Kemi AB
*Chemical Description:* Fatty alcohol ethoxylate
*Category:* Detergent, nonionic
*Applications:* Low foam detergent for textile industry for washing and dyeing/washing in Jet or HP machines
*Form:* Clear liquid; Color: < 100 (Hazen)
*Properties:* Solubility: Sol. in water, ethanol, xylene, trichloroethylene, white spirit, light fuel oil; Density: 1.00 g/cm$^3$ @ 20 C; Visc.: ≈ 50 cps @ 20 C; Cloud Pt.: ≈ 45 C; Clear Pt.: < 5 C; Flash Pt.: > 100 C; Pour Pt.: < 0 C; Stability: Stable to weak acids, bases, oxidizing and reducing agents; Surface Tension: 32 dynes/cm (0.1% sol'n.); pH: 6–7 (1% aq.); Biodegradable
*Composition:* 90% active ingred.
*Toxicity:* Degreasing effect on skin in concentrated form
*Containers:* 200 kg net drums

**Berol Lanco.** Berol Kemi AB
*Chemical Description:* Alkyl phenol ethylene oxide adduct type
*Category:* Detergent, wetting agent, emulsifier, dispersant, nonionic
*Applications:* Textile wetting and cleaning agent for natural and synthetic fibers; esp. for washing of virgin wool; emulsifier for waxes, oils, solvents, and paint resins; emulsion polymerization; dispersant for pigment and tinting pastes in paint industry
*Form:* Clear liquid; Color: < 150 (Hazen)
*Properties:* Solubility: Sol. in water, ethanol, xylene, trichloroethylene; Density: 1.05 g/cm$^3$ @ 20 C; Visc.: ≈ 350 cps @ 20 C; HLB: 13.3; Cloud Pt.: 52–58 C (1% aq.); Clear Pt.: ≈ 7 C; Flash Pt.: ≫ 100 C; Pour Pt.: ≈ 5 C; Stability: Stable to weak acids, bases, oxidizing and reducing agents; Surface Tension: 33 dynes/cm; pH: 6–7 (1% aq.); Biodegradable under normal biological sewage purification conditions
*Composition:* 100% active
*Toxicity:* Degreasing effect on skin in concentrated form
*Containers:* 200 kg net drums

**Berol Nonylphenol.** Berol Kemi AB
*Chemical Description:* Nonylphenol
*Category:* Raw material
*Applications:* Raw material for detergents, dispersants, wetting agents, additives for lubricating oils, corrosion inhibitors, chemicals for the rubber industry, plasticizers, stabilizers for plastics, germicides, insecticides, oil-soluble resins, tanning agents, dyestuffs and pharmaceutical preparations
*Form:* Viscous liquid; Color: Almost colorless; Odor: Faint phenol
*Properties:* m.w.: 220; Ref. Index: 1.513; Solubility: Sol. in ethanol, acetone, trichloroethylene, butyl glycol, white spirit, mineral oil; Density: 0.949 g/cm$^3$ @ 20 C; Visc.: ≈ 2750 cps @ 20 C; Flash Pt.: 145 C; Pour Pt.: < 0 C; Hydroxyl No. 252
*Reaction:* Reacts with ethylene oxide to form polyoxyethylene

adducts; with hexamethylenetetramine, etc. to form polymers; can be sulfonated, halogenated, nitrated, etc. to yield various products
*Toxicity:* Severe irritation and burning to skin, eyes, etc.; respiratory irritation on prolonged exposure to vapor
*Warnings and Caution:* Protective clothing, goggles, gloves should be worn

**Berol WASC.** Berol Kemi AB
*Chemical Description:* Alkyl phenol ethylene oxide adduct
*Category:* Detergent, rewetting agent, dispersant, nonionic
*Applications:* Household and industrial detergents, dishwashing liquids; textile washing agent; rewetting agent for cotton and cellulose; dispersing agent for pigment pastes in paint and varnish industry
*Form:* Clear liquid; Color: $<$ 150 (Hazen)
*Properties:* Solubility: Sol. in water, ethanol; disp. in xylene, trichloroethylene, paraffin oil, white spirit; Density: 1.06 g/cm$^3$ @ 20 C; Visc.: $\approx$ 375 cps @ 20 C; HLB: 14.1; Cloud Pt.: 75–80 C (1% aq.); Clear Pt.: $\approx$ 5 C; Flash Pt.: $\gg$ 100 C; Pour Pt.: $\approx$ 0 C; Stability: Stable to weak acids, bases, oxidizing and reducing agents; Surface Tension: 34 dynes/cm (0.1% sol'n.); pH: 6–7 (1% aq.); Biodegradable under normal biological sewage purification conditions
*Composition:* 95% active ingred.
*Toxicity:* Degreasing effect on skin in concentrated form
*Containers:* 200 kg net drums

**Biopal NR-20.** GAF Corp.
*Chemical Description:* Nonylphenoxy poly (ethyleneoxy) ethanol-iodine complex
*Form:* Viscous free flowing liquid; Color: Dark, reddish brown

**Biopal VRO-20.** GAF Corp.
*Chemical Description:* Nonylphenoxypoly(ethyleneoxy)-ethanol iodine complex providing 20% available iodine
*Category:* Iodophor concentrate
*Applications:* Iodine complexed with a nonionic surfactant for formulation of sanitizing rinses, germicidal rinses and cleaners, sanitizing cleaners, detergent-sanitizers, disinfecting agents
*Form:* Pourable liquid; Color: Dark brown
*Properties:* Solubility: Water-sol.; Sp. gr.: 1.34; Visc.: 1700 cps; Pour Pt.: 10.4 C; Solidification Pt.: 5.2 C; Stability: Excellent (storage)
*Warnings and Caution:* Mixing with concentrated forms of oxidizing or reducing agents may be hazardous

**Bio Soft D-35X.** Stepan Chem.
*Chemical Description:* Sodium linear alkylate sulfonates
*Category:* Detergent with good foaming, wetting, emulsifying properties
*Applications:* Liquid detergents
*Form:* Liquid; Color: Light straw

*Properties:* Solubility: Water-sol.; Density: 8.9 lb/gal; Cloud Pt.: 10 C; Stability: Good; Biodegradable
*Composition:* 34% min. active ingred.
*Reaction:* Excellent compatability with other anionics, nonionics and alkaline builders

**Bio Soft D-40.** Stepan Chem.
*Chemical Description:* Sodium linear alkylate sulfonate
*Category:* Detergent with good foaming, wetting and emulsifying properties
*Applications:* Liquid detergent
*Form:* Clear liquid; Odor: Bland
*Properties:* Solubility: Water-sol.; Density: 8.7 lb/gal; Cloud Pt.: 29 C; Biodegradable
*Composition:* 40% active ingred.
*Reaction:* Compatible with other anionics, nonionics and alkaline builders

**Bio Soft D-60.** Stepan Chem.
*Chemical Description:* Linear sodium alkyl aryl sulfonate
*Category:* Detergent, foaming and wetting agent
*Form:* Slurry; Color: Light
*Properties:* Solubility:Water-sol.; Surface Tension: 37.4 dynes/cm (0.1% conc.); Biodegradable
*Composition:* 50% active ingred.; 21% water

**Bio Soft D-62.** Stepan Chem.
*Chemical Description:* Sodium linear alkylate sulfonates
*Category:* Detergent with good foaming, wetting, emulsifying properties
*Form:* Slurry; Color: Straw yellow; Odor: Bland
*Properties:* Solubility: Water sol.; Density: 9.1 lb/gal; Biodegradable
*Composition:* 60% active ingred.
*Reaction:* Compatible with other anionics, nonionics and alkaline builders

**Bio Soft EA-8.** Stepan Chem.
*Chemical Description:* Linear alkoxypolyoxethylene ethanols
*Category:* Detergent, nonionic
*Applications:* Chemical specialties: heavy duty laundry detergents, agricultural emulsions, sanitizers, metal cleaning and pickling, shampoo ingredient; Textiles: bleaching and dyeing assistant, Kier boiling, scouring of wool and cotton, soaping and washing, wool carbonizing; Petroleum: drilling muds, oil well acidizing; Paper: de-inking, de-pitching, felt washing, re-wetting
*Form:* Hazy white liquid
*Properties:* Solubility: Sol. in various organic solvents; Density: 8.42 lb/gal; Sp. gr.: 1.01; Cloud Pt.: 46 C; Stability: Stable to acid or alkaline hydrolysis, electrolytes, oxidizing or reducing agents, cationics, anionics and hard water; Biodegradable
*Composition:* 100% active ingred.

**Bio Soft EA-10.** Stepan Chem.
*Chemical Description:* Linear alkoxypolyoxethylene ethanols
*Category and Applications:* See Bio Soft EA-8
*Form:* Hazy, white liquid

*Properties:* Solubility: Sol. in various organic solvents, oil, water; Density: 8.42 lb/gal; Sp. gr.: 1.01; Cloud Pt.: 58.5 C; Stability: See Bio Soft EA-8; Biodegradable
*Composition:* 100% active ingred.

**Bio Soft HD-100.** Stepan Chem.
*Chemical Description:* Blend of nonionic and anionic surfactants
*Category:* Intermediate
*Applications:* Intermediate for the formulation of built and unbuilt liquid laundry detergents
*Form:* Clear liquid; Color: Light yellow
*Properties:* Solubility: Water-sol.; Density: 8.9 lb/gal; Visc.: 600 cps; Pour Pt.: 5 C
*Composition:* 100% active ingred.

**Bio Soft LD-32.** Stepan Chem.
*Chemical Description:* Anionic-nonionic blend with sodium cation
*Category:* Detergent
*Applications:* Liquid dishwashing detergent concentrate; bubble baths, car wash detergent, rug shampoos, textile scouring, aircraft cleaners, air entraining agent in concrete
*Form:* Clear liquid; Color: Pale straw
*Properties:* Solubility: Water-sol.; Density: 9.23 lb/gal; Visc.: 350 cps; Cloud Pt.: 5 C; Biodegradable
*Composition:* 60% active ingred.

**Bio Soft LD-68.** Stepan Chem.
*Category:* Detergent
*Applications:* Liquid detergent concentrate for dishwashing, car wash, fine fabrics
*Form:* Clear liquid; Color: Pale straw
*Properties:* Density: 9.0 lb/gal; Visc.: 800 cps; Biodegradable
*Composition:* 60% solids contains no phosphates

**Bio Soft LD-70.** Stepan Chem.
*Chemical Description:* Sodium olefin sulfonate
*Category:* Detergent, foaming agent, nonionic
*Applications:* Shampoos, bubble baths, dishwashing, car washing, fine fabric detergents
*Form:* Clear liquid; Color: Amber
*Properties:* Density: 8.7 lb/gal; Visc.: 300 cps; Cloud Pt.: $>$ 40 F; Biodegradable
*Composition:* 35% active ingred.; contains no phosphates

**Bio Soft LD-80.** Stepan Chem.
*Chemical Description:* Sodium olefin sulfonate
*Category:* Detergent
*Applications:* Dishwashing detergent concentrate
*Form:* Liquid; Color: Amber
*Properties:* Solubility: Water-sol.; Density: 8.8 lb/gal; Visc.: 225 cps; Cloud Pt.: $<$ 5 C; Biodegradable
*Composition:* 41% solids

**Bio Soft LD-95.** Stepan Chem.
*Chemical Description:* Anionic-nonionic blend with sodium cation
*Applications:* Liquid dishwashing detergent concentrate
*Form:* Clear liquid; Color: Pale straw
*Properties:* Solubility: Water-sol.; Density: 9.11 lb/gal; Visc.: 600

cps; Cloud Pt.: 5 C; Biodegradable
*Composition:* 60% active ingred.

**Bio Soft N-21.** Stepan Chem.
*Category:* Detergent
*Applications:* Liquid dishwashing detergent concentrate
*Form:* Clear liquid; Color: Amber
*Properties:* Solubility: Water-sol.; Density: 8.6 lb/gal; Visc.: 220 cps; Biodegradable
*Composition:* 60% active ingred.

**Bio Soft N-300.** Stepan Chem.
*Chemical Description:* Triethanolamine linear alkyl aryl sulfonate
*Category:* Detergent, wetting and foaming agent, anionic
*Applications:* Liquid dishwashing, car washing detergent, shampoos, textile scouring agents, metal cleaners, latex paints
*Form:* Clear liquid; Color: Pale yellow
*Properties:* Solubility: Water-sol.; Density: 9.0 lb/gal; Sp. gr.: 1.08; Visc.: 3200 cps; Surface Tension: 38.8 dynes/cm (0.1% conc.); Biodegradable
*Composition:* 60% active ingred.

**Bio Soft S-100.** Stepan Chem.
*Chemical Description:* Linear alkyl aryl sulfonic acid
*Category:* Intermediate
*Applications:* Intermediate for formulation of built detergents by dry neutralization
*Form:* Dark viscous liquid
*Properties:* Solubility: Water-sol.; Sp. gr.: 1.04; Biodegradable
*Composition:* 97% active ingred.; 0.5% sulfuric acid; 1.8% unsulfonated oil

**Bio Terge AS-40.** Stepan Chem.
*Chemical Description:* 40% aqueous sol'n. of sodium alpha olefin sulfonate
*Category:* Detergent, foaming agent
*Applications:* Mild liquid dishwash, car wash; commercial and industrial formulations
*Form:* Liquid; Color: Yellow
*Properties:* Solubility: Water-sol.; Density: 8.9 lb/gal; Biodegradable
*Composition:* 40% active ingred.; contains no phosphates

**Bio Terge AS-90F.** Stepan Chem.
*Chemical Description:* Sodium alpha olefin
*Category:* Detergent
*Applications:* Powdered detergents
*Form:* Cream flake
*Properties:* Solubility: Water-sol.; Biodegradable
*Composition:* 90% active ingred.

**Blancol.** GAF Corp.
*Chemical Description:* Sodium naphthalenesulfonate–formaldehyde condensate
*Category:* Thinning, dispersing, dye-leveling agents; anionic
*Applications:* Paper industry for slime control, reduction in two-sidedness, improved retention of fillers or fines, improved sizing, lowered viscosity of coatings
*Form:* Free-flowing coarse granules; Color: Tan to brown; Odor: None
*Properties:* Solubility: Sol. in warm or cold water; Density:

0.77-0.82 g/ml; Stability: Good
*Composition:* 88-90% active
*Containers:* C/L bags; 10 lb pkgs

**Blancol N.** GAF Corp.
*Chemical Description:* Sodium naphthalenesulfonate - formaldehyde condensate
*Category and Applications:* See Blancol
*Form:* Free-flowing granular powder; Color: Tan to brown; Odor: None
*Properties:* Solubility: Sol. in warm or cold water; Density: 0.65-0.75 g/ml; Stability: Good
*Composition:* 88-90% active
*Containers:* C/L bags, 10 lb pkgs

**Brij 30.** I.C.I. United States
*Chemical Description:* Polyoxyethylene (4) lauryl ether
*Category:* Nonionic surfactant
*Form:* Liquid; Color: Colorless to light yellow (may become hazy or form a precipitate)
*Properties:* Solubility: Sol. in alcohol, propylene glycol, cottonseed oil; Sp. gr.: ≈ 0.95; Visc.: ≈ 30 cps; HLB: 9.7; Flash Pt.: > 300 F; Fire Pt.: > 300 F

**Brij 30 SP.** I.C.I. United States
*Chemical Description:* Polyoxyethylene (4) lauryl ether with 0.01% BHA and 0.005% citric acid added as preservatives
*Category:* Nonionic surfactant
*Form:* Liquid; Color: Colorless to light yellow
*Properties:* Solubility: Sol. in alcohol, propylene glycol, cottonseed oil; Sp. gr.: ≈ 0.95; Visc.: ≈ 30 cps; HLB: 9.5; Flash Pt.: > 300 F; Fire Pt.: > 300 F

**Brij 35.** I.C.I. United States
*Chemical Description:* Polyoxyethylene (23) lauryl ether
*Category:* Nonionic surfactant
*Form:* Waxy solid; Color: White
*Properties:* Solubility: Sol. in water, alcohol, propylene glycol; Sp. gr.: ≈ 1.05; HLB: 16.9; Flash Pt.: > 300 F; Fire Pt.: > 300 F; Pour Pt.: ≈ 33 C

**Brij 35 SP.** I.C.I. United States
*Chemical Description:* Polyoxyethylene (23) lauryl ether with 0.01% BHA and 0.005% citric acid added as preservatives
*Category:* Nonionic surfactant
*Form:* Waxy solid; Color: White
*Properties:* Solubility: Sol. in water, alcohol, propylene glycol; Sp. gr.: ≈ 1.05; HLB: 16.9; Flash Pt.: > 300 F; Fire Pt.: > 300 F; Pour Pt.: ≈ 33 C

**Brij 52.** I.C.I. United States
*Chemical Description:* Polyoxyethylene (2) cetyl ether (antioxidants added)
*Category:* Nonionic surfactant
*Form:* Waxy solid; Color: White
*Properties:* Solubility: Sol. in alcohol, cottonseed oil, mineral oil; HLB: 5.3; Flash Pt.: > 300 F; Fire Pt.: > 300 F; Pour Pt.: ≈ 33 C

**Brij 56.** I.C.I. United States
*Chemical Description:* Polyoxyethylene (10) cetyl ether (antioxidants added)
*Category:* Nonionic surfactant

*Form:* Waxy solid; Color: White
*Properties:* Solubility: Sol. in alcohol; HLB: 12.9; Flash Pt.: > 300 F; Fire Pt.: > 300 F; Pour Pt.: ≈ 31 C

**Brij 58.** I.C.I. United States
*Chemical Description:* Polyoxyethylene (20) cetyl ether with 0.01% BHA and 0.005% citric acid added as preservatives
*Category:* Nonionic surfactant
*Form:* Waxy solid; Color: White
*Properties:* Solubility: Sol. in water and alcohol; HLB: 15.7; Flash Pt.: > 300 F; Fire Pt.: > 300 F; Pour Pt.: ≈ 38 C
*Containers:* 55 gal open head steel drum

**Brij 72.** I.C.I. United States
*Chemical Description:* Polyoxyethylene (2) stearyl ether with 0.01% BHA and 0.005% citric acid added as preservatives
*Category:* Nonionic surfactant
*Form:* Waxy solid; Color: White
*Properties:* Solubility: Sol. in alcohol, cottonseed oil; HLB: 4.9; Flash Pt.: > 300 F; Fire Pt.: > 300 F; Pour Pt.: ≈ 43 C
*Containers:* 55 gal open head steel drum

**Brij 76.** I.C.I. United States
*Chemical Description:* Polyoxyethylene (10) stearyl ether with 0.01% BHA and 0.005% citric acid added as preservatives
*Category:* Nonionic surfactant
*Form:* Waxy solid; Color: White
*Properties:* Solubility: Sol. in ethanol, propylene glycol; HLB: 12.4; Flash Pt.: > 300 F; Fire Pt.: > 300 F; Pour Pt.: ≈ 38 C

**Brij 78.** I.C.I. United States
*Chemical Description:* Polyoxyethylene (20) stearyl ether with 0.01% BHA and 0.005% citric acid added as preservatives
*Category:* Nonionic surfactant
*Form:* Waxy solid; Color: White
*Properties:* Solubility: Sol. in alcohol; HLB: 15.3; Flash Pt.: > 300 F; Fire Pt.: > 300 F; Pour Pt.: ≈ 38 C

**Brij 92.** I.C.I. United States
*Chemical Description:* Polyoxyethylene (2) oleyl ether with 0.01% BHA and 0.005% citric acid added as preservatives
*Category:* Nonionic surfactant
*Form:* Liquid; Color: Yellow
*Properties:* Solubility: Sol. in alcohol, propylene glycol, cottonseed and mineral oil; Visc.: ≈ 30 cps; HLB: 4.9; Flash Pt.: > 300 F; Fire Pt.: > 300 F; Pour Pt.: ≈ 10 C

**Brij 93.** I.C.I. United States
*Chemical Description:* Polyoxyethylene (2) oleyl ether with 0.01% BHA and 0.005% citric acid added as preservatives
*Category:* Nonionic surfactant
*Form:* Liquid; Color: Pale yellow
*Properties:* Solubility: Sol. in alcohol, propylene glycol, cottonseed and mineral oils; Visc.: ≈ 30 cps; HLB: 4.9; Flash Pt.: > 300 F; Fire Pt.: > 300 F; Pour Pt.: ≈ 10 C

**Brij 96.** I.C.I. United States
*Chemical Description:* Polyoxyethylene (10) oleyl ether (antioxidants added)
*Category:* Nonionic surfactant
*Form:* Yellow liquid with some solids
*Properties:* Solubility: Sol. in water and alcohol; Visc.: ≈ 100 cps; HLB: 12.4; Flash Pt.: > 300 F; Fire Pt.: > 300 F; Pour Pt.: ≈ 16 C

**Brij 97.** I.C.I. United States
*Chemical Description:* Polyoxyethylene (10) oleyl ether with 0.01% BHA and 0.005% citric acid added as preservatives
*Category:* Nonionic surfactant
*Form:* Liquid; Color: Pale yellow
*Properties:* Solubility: Sol. in water and alcohol; Visc.: ≈ 100 cps; HLB: 12.4; Flash Pt.: > 300 F; Fire Pt.: > 300 F; Pour Pt.: ≈ 16 C
*Containers:* 55 gal closed head steel drum

**Brij 98.** I.C.I. United States
*Chemical Description:* Polyoxyethylene (20) oleyl ether (antioxidants added)
*Category:* Nonionic surfactant
*Form:* Soft waxy solid; Color: Cream
*Properties:* Solubility: Sol. in water, alcohol, propylene glycol; HLB: 15.3; Flash Pt.: > 300 F; Fire Pt.: > 300 F; Pour Pt.: ≈ 30 C

**Brij 99.** I.C.I. United States
*Chemical Description:* Polyoxyethylene (20) oleyl ether (antioxidants added)
*Category:* Nonionic surfactant
*Form:* Solid; Color: Cream
*Properties:* Solubility: Sol. in water, alcohol, propylene glycol; HLB: 15.3; Flash Pt.: > 300 F; Fire Pt.: > 300 F; Pour Pt.: ≈ 33 C

**BTC 50, 50 USP.** Onyx Chem. Co.
*Chemical Description:* *N*-alkyl dimethyl benzyl ammonium chlorides
*Category:* Disinfectant, cationic
*Applications:* Hard surface disinfection, sanitization and deodorization
*Form:* Liquid
*Properties:* Sp. gr.: 0.96 (25/20 C); Flash Pt.: 126 F
*Composition:* 50% min. quaternary

**BTC 65, 65 USP.** Onyx Chem. Co.
*Chemical Description:* *N*-alkyl dimethyl benzyl ammonium chlorides
*Category:* Disinfectant, cationic
*Applications:* Hard surface disinfection, sanitization and deodorization
*Form:* Liquid
*Properties:* Sp. gr.: 0.97 (25/20 C); Flash Pt.: 132 F
*Composition:* 50% min. quaternary

**BTC 100.** Onyx Chem. Co.
*Chemical Description:* *N*-alkyl dimethyl benzyl ammonium chlorides
*Category:* Disinfectant, cationic

*Applications:* Hard surface disinfection, sanitization and deodorization
*Form:* Solid
*Properties:* Sp. gr.: 0.95 (25/20 C); Flash Pt.: > 200 F
*Composition:* 95% min. quaternary

**BTC 776.** Onyx Chem. Co.
*Chemical Description:* *N*-alkyl dimethyl benzyl and dialkyl methyl benzyl ammonium chlorides
*Category:* Disinfectant, cationic
*Applications:* Algaecidal and slimicidal agents for swimming pool and industrial water treatment
*Form:* Liquid
*Properties:* Sp. gr.: 0.96 (25/20 C); Flash Pt.: 102 F
*Composition:* 50% min. quaternary

**BTC 812.** Onyx Chem. Co.
*Chemical Description:* *N*-octyl dodecyl dimethyl ammonium chlorides
*Category:* Disinfectant, cationic
*Applications:* Hard surface disinfection and sanitization fungicide
*Form:* Liquid
*Properties:* Sp. gr.: 0.93 (25/20 C); Flash Pt.: 126 F
*Composition:* 50% min. quaternary

**BTC 824.** Onyx Chem. Co.
*Chemical Description:* *N*-alkyl dimethyl benzyl ammonium chlorides
*Category:* Disinfectant, cationic
*Applications:* Hard surface disinfection and sanitization, algaecide in swimming pool and industrial water treatment
*Form:* Liquid
*Properties:* Sp. gr.: 0.96 (25/20 C); Flash Pt.: 120 F
*Composition:* 50% min. quaternary

**BTC 824 P-100.** Onyx Chem. Co.
*Chemical Description:* *N*-tetradecyl dimethyl benzyl ammonium chloride monohydrate
*Category:* Disinfectant, cationic
*Applications:* Nondusting water soluble powder for manufacturing disinfectants, sanitizers, and deodorizers
*Form:* Powder
*Properties:* Sp. gr.: 0.40 (25/20 C); Flash Pt.: > 200 F
*Composition:* 95% min. quaternary

**BTC 835.** Onyx Chem. Co.
*Chemical Description:* *N*-alkyl dimethyl benzyl ammonium chlorides
*Category:* Disinfectant, cationic
*Applications:* Hard surface disinfection and sanitization, algaecide in swimming pool and industrial water treatment
*Form:* Liquid
*Properties:* Sp. gr.: 0.97 (25/20 C); Flash Pt.: 130 F
*Composition:* 50% min. quaternary

**BTC 1100.** Onyx Chem. Co.
*Chemical Description:* *N*-alkyl dimethyl 1-naphthylmethyl ammonium chlorides

*Category:* Disinfectant, cationic
*Applications:* High active nondusting powder used as a disinfectant, sanitizer, algaecide, and deodorizer
*Form:* Powder
*Properties:* Sp. gr.: 0.50 (25/20 C); Flash Pt.: > 200 F
*Composition:* 99% min. quaternary

**BTC 2125.** Onyx Chem. Co.
*Chemical Description:* *N*-alkyl dimethyl benzyl and dimethyl ethylbenzyl ammonium chlorides
*Category:* Disinfectant, cationic
*Applications:* Hard surface disinfection and sanitization, algaecide in swimming pool and industrial water treatment
*Form:* Liquid
*Properties:* Sp. gr.: 0.97 (25/20 C); Flash Pt.: > 200 F
*Composition:* 50% min. quaternary

**BTC 2125 80%.** Onyx Chem. Co.
*Chemical Description:* *N*-alkyl dimethyl benzyl and dimethyl ethylbenzyl ammonium chlorides
*Category:* Disinfectant, cationic
*Applications:* Hard surface disinfection and sanitization, algaecide in swimming pool and industrial water treatment
*Form:* Liquid
*Properties:* Sp. gr.: 0.94 (25/20 C); Flash Pt.: 130 F
*Composition:* 80% min. quaternary

**BTC 2125M.** Onyx Chem. Co.
*Chemical Description:* *N*-alkyl dimethyl benzyl and dimethyl ethylbenzyl ammonium chlorides
*Category:* Disinfectant, cationic
*Applications:* Used in disinfectant and sanitizer formulations
*Form:* Liquid
*Properties:* Sp. gr.: 0.97 (25/20 C); Flash Pt.: > 200 F
*Composition:* 50% min. quaternary

**BTC 2125M 80%.** Onyx Chem. Co.
*Chemical Description:* *N*-alkyl dimethyl benzyl and dimethyl ethylbenzyl ammonium chlorides
*Category:* Disinfectant, cationic
*Applications:* Used in disinfectant and sanitizer formulations
*Form:* Liquid
*Properties:* Sp. gr.: 0.94 (25/20 C); Flash Pt.: 130 F
*Composition:* 80% min. quaternary

**BTC 2125 P-40.** Onyx Chem. Co.
*Chemical Description:* *N*-alkyl dimethyl benzyl and dimethyl ethylbenzyl ammonium chlorides
*Category:* Disinfectant, cationic
*Applications:* Hard surface disinfection and sanitization, algaecide in swimming pool and industrial water treatment
*Form:* Powder
*Properties:* Sp. gr.: 0.55 (25/20 C); Flash Pt.: > 200 F
*Composition:* 40% min. quaternary

**BTC 2565.** Onyx Chem. Co.
*Chemical Description:* *N*-alkyl dimethyl benzyl ammonium chlo-

rides
*Category:* Disinfectant, cationic
*Applications:* Algaecidal and slimicidal agents for swimming pool and industrial water treatment
*Form:* Liquid
*Properties:* Sp. gr.: 0.96 (25/20 C); Flash Pt.: 110 F
*Composition:* 50% min quaternary

**BTC 8248.** Onyx Chem. Co.
*Chemical Description:* *N*-alkyl dimethyl benzyl ammonium chlorides
*Category:* Disinfectant, cationic
*Applications:* Hard surface disinfection and sanitization, algaecide in swimming pool and industrial water treatment
*Form:* Liquid
*Properties:* Sp. gr.: 0.95 (25/20 C); Flash Pt.: 110 F
*Composition:* 80% min. quaternary

**BTC 8249.** Onyx Chem. Co.
*Chemical Description:* *N*-alkyl dimethyl benzyl ammonium chlorides
*Category:* Disinfectant, cationic
*Applications:* Hard surface disinfection and sanitization, algaecide in swimming pool and industrial water treatment
*Form:* Liquid
*Properties:* Sp. gr.: 0.94 (25/20 C); Flash Pt.: 118 F
*Composition:* 90% min. quaternary

**BTC E-8358.** Onyx Chem. Co.
*Chemical Description:* *N*-alkyl dimethyl benzyl ammonium chlorides
*Category:* Disinfectant, cationic
*Applications:* Hard surface disinfection and sanitization, algaecide in swimming pool and industrial water treatment
*Form:* Liquid
*Properties:* Sp. gr.: 0.93 (25/20 C); Flash Pt.: 100 F
*Composition:* 80% min. quaternary

**BTCO 1010.** Onyx Chem. Co.
*Chemical Description:* *N*-didecyl dimethyl ammonium chloride
*Category:* Disinfectant, cationic
*Applications:* Hard surface disinfection and sanitization fungicide
*Form:* Liquid
*Properties:* Sp. gr.: 0.89 (25/20 C); Flash Pt.: 86 F
*Composition:* 50% min. quaternary

**Bubble Breaker 259.** Witco Chem.
*Chemical Description:* Blend of organic, nonsilicone compounds
*Category:* Defoamer
*Applications:* Manufacture of water-based paper coatings, textile-processing formulations, paints, adhesives, and inks

**Bubble Breaker 260.** Witco Chem.
*Chemical Description:* Blend of organic, nonsilicone compounds
*Category:* Defoamer
*Applications:* Latex manufacturing operations including polymerization, transfer and monomer stripping; formulation of agricultural chemical products

**Bubble Breaker 613-M.** Witco Chem.
*Chemical Description:* Blend of organic, nonsilicone compounds
*Category:* Defoamer
*Applications:* Processing of beet sugar; formulation of agricultural chemical products

**Bubble Breaker 622.** Witco Chem.
*Chemical Description:* Blend of organic, nonsilicone compounds
*Category and Applications:* See Bubble Breaker 613-M

**Bubble Breaker 730.** Witco Chem.
*Chemical Description:* Blend of organic, nonsilicone compounds
*Category and Applications:* See Bubble Breaker 259

**Bubble Breaker 737.** Witco Chem.
*Chemical Description:* Blend of organic, nonsilicone compounds
*Category and Applications:* See Bubble Breaker 259

**Bubble Breaker 746.** Witco Chem.
*Chemical Description:* Blend of organic, nonsilicone compounds
*Category:* Defoamer
*Applications:* Manufacture of water-based paper coatings, textile-processing formulations, paints, adhesives, inks; formulation of agricultural chemical products

**Bubble Breaker 748.** Witco Chem.
*Chemical Description:* Blend of organic, nonsilicone compounds
*Category and Applications:* See Bubble Breaker 746

**Bubble Breaker 776.** Witco Chem.
*Chemical Description:* Dispersion of reacted silica in hydrocarbon solvent
*Category:* Defoamer
*Applications:* Latex manufacturing operations including polymerization, transfer, monomer stripping and effluent treatment; formulation of water-based paints and adhesives

**Bubble Breaker 900.** Witco Chem.
*Chemical Description:* Blend of organic, nonsilicone compounds
*Category and Applications:* See Bubble Breaker 746

**Bubble Breaker 913.** Witco Chem.
*Chemical Description:* Blend of organic, nonsilicone compounds
*Category and Applications:* See Bubble Breaker 259

**Bubble Breaker 917.** Witco Chem.
*Chemical Description:* Blend of organic, nonsilicone compounds
*Category:* Defoamer
*Applications:* Formulation of semi- and high-gloss latex paints, agricultural chemical products

**Bubble Breaker 1840X.** Witco Chem.
*Chemical Description:* Sodium salt of sulfonated fatty acid
*Category:* Defoamer
*Applications:* Defoamer for "wet-process" production of phosphoric acid

**Bubble Breaker 3009-F.** Witco Chem.
*Chemical Description:* Blend of organic, nonsilicone compounds
*Category:* Defoamer
*Applications:* Defoamer for wash water of sliced potatoes

**Bubble Breaker 3017-A.** Witco Chem.
*Chemical Description:* Dispersion of reacted silica in hydrocarbon solvent
*Category and Applications:* See Bubble Breaker 776

**Bubble Breaker 3056-A.** Witco Chem.
*Chemical Description:* Dispersion of reacted silica in hydrocarbon solvent
*Category:* Defoamer
*Applications:* Latex manufacturing operations including polymerization, transfer, monomer stripping and effluent treatment; formulation of water-based paints and adhesives, water-based drilling fluids for the petroleum industry

**Bubble Breaker 3073-7.** Witco Chem.
*Chemical Description:* Blend of organic, nonsilicone compounds
*Category and Applications:* See Bubble Breaker D

**Bubble Breaker D.** Witco Chem.
*Chemical Description:* Blend of organic, nonsilicone compounds
*Category:* Defoamer
*Applications:* For water-based drilling fluids for the petroleum industry

**Bubble Breaker PR.** Witco Chem.
*Chemical Description:* Sodium salt of sulfonated fatty acid
*Category and Applications:* See Bubble Breaker 1840X

# C

**Calcium Petronate.** Witco Chem. Sonneborn Div.
*Category:* Detergent
*Applications:* Detergent and rust inhibitor component in lube oil additives, in rust preventive formulations and emulsifier for w/o systems
*Form:* Viscous liquid; Color: Dark brown
*Properties:* Solubility: Sol. in organic solvents; Sp. gr.: 0.98
*Composition:* 44% active ingred; 55% mineral oil; 1.0% water
*Reaction:* Acid
*Toxicity:* May cause eye burn
*Containers:* Drums, tank wagons, tank cars

**Calsolene Oil HSA.** I.C.I. United States
*Chemical Description:* Sulfated fatty acid ester
*Category:* Anionic surfactant
*Form:* Clear liquid (may be turbid at room temp.); Color: Gold yellow
*Properties:* Solubility: Dissolves in hard and soft water and in salt solutions; Flash Pt.: 200 F (open cup)
*Toxicity:* Causes eye burns and skin irritation
*Warnings and Caution:* Avoid contact; use with forced ventilation

**Carbonox.** NL Baroid/NL Industries
*Category:* Thinner, dispersant
*Applications:* Thinner and dispersant used to control viscosity and gel strengths of drilling fluids, to aid in filtration control, to emulsify oil into water in emulsion muds
*Containers:* Multiwall paper bags containing 50 lb

**Cardene.** Cincinnati Milacron
*Category:* Wetting agent
*Applications:* Secondary surfactant improving o/w emulsions
*Form:* Liquid; Color: Light amber
*Properties:* Solubility: Oil sol.; Sp. gr.: 0.979 @ 60 F; Visc.: 2200 SUS @ 100 F
*Composition:* 100% active ingred.
*Containers:* 55 gal drums

**Carnapol 77X.** Reilly-Whiteman Inc.
*Chemical Description:* Fatty dia-

mide
*Category:* Release agent, nonionic
*Applications:* High melting point wax for use as release agents and wax finishes
*Form:* Solid
*Composition:* 100% active

**Carsamide 7644.** Carson Chem.
*Chemical Description:* Modified coconut diethanolamide
*Category:* Emulsifier, thickening agent; nonionic/anionic
*Applications:* Controlled suds detergent, textile fulling and scouring agent
*Form:* Liquid; Color: 6 (Gardner)
*Properties:* Solubility: Sol. in water, alcohol, chlorinated and/or aromatic hydrocarbons; dispersible in mineral oil, kerosene, natural fats and oils, mineral spirits; Density: 8.3 lb/gal

**Carsamide C-3.** Carson Chem.
*Category:* Intermediate, emulsifier
*Applications:* Intermediate for liquid detergent formulation; car and hair shampoos; improves the foaming, detergency and hard water resistance of soaps; solvent cleaners and dry cleaning formulations; emulsifier for aromatic solvents
*Containers:* 55 gal steel drums

**Carsamide C-7944.** Carson Chem.
*Form:* Amber liquid free of foreign matter

**Carsamide CA.** Carson Chem.
*Chemical Description:* High activity coco diethanolamide
*Category:* Detergent, emulsifier, dispersant, wetting agent, nonionic
*Applications:* Industrial, institutional, cosmetic and household cleaners
*Properties:* Solubility: Sol. in water, alcohol, chlorinated hydrocarbons, aromatic hydrocarbons, "Polysolve," "Cellosolves," "Carbitols" and natural fats and oils; Density: 8.3 lb/gal; Sp. gr.: 1.0; Biodegradable
*Composition:* 100% active ingred.

**Carsamide LE.** Carson Chem.
*Chemical Description:* Super coco fatty diethanolamide
*Form:* Liquid; Color: Light amber

**Carsamide SAC.** Carson Chem.
*Chemical Description:* Coco super diethanolamide
*Category:* Nonionic detergent, emulsifier, dispersant, wetting agent, viscosity builder
*Applications:* Detergent for industrial, institutional, cosmetic and household cleaners; foam stabilizer for shampoos, bubble baths, detergents
*Form:* Viscous liquid; Color: Pale amber; Odor: Mild, fatty
*Properties:* Solubility: Sol. in water, alcohol, chlorinated and aromatic hydrocarbons, "Polysolvs," "Cellosolves," "Carbitols," natural fats and oils; Density: 8.2 lb/gal
*Composition:* 100% active ingred.

**Carsamide SAL-7.** Carson Chem.
*Chemical Description:* High activity lauric diethanolamide
*Category:* Nonionic detergent,

emulsifier, foaming agent and foam stabilizer
*Applications:* Foam stabilizer, thickener, detergent and foaming agent in household, institutional and industrial cleaners; car washes, rug and upholstery cleaners, shampoos, bubble baths, lotions, creams
*Form:* Solid; Color: 2 (Gardner); Odor: Mild, fatty
*Properties:* Solubility: Sol. in alcohol, chlorinated and aromatic hydrocarbons; Density: 8.2 lb/gal; M.P.: 104–114 F; Biodegradable
*Containers:* 55 gal open top steel drums

**Carsamide SAL-9.** Carson Chem.
*Chemical Description:* High activity lauric diethanolamine
*Category:* Detergent, emulsifier foaming agent, nonionic
*Applications:* Foam stabilizer thickener, detergent and foaming agent in household, institutional and industrial cleaners, car washes rug and upholstery cleaners, shampoos, bubble baths, lotions, creams
*Form:* Solid; Color: 2 (Gardner); Odor: Mild, fatty
*Properties:* Solubility: Sol. in alcohol, chlorinated and aromatic hydrocarbons; Density: 8.1 lb/gal; Biodegradable
*Containers:* 55 gal open top steel drums

**Carsamide SAL-82.** Carson Chem.
*Form:* Clear yellow liquid free from foreign matter
*Properties:* Visc.: 500 cps max.

**Carsamine 2414.** Carson Chem.
*Chemical Description:* A mixture of aliphatic and heterocyclic mono and diamines
*Properties:* Solubility: Complete in water; Density: 9.0 lb/gal; Boiling range: 242–310 C; decomposes at 310 C

**Carsamine CB.** Carson Chem.
*Form:* Clear liquid; Color: Amber
*Properties:* Density: 7.6 lb/gal

**Carsamine Oxide LM.** Carson Chem.
*Chemical Description:* Lauryl/myristyl-dimethyl amine oxide
*Category:* Foaming and conditioning agent
*Applications:* Foaming and conditioning agent in household detergents, industrial detergents, shampoos, bubble baths; foam stabilizer; wetting agent in concentrated electrolyte sol'ns.; mildly cationic under acidic conditions
*Form:* Clear liquid; Color: 2 (Gardner)
*Composition:* 29% active ingred.
*Reaction:* Compatible with anionic and nonionic detergents

**Carsemol A-500.** Carson Chem.
*Chemical Description:* Ethyl palmitate, ethyl myristate, ethyl stearate and ethyl pelargonate
*Applications:* Emollient creams and lotions, moisturizers, toners, eye make-up, hair products, lipsticks, nail enamels, bath products
*Properties:* Sp. gr.: 0.865; Cloud Pt.: 11 C

**Carsodet TD-7C.** Carson Chem.
*Chemical Description:* Trideceth-7

carboxylic acid supplied as free carboxylic acid
*Applications:* Neutralized with other materials to produce mild anionic surfactants; topical cleansing products; shampoos
*Form:* Colorless liquid
*Properties:* Density 8.5 lb/gal; pH: 1 (as is)

**Carsofoam 211.** Carson Chem.
*Chemical Description:* Blend of sulfonates, nonionics, and coconut derived alkanolamides
*Category:* Detergent concentrate
*Applications:* All-purpose cleaners, solvent emulsion cleaners, paint strippers, high foam light-duty liquids; base for aircraft cleaners, dairy cleaners, dishwashing liquids, car washes, metal cleaners, laundry cleaners, textile processing
*Form:* Liquid; Color: Amber; Odor: Mild
*Properties:* Density: 8.8 lb/gal; Clear Pt.: 1 C; pH: 8.2 ± 0.2; Biodegradable
*Composition:* 100% active ingred.

**Carsofoam MS Concentrate.** Carson Chem.
*Category:* Detergent
*Applications:* Shampoo concentrate esp. for mild shampoos and baby shampoos
*Form:* Hazy liquid; Color: Light yellow
*Properties:* pH ≅ 7.3 (10% sol'n.)
*Composition:* 50% active ingred.; 2% sodium chloride

**Carsofoam PS-1.** Carson Chem.
*Category:* Shampoo concentrate
*Applications:* Manufacture of liquid cream and pearlized shampoos
*Form:* Pearlized lotion; Color: White; Odor: Mild, pleasant
*Composition:* ≈ 38% active

**Carsofoam PS-3.** Carson Chem.
*Category:* Shampoo concentrate
*Applications:* Manufacture of liquid pearlized and cream shampoos
*Form:* Pearlized lotion; Color: White; Odor: Mild, pleasant
*Properties:* pH ≅ 7 (10% sol'n.)
*Composition:* ≈ 38% active

**Carsofoam SC.** Carson Chem.
*Chemical Description:* Sodium laureth sulfate, cocamide DEA, DEA lauryl sulfate and tetrasodium EDTA
*Category:* Detergent concentrate
*Applications:* High quality shampoos and bubble baths
*Form:* Thick, flowable gel; Color: 4 (Gardner)
*Properties:* pH ≅ 7 (10% sol'n.)
*Composition:* 65% solids

**Carsofoam T-60-L.** Carson Chem.
*Chemical Description:* Triethanolamine salt of linear alkyl benzene sulfonic acid
*Category:* High foaming detergent, wetting agent
*Applications:* Liquid cleaning compounds for cosmetic, industrial and institutional usage
*Form:* Clear, viscous liquid; Color: < 5 (Gardner); Odor: Bland
*Properties:* Density: 9.0 lb/gal; pH: 7.0–7.6 (as is); Biodegradable

*Composition:* 60% solids
*Containers:* 55 gal lined steel or liquipak drums

**Carsofos 4326.** Carson Chem.
*Chemical Description:* Anionic/nonionic blend
*Category:* Anionic detergent, wetting agent, emulsifier, penetrant
*Applications:* All-purpose cleaners
*Form:* Liquid; Color: Pale amber; Odor: Mild, fatty
*Properties:* Density: 8.25 lb/gal; Sp. gr.: 0.989; pH: 7.5–8.5 (as is)
*Composition:* 92% solids

**Carsofos N.** Carson Chem.
*Applications:* Formulation of aqueous degreasers
*Form:* Clear viscous liquid; Color: 5 (Gardner)
*Properties:* Density: 8.39 lb/gal; Visc.: 300 cps (10% sol'n. in water); pH: 9.2 (1% sol'n.)

**Carsofos NP-6.** Carson Chem.
*Form:* Viscous liquid; Color: Amber, 5 (Gardner)
*Properties:* Acid Value: 105; pH: 2.5 (1% aq.)

**Carsofos NP-9.** Carson Chem.
*Form:* Slightly hazy viscous liquid; Color: Yellow, 4 (Gardner)
*Properties:* Acid Value: 120; pH: 2.0 (1% aq.)

**Carsonam 3.** Carson Chem.
*Category:* Amphoteric surfactant, detergent, wetting agent, emulsifier, foaming and coupling agents
*Applications:* Shampoos, lotions, antidandruff shampoos and germicides; chemical specialties such as liquid hand cleaners, rug and upholstery shampoos, dishwashing compounds
*Form:* Liquid; Color: 5
*Properties:* Density: 8.45 lb/gal; pH: 5.5–7.5 (10% sol'n.); Biodegradable
*Composition:* 30% active ingred; 64% water; 6% max. chloride

**Carsonam 32.** Carson Chem.
*Chemical Description:* Coco amido betaine (industrial grade)
*Applications:* Industrial use
*Form:* Clear liquid; Color: Amber
*Properties:* Solubility: Solvent—Proprietary ethanol; pH: 6.0–7.0 (as is)
*Composition:* 58–62% solids; 4% max. sodium chloride

**Carsonam 33-S.** Carson Chem.
*Category and Applications:* See Carsonam 3
*Form:* Liquid; Color: 6 max.
*Properties:* Density: 8.45 lb/gal; pH: 5.5–7.5 (10% sol'n.); Biodegradable
*Composition:* 30% active ingred.; 63% water; 7% max. chloride

**Carsonam 3147.** Carson Chem.
*Category and Applications:* See Carsonam 3
*Form:* Liquid; Color: 5 max.
*Properties:* Density: 8.45 lb/gal; Biodegradable
*Composition:* 33% active ingred.; 64% water; 2% max. chloride

**Carsonam BCW.** Carson Chem.
*Chemical Description:* Coco fatty

betaine
*Category:* Foaming and conditioning agent
*Applications:* Used in household and industrial detergents; also hair shampoos, bubble baths
*Form:* Liquid; Color: 2 (Gardner)
*Properties:* Density: 8.5 lb/gal: Sp. gr.: 1.044; pH: 8.0 (as is); Biodegradable
*Composition:* 44% solids; 56% water; 8% sodium chloride

**Carsonam C.** Carson Chem.
*Chemical Description:* Coco imidazoline betaine
*Category:* Emulsifier, detergent, sequestrant, mild germistat and fungistat, amphoteric
*Applications:* Human, rug, upholstery shampoos; heavy duty liquid cleaners, dishwashing compounds, liquid steam cleaners
*Form:* Liquid; Color: 5 (Gardner)
*Properties:* pH: 9.7 (as is); Biodegradable
*Composition:* 37% solids content

**Carsonam C-SF.** Carson Chem.
*Chemical Description:* Coco imidazoline betaine—salt free
*Category and Applications:* See Carsonam C
*Form:* Liquid; Color: 5 (Gardner)
*Properties:* pH: 9.7 (as is); Biodegradable
*Composition:* 37% solids content

**Carsonam C-SPCL.** Carson Chem.
*Chemical Description:* Coco imidazoline betaine
*Category and Applications:* See Carsonam C
*Form:* Color: 7 (Gardner)
*Properties:* pH: 6.0 (as is); Biodegradable
*Composition:* 48% solids content

**Carsonam DC.** Carson Chem.
*Chemical Description:* Coco imidazoline betaine dicarboxylate
*Category and Applications:* See Carsonam C
*Form:* Gel; Color: 9 (Gardner)
*Properties:* pH: 8.5 (5% sol'n.); Biodegradable
*Composition:* 50% solids; 11% sodium chloride; 2.4% nitrogen

**Carsonam DC-SF.** Carson Chem.
*Chemical Description:* Coco imidazoline betaine dicarboxylate—salt free
*Category and Applications:* See Carsonam C
*Form:* Gel; Color: 8 (Gardner)
*Properties:* pH: 9.6 (10% sol'n.); Biodegradable
*Composition:* 38% solids; 2.5% nitrogen

**Carsonam DC-70%-SF.** Carson Chem.
*Chemical Description:* Coco imidazoline betaine dicarboxylate—salt free
*Category and Applications:* See Carsonam C
*Form:* Color: 11 (Gardner)
*Properties:* pH: 9.6 (10% sol'n.); Biodegradable
*Composition:* 70% solids; 4.35% nitrogen

**Carsonam L.** Carson Chem.
*Chemical Description:* Lauryl

imidazoline betaine
*Category and Applications:* See Carsonam C
*Form:* Liquid; Color: 6 (Gardner)
*Properties:* Sp. gr.: 1.1; pH: 9.3 (as is); Biodegradable
*Composition:* 43% solids; 6.5% sodium chloride; 2.6% nitrogen

**Carsonol ALS.** Carson Chem.
*Chemical Description:* Ammonium lauryl sulfate
*Category:* Detergent with high foam, good wetting and emulsifying properties, anionic
*Applications:* Cosmetics–shampoos, bubble baths, shaving and cleansing creams, skin lotions; Chemical Specialties–industrial cleaners, rug and upholstery cleaners, liquid household detergents, vegetable scrubbing, emulsion polymerization, automobile and pet shampoos
*Form:* Viscous liquid; Color and Odor: Low
*Properties:* Solubility: Water sol.; Density: 8.4 lb/gal; Stability: Good > pH 4; pH: 6.5–7.0 (10% sol'n.); Biodegradable
*Composition:* 30% active ingred.

**Carsonol ALS Special.** Carson Chem.
*Chemical Description:* Ammonium lauryl sulfate
*Category and Applications:* See Carsonol ALS
*Form:* Clear liquid
*Properties:* Solubility: Water sol.; Density: 8.4 lb/gal; Stability: Good > pH 4; pH: 6.5–7.0 (10% sol'n.); Biodegradable
*Composition:* 30% active ingred.

**Carsonol BD.** Carson Chem.
*Chemical Description:* Blend of carboxybetaine, sulfate and amine groups
*Category:* Detergent, foaming agent, anionic/amphoteric
*Applications:* Shampoo concentrate with good hair manageability, foaming, cleansing and hard water stability; bubble baths
*Form:* Moderately viscous liquid; Color: Light amber
*Properties:* Solubility:Excellent water sol.; Cloud Pt.: 50 F; pH: 7.5–8.5
*Composition:* 30–32% active

**Carsonol DLS.** Carson Chem.
*Chemical Description:* Diethanolamine lauryl sulfate
*Category and Applications:* See Carsonol ALS
*Form:* Clear liquid; Color and Odor: Low
*Properties:* Solubility: Water sol.; Density: 8.49 lb/gal; pH: 7.5–8.5 (10% sol'n.); Biodegradable
*Composition:* 35% active ingred.

**Carsonol ILS.** Carson Chem.
*Chemical Description:* Mixed isopropanolamine lauryl sulfate
*Category and Applications:* See Carsonol ALS
*Form:* Viscous liquid; Color and Odor: Low
*Properties:* Solubility: Water sol.; Density: 8.63 lb/gal; pH: 7.0–7.4 (10% sol'n.); Biodegradable
*Composition:* 75% active ingred.

**Carsonol MLS.** Carson Chem.
*Chemical Description:* Magnesium lauryl sulfate
*Category and Applications:* See Carsonol ALS
*Form:* Viscous liquid; Color and Odor: Low
*Properties:* Solubility: Water sol.; Density: 8.57 lb/gal; pH: 6.5–7.5 (10% sol'n.); Biodegradable
*Composition:* 30% active ingred.

**Carsonol SES-A.** Carson Chem.
*Chemical Description:* Ammonium lauryl ether sulfate
*Category:* Anionic surfactants with excellent foaming in hard and soft water
*Applications:* Shampoos, bubble baths, liquid detergents, automobile shampoos, household chemical specialties, industrial foaming agents
*Form:* Liquid
*Properties:* Density: 8.4 lb/gal; pH: 6.5–7.0 (10% sol'n.); Biodegradable
*Composition:* 60% active ingred.

**Carsonol SES-S.** Carson Chem.
*Chemical Description:* Sodium lauryl ether sulfate
*Category and Applications:* See Carsonol SES-A
*Form:* Liquid
*Properties:* Density: 8.8 lb/gal; Stability: Good in acidic and alkaline media; pH: 7.0–8.0 (10% sol'n.); Biodegradable
*Composition:* 60% active ingred.

**Carsonol SHS.** Carson Chem.
*Chemical Description:* Sodium 2-ethylhexyl sulfate
*Category:* Low foam detergent, wetting agent, penetrant
*Applications:* Used in caustic solutions for peeling of fruits and vegetables; very stable to high electrolyte and in acid and basic solutions
*Form:* Clear liquid; Color: 3 (Gardner)
*Properties:* Density: 9.6 lb/gal; Sp. gr.: 1.15; Visc.: 35 cps; pH: 10.3 (10% sol'n.)
*Composition:* 40% active ingred.
*Containers:* 480 lb net

**Carsonol SLES.** Carson Chem.
*Chemical Description:* Sodium lauryl ether sulfate
*Category and Applications:* See Carsonol SES-A
*Form:* Liquid
*Properties:* Density: 8.7 lb/gal; Stability: Good in acidic and alkaline media; pH: 7.5–8.5 (10% sol'n.); Biodegradable
*Composition:* 30% active ingred.

**Carsonol SLES-2.** Carson Chem.
*Chemical Description:* Sodium lauryl ether sulfate
*Category and Applications:* See Carsonol SES-A
*Form:* Liquid
*Properties:* Density: 8.63 lb/gal; Stability: Good in acidic and alkaline media; pH: 7.0–8.0 (10% sol'n.); Biodegradable
*Composition:* 30% active ingred.

**Carsonol SLS.** Carson Chem.
*Chemical Description:* Sodium lauryl sulfate

*Category and Applications:* See Carsonol ALS
*Form:* Clear liquid; Color and Odor: Low
*Properties:* Solubility: Water sol.; Density: 8.7 lb/gal; pH: 7.5–8.5 (10% sol'n.); Biodegradable
*Composition:* 30% active ingred.

**Carsonol SLS Paste B.** Carson Chem.
*Chemical Description:* Sodium lauryl sulfate
*Category and Applications:* See Carsonol ALS
*Form:* Paste; Color and Odor: Low
*Properties:* Solubility: Water sol.; Density: 8.8 lb/gal; pH: 8.4–8.8 (10% sol'n.); Biodegradable
*Composition:* 30% active ingred.

**Carsonol SLS Special.** Carson Chem.
*Chemical Description:* Sodium lauryl sulfate
*Category and Applications:* See Carsonol ALS
*Form:* Liquid; Color and Odor: Low
*Properties:* See Carsonol SLS
*Composition:* 30% active ingred.

**Carsonol TLS.** Carson Chem.
*Chemical Description:* Triethanolamine lauryl sulfate
*Category and Applications:* See Carsonol ALS
*Form:* Clear liquid; Color and Odor: Low
*Properties:* Solubility: Water sol.; Density: 8.7 lb/gal; pH: 7.0–7.5 (10% sol'n.); Biodegradable
*Composition:* 40% active ingred.

**Carsonon L-985.** Carson Chem.
*Chemical Description:* 9–10 mole ethoxylated linear primary alcohol
*Category:* Wetting agent, detergent, emulsifier, dispersant, nonionic
*Applications:* Maintenance and institutional cleaners; textile processing; metal cleaning; paper and paint industries
*Form:* Liquid; Color: Colorless
*Properties:* Solubility: Sol. in acid and alkaline solutions; Density: 8.32 lb/gal; Cloud Pt.: ≈ 137 F; Pour Pt.: 39–43 F; Solidification Pt.: 30–34 F; Stability: Stable in acid and alkaline solutions; pH: neutral (3% sol'n.); Biodegradable
*Composition:* 85% active ingred.

**Carsonon LF-5.** Carson Chem.
*Chemical Description:* Linear alcohol alkoxylate
*Category:* Low foam surfactant, nonionic
*Applications:* Pulp and paper additives, mechanical dishwashing, metal cleaning, dairy equipment cleaning
*Form:* Color: 100 (APHA); Odor: Mild
*Properties:* Solubility: Sol. in aromatic solvents, polar solvents, water below cloud pt. and in the common chlorinated solvents; Sp. gr.: 1.02 (60 F/60 F); Visc.: 125 cts; Cloud Pt.: 43–49 C; Flash Pt.: 460 F (COC); Stability: Stable to dry alkalis; Surface Tension: 33.5 dynes/cm (0.1% conc.); pH: 5.5–7.0 (1% aq.); Biodegradable

**Carsonon LF-46.** Carson Chem.
*Chemical Description:* Polyoxyalkalene glycol
*Category:* Low foam surfactant, nonionic
*Applications:* Low foam use as pulp and paper additives, mechanical dishwashing, metal cleaning, dairy equipment cleaning
*Form:* Color: 1 (Gardner); Odor: Mild
*Properties:* Solubility: Sol. in aromatic solvents, polar solvents, water below cloud pt., and in the common chlorinated solvents; Visc.: 200 cps; Cloud Pt.: 44 C; Flash Pt.: > 230 F; pH: 5.5–7.0 (1% aq.); Biodegradable

**Carsonon N-4.** Carson Chem.
*Chemical Description:* 4-mole ethoxylated nonylphenol
*Category:* Emulsifier, detergent, dispersant, nonionic
*Applications:* Oil soluble emulsifier; oil soluble detergent and dispersing agent for petroleum oils; intermediate in the synthesis of high foaming, water soluble sulfate esters; drycleaning detergent
*Form:* Liquid; Color: Pale
*Properties:* Solubility: Sol. in kerosene, Stoddard solvent, alcohol, chlorinated and aromatic solvents; Density: 8.6 lb/gal; Sp. gr.: 1.02; Visc.: 175–250 cps; Flash Pt.: 500–600 F; Fire Pt.: 500–600 F; Pour Pt.: -5 ± 2 F; Solidification Pt.: -20 ± 2 F
*Composition:* 100% active ingred.

**Carsonon N-6.** Carson Chem.
*Chemical Description:* 6-mole ethoxylated nonylphenol
*Category:* Emulsifier, wetting agent, nonionic
*Applications:* Used for foam control in household detergent, industrial cleaning compounds; low foaming wetting agent; used with Carsonon N-9 to improve oil solubility and detergency in solvent systems
*Form:* Liquid; Color: Pale; Odor: Mild, pleasant
*Properties:* Solubility: Sol. in alcohols, aromatic, chlorinated and aliphatic solvents, fatty oil; dispersible in water; Density: 8.68 lb/gal; Visc.: 180–240; Cloud Pt.: 90–102 F; Flash Pt.: 500 F; Solidification Pt.: –33 to –29 F
*Composition:* 100% active ingred.

**Carsonon N-8.** Carson Chem.
*Chemical Description:* 8–8½ mole ethoxylated nonylphenol
*Category:* Wetting agent, detergent, emulsifier, dispersant, nonionic
*Applications:* Household and industrial cleaners, textile processing, paper industry, wetting agent for use with mineral acids and corrosion inhibitors
*Form:* Liquid; Color: Pale; Odor: Mild, pleasant
*Properties:* Solubility: Sol. in water, alcohol, chlorinated and aromatic solvents; Density: 8.75 lb/gal; Visc.: 200–300 cps; Cloud Pt.: 82–86 F; Flash Pt.: ≈ 435 F; Fire Pt.: ≈ 465 F; Pour Pt.: 41–45 F; Solidification Pt.: 30–34 F; pH: neutral (1% sol'n.)
*Composition:* 100% active ingred.

**Carsonon N-9.** Carson Chem.
*Chemical Description:* 9–10 mole ethoxylated nonylphenol
*Category:* Wetting agent, detergent, emulsifier, dispersant, nonionic
*Applications:* Household and industrial cleaners; textile processing; paper industry; wetting agent for use with mineral acids and corrosion inhibitors
*Form:* Liquid; Color: Pale; Odor: Mild, pleasant
*Properties:* Solubility: Sol. in water, alcohol, chlorinated and aromatic solvents; Density: 8.83 lb/gal; Visc.: 180–250 cps; Flash Pt.: ≈ 540 F; Fire Pt.: ≈ 600 F; Cloud Pt.: ≈ 130 F; Pour Pt.: 35–39 F; Solidification Pt.: 24–28 F; pH: neutral (1% sol'n.)
*Composition:* 100% active ingred.

**Carsonon N-10.** Carson Chem.
*Chemical Description:* 10–11 mole ethoxylated nonylphenol
*Category:* Wetting agent, detergent, emulsifier, dispersant, nonionic
*Applications:* Household and industrial cleaners, textile processing, paper industry, wetting agent for use with mineral acids and corrosion inhibitors
*Form:* Liquid; Color: Pale; Odor: Mild, pleasant
*Properties:* Solubility: Sol. in water, alcohol, chlorinated and aromatic solvents; Density: 8.84 lb/gal; Visc.: 180–250 cps; Cloud Pt.: 143–146 F; Flash Pt.: ≈ 540 F; Fire Pt.: ≈ 600 F; Pour Pt.: 35–39 F; Solidification Pt.: 24–28 F; pH: neutral (1% sol'n.)
*Composition:* 100% active ingred.

**Carsonon N-30.** Carson Chem.
*Chemical Description:* 30 mole ethoxylated nonylphenol
*Category:* Wetting agent, detergent, emulsifier, nonionic
*Applications:* Emulsion polymerization; emulsifier for fats, oils, waxes and essential oils
*Form:* Solid; Odor: Mild, pleasant
*Properties:* Density: 9.75 lb/gal; Sp. gr.: 1.17; Cloud Pt.: Clear @ 212 F (1% sol'n.); Solidification Pt.: 104 ± 2 F; pH: neutral (1% sol'n.)
*Composition:* 100% active ingred.

**Carsonon N-30, 70%.** Carson Chem.
*Category and Applications:* See Carsonon N-30
*Form:* Liquid to semisolid; Color: Pale yellow
*Properties:* Sp. gr.: 1.09; Cloud Pt.: 167–171 F; pH: 6.0–7.0 (3% sol'n.)
*Composition:* 70% active ingred.; 30% water

**Carsonon N-40, 70%.** Carson Chem.
*Form:* Semisolid; Color: White to yellow
*Properties:* Cloud Pt.: Clear @ 212 F (1% sol'n.); pH: 6.0–7.0 (3% sol'n.)
*Composition:* 70% active, 30% water

**Carsonon N-50, 70%.** Carson Chem.
*Form:* Semisolid; Color: White to yellow
*Properties:* Cloud Pt.: Clear @ 212 F (1% sol'n.); pH: 6.0–7.0

(3% sol'n.)
*Composition:* 70% active ingred.; 30% water

**Carsonon N-100, 70%.** Carson Chem.
*Chemical Description:* Polyethoxylated nonylphenol
*Category:* Emulsifier, stabilizer
*Applications:* Effective at high temperatures and in concentrated electrolyte solutions; very high water solubility
*Form:* Semisolid; Color: White to off white
*Properties:* Solubility: Very high water sol.; Cloud Pt.: Clear @ 212 F (1% sol'n.); pH: 6.0–7.0 (3% sol'n.)
*Composition:* 70% active, 30% water

**Carsonon ND-317.** Carson Chem.
*Chemical Description:* Polyethoxylated nonylphenol
*Category:* Nonionic surfactant
*Applications:* Base for solid toilet bowl cleaners
*Form:* Opaque, viscous liquid; Color: 2 (Gardner)
*Properties:* Sp. gr.: 1.11; M.P.: High; Cloud Pt.: 162 ± 2 F (1% in 10% NaCl sol'n.)

**Carsonon P-30M.** Carson Chem.
*Form:* Color: 5 max (Molten)
*Properties:* Cloud Pt.: 30–32 ml $H_2O$; pH: 7.5–8.0 (3% aq.)
*Composition:* 1% max water

**Carsonon TD-10.** Carson Chem.
*Chemical Description:* 10 mole ethoxylated tridecyl alcohol
*Category:* Emulsifier, detergent, wetting agent, foam builder, solubilizer, nonionic
*Applications:* Household industrial cleaners; textile processing; paper industry; emulsifier for essential oils, aromatic solvents, fats; wetting agent for use with mineral acids and corrosion inhibitors
*Form:* Liquid; Color: 1 (Gardner)
*Properties:* Solubility: Water sol.; Density: 8.5 lb/gal; HLB: 13.7; Cloud Pt.: 162–166 F (1% in distilled water); Flash Pt.: > 300 F (COC); Pour Pt.: 67 F; pH: 5.0–7.0 (3% aq.)
*Composition:* 100% active ingred.

**Carsonon TD-11.** Carson Chem.
*Chemical Description:* 11 mole ethoxylated tridecyl alcohol; available as 100% material and 70% solution in water
*Category:* Detergent, emulsifier, wetting agent, nonionic
*Applications:* Household and industrial cleaners; textile processing; paper industry; emulsifier for essential oils, aromatic solvents, fats; wetting agent for use with mineral acids and corrosion inhibitors; high foaming properties; high temperature detergency
*Form:* Waxy solid; Color: 1 (Gardner)
*Properties:* Density: 8.5 lb/gal; HLB: 14.6; Cloud Pt.: 137–141 F (1% in 10% NaCl sol'n.); Flash Pt.: > 300 F (COC); Pour Pt.: 92 F; pH: 5.0–7.0 (3% aq.)
*Composition:* Available as 100% material or 70% solution in water
*Reaction:* Compatible with electrolytes, strong alkalis

**Carsoquat 621.** Carson Chem.
*Chemical Description:* n-alkyl dimethyl benzyl ammonium chloride
*Category:* Corrosion inhibitor, cationic surfactant
*Form:* Liquid; Color: Almost water-white to pale yellow; Odor: Mild
*Properties:* Solubility: Miscible with water, lower alcohols, most polar solvents; Density: 8 lb/gal; Sp. gr.: 0.96; Stability: Good
*Composition:* 50% active
*Reaction:* Compatible with nonionic and amphoteric detergents
*Containers:* 55 gal drums
*Warnings and Cautions:* Avoid contact with eyes, skin, clothing; wash affected areas with plenty of soap and water

**Carsoquat 621 (80%).** Carson Chem.
*Chemical Description:* Alkyl dimethyl ammonium chloride
*Category and Applications:* See Carsoquat 621
*Form:* Clear liquid; Color: Straw
*Properties:* m.w.: 361; pH: 7–8 (1% sol'n.)
*Composition:* 80% active ingred.; 10% water; 10% alcohol

**Carsoquat 816-C.** Carson Chem.
*Chemical Description:* Blended cetearyl alcohol, PEG-40 castor oil and stearalkonium chloride
*Category:* Creme rinse concentrate
*Applications:* Finished creme rinse requires only dispersion in water; compatible with hydrolized animal protein, lanolin, lanolin derivatives
*Form:* White waxy flakes; Odor: Mild, fatty
*Properties:* Solubility: Disperses readily in water; pH: 6.3 (5% disp.)
*Composition:* 96% min active

**Carsoquat CTM-29.** Carson Chem.
*Chemical Description:* Cetyl trimethyl ammonium chloride
*Category:* Cationic surfactant
*Applications:* Conditioners in specialty hair preparations; coagulating agent in the manufacture of antibiotics and other pharmaceuticals
*Form:* Liquid; Color: Almost clear to pale yellow
*Properties:* m.w.: 319; Density: 8.1 lb/gal; Sp. gr.: 0.968; pH: 7.0–9.0 (10% active)
*Composition:* 29% active ingred.; 71% water
*Reaction:* Compatible with nonionic and cationic polymers, organic chelates, most polar solvents and inorganic detergent builders

**Carsoquat CTM-429.** Carson Chem.
*Chemical Description:* Cetyl trimethyl ammonium chloride
*Category and Applications:* See Carsoquat CTM-29
*Form and Properties:* Identical to Carsoquat CTM-29 except for pH: 3.5–4.0 (2% active)
*Composition:* 29% active; 71% water

**Carsoquat SDQ-25.** Carson Chem.
*Chemical Description:* Stearyl dimethyl benzyl ammonium chloride
*Category:* Cationic surfactant
*Applications:* Softener, creme hair

rinse, hair conditioner; for cosmetics and natural fibers
*Form:* Thick creamy paste; Color: White; Odor: Mild and sweet
*Properties:* pH: 3.0–4.0 (1% disp.)
*Composition:* 20–21 % active

**Carsoquat SDQ-85.** Carson Chem.
*Chemical Description:* Stearyl dimethyl benzyl ammonium chloride
*Category:* Hair conditioner and softener, cationic
*Form:* Waxy flake; Color: White
*Properties:* M.P.: ≈ 140 F; pH: 5–7 (0.5 sol'n.)
*Composition:* 85% active ingred.

**Carsosoft S-75.** Carson Chem.
*Chemical Description:* Alkyamidoethyl alkyl imidazolinium methyl methosulfate
*Category:* Fabric softener base, cationic
*Applications:* Fabric softener base with lubricity, softness and antistatic and antilinting properties, excellent rewet properties and nonyellowing
*Form:* Liquid; Color: 4 (Gardner)
*Properties:* m.w.: 720; Stability: Excellent thermal stability; pH: 5–7 (5% sol'n.)
*Composition:* 75% active ingred.; diluent: isopropanol

**Carsosoft S-90.** Carson Chem.
*Form:* Semisold; Color: Yellow
*Properties:* m.w.: 723; pH: 5.0–7.0 (5% aq.)
*Composition:* 85–88% active

**Carsosoft S-90-M.** Carson Chem.
*Chemical Description:* Modified alkyamidoethyl alkyl imidazolinium methyl methosulfate
*Category and Applications:* See Carsosoft S-75
*Form:* Liquid; Color: 4 (Gardner)
*Properties:* Solubility: Readily dispersible in warm and cold water; Stability: Excellent thermal stability; pH: 5.0–7.0 (5% sol'n.)
*Composition:* 88–91% active ingred.; diluent: isopropanol

**Carsosoft T-90.** Carson Chem.
*Chemical Description:* Quaternary ditallow diamido
*Category:* Fabric softener, cationic
*Applications:* Home and commercial laundry fabric softener with excellent softening, rewet properties, nonyellowing, antistat and antilint properties
*Form:* Liquid; Color: Yellow
*Properties:* pH: 5–7 (5% sol'n.)
*Composition:* 88–90% active ingred.; Diluent: isopropanol

**Carsosoft V-90.** Carson Chem.
*Chemical Description:* Dihydrogenated tallow dimethyl ammonium methosulfate
*Form:* Paste; Color: White, 4 (Gardner)
*Properties:* pH: 5.5 (10% in 1:1 IPA/aq.)
*Composition:* 91% solids

**Carsosoft V-100.** Carson Chem.
*Chemical Description:* Dihydrogenated tallow dimethyl ammonium sulfate
*Form:* Flake or powder; Color: White
*Properties:* pH: 5.5–6.5 (10% in

1:1 IPA/aq.)
*Composition:* 95% solids

**Carsosulf SXS-Liquid.** Carson Chem.
*Chemical Description:* Sodium xylene sulfonate
*Category:* Coupling agent, solubilizer, anionic
*Applications:* Nonflammable coupling agent and solubilizer in liquid detergent systems; esp. for coupling inorganics into cleaning compounds
*Form:* Liquid
*Properties:* Sp. gr.: 1.18–1.22; pH: 7.5–10.5 (as is)
*Composition:* 40% active ingred.

**Carsosulf UL-100 Acid.** Carson Chem.
*Chemical Description:* Linear alkyl benzene sulfonic acid
*Properties:* m.w.: 316; Density: 8.84 lb/gal; Sp. gr.: 1.06; Biodegradable
*Composition:* 97 ± 1% active
*Containers:* 55 gal steel lined drum

**Cartaretin F-4.** Sandoz Colors and Chem.
*Category:* Cationic polymer
*Applications:* Conditioning agent for hair and skin; as wetting agent for sizing
*Form:* Liquid; Color: Light yellow; Odor: Natural
*Properties:* Solubility: Sol. in water and alcohol @ 20 C; Sp. gr.: 1.07; Visc.: 400 cps; F.P.: –4 C; B.P.: 220 F; Stability: Very stable
*Composition:* 30% active ingred.
*Containers:* 500 lb fiber drum
*References:* Bulletin 7-159/78

**Cartaretin F-8.** Sandoz Colors and Chem.
*Category:* Cationic polymer
*Applications:* Conditioning agent for hair and skin; wetting agent in sizing
*Form:* Liquid; Color: Amber; Odor: Natural
*Properties:* Solubility: Sol. in water and alcohol @ 20 C; Sp. gr.: 1.07; Visc.: 800 cps; F.P.: –4 C; B.P.: 220 F; Stability: Very stable
*Composition:* 30% active ingred.
*Containers:* 500 lb fiber drum
*References:* Bulletin 7-159/78
*Patents:* U.S. 3,632,559

**Casul 70 HF.** Thompson-Hayward Chemical Co.
*Chemical Description:* Calcium salt of a branched chain dodecylbenzene alkylate
*Category:* Emulsifier, anionic
*Applications:* Co-emulsifier for oil-in-water and water-in-oil formulations
*Form:* Liquid; Color: Amber
*Properties:* Density: 8.50 lb/gal; Sp. gr.: 1.02; Visc.: 18,400 cps; Flash Pt.: 104 F (PMCC); Pour Pt.: 46 F; pH: 5.0–7.0
*Composition:* 70% active ingred.
*Containers:* 55 gal. drums

**Cation Softener X.** Onyx Chem. Co.
*Chemical Description:* Alkyl imidazoline derivative
*Category:* Softener, lubricant, mordant
*Applications:* Softener, lubricant and glass fiber mordant with good substantivity

*Form:* Paste
*Properties:* Sp. gr.: 0.92 (25/20 C); Flash Pt.: 60 F

**CC-16.** NL Baroid Petroleum Services
*Chemical Description:* Sodium salt of a humic acid material
*Category:* Drilling fluid additive, emulsifier, gel dispersing agent, anionic
*Applications:* Thinner and emulsifier for oil-in-water emulsion muds esp. in normal-to-high pH drilling fluids
*Form:* Free flowing powder
*Properties:* Solubility: Sol. in water-based drilling fluids; Stability: Stable at high temperature
*Composition:* 100% active ingred.
*Containers:* 50 lb multiwall paper bags

**CC-603.** PVO International
*Chemical Description:* Blend of carrageenan, guar gum and dextrose
*Category:* Stabilizer
*Applications:* Food grade stabilizer for use in whipped toppings, fillings, coffee whiteners

**Centrol Series.** Central Soya Co.
*Chemical Description:* Natural soybean lecithin
*Category:* Wetting agent, emulsifier, emollient, softening agent, lubricant, solubilizer
*Applications:* Cosmetics; food processing; pharmaceuticals; coatings manufacture; plastic and rubber industry; glass and ceramics processing; paper and printing industry; masonry and asphalt products; petroleum industry as specialty oils and greases; metal processing; pesticides; textiles and leather
*Form:* Liquid
*Composition:* 100% active ingred.

**Centrolene Series.** Central Soya Co.
*Chemical Description:* Modified soybean lecithin
*Category:* Wetting agent, emulsifier, release agent, dispersing agent, lubricant, anticorrosive agent, emollient
*Applications:* Food processing; cosmetics; pharmaceuticals; coatings manufacture; plastic and rubber industries; glass and ceramics; paper and printing; masonry and asphalt products; petroleum industry as specialty greases and oils; metal processing; textiles and leather
*Form:* Liquid
*Composition:* 100% active ingred.
*Toxicity:* Food approved and nonfood approved

**Centrolex Series.** Central Soya Co.
*Chemical Description:* Refined "oil free" phosphatides (soybean lecithin)
*Category:* Wetting agent, emulsifier, dietary supplement, dispersant, emollient
*Applications:* Food processing; cosmetics; pharmaceuticals; plastics and rubber industries
*Form:* Granular
*Composition:* 100% active ingred.

**Centromix Series.** Central Soya Co.
*Chemical Description:* Standard

soybean lecithin with selected surfactants
*Category:* Emulsifier, wetting agent, dispersant, penetrant
*Applications:* Food processing; cosmetics; coatings manufacture and finishes; plastic and rubber industries; glass and ceramic processing; paper and printing; masonry and asphalt; metal processing; pesticides; adhesives; textiles and leather
*Form:* Liquid, paste
*Composition:* 100% active ingred.

**Centrophil Series.** Central Soya Co.
*Chemical Description:* Refined soybean lecithin with selected carriers
*Category:* Wetting agent, emulsifier, dispersant, release agent, dietary supplement
*Applications:* Food processing; cosmetics; pharmaceuticals; plastics and rubber industries; masonry and asphalt products; metal processing
*Form:* Liquid, solid, other
*Composition:* 100% active ingred.
*Toxicity:* Most are food approved; some for industrial use only

**Ceramol.** Aceto Chem.
*Chemical Description:* Partially sulfonated fatty alcohol from hydrogenated vegetable oil
*Category:* Emulsifier
*Applications:* Cosmetics
*Form:* Flake

**Ceranine AT.** Sandoz Ltd.
*Chemical Description:* Fatty acid derivative
*Category:* Softener, antistatic, cationic
*Applications:* Softener and antistatic agent for synthetic fibers
*Form:* Paste

**Ceranine AW.** Sandoz Ltd.
*Chemical Description:* Fatty acid derivative
*Category:* Softener, antistatic, cationic
*Applications:* Softener and antistatic agent for synthetic fibers
*Form:* Paste/solid

**Ceranine HC.** Sandoz Ltd.
*Chemical Description:* Fatty acid derivative
*Category:* Softener, cationic
*Applications:* Softener for all fibers
*Form:* Paste

**Ceranine HC Hi Conc.** Sandoz Colors and Chem.
*Category:* Emulsifier
*Applications:* Skin care products
*Form:* Solid; Color: Tan; Odor: Slight
*Properties:* Solubility: Dispersible in water; M.P.: 55–65 C; Stability: Good
*Composition:* 100% active ingred.
*Containers:* 200 lb fiber drum

**Ceranine HCA.** Sandoz Colors and Chem.
*Chemical Description:* A fatty amide derivative
*Category:* Emulsifier, cationic
*Applications:* Skin care products; softener for textiles
*Form:* Granule; Color: Cream;

Odor: Slight
*Properties:* Solubility: Dispersible in water; M.P.: 71–81 C; Stability: Good
*Composition:* 100% active ingred.
*Containers:* 250 lb fiber drum

**Ceranine HCL.** Sandoz Ltd.
*Chemical Description:* Fatty acid derivative
*Category:* Softener, cationic
*Applications:* Softener for all fibers
*Form:* Liquid/flakes

**Ceranine HCS.** Sandoz Ltd.
*Chemical Description:* Fatty acid derivative
*Category:* Softener, cationic
*Applications:* Softener for all fibers
*Form:* Liquid

**Ceranine L.** Sandoz Ltd.
*Chemical Description:* Polyethylene dispersion
*Category:* Softener, nonionic
*Applications:* Softener for cellulosic fibers and their blends with synthetic fibers, especially for high-grade finishing
*Form:* Liquid

**Ceranine PAP.** Sandoz Ltd.
*Chemical Description:* Polymerization product
*Category:* Softener, cationic
*Applications:* Permanent softener for polyamide fibers
*Form:* Liquid

**Ceranine PNA.** Sandoz Ltd.
*Chemical Description:* Fatty acid derivative
*Category:* Softener, cationic
*Applications:* Softener for synthetic fibers
*Form:* Paste/powder

**Ceranine PNL.** Sandoz Ltd.
*Chemical Description:* Substituted fatty acid amide
*Category:* Softener, cationic
*Applications:* Softener for synthetic fibers
*Form:* Liquid

**Ceranine PNS.** Sandoz Ltd.
*Chemical Description:* Fatty acid derivative
*Category:* Softener, cationic
*Applications:* Softener for all fibers, especially for synthetic fibers
*Form:* Flakes/paste

**Ceranine RW.** Sandoz Ltd.
*Chemical Description:* Fatty acid derivative
*Category:* Softener, cationic
*Applications:* Softener with rewetting properties for cellulosic fibers and their blends with synthetic fibers
*Form:* Liquid/paste

**Ceranine RWN.** Sandoz Ltd.
*Chemical Description:* Fatty acid derivative
*Category:* Softener, cationic
*Applications:* Softener with rewetting properties for cellulosic fibers and their blends with synthetic fibers
*Form:* Liquid

**Ceranine SG.** Sandoz Ltd.
*Chemical Description:* Fatty acid derivative
*Category:* Softener, nonionic/anionic
*Applications:* Softener for all fibers, particularly for scrooping
*Form:* Paste/solid

**Ceranine SIT.** Sandoz Ltd.
*Chemical Description:* Fatty acid derivative/silicones
*Category:* Softener, cationic
*Applications:* Softener for all fibers with antistatic effect
*Form:* Paste/solid

**Ceranine SWP.** Sandoz Ltd.
*Chemical Description:* Fatty acid derivative
*Category:* Softener, cationic
*Applications:* Permanent softener for polyester and polyacrylic fibers
*Form:* Liquid

**Ceranine VR.** Sandoz Ltd.
*Chemical Description:* Fatty alcohol sulfates
*Category:* Softener, anionic
*Applications:* Softener for all fibers
*Form:* Paste

**Ceranine VS.** Sandoz Ltd.
*Chemical Description:* Fatty acid condensation product in aqueous dispersion
*Category:* Softener, cationic
*Applications:* Softener with hydrophobic properties for cellulosic fibers and their blends with synthetic fibers
*Form:* Liquid

**Ceranine WN.** Sandoz Ltd.
*Chemical Description:* Fatty acid ester
*Category:* Softener, nonionic
*Applications:* Softener for cellulosic and synthetic fibers
*Form:* Liquid/flakes

**Cerol A Liquid.** Sandoz Colors and Chem.
*Chemical Description:* Cationic zirconium/wax emulsion
*Category:* Water repellent, cationic
*Applications:* Water repellent for use on cotton and polyester/cotton blends

**Cerol C.** Sandoz Ltd.
*Chemical Description:* Fatty acid chrom-complex
*Category:* Waterproof agent, cationic
*Applications:* Waterproofing agent, fast to washing and dry cleaning
*Form:* Liquid

**Cerol EWL.** Sandoz Ltd.
*Chemical Description:* Modified synthetic resin
*Category:* Waterproof agent, cationic
*Applications:* Waterproofing agent, wash fast
*Form:* Liquid

**Cerol ZN.** Sandoz Ltd.
*Chemical Description:* Paraffin dispersion containing zirconium
*Category:* Waterproof agent, cationic
*Applications:* Waterproofing agent
*Form:* Liquid

**Charlab Condensate-K.** Catawba-Charlab, Inc.
*Chemical Description:* Alkanolamide
*Category:* Wetting agent, emulsifier, nonionic
*Applications:* Wetting agent, emulsifier, scouring agent
*Form:* Liquid; Color: Amber
*Properties:* Solubility: Water sol.; Sp. gr.: 1.01
*Composition:* 100% active ingred.

**Charlab Leveler AT Special.** Catawba-Charlab, Inc.
*Chemical Description:* Sodium hydrocarbon sulfonate
*Category:* Detergent, emulsifier, anionic
*Applications:* Textile dyeing-leveling agent, softener, lubricant
*Form:* Solid; Color: Amber
*Properties:* Solubility: Sol. in water @ 160 F; Sp. gr.: 1.0; M.P.: 110–125 F
*Composition:* 80% active ingred.

**Charlab Leveler DSL.** Catawba-Charlab, Inc.
*Chemical Description:* Sodium hydrocarbon sulfonate
*Category:* Detergent, emulsifier, anionic
*Applications:* Textile leveling agent for dyeing, softener, dispersant
*Form:* Paste; Color: Light brown
*Properties:* Solubility: Sol. in hot water @ 160 F; Density: 1.03 lb/gal
*Composition:* 40% active ingred.

**Charlab LPC.** Catawba-Charlab Inc.
*Chemical Description:* Lauryl pyridinium chloride
*Category:* Cationic surfactant
*Form:* Solid; Color: Light brown; Odor: Pyridine
*Properties:* Solubility: Water sol.; F.P.: 45 C
*Composition:* 97–100% active ingred.; other components: small amount pyridine; lauryl alcohol; lauryl ether
*Toxicity:* Skin irritant; solid material or powder may cause a rash with certain individuals

**Chemannate C-60B.** Chemform Corp.
*Chemical Description:* Calcium salt of dodecyl benzene sulfonic acid
*Category:* Anionic base for emulsifiers
*Form:* Viscous liquid; Color: Amber
*Properties:* Solubility: Oil sol.; Density: 8.45 lb/gal; Visc.: 600 cps @ 20 C (Brookfield); Flash Pt.: > 100 F; pH: 4–7 (5% in 25% isopropanol)
*Composition:* 60% active ingred.

**Chemannate C-70B.** Chemform Corp.
*Chemical Description:* Calcium salt of dodecyl benzene sulfonic acid
*Category:* Emulsifier base, anionic
*Applications:* Agricultural chemicals
*Form:* Viscous liquid; Color: Dark amber
*Properties:* Visc.: 3500 cps @ 140 F; Flash Pt.: > 100 F; pH:

3.0–6.0 (5% in 25% IPA)
*Composition:* 70% active ingred.

**Chemcol 14-Z.** Chemform Corp.
*Chemical Description:* Anionic/nonionic blend
*Category:* Emulsifier, anionic/nonionic
*Applications:* Emulsifier for aromatic solvent aquatic weed control esp. for xylene or aromatic hydrocarbon-xylene solvents
*Form:* Color: Light amber
*Properties:* Solubility: Sol. in xylene, miscible with water; Pour Pt.: < 25 F; Stability: Good
*Containers:* 55 gal drums

**Chemcol 610.** Chemform Corp.
*Category:* Emulsifier
*Applications:* Low foaming, oil sol. emulsifier for agricultural sprays, esp. petroleum oils for citrus sprays
*Form:* Clear liquid; Color: < 1 (Gardner); Odor: Mild
*Properties:* Solubility: Oil sol.; Pour Pt.: < –10 F; pH: 6–8 (1% sol'n.)
*Containers:* 55 gal drums; tank cars and tank trucks

**Chemcol Agwet 110.** Chemform Corp.
*Chemical Description:* Alkyl aryl polyoxyethylene ether
*Category:* Agricultural surfactant, nonionic
*Applications:* Reduces surface tension, improves wetting and plant coverage of herbicide and pesticide sprays
*Form:* Clear liquid
*Properties:* Solubility: Water sol.; Density: 8.2 lb/gal; Flash Pt.: near 80 F
*Composition:* 80% active ingred.; 20% isopropyl alcohol
*Toxicity:* Skin and eye irritant
*Containers:* Bulk and drum
*Warnings and Caution:* Avoid skin and eye contact; flammable

**Chemcol Agwet 115-M.** Chemform Corp.
*Chemical Description:* Alkyl polyoxyethylene glycol ether
*Category:* Wetting agent, nonionic
*Applications:* Agricultural surfactant to improve wetting properties, reduce surface tension, improve plant coverage of pesticide and herbicide sprays
*Form:* Liquid; Color: Water white to pale yellow
*Properties:* Solubility: Water sol.; Density: 7.8 lb/gal; Flash Pt.: 108 F; Biodegradable
*Composition:* 80% active ingred.; 20% methyl propanol
*Toxicity:* Skin and eye irritant
*Containers:* Bulk and 55 gal drums

**Chemcol CO-28.** Chemform Corp.
*Chemical Description:* Polyethoxylated triglyceride (28-mole ethylene oxide adduct of castor oil)
*Category:* Emulsifier, nonionic
*Applications:* Textiles and agricultural chemicals
*Form:* Liquid; Color: Pale yellow
*Properties:* Density: 8.6 lb/gal; Cloud Pt.: < 25 C; Pour Pt.: < 0 C; pH: 6–8
*Composition:* 100% active ingred.

**Chemcol CO-30.** Chemform Corp.
*Chemical Description:* Polyethoxylated triglyceride (reaction of a castor oil with 30 moles of ethylene oxide
*Category and Applications:* See Chemcol CO-28
*Form:* Liquid; Color: Yellow
*Properties:* Density: 8.7 lb/gal; HLB: 11.7; Cloud Pt.: > 100 C; Pour Pt.: < 0 C; pH: 6–8
*Composition:* 100% active ingred.

**Chemcol CO-40.** Chemform Corp.
*Chemical Description:* Polyethoxylated triglyceride (40-mole ethylene oxide adduct of castor oil)
*Category:* Emulsifier, lubricant, dispersant, antistat, nonionic
*Applications:* Agricultural chemicals; textiles; promotes dye leveling
*Form:* Viscous liquid; Color: Yellow
*Properties:* Density: 8.8–8.9 lb/gal; HLB: 13.1; Cloud Pt.: > 100 C; Pour Pt.: 10 C; pH: 6.5–7.5
*Composition:* 97% active ingred.

**Chemcol HE-1.** Chemform Corp.
*Chemical Description:* Anionic/nonionic blended surfactant
*Category:* Agricultural emulsifiers, anionic/nonionic
*Applications:* Improves soft water performance of emulsifiable concentrates; used to couple polar materials into nonpolar formulations
*Form:* Slightly viscous liquid; Color: Amber
*Properties:* Density: 8.3 lb/gal; Visc.: 80 cps @ 20 C; Flash Pt.: > 100 F; Pour Pt.: < 5 C
*Toxicity:* Skin and eye irritant on prolonged contact
*Containers:* Bulk and 55 gal drums

**Chemcol HE-2.** Chemform Corp.
*Category:* Agricultural emulsifier, nonionic/anionic
*Applications:* Improves hard water performance of many toxicant formulations
*Form:* Viscous liquid; Color: Light amber
*Properties:* Density: 8.6 lb/gal; Visc.: 146 cps @ 20 C; Flash Pt.: > 100 F; Pour Pt.: 17 C; pH: 4–7 (5% in 25% IPA)
*Toxicity:* Skin and eye irritant on prolonged contact
*Containers:* Bulk and 55 gal drums

**Chemcol HE-8.** Chemform Corp.
*Category:* Agricultural emulsifier, nonionic/anionic
*Applications:* Improves spontaneity and emulsion stability of organo-phosphate insecticides
*Form:* Viscous liquid; Color: Light amber
*Properties:* Density: 8 lb/gal; Visc.: 1850 cps @ 20 C; Flash Pt.: > 100 F; Pour Pt.: 15 C; pH: 4–7 (5% in 25% IPA)
*Toxicity:* Skin and eye irritant on prolonged contact
*Containers:* Bulk and 55 gal drums

**Chemcol JL-60.** Chemform Corp.
*Chemical Description:* Polyoxyethylene ether alcohol
*Category:* Detergent, dispersant, scouring agent, leveling agent, nonionic

*Applications:* Cosmetics, textiles
*Form:* Cloudy liquid; Color: White
*Properties:* Density: 8.2 lb/gal; Cloud Pt.: < 25 C; Pour Pt.: 4.5–7.5 C; pH: 6–8
*Composition:* 100% active ingred.

**Chemcol JL-120.** Chemform Corp.
*Chemical Description:* Polyoxyethylene ether alcohol
*Category:* Detergent, dispersant, scouring and leveling agents, nonionic
*Applications:* Textiles, detergents, cosmetics, agricultural chemicals
*Form:* Semisolid; Color: White
*Properties:* Density: 8.4 lb/gal; Pour Pt.: Pourable liquid @ 50 C; pH: 6–8
*Composition:* 100% active ingred.

**Chemcol JL-180.** Chemform Corp.
*Chemical Description:* Polyoxyethylene ether alcohol
*Category:* Detergent, wetting agent, dispersant, scouring and leveling agent, nonionic
*Applications:* Textiles
*Form:* White waxy solid
*Properties:* Cloud Pt.: > 100 C; Pour Pt.: Pourable liquid @ 50 C; pH: 6–8
*Composition:* 100% active ingred.

**Chemcol MS-910.** Chemform Corp.
*Category:* Agricultural chemical, nonionic/anionic
*Applications:* Miscible surfactant for organic arsenicals
*Form:* Clear liquid; Color: < 6 (Gardner); Odor: Mild, pleasant
*Properties:* Density: 9.1 lb/gal; Sp. gr.: 1.09 @ 20 C; pH: 5.5–6.5 (10% sol'n.)
*Composition:* 100% active ingred.
*Toxicity:* Skin and eye irritant on prolonged contact
*Containers:* Bulk and drum; corrosive to ferrous surfaces; store in stainless steel, fiberglass lined tanks

**Chemcol NPE-40.** Chemform Corp.
*Chemical Description:* Nonyl phenol ethoxylate (the reaction product of nonyl phenol with 4 moles of ethylene oxide)
*Category:* Detergent, stabilizer, coemulsifier, dispersant, nonionic
*Applications:* Textile, household cleaners, agricultural chemicals, petroleum oil dispersant
*Form:* Liquid; Color: Pale yellow
*Properties:* Solubility: Oil sol.; Density: 8.5 lb/gal; HLB: 8.8; Cloud Pt.: Insoluble (1% in water); Flash Pt.: > 400 F (TOC); Pour Pt.: < 0 C; pH: 6–8 (5% in 25% IPA)
*Composition:* 100% active ingred.

**Chemcol NPE-60.** Chemform Corp.
*Chemical Description:* Nonyl phenol ethoxylate (reaction product of nonyl phenol with 6 moles of ethylene oxide)
*Category and Applications:* See Chemcol NPE-40
*Form:* Liquid; Color: Pale yellow
*Properties:* Solubility: Water sol.; Density: 8.7 lb/gal; HLB: 10.8; Cloud Pt.: Insoluble (1% in water); Flash Pt.: > 400 F (TOC); Pour Pt.: < 10 C; pH: 6–8 (5% in 25% IPA)
*Composition:* 100% active ingred.

**Chemcol NPE-100.** Chemform Corp.
*Chemical Description:* Nonyl phenol ethoxylate (reaction of nonyl phenol with 10 moles ethylene oxide)
*Category and Applications:* See Chemcol NPE-40
*Form:* Liquid; Color: Colorless to pale yellow
*Properties:* HLB: 13.4; Cloud Pt.: 58–64 C (1% in water); pH: 6–8 (5% in 25% IPA)
*Composition:* 100% active ingred.

**Chemcol NPE-140.** Chemform Corp.
*Chemical Description:* Nonyl phenol ethoxylate (reaction product of nonyl phenol with 14 moles of ethylene oxide)
*Category and Applications:* See Chemcol NPE-40
*Form:* Viscous liquid to off-white solid
*Properties:* Density: 8.9 lb/gal; HLB: 14.7; Cloud Pt.: 92–96 C (1% in water); Flash Pt.: > 450 F (TOC); Pour Pt.: ≈ 24 C; pH: 3.5–4.5 (10% sol'n.)
*Composition:* 100% active ingred.

**Chemcol NPE-200.** Chemform Corp.
*Chemical Description:* Nonyl phenol ethoxylate (reaction product of nonyl phenol with 20 moles of ethylene oxide)
*Category and Applications:* See Chemcol NPE-40
*Form:* Off-white solid
*Properties:* Density: Solid; HLB: 16.0; Cloud Pt.: 67–73 C (1% in 10% NaCl); Flash Pt.: > 450 F (TOC); Pour Pt.: ≈ 36 C; pH: 3.5–4.5 (10% sol'n.)
*Composition:* 100% active ingred.

**Chemcol NPE-400.** Chemform Corp.
*Chemical Description:* Nonyl phenol ethoxylate (reaction product of nonyl phenol with 40 moles of ethylene oxide)
*Category and Applications:* See Chemcol NPE-40
*Form:* Off-white solid
*Properties:* Density: Solid; HLB: 17.7; Cloud Pt.: > 100 C (1% in water); Flash Pt.: > 450 F (TOC); Pour Pt.: ≈ 48 C; pH: 3.5–4.5 (10% sol'n.)
*Composition:* 100% active ingred.

**Chemcol PE-34.** Chemform Corp.
*Chemical Description:* Blended anionic/nonionic surfactant
*Category:* Emulsifier, anionic/nonionic
*Applications:* Emulsifier for agricultural chemicals esp. Propanil and where mesityl oxide is the principal solvent; used in soft and hard waters
*Form:* Viscous liquid; Color: Dark amber; Odor: Mild, aromatic
*Properties:* Density: 8.9 lb/gal @ 40 C; Sp. gr.: 1.067 @ 40 C; Visc.: 550 cps @ 40 C; Pour Pt.: < 0 C; pH: 3.5–4.5 (10% sol'n.)
*Composition:* 100% active ingred.
*Toxicity:* Skin and eye irritant on prolonged contact
*Containers:* Bulk or drum quantities; store in phenolic lined equipment
*Warnings and Caution:* Flush with water in case of contact with skin or eyes; flammable liquid

**Chemcol PS-160.** Chemform Corp.
*Chemical Description:* Complex phosphate ester of an ethylene oxide adduct
*Category:* Detergent, emulsifier
*Applications:* Agricultural chemicals; emulsion polymerization; good electrolyte tolerance
*Form:* Clear liquid; Color: Straw; Odor: Mild
*Properties:* Density: 9.2 lb/gal; Sp. gr.: 1.12; Pour Pt.: < 0 C
*Composition:* 100% active ingred.

**Chemcol T-10.** Chemform Corp.
*Chemical Description:* Ethoxylated fatty amine (reaction of tallow-amine with 10 moles of ethylene oxide)
*Category:* Emulsifier, antistat, dispersant, cationic
*Applications:* Textiles
*Form:* Liquid; Color: 8 (Gardner)
*Properties:* Density: 8.2 lb/gal; Cloud Pt.: 80 ± 10 C; Pour Pt.: 54 F
*Composition:* 100% active ingred.

**Chemcol T-15.** Chemform Corp.
*Chemical Description:* Ethoxylated fatty amine (reaction of tallow-amine with 15 moles of ethylene oxide)
*Category:* Emulsifier, antistat, dispersant, cationic
*Applications:* Textiles
*Form:* Liquid; Color: Amber
*Properties:* Density: 8.5 lb/gal; Cloud Pt.: > 100 C; Pour Pt.: < 5 C; pH: 11.74
*Composition:* 100% active ingred.

**Chemcol T-20.** Chemform Chem.
*Chemical Description:* Ethoxylated fatty amine (reaction of tallow-amine with 20 moles of ethylene oxide)
*Category:* Emulsifier, antistat, dispersant, cationic
*Applications:* Textiles
*Form:* Liquid; Color: Amber
*Properties:* Density: 8.7 lb/gal; Cloud Pt.: > 100 C; Pour Pt.: 10 C; pH: 11.59
*Composition:* 100% active ingred.

**Chemcol T-25.** Chemform Corp.
*Chemical Description:* Ethoxylated fatty amine (reaction of tallow-amine with 25 moles of ethylene oxide)
*Category:* Emulsifier, cationic
*Applications:* Textiles
*Form:* Solid; Color: 6 (Gardner)
*Properties:* Density: 8.8 lb/gal; Cloud Pt.: > 100 C; Pour Pt.: 110 F
*Composition:* 100% active ingred.

**Chemcol TO-10.** Chemform Corp.
*Chemical Description:* Polyethoxylated mixed rosin and fatty acid (reaction of tall-oil with 10 moles of ethylene oxide)
*Category:* Emulsifier, lubricant, dye assistant, nonionic
*Applications:* Textiles
*Form:* Liquid; Color: Yellow
*Properties:* Sp. gr.: 0.872; Pour Pt.: < 50 F; Saponification No.: 77 ± 5
*Composition:* 100% active ingred.

**Chemcol TO-15.** Chemform Corp.
*Chemical Description:* Polyethoxylated mixed rosin and fatty acid

(reaction of tall-oil with 15 moles of ethylene oxide)
*Category:* Emulsifier, lubricant, dye assistant, nonionic
*Applications:* Textiles
*Form:* Solid; Color: 8 (Gardner)
*Properties:* Sp. gr.: 0.866; Pour Pt.: 80 F; Saponification No.: 59 ± 5; pH: 6–8
*Composition:* 100% active ingred.

**Chemcol TO-16.** Chemform Corp.
*Chemical Description:* Polyethoxylated mixed rosin and fatty acid (reaction of tall-oil with 16 moles of ethylene oxide)
*Category:* Emulsifier, lubricant, dye assistant, nonionic
*Applications:* Textiles
*Form:* Liquid; Color: Light amber
*Properties:* Density: 9 lb/gal; HLB: 13.2; Pour Pt.: 68 F; pH: 7–8
*Composition:* 97% active ingred.

**Chemcol VCC-5.** Chemform Corp.
*Chemical Description:* Anionic/nonionic blend of metal sulfonates and polyethoxylates
*Category:* Emulsifier, anionic/nonionic
*Applications:* Multi-purpose pesticide emulsifier
*Form:* Liquid; Color: Amber
*Properties:* Density: 8.49 lb/gal; Sp. gr.: 1.019; Flash Pt.: 170 F (TOC); pH: 4–6 (5% sol'n. in 3:1 water:IPA)
*Composition:* 100% active ingred.
*Toxicity:* Avoid contact with skin and eyes; flush with water in case of contact
*Containers:* Store and transfer using carbon steel equipment
*Warnings and Caution:* Contains flammable solvents

**Chemical Base 6532.** Sandoz Colors and Chem.
*Chemical Description:* Stearamido ethyl diethalamine
*Category:* Emulsifier, cationic
*Applications:* Conditioning agent for skin and hair
*Form:* Solid; Color: Yellowish tan; Odor: Slight
*Properties:* M.P.: 35–45 C; Stability: Good
*Composition:* 100% active ingred.
*Containers:* 250 lb fiber drum
*References:* Bulletin 7-254/78

**Chemstick 56.** Chemform Corp.
*Applications:* Spray sticker designed to increase the effectiveness of wettable powder and emulsifiable concentrate for agricultural sprays
*Form:* Liquid; Color: Amber
*Properties:* Density: 7.56 lb/gal

**Cholesteryl Acetate.** Van Schuppen chemie
*Chemical Description:* Acetic ester of a high quality cholesterol
*Category:* Raw material
*Applications:* As a starting material for the synthesis of various steroid derivatives, esp. vitamin $D_3$; in liquid crystal compositions
*Form:* Microcrystalline powder; Color: White or faintly yellow; Odor: Odorless
*Properties:* Solubility: Sol. in acetic acid, alcohol, chloroform, dioxane, ether, ethyl acetate, hexane, pyridine; insol. in water; M.P.:

111.5–115 C; Acid No.: 0.3 max.
*References:* Product information No. 3A

**Cholesteryl Adipate.** Van Schuppen chemie
*Applications:* Production of thermochromic liquid crystals and temperature-insensitive liquid crystals for industrial and medical fields and in displays; as a starting material for pharmaceuticals (e.g., vitamin $D_3$ and hormones) and for cosmetic preparations
*Form:* Crystalline solids; Color: White to faint yellow
*Properties:* M.P.: 184 C

**Cholesteryl Benzoate.** Van Schuppen chemie
*Applications:* See Cholesteryl Adipate
*Form:* Crystalline solids; Color: White to faint yellow
*Properties:* M.P.: 149 C

**Cholesterol B.P.** Van Schuppen chemie
*Category:* Raw material, emulsifier
*Applications:* A basic material for many pharmaceuticals, e.g., vitamin $D_3$, steroid hormones and for liquid crystals; also as a natural emulsifier in cosmetic preparations (characteristic emolleint feel)
*Form:* Powder; Color: White; Odor: Almost odorless
*Properties:* Solubility: Sol. in acetic acid, acetone, alcohol, benzene, chloroform, dioxane, ether, ethyl acetate, hexane, pyridine, toluene and vegetable oils; insol. in water; M.P.: 145–149 C; Acid No.: 0.3 max.
*References:* Product information No. 1 B

**Cholesteryl Butoxy Ethyl Carbonate.** Van Schuppen chemie
*Applications:* See Cholesteryl adipate
*Form:* Crystalline solids; Color: White to faint yellow
*Properties:* M.P.: 65 C

**Cholesteryl Butyrate.** Van Schuppen
*Applications:* See Cholesteryl adipate
*Form:* Crystalline solids; Color: White to faint yellow
*Properties:* M.P.: 97 C

**Cholesterol Techn.** Van Schuppen Chemie
*Category:* Raw material, emulsifier, nonionic
*Applications:* Raw material for alkoxylated compounds; W/O emulsifier for cosmetics
*Form:* Powder; Color: White to faintly yellow; Odor: Odorless
*Properties:* Solubility: Sol. in isopropyl myristate, isostearic acid, hexadecyl alcohol, chlorinated hydrocarbons and common organic solvents; insol. in water; M.P.: 135–140 C; Acid No.: 1.0 max.
*References:* Product information No. 1D

**Cholesterol USP.** Van Schuppen chemie
*Category:* Raw material, emulsifier
*Applications:* Raw material for

pharmaceuticals, e.g., vitamin $D_3$, steroid hormones, for liquid crystals; natural emulsifier in cosmetic preparations (emollient feel)
*Form:* Crystalline powder; Color: White; Odor: Almost odorless
*Properties:* Solubility: Sol. in acetic acid, acetone, alcohol, benzene, chloroform, dioxane, ether, ethyl acetate, hexane, pyridine, toluene and vegetable oils; insol. in water; M.P.: 147–150 C; Acid No.: 0.3 max.
*References:* Product information No. 1A

**Cindet 201.** Cindet Chemicals
*Chemical Description:* Blend of ethoxylated alcohols and aryl sulfonates
*Category:* Detergent, dyeing assistant, dispersing agent, penetrant, anionic
*Applications:* Detergent for scouring wool, cotton, synthetics; in dyeing procedures at neutral or alkaline pH
*Form:* Liquid; Color: Light amber
*Properties:* Stability: Stable in hard water, alkali media and moderate concentrations of acids; pH: 7.5–8.5 (1% sol'n.)
*Composition:* 25% active ingred.

**Cindet 542.** Cindet Chemicals
*Chemical Description:* Polyoxyethylene compound
*Category:* Dye assistant, scouring and soaping off agent, detergent, retarding and leveling agent, nonionic
*Applications:* Preparation, dyeing and finishing of textiles; wet processing of natural and synthetic fibers; retarding and leveling agent in dyeing of polyamide, polyester, acrylic fibers; detergent as afterscour; thermosol dye assistant
*Form:* Liquid; Color: Water white
*Properties:* Stability: Stable in hard water and acid and alkali media; Fast wetting speed; medium foaming; excellent rinsability; pH: 6.5–7.5 (1% sol'n.)
*Composition:* 50% active ingred.
*References:* Bulletin TB-270-2

**Cindet AL.** Cindet Chemicals
*Chemical Description:* Sodium alcohol ether sulfate
*Category:* Dye assistant, dispersant, penetrant, scouring agent, anionic
*Applications:* Scouring agent for wool, cotton, synthetics; dyeing procedures at neutral or alkaline pH; desizing
*Form:* Liquid; Color: Light amber
*Properties:* Stability: Stable to alkali media; Very free rinsing, fast wetting speed, medium foaming; pH: 7.0–8.0 (1% sol'n.)
*Composition:* 60% active ingred.

**Cindet AP.** Cindet Chemicals
*Chemical Description:* Modified blend of fatty nonionic and sulfonated hydrocarbons
*Category:* Wetting agent, scouring agent, afterscour, anionic
*Applications:* Wetting agent for causticizing and bleach operations; improves wetting speed and levelness in hydrogen peroxide bleach systems; soaping-off aid on con-

tinuous dyeing operations
*Form:* Transparent liquid; Color: Amber
*Properties:* Solubility: Water sol.; pH: 7.5–8.5 (1% sol'n.)

**Cindet C-12.** Cindet Chemicals
*Chemical Description:* Coconut amine condensate
*Category:* Detergent, emulsifier, wetting agent, scouring agent, lubricant, fulling agent, dyeing assistant, anionic
*Applications:* Scouring baths, fulling formulations for wool, scouring agent and detergent as afterscour
*Form:* Viscous liquid; Color: Amber
*Properties:* Stability: Stable in alkaline media; Medium wetting and foaming properties; good rinsability; pH: 8.5–9.5 (1% sol'n.)
*Composition:* 99% active ingred.
*References:* Bulletin TB-270-10

**Cindet GE.** Cindet Chemicals
*Chemical Description:* Isopropylamine salt of dodecyl benzene sulfonic acid
*Category:* Emulsifier, anionic
*Applications:* Emulsifier in polyester and triacetate dye carriers; emulsifier with varsol, xylene, etc. in solvent scours
*Form:* Viscous liquid; Color: Amber
*Properties:* Solubility: Sol. in low concentrations in water, completely sol. in many solvents; Medium foaming; pH: 7.2–7.8 (1% sol'n.)
*Composition:* 94% active ingred.

**Cindet JS.** Cindet Chemicals
*Chemical Description:* Polyoxyethylene compound
*Category:* Dye assistant, scouring agent, retarding and leveling agent, soaping off agent, detergent, nonionic
*Applications:* Nonfoaming surfactant in scouring and dyeing in the Gaston County Jet Machine; desizing woven goods; retarding and leveling agent in dyeing of polyamide, polyester, acrylic fibers; afterscour
*Form:* Liquid; Color: Light amber
*Properties:* Stability: Stable in acid and alkaline media; Excellent rinsability; pH: 6.5–7.5 (1% sol'n.)

**Cindet K-1.** Cindet Chemicals
*Chemical Description:* Blend of synthetic detergents, phosphates, and other alkalis
*Category:* Detergent, scouring agent, anionic
*Applications:* Scouring agent for cotton, rayon, nylon, polyester, polyolefin fibers; kier boiling of cotton prior to bleaching; free rinsing; in hard water conditions
*Form:* Powder; Color: White
*Properties:* Stability: Stable in alkaline media; Medium wetting speed and foaming; excellent rinsability; pH: 10–11 (1% sol'n.)
*Composition:* 100% active ingred.

**Cindet NAK.** Cindet Chemicals
*Chemical Description:* Blend of organic compounds
*Category:* Detergent; fulling, scouring and soaping agent; dye dispersant, anionic

*Applications:* Dye assistant as detergent, dispersant, fulling, scouring and soaping agent; removes dye carriers after dyeing hydrophobic fibers; oil, grease, soil removal
*Form:* Viscous liquid; Color: Light amber
*Properties:* Stability: Stable to alkali, hard water and peroxides; low foaming, medium wetting speed, excellent rinsability; pH: 7.5–8.5 (1% sol'n.)
*Composition:* 50% active ingred.
*References:* Bulletin TB-270-5

**Cindet R.** Cindet Chemicals
*Chemical Description:* Fatty polyoxethylene compound
*Category:* Dyeing assistant, scouring agent, retarder, detergent as afterscour, emulsifier, nonionic
*Applications:* Preparation, dyeing, and finishing of textiles; used in neutral, acid or alkaline media for wet processing of natural or synthetic fibers; oil and grease emulsifier; soaping off after dyeing
*Form:* Liquid; Color: Light amber
*Properties:* Stability: Stable in acid and alkaline media; Medium wetting speed, medium to low foaming, excellent rinsability; pH: 6.5–7.5 (1% sol'n.)
*Composition:* 70% active ingred.
*References:* Bulletin TB-270-1

**Cinwet A-35.** Cindet Chemicals
*Chemical Description:* Blend of anionic and nonionic surfactants
*Category:* Detergent, dyeing assistant, dispersant, penetrant
*Applications:* Detergent of superior wetting speed for scouring wool, cotton, synthetics; oil and grease removal
*Form:* Clear liquid; Color: Slight yellow
*Properties:* Stability: Stable in acid and alkaline media in moderate concentration; Fast wetting speed, medium foaming, excellent rinsability; pH: 7–8 (1% sol'n.)
*Composition:* 35% active ingred.

**Cinwet BOS.** Cindet Chemicals
*Chemical Description:* Sulfated propyl oleate
*Category:* Dyeing assistant, dispersant, leveling agent, wetting agent, lubricant, anionic
*Applications:* Dispersant and leveling agent in dyeing of polyester and polyamide; wetting agent in bleaching and dyeing of cellulosic fibers; coning emulsions; rewetting properties
*Form:* Liquid; Color: Amber
*Properties:* Stability: Stable in alkaline media and moderately acid media; Medium to fast wetting speed; low foaming; excellent rinsability; pH: 6–7 (1% sol'n.)
*Composition:* 65% active ingred.

**Cithrol 2ML.** Croda Chemicals
*Chemical Description:* Polyethylene glycol 200 monolaurate
*Category:* Wetting agent, emulsifier, detergent, thickening agent, solubilizer, dispersant, softening agent, lubricant, antistat
*Applications:* Cosmetics, textiles, glass fiber, metal treatment
*Properties:* m.w.: 200; HLB: 8.8; Saponification Value: 158–168

**Cithrol 4ML.** Croda Chemicals
*Chemical Description:* Polyethylene glycol 400 monolaurate
*Category:* Emulsifier, dye assistant, penetrant, detergent, nonionic
*Applications:* Component of paper and paperboard; in closures for food containers; as viscosity stabilizer and antistat in plastics industry; emulsifier for DDT and Chordane insecticides; cotton and rayon dyeing; scouring, desizing and dewashing of textiles
*Properties:* m.w.: 400; Solubility: Water sol.; HLB: 19; Saponification Value: 92–98

**Cithrol 6ML.** Croda Chemicals
*Chemical Description:* Polyethylene glycol 600 monolaurate
*Category:* See Cithrol 2ML
*Applications:* See Cithrol 2ML; also particularly suited for preparation of clear cod oil concentrates for dispersion in water to form fat liquor emulsions for leather treatment
*Properties:* m.w.: 600; HLB: 15.8; Saponification Value: 64–74

**Cithrol 10ML.** Croda Chemicals
*Chemical Description:* Polyethylene glycol 1000 monolaurate
*Category and Applications:* See Cithrol 2ML; esp. as emulsifier for oleic acid and colloidal silica; low foam detergent for mechanical washing machines; in leather industry as emulsifier and penetrant in straight sperm oil emulsions
*Properties:* m.w.: 1000; Solubility: Sol. in water, alcohol, polar solvents; HLB: 18; Saponification Value: 51–55

**Cithrol 60ML.** Croda Chemicals
*Chemical Description:* Polyethylene glycol 6000 monolaurate
*Category and Applications:* See Cithrol 2ML
*Properties:* m.w.: 6000; HLB: 19.2; Saponification Value: 8–13

**Cithrol 2MO.** Croda Chemicals
*Chemical Description:* Polyethylene glycol 200 monooleate
*Category:* Wetting agent, penetrant, detergent, emulsifier, solubilizer, thickening agent, dispersant, textile auxiliary, lubricant
*Applications:* O/W emulsifier for cosmetics; textile industry as dyeing, dispersing and softening agents; lubricants in metal working processes; emulsifier for soluble oils and insecticides
*Properties:* m.w.: 200; HLB: 6.2; Saponification Value: 110–112

**Cithrol 4MO.** Croda Chemicals
*Chemical Description:* Polyethylene glycol 400 monooleate
*Category and Applications:* See Cithrol 2MO
*Properties:* m.w.: 400; HLB: 11.4; Saponification Value: 85–93

**Cithrol 6MO.** Croda Chemicals
*Chemical Description:* Polyethylene glycol 600 monooleate
*Category and Applications:* See Cithrol 2MO
*Properties:* m.w.: 600; HLB: 13.1;

Saponification Value: 65–75

**Cithrol 10MO.** Croda Chemicals
*Chemical Description:* Polyethylene glycol 1000 monooleate
*Category and Applications:* See Cithrol 2MO
*Properties:* m.w.: 1000; HLB: 16.2; Saponification Value: 35–40

**Cithrol 40MO.** Croda Chemicals
*Chemical Description:* Polyethylene glycol 4000 monooleate
*Category and Applications:* See Cithrol 2MO
*Properties:* m.w.: 4000; HLB: 18.7; Saponification Value: 10–15

**Cithrol 60MO.** Croda Chemicals
*Chemical Description:* Polyethylene glycol 6000 monoleate
*Category and Applications:* See Cithrol 2MO
*Properties:* m.w.: 6000; HLB: 19.0; Saponification Value: 5–10

**Cithrol 2MS.** Croda Chemicals
*Chemical Description:* Polyethylene glycol 200 monostearate
*Category:* Nonionic emulsifier, detergent, wetting agent, solubilizer, thickening agent, anti-frothing agent, foaming agent
*Applications:* O/W and W/O emulsifiers; cosmetics as O/W emulsifiers and wetting agents; solubilizers and thickeners in perfumery; emulsifiers for insecticides; esp. useful in preparing emulsions of high electrolyte content
*Properties:* m.w.: 200; M.P.: 39–41 C; HLB: 6.3; Saponification Value: 110–120

**Cithrol 3MS.** Croda Chemicals
*Chemical Description:* Polyethylene glycol 300 monostearate
*Category and Applications:* See Cithrol 2MS
*Properties:* m.w.: 300; M.P.: 30–33 C; HLB: 10.7; Saponification Value: 95–105

**Cithrol 4MS.** Croda Chemicals
*Chemical Description:* Polyethylene glycol 400 monostearate
*Category and Applications:* See Cithrol 2MS
*Properties:* m.w.: 400; M.P.: 31–34 C; HLB: 11.0; Saponification Value: 95–105

**Cithrol 6MS.** Croda Chemicals
*Chemical Description:* Polyethylene glycol 600 monostearate
*Category and Applications:* See Cithrol 2MS
*Properties:* m.w.: 600; M.P.: 33–35 C; HLB: 14.0; Saponification Value: 68–76

**Cithrol 10MS.** Croda Chemicals
*Chemical Description:* Polyethylene glycol 1000 monostearate
*Category and Applications:* See Cithrol 2MS
*Properties:* m.w.: 1000; M.P.: 34–40 C; HLB: 16.0; Saponification Value: 36–50

**Cithrol 15MS.** Croda Chemicals
*Chemical Description:* Polyethylene glycol 1500 monostearate
*Category and Applications:* See

Cithrol 2MS
*Properties:* m.w.: 1500; HLB: 17.0; Saponification Value: 30–55

**Cithrol 40MS.** Croda Chemicals
*Chemical Description:* Polyethylene glycol 4000 monostearate
*Category and Applications:* See Cithrol 2MS
*Properties:* m.w.: 4000; HLB: 18.8; Saponification Value: 12–17

**Cithrol 'A'.** Croda Chemicals
*Chemical Description:* Poloxyethylene glycol oleate
*Category:* Emulsifier, dispersant, antistat, nonionic
*Applications:* Antistatic agent for textile processing; emulsions of mineral and vegetable oils, waxes, and fatty acids; emulsion-type metal cleaners, polishers, cutting oils; emulsions for leather and paper trades; emulsions of casein for the lacquer industry; oil emulsions for rubber latex; rag pulling oils and wool lubricants
*Form:* Liquid; Color: Amber; Odor: Slight agreeable
*Properties:* Solubility: Sol. in white spirit and liquid paraffin; completely sol. in xylol, ethyl acetate, methylated spirits, oleic acid; dispersible in water; Sp. gr.: 1.013; Flash Pt.: ≈ 215 C; Saponification Value: 85–93; pH: 6.5–7.5 (10% sol'n.)
*Containers:* Mild steel drums

**Cithrol GMS A/S ES 0743.** Croda Chemicals
*Chemical Description:* Glyceryl monostearate
*Category:* Emulsifier
*Applications:* Emulsifier and opacifier for cosmetics, esp. antiperspirants based on astringent aluminum salts, acidic skin creams, hair rinses and bleaches; base for pharmaceutical creams and ointments
*Form:* Color: 3 (Gardner)
*Properties:* M.P.: ≈ 52 C; Saponification Value: ≈ 96; Stability: Stable in acids, solutions of electrolytes and with anionics and cationics

**Clindrol 101CG.** Clintwood Chem.
*Chemical Description:* Alkanolamide
*Category:* Emulsifier, detergent, thickener, nonionic
*Applications:* Emulsifier of aliphatic hydrocarbons; for hydrocarbon propellents and aerosol packaged products; household uses as detergent
*Form:* Liquid; Color: Dark amber; Odor: Typical alkanolamide
*Properties:* Solubility: Sol. in alcohols, glycols, ketones, esters, aromatic and aliphatic hydrocarbons; Sp. gr.: 0.99; F.P.: 6 C
*Composition:* 100% active ingred.
*Reaction:* Alkaline
*Toxicity:* Skin and eye irritant
*Containers:* 55 gal tight head steel drums (450# net each)

**Clindrol 101LI.** Clintwood Chem.
*Chemical Description:* Lauric isopropanolamide
*Category:* Foam stabilizer and

thickener, nonionic
*Applications:* Foam stabilizer and thickener in rug shampoo, hand cleaners, hair shampoo; thickener for alpha olefin sulfonate solutions
*Form:* Solid flakes; Color: Tan
*Properties:* Solubility: Sol. in alcohols, glycols, ketones, esters, aromatic and chlorinated hydrocarbons; F.P.: 53–60 C; Stability: Stable
*Composition:* 100% active ingred.
*Reaction:* Alkaline
*Toxicity:* Skin and eye irritant
*Containers:* 55 gal open head fiber drums 200# net each

**Clindrol 200CG.** Clintwood Chem.
*Chemical Description:* Coconut oil alkanolamide
*Category:* Detergent, wetting agent, foam stabilizer, thickener, nonionic
*Applications:* Detergent; general purpose cleaners, metal cleaners, hard surface cleaners for household use, textiles; Kier boiling
*Form:* Liquid; Color: Dark amber; Odor: Typical alkanolamide
*Properties:* Solubility: Sol. in water, alcohols, glycols, ketones, aromatic and chlorinated hydrocarbons; Sp. gr.: 1.00; F.P.: –8 C
*Composition:* 100% active ingred.
*Reaction:* Alkaline
*Toxicity:* Skin and eye irritant
*Containers:* 55 gal tight head steel drums 450# net each

**Clindrol 200 CGN.** Clintwood Chem.
*Chemical Description:* Coconut oil diethanolamide
*Category:* Detergent, wetting agent, emulsifier, foam stabilizer, thickener, nonionic/anionic
*Applications:* Built hard surface cleaners, hand cleaners, metal degreasers, burnishing compounds, floor cleaners, wax strippers; textiles, laundering and household uses; desizing and boiling off cottons; Kier boiling, making of cutting and soluble oils, rinseable degreasers
*Form:* Liquid; Color: Dark amber; Odor: Typical alkanolamide
*Properties:* Solubility: Sol. in alcohols, water, ketones, aromatic and chlorinated hydrocarbons; Sp. gr.: 0.99; F.P.: 0 C
*Composition:* 100% active ingred.
*Reaction:* Alkaline
*Toxicity:* Skin and eye irritant
*Containers:* 55 gal tight head steel drums 450# net each

**Clindrol 200HC.** Clintwood Chem.
*Chemical Description:* Hydrogenated coconut fatty acid alkanolamide
*Category:* Detergent, emulsifier, foam stabilizer, thickener, nonionic
*Applications:* Hair shampoos, metal cleaners; household and cosmetic uses
*Form:* Liquid; Color: Straw; Odor: Typical alkanolamide
*Properties:* Solubility: Sol. in alcohols, water, aromatic and chlorinated hydrocarbons, ketones; Sp. gr.: 1.02; F.P.: 14–15 C
*Composition:* 100% active ingred.
*Reaction:* Alkaline
*Toxicity:* Skin and eye irritant

*Containers:* 55 gal tight head steel drums 450# net each

**Clindrol 200L.** Clintwood Chem.
*Chemical Description:* Lauric acid diethanolamide
*Category:* Detergent, wetting agent, emulsifier, foam stabilizer and thickener, nonionic
*Applications:* Foam stabilizer and thickener in hair shampoos, bubble bath and liquid dishwash detergent; hard surface cleaners; household and cosmetic uses
*Form:* Paste; Color: White; Odor: Typical alkanolamide
*Properties:* Solubility: Sol. in alcohols, water, ketones, glycols, aromatic hydrocarbons; Sp. gr.: 1.02; M.P.: 25–30 C
*Composition:* 100% active ingred.
*Reaction:* Alkaline
*Toxicity:* Skin and eye irritant
*Containers:* 55 gal open head steel drums 425# net each

**Clindrol 200-O.** Clintwood Chem.
*Chemical Description:* Diethanolamide of oleic acid
*Category:* Emulsifier, thickening agent, nonionic
*Applications:* Emulsifier for higher boiling aliphatic hydrocarbons; rinseable degreasers, fire safe fuels, thickener for amphoteric surfactants (shampoos, bubble baths); making of cutting and soluble oils; household and cosmetic uses
*Form:* Liquid; Color: Amber
*Properties:* Solubility: Sol. in alcohols, glycols, aliphatic and aromatic hydrocarbons, ketones, chlorinated hydrocarbons; Sp. gr.: 0.99; F.P.: –8 C
*Composition:* 100% active ingred.
*Reaction:* Alkaline
*Toxicity:* Skin and eye irritant
*Containers:* 55 gal tight head steel drums 450# net each

**Clindrol 200MS.** Clintwood Chem.
*Chemical Description:* Alkanolamide
*Category:* Pearl producing agent
*Applications:* Hair shampoos, hand cleansers, skin care products; household and cosmetics uses
*Form:* Flakes; Color: Tan; Odor: Ammoniacal
*Properties:* Solubility: Sol. in alcohols, glycols, esters, ketones, aromatic hydrocarbons; Sp. gr.: 0.998; M.P.: 98 C
*Composition:* 100% active ingred.
*Reaction:* Alkaline
*Toxicity:* Skin and eye irritants
*Containers:* 55 gal open head fiber drums 200# net each

**Clindrol 200RC.** Clintwood Chem.
*Chemical Description:* Coconut fatty acid alkanolamide
*Category:* Detergent, wetting agent, emulsifier, solubilizing agent, nonionic
*Applications:* Solubilizing agent, emulsifier in insecticide compositions, dye solvent for use in inks in printing industry, foam stabilizer in hair shampoos and other cosmetics; household detergents
*Form:* Liquid; Color: Amber
*Properties:* Solubility: Sol. in alcohols, glycols, water, ketones, aromatic and chlorinated hydrocarbons; Sp. gr.: 1.02

*Composition:* 100% active ingred.
*Reaction:* Alkaline
*Toxicity:* Skin and eye irritant
*Containers:* 55 gal tight head steel drums, 450# net each

**Clindrol 200S.** Clintwood Chem.
*Chemical Description:* Stearic acid derived alkanolamide
*Category:* Emulsifier, thickener and dispersing agent, nonionic
*Applications:* Emulsifier for skin, hair care products, other cosmetics (creams and lotions); hand cleaners, waterless hand cleaners; gel shampoos; porcelain cleaners; abrasive metal cleaners
*Form:* Solid; Color: White
*Properties:* Solubility: Sol. in alcohols, glycols, ketones, esters, aromatic and chlorinated hydrocarbons; M.P.: 36–40 C
*Composition:* 100% active ingred.
*Reaction:* Alkaline
*Toxicity:* Skin and eye irritant
*Containers:* 55 gal open head steel drums, 400# net each

**Clindrol 201CGN.** Clintwood Chem.
*Chemical Description:* Coconut alkanolamide
*Category:* Detergent, emulsifier, wetting agent, foam stabilizer, thickener
*Applications:* Built hard surface cleaners, hand cleaners, metal degreasers, burnishing compounds, floor cleaners, wax strippers; textiles, laundering, household uses; desizing and boiling off cottons; Kier boiling; making of cutting and soluble oils; rinseable degreasers
*Form:* Liquid; Color: Dark amber; Odor: Typical alkanolamide
*Properties:* Solubility: Sol. in alcohols, water, ketones, aromatic and chlorinated hydrocarbons; Sp. gr.: 0.99; F.P.: 0 C
*Composition:* 100% active ingred.
*Reaction:* Alkaline
*Toxicity:* Skin and eye irritant
*Containers:* 55 gal tight head steel drums, 450# net each

**Clindrol 202CGN.** Clintwood Chem.
*Chemical Description:* Coconut oil alkanolamide
*Category:* Detergent, emulsifier, wetting agent, foam stabilizer and thickener, nonionic/anionic
*Applications:* Built hard surface cleaners, hand cleaners, metal degreasers, burnishing compounds, floor cleaners, wax strippers, steam cleaners; textiles, laundering, transport and household uses as detergents; desizing and boiling off cottons; Kier boiling; making of cutting and soluble oils; rinseable degreasers
*Form:* Liquid; Color: Dark amber; Odor: Typical alkanolamide
*Properties:* Solubility: Sol. in alcohols, water, ketones, aromatic and chlorinated hydrocarbons; Sp. gr.: 0.99; F.P.: 0 C
*Composition:* 100% active ingred.
*Reaction:* Alkaline
*Toxicity:* Skin and eye irritant
*Containers:* 55 gal tight head steel drums, 450# net each

**Clindrol 203CG.** Clintwood Chem.
*Category:* Detergent, wetting a-

gent, foam stabilizer, thickener
*Applications:* Detergent, general purpose cleaners, metal cleaners, hard surface cleaners, cleaner-sanitizers with quaternary ammonium compounds; textiles, household and cosmetic uses; Kier boiling
*Form:* Liquid; Color: Straw to light amber; Odor: Typical alkanolamide
*Properties:* Solubility: Sol. in alcohols, water, glycols, ketones, aromatic and chlorinated hydrocarbons; Sp. gr.: 1.00; F.P.: –8 C
*Composition:* 100% active ingred.
*Reaction:* Alkaline
*Toxicity:* Skin and eye irritant
*Containers:* 55 gal tight head steel drums 450# net each

**Clindrol 204CGN.** Clintwood Chem.
*Chemical Description:* Coconut oil alkanolamide
*Category:* Detergent, emulsifier, wetting agent, foam stabilizer and thickener
*Applications:* Built hard surface cleaners, hand cleaners, metal degreasers, burnishing compounds, floor cleaners, wax strippers; textiles, laundering and household uses as detergent; desizing and boiling off cottons; Kier boiling; making of cutting and soluble oils; rinseable degreasers
*Form:* Liquid; Color: Straw to light amber; Odor: Typical alkanolamide
*Properties:* Solubility: Sol. in alcohols, water, ketones, aromatic and chlorinated hydrocarbons; Sp. gr.: 0.99; F.P.: 0 C
*Composition:* 100% active ingred.
*Reaction:* Alkaline
*Toxicity:* Skin and eye irritant
*Containers:* 55 gal tight head steel drums, 450# net each

**Clindrol 206CGN.** Clintwood Chem.
*Chemical Description:* Coconut alkanolamide
*Category:* Detergent, emulsifier, wetting agent, conveyor chain lubricant, anionic
*Applications:* Hard surface cleaner and fabric cleaner; household and industrial hard surface cleaners, car wash, dishwash, food machinery cleaner, fine fabric cleaner; detegent in transport and cosmetics industries; wetting agent for removal of spots, insecticides, herbicides, fungicides
*Form:* Liquid; Color: Dark amber
*Properties:* Solubility: Sol. in water, alcohols, glycols, esters, ketones, aromatic and chlorinated hydrocarbons; Sp. gr.: 1.02; F.P.: –5 C
*Composition:* 100% active ingred.
*Reaction:* Alkaline
*Toxicity:* Skin and eye irritant
*Containers:* 55 gal tight head steel drums, 450# net each

**Clindrol 207CGN.** Clintwood Chem.
*Chemical Description:* Coconut alkanolamide
*Category:* Detergent, emulsifier, wetting agent, conveyor chain lubricant, anionic
*Applications:* Household and industrial hard surface cleaners, car wash, dishwash, food machinery cleaner, fine fabric cleaner, con-

veyor chain lubricant; detergent in laundering, foods, transport, cosmetics industries; as wetting agent in removal of spots and in insecticides, herbicides, fungicides
*Form:* Liquid; Color: Amber
*Properties:* Solubility: Sol. in water, alcohols, glycols, esters, ketones, aromatic and chlorinated hydrocarbons; Sp. gr.: 1.02; F.P.: –5 C
*Composition:* 100% active ingred.
*Reaction:* Alkaline
*Toxicity:* Skin and eye irritant
*Containers:* 55 gal tight head steel drums, 450# net each

**Clindrol 209CGN.** Clintwood Chem.
*Chemical Description:* Coconut alkanolamide
*Category:* Detergent, wetting agent, emulsifier, solubilizer, nonionic/anionic
*Applications:* Solubilizing agent and emulsifier for pesticides; agricultural wetting agent; household detergents; degreasers
*Form:* Liquid; Color: Amber
*Properties:* Solubility: Sol. in water, alcohols, glycols, glycol ethers, aromatic and chlorinated hydrocarbons; Sp. gr.: 1.06; M.P.: –7 to –3 C
*Composition:* 100% active ingred.
*Reaction:* Alkaline
*Toxicity:* Skin and eye irritant
*Containers:* 55 gal tight head steel drums, 450# net each

**Clindrol 210CGN.** Clintwood Chem.
*Chemical Description:* Coconut alkanolamide
*Category:* Detergent, wetting agent, emulsifier, conveyor chain lubricant, nonionic/anionic
*Applications:* Conveyor chain lubricants, rinseable degreasers, making of cutting and soluble oils
*Form:* Liquid; Color: Amber
*Properties:* Solubility: Sol. in water, alcohols, glycols, esters, ketones, aromatic and aliphatic hydrocarbons; Sp. gr.: 0.99; M.P.: 0–5 C
*Composition:* 100% active ingred.
*Reaction:* Alkaline
*Toxicity:* Skin and eye irritant
*Containers:* 55 gal tight head steel drums, 450# net each

**Clindrol 868.** Clintwood Chem.
*Chemical Description:* Fatty acid alkanolamide
*Category:* Emulsifier, thickener, nonionic
*Applications:* Gel shampoos and hand cleaners; household and cosmetic applications
*Form:* Solid; Color: White
*Properties:* Solubility: Sol. in alcohols, ketones, chlorinated and aromatic solvents; M.P.: 58–63 C
*Composition:* 100% active ingred.
*Reaction:* Alkaline
*Toxicity:* Skin and eye irritant
*Containers:* Cubitainers, 40# net each

**Clindrol SDG.** Clintwood Chem.
*Chemical Description:* Diethylene glycol monostearate
*Category:* Emulsifier, opacifying agent, nonionic
*Applications:* Auxiliary emulsifier and opacifier in skin and hair care products

*Form:* Solid; Color: White
*Properties:* Solubility: Sol. in alcohols, glycols, chlorinated, aromatic and aliphatic hydrocarbons; Sp. gr.: 1.01; M.P.: 42–45 C
*Composition:* 100% active ingred.
*Reaction:* Neutral
*Containers:* 40 lb net cubitainers

**Clindrol SEG.** Clintwood Chem.
*Chemical Description:* Ethylene glycol monostearate
*Category:* Opacifier and pearl producing agent, nonionic
*Applications:* Hair shampoos, liquid dishwash detergent, hair and skin care products, hand cleaners
*Form:* Flakes; Color: White; Odor: Slightly fatty
*Properties:* Solubility: Sol. in alcohols, glycols, esters, ketones, aromatic, aliphatic and chlorinated hydrocarbons; Sp. gr.: 0.998; M.P.: 52–60 C
*Composition:* 100% active ingred.
*Reaction:* Neutral
*Containers:* 55 gal open head fiber drums, 200# net each

**Clindrol Superamide 100C.** Clintwood Chem.
*Chemical Description:* Coconut diethanolamide
*Category:* Foam stabilizer and thickener, nonionic
*Applications:* Hair shampoos, bubble bath, liquid dishwash detergent, hand cleaners
*Form:* Liquid; Color: Straw
*Properties:* Solubility: Sol. in alcohols, glycols, ketones, aromatic and chlorinated hydrocarbons; Sp. gr.: 0.99; F.P.: 8 C; Stability: Stable
*Composition:* 100% active ingred.
*Reaction:* Alkaline
*Toxicity:* Skin and eye irritant
*Containers:* 55 gal open head drums, 425# net each

**Clindrol Superamide 100CG.** Clintwood Chem.
*Chemical Description:* Coconut diethanolamide
*Category:* Foam stabilizer and thickener, nonionic
*Applications:* Foam stabilizer and thickener in hair shampoos, bubble bath, liquid dishwash detergents, hand cleaners
*Form:* Liquid; Color: Straw to amber
*Properties:* Solubility: Sol. in alcohols, glycols, water, aromatic and chlorinated hydrocarbons; Sp. gr.: 0.99; F.P.: –3 C
*Composition:* 100% active ingred.
*Reaction:* Alkaline
*Toxicity:* Skin and eye irritant
*Containers:* 55 gal tight head steel drums, 450# net each

**Clindrol Superamide 100L.** Clintwood Chem.
*Chemical Description:* Lauric diethanolamide
*Category:* Foam stabilizer and thickener, nonionic
*Applications:* Foam stabilizer in aqueous systems; in shampoos, bubble bath, dishwash detergent; thickener in gel shampoos
*Form:* Solid; Color: White; Odor: Typical alkanolamide
*Properties:* Solubility: Sol. in alco-

hols, ketones, chlorinated and aromatic solvents; Sp. gr.: 0.99; M.P.: 42–46 C; Stability: Stable
*Composition:* 100% active ingred.
*Reaction:* Alkaline
*Toxicity:* Skin and eye irritant
*Containers:* 55 gal open head steel drums, 435# net; cubitainers, 40# net

**Clindrol Superamide 100LM.** Clintwood Chem.
*Chemical Description:* Diethanolamide of 70% lauric–30% myristic acid
*Category:* Foam stabilizer and thickener, nonionic
*Applications:* Foam stabilizer in aqueous systems; for hair shampoos, bubble bath, liquid dishwash detergent; thickener in gel shampoos
*Form:* Paste, solid; Color: White
*Properties:* Solubility: Sol. in alcohols, ketones, chlorinated and aromatic solvents; Sp. gr.: 0.99; M.P.: 29–33 C; Stability: Stable
*Composition:* 100% active ingred.
*Reaction:* Alkaline
*Toxicity:* Skin and eye irritant
*Containers:* 55 gal open head steel drums, 425# net each

**Clindrol 100MC.** Clintwood Chem.
*Chemical Description:* Monoethanolamide of coconut fatty acid
*Category:* Foam stabilizer, nonionic
*Applications:* Foam stabilizer in shampoo and liquid dishwash detergent
*Form:* Flakes; Color: Tan; Odor: Ammoniacal
*Properties:* Solubility: Sol. in ethanol, isopropanol, aliphatic and aromatic hydrocarbons; Sp. gr.: 1.0; M.P.: 63 C
*Composition:* 100% active ingred.
*Reaction:* Alkaline
*Toxicity:* Skin and eye irritant
*Containers:* 55 gal open head fiber drums, 200# net each

**CM-88.** Georgia-Pacific
*Chemical Description:* Aqueous alkaline solution of a lignoamine
*Category:* Modifier, cationic
*Applications:* Replaces 40 to 60% of a primary emulsifier thus providing a lower cost emulsifier system
*Form:* Solid
*Properties:* Density: 1.25 g/ml; Solidification: 46%

**CMTC-T.** Hercules Inc.
*Chemical Description:* Sodium carboxymethylcellulose
*Category:* Cellulose gum, ionic
*Applications:* Used as a synthetic detergent to prevent redeposition of soil that has been removed from fabrics, built soaps, laundry sizes, imparting soil resistance to cotton carpeting, adhesions to cellulose yarns, binding and adhesive properties that make it useful alone in combination with other ingredients as an adhesive
*Form:* Powder
*Properties:* Solubility: Sol. in water
*Containers:* 50 lbs/bag

**Cobee 76.** PVO International
*Chemical Description:* Refined,

bleached, deodorized coconut oil
*Category:* Emollient, superfatting agent, clouding agent
*Applications:* Food industry as frying oil, ice cream coatings, cracker spray, chewy candies; hair conditioners; pharmaceutical vehicles
*Form:* Soft solid; Color: 20/2 (Lovibond); Odor and Taste: None
*Properties:* M.P.: 76–80 F

**Cobee 92.** PVO International
*Chemical Description:* Refined, bleached, deodorized, hydrogenated coconut oil
*Category:* Emollient, superfatting agent, clouding agent
*Applications:* Cream fillings, nonchewy candies, icings, sandwich fillings, snack roasting, frozen dessert, peanut roasting; vegetable dairy industry; soaps and shampoos as emollients and superfatting agents
*Form:* Solid; Color: 20/2 (Lovibond); Odor and Taste: None
*Properties:* M.P.: 98–101 F

**Cobee 110.** PVO International
*Chemical Description:* Refined, bleached, deodorized, hydrogenated coconut oil
*Category and Applications:* See Cobee 92
*Form:* Solid; Color: 20/2 (Lovibond); Odor and Taste: None
*Properties:* M.P.: 110–114 F

**Complemix 50.** American Cyanamid Co.
*Category:* Wetting agent
*Applications:* Food and beverage;
*Form:* Liquid; Color: Nearly colorless
*Properties:* Sp. gr.: 0.93; B.P.: 173 F
*Composition:* 50% active ingred.; other components: beverage grade ethanol
*Toxicity:* Eye irritant

**Complemix 100.** American Cyanamid Co.
*Category:* Wetting agent
*Applications:* Food and beverage
*Form:* Solid; Color: White
*Properties:* Solubility: 1.5 g in 100 ml water @ 25 C; Sp. gr.: 1.1
*Composition:* 100% active ingred.
*Toxicity:* Eye irritant
*Warnings and Caution:* Avoid eye contact

**Compound 8-S.** DuPont Co.
*Chemical Description:* Complex alkylaryl sulfonate
*Category:* Dispersing agent, anionic
*Applications:* Nonfoaming dispersant for disperse- and vat-type dyes; application to clays; dispersing agent for carbon black
*Form:* Liquid
*Properties:* Solubility: Water-sol.; Density: 9.6 lb/gal; Sp. gr.: 1.152; Visc.: Readily pourable; Stability: Freezing affects but does not damage physical form; thaw and mix before using
*References:* Bulletin

**Compound 9264B.** PVO Internat'l.
*Chemical Description:* Ethoxylated glycerol monostearate

*Category:* Softener, nonionic
*Form:* Solid

**Conco AAS-98S.** Continental Chem. Co.
*Chemical Description:* Dodecyl benzene sulfonic acid
*Category:* Detergent intermediate, anionic
*Applications:* For manufacturing detergents by neutralizing dodecyl benzene sulfonic acid; manufacture of shampoos, detergent, metal cleaners
*Form:* Liquid; Color: Brown
*Properties:* Solubility: Water sol.; Sp. gr.: 1.04; Biodegradable
*Composition:* 98+% active ingred.; other components: water and dodecyl benzene
*Reaction:* Acid
*Containers:* 55 gal open head drums (450 lb net), tankwagons or tankcars
*Warnings and Caution:* Acid is corrosive (salts of the acid are noncorrosive)
*References:* Technical bulletin

**Conco ATR-98S.** Continental Chem. Co.
*Chemical Description:* Tridecyl benzene sulfonic acid
*Category:* Detergent intermediate, anionic
*Applications:* Detergent intermediate; neutralized with wide range of alkalies
*Form:* Liquid
*Properties:* Sp. gr.: 1.04; Biodegradable
*Composition:* 98+% active ingred.
*Reaction:* Acid
*Containers:* 55 gal drums
*Warnings and Caution:* Acid is corrosive
*References:* Technical bulletin

**Condanol 42/FAS.** Dutton and Reinisch Ltd.
*Chemical Description:* Surfactant blend
*Category:* Detergent
*Applications:* Shampoo base
*Form:* Liquid; Color: Pale yellow; Odor: Faint, typical
*Properties;* Sp. gr.: ≈ 1.0; Visc.: Medium
*Composition:* 25% active ingred.; other components: water
*Reaction:* Neutral
*Containers:* Lacquered drums

**Condanol B 2003.** Dutton and Reinisch Ltd.
*Chemical Description:* Tetra-sodium (1,2-dicarboxyethyl)–N-alkyl sulfosuccinamide
*Category:* Detergent, emulsifier, foaming agent, foam stabilizer, solubilizer, anionic
*Applications:* Industrial and metal cleaners; dishwashing liquids, household and general purpose cleaners; carpet and upholstery shampoos; soldering aid; reduction of gel formation; dispersion of pigments; emulsifier for wax and oil; cosmetic preparations
*Form:* Clear liquid; Color: Amber
*Properties:* Solubility: Sol. in water and electrolytes; Stability: Good esp. in presence of inorganic salts
*Composition:* 35% solids

**Condanol CMS.** Dutton and Reinisch Ltd.

*Chemical Description:* Disodium coconut sulphosuccinamate
*Category:* Emulsifier, foaming agent
*Form:* Liquid; Color: Pale yellow; Odor: Slight typical
*Properties:* Sp. gr.: ≈ 1.0
*Composition:* 40% active ingred.; other components: water
*Reaction:* Slight alkaline
*Containers:* Lacquered drums

**Condanol CS 30.** Dutton and Reinisch Ltd.
*Chemical Description:* Surfactant blend
*Category:* Detergent
*Applications:* Carpet shampoo base
*Form:* Liquid; Color: Pale yellow; Odor: Of formalin
*Properties:* Sp. gr.: ≈ 1.0; Visc.: Low
*Composition:* 80% active ingred.; other components: water
*Reaction:* Slight acid
*Containers:* Lacquered drums

**Condanol DLS 35.** Dutton and Reinisch Ltd.
*Chemical Description:* Diethanolamine lauryl sulfate
*Category:* Detergent
*Applications:* Shampoo base
*Form:* Liquid; Color: Pale yellow; Odor: Slight typical
*Properties:* Sp. gr.: ≈ 1.0; Visc.: Low
*Composition:* 35% active ingred.; other components: water
*Reaction:* Neutral
*Containers:* Lacquered drums

**Condanol Emulsifier BW.** Rewo Chem.
*Chemical Description:* Modified nonionic fatty acid derivative
*Category:* Emulsifier, nonionic
*Applications:* Cleansing liquids, hand cleaners, cleaning of oily and greasy surfaces
*Form:* Clear viscous liquid; Color: Yellow
*Properties:* Solubility: Oil sol.
*Composition:* 97% solids; 3% water

**Condanol Emulsifier BWA.** Dutton and Reinisch Ltd.
*Chemical Description:* Modified nonionic fatty acid derivative
*Category:* Emulsifier, nonionic
*Applications:* Metal cleaning formulations as emulsifier for engine cleaners, solvent degreasers; emulsifying paraffin wax; emulsifier for cold cleaners to de-wax cars
*Form:* Clear viscous liquid; Color: Yellow
*Composition:* 97% solids; 3% water

**Condanol FBR.** Dutton and Reinisch Ltd.
*Chemical Description:* Detergent combination
*Category:* Detergent, anionic
*Applications:* Raw material for weakly alkaline household and general-purpose cleaners; removal of latex based floor waxes and polishes; corrosion-inhibitors
*Form:* Clear viscous liquid
*Properties:* Density: 0.98 g/cm$^3$ @ 20 C; Visc.: 350 cp (Brookfield spindle 3)
*Composition:* 98% solids; 2% water

**Condanol FLM.** Dutton and Reinisch Ltd.
*Category:* Anionic foam fire-extinguisher base
*Applications:* Raw material for low-density foam fire-extinguisher
*Form:* Clear liquid; Color: Light yellow
*Properties:* Density @ 20 C: 1.0–1.1 g/cm$^3$; Clear Pt.: –12 C; Biodegradable

**Condanol M 5047.** Dutton and Reinisch Ltd.
*Chemical Description:* Detergent combination
*Category:* Detergent
*Applications:* Surface cleaner preconcentrate
*Form:* Clear liquid; Color: Colorless
*Composition:* 45% solids content

**Condanol MLS 35.** Dutton and Reinisch Ltd.
*Chemical Description:* Monoethanolamine lauryl sulfate
*Category:* Detergent
*Applications:* Shampoo base
*Form:* Liquid; Color: Pale yellow; Odor: Slight typical
*Properties:* Sp. gr.: ≈ 1.0; Visc.: High
*Composition:* 35% active ingred.; other components: water
*Reaction:* Neutral
*Containers:* Lacquered drums

**Condanol NL2.** Dutton and Reinisch Ltd.
*Chemical Description:* Sodium lauryl alcohol polyglycol ether sulfate (approx. 2 mole ethylene oxide)
*Category:* Detergent
*Applications:* Shampoo base
*Form:* Liquid; Color: Water white; Odor: Slight typical
*Properties:* Sp. gr.: ≈ 1.0; Visc.: Medium
*Composition:* 30% active ingred.; other components: water
*Reaction:* Neutral
*Containers:* Lacquered drums

**Condanol NL3.** Dutton and Reinisch Ltd.
*Chemical Description:* Sodium lauryl polyglycol ether sulfate (3 moles ethylene oxide)
*Category:* Detergent
*Applications:* Shampoo base
*Form:* Liquid; Color: Water white; Odor: Slight typical
*Properties:* Sp. gr.: ≈ 1.0; Visc.: Low
*Composition:* 30% active ingred.; other components: water
*Reaction:* Neutral
*Containers:* Lacquered drums

**Condanol NL 3, 28%.** Dutton and Reinisch Ltd.
*Chemical Description:* Sodium lauryl ether sulfate
*Category:* Detergent, foaming agent, anionic
*Applications:* Cleaning formulations, high quality shampoos and foam bath preparations
*Form:* Clear liquid; Color: Almost colorless; Odor: Low
*Properties:* Visc.: 50–150 cp; Stability: Good temperature stability
*Composition:* 29% active ingred.

**Condanol NLS 28.** Dutton and Reinisch Ltd.
*Chemical Description:* Sodium lauryl sulfate

*Category:* Detergent, emulsifier, anionic
*Applications:* Washing and cleaning formulations; cosmetic preparations: foam baths and shampoos; emulsion polymerization
*Form:* Liquid; Color: Light yellow
*Properties:* m.w.: 300
*Composition:* 28% active ingred.

**Condanol NLS 30.** Dutton and Reinisch Ltd.
*Chemical Description:* Sodium lauryl sulfate
*Category:* Detergent
*Applications:* Shampoo base, emulsion polymerization
*Form:* Paste; Color: Off white; Odor: Slight typical
*Properties:* Sp. gr.: ≈ 1.0
*Composition:* 30% active ingred.; other components: water
*Reaction:* Alkaline
*Containers:* Lacquered drums

**Condanol NLS 90.** Dutton and Reinisch Ltd.
*Chemical Description:* Sodium lauryl sulfate
*Category and Applications:* See Condanol NLS 28
*Form:* Powder; Color: White to light yellow
*Composition:* 85% active ingred.

**Condanol OFR.** Dutton and Reinisch Ltd.
*Category:* Surface cleaner
*Applications:* Cleaning of synthetic car seats and furniture coverings, white-walled tires, baths, sinks, table-tops, floors, mirrors
*Form:* Clear, colorless liquid
*Composition:* 16% solids content

**Condanol S 1590.** Dutton and Reinisch Ltd.
*Chemical Description:* Detergent combination
*Category:* Detergent, anionic
*Applications:* Wash-active preconcentrate; dishwashing liquids
*Form:* Clear viscous liquid; Color: Yellow
*Properties:* Stability: Temperature stable down to 3 C
*Composition:* 44% solids content

**Condanol S 1605.** Dutton and Reinisch Ltd.
*Chemical Description:* Detergent combination
*Category:* Detergent, anionic; good wetting and emulsifying properties; foam stability
*Applications:* Liquid washing, rinsing, cleaning agents for household and industrial uses; skin-protecting properties
*Form:* Clear viscous liquid; Color: Yellow
*Composition:* 48% solids content

**Condanol SB3OL.** Dutton and Reinisch Ltd.
*Chemical Description:* Surfactant blend
*Category:* Detergent
*Applications:* Shampoo base, cream shampoos
*Form:* Liquid; Color: Pale yellow; Odor: Nil
*Properties:* Sp. gr.: ≈ 1.0; Visc.: Medium
*Composition:* 30% active ingred.; other components: water
*Reaction:* Neutral
*Containers:* Lacquered drums

**Condanol SBA/1.** Dutton and Reinisch Ltd.
*Chemical Description:* Sodium salt of the monosulfosuccinate of lauryl alcohol polyglycol ether (short chain)
*Category:* Detergent
*Applications:* Hair shampoos, foam baths, metal cleaning
*Form:* Paste; Color: Off white; Odor: Slight typical
*Properties:* Sp. gr.: ≈ 1.0
*Composition:* 40% active ingred.; other components: water
*Reaction:* Slight acid
*Containers:* Lacquered drums

**Condanol SBD060.** Dutton and Reinisch Ltd.
*Chemical Description:* Dioctyl sodium sulfosuccinate based on 2-ethyl hexanol
*Category:* Wetting agent
*Applications:* Super wetting agent with low foaming properties
*Form:* Liquid; Color: Pale yellow; Odor: Strong typical
*Properties:* Sp. gr.: ≈ 1.0; Visc.: Medium
*Composition:* 60% active ingred.; other components: water and isopropanol
*Reaction:* Neutral
*Containers:* Lacquered drums

**Condanol SB-DO 70.** Dutton and Reinisch Ltd.
*Chemical Description:* Sodium diisooctyl sulfosuccinate
*Category:* Wetting agent, solubilizer, anionic
*Applications:* Wetting agent in cosmetic preparations and household cleaners; cleaning lotions; nail-varnish removers; shampoos; foam baths; dishwashing agents; window and glass cleaners technical applications: metal cleaning, anti-smear agent for car windscreens; paper, textile and paint industries; pickling agents; dye industry; dry-cleaning process
*Form:* Almost colorless liquid
*Properties:* Stability: Good in pH range 3–9; Surface Tension: 30 dynes/cm (0.1% sol'n. @ 20 C); Biodegradable
*Composition:* 64% active ingred.

**Condanol SBF 12.** Dutton and Reinisch Ltd.
*Chemical Description:* Disodium lauryl alcohol sulfosuccinate
*Category:* Detergent, foam-improver, anionic
*Applications:* Cosmetic preparations; carpet and upholstery shampoos; synthetic bar soap formulations
*Form:* White paste
*Properties:* Biodegradable
*Composition:* 39% solids content

**Condanol SBF 12-Powder.** Dutton and Reinisch Ltd.
*Chemical Description:* Disodium lauryl alcohol sulfosuccinate
*Category:* Detergent, anionic
*Applications:* Synthetic soaps, normal soap bars, dry shampoos, enzyme washing powders, moisture sensitive technical products; used to improve the spray-drying properties of softer materials
*Form:* Powder; Color: White or light yellow

*Properties:* Solubility: Water-sol.; Density: 250 g/l; Stability: Good in weakly acid and weakly alkaline media; Biodegradable
*Composition:* 95% solids content

**Condanol SBFA/3.** Dutton and Reinisch Ltd.
*Chemical Description:* Sodium salt of the monosulfosuccinate of lauryl alcohol polyglycol ether (medium chain)
*Category:* Detergent
*Applications:* Mild hair shampoos and foam baths
*Form:* Liquid; Color: Water white
*Properties:* Sp. gr.: ≈ 1.0; Visc.: Low
*Composition:* 40% active ingred.; other components: water
*Reaction:* Neutral
*Containers:* Lacquered drums

**Condanol SBFA 30, 40%.** Dutton and Reinisch Ltd.
*Chemical Description:* Disodium lauryl alcohol polyglycol-ether sulfosuccinate
*Category:* Detergent raw material, anionic
*Applications:* Mild detergent base with good foaming properties; high quality shampoos, baby preparations, intimate hygiene preparations
*Form:* Clear liquid; Color: Almost colorless
*Properties:* Visc.: 100–200 cp; Biodegradable
*Composition:* 39% solids content

**Condanol SBIL/3.** Dutton and Reinisch Ltd.
*Chemical Description:* Sodium salt of the monosulfosuccinate of lauric isopropanolamide polyglycol ether (medium chain)
*Category:* Detergent
*Applications:* Hair shampoos
*Form:* Gel; Color: Pale yellow; Odor: Typical slight
*Properties:* Sp. gr.: ≈ 1.0; Visc.: High
*Composition:* 50% active ingred.; other components: water
*Reaction:* Slight acid
*Containers:* Lacquered drums

**Condanol SBL.** Dutton and Reinisch Ltd.
*Chemical Description:* Sodium salt of the monosulfosuccinate of lauric monoethanolamide
*Category:* Detergent, wetting agent
*Applications:* Carpet shampoo base; fire fighting formulations
*Form:* Liquid, paste; Color: Pale yellow; Odor: Slight typical
*Properties:* Sp. gr.: ≈ 1.0; Visc.: Low
*Composition:* 35% active ingred.; other components: water
*Reaction:* Slight acid
*Containers:* Lacquered drums

**Condanol SBL 203.** Dutton and Reinisch Ltd.
*Chemical Description:* Disodium lauric acid ethanolamide sulfosuccinate
*Category:* Detergent, anionic
*Applications:* Mild detergent, foam improver; cosmetics; aids spray-drying; carpet and uphols-

tery shampoos; synthetic soaps
*Form:* Turbid liquid; Color: Yellow
*Properties:* Biodegradable
*Composition:* 39% solids content

**Condanol SBV.** Dutton and Reinisch Ltd.
*Chemical Description:* Modified sulfosuccinate
*Category:* Detergent, foaming agent, anionic
*Applications:* Mild detergent, nonirritant shampoos and foam baths, rinsing and cleaning agents with skin-protecting properties
*Form:* Clear liquid; Color: Light yellow
*Properties:* Visc.: 2500 cp; Biodegradable
*Composition:* 28% solids content

**Condanol SBZ.** Dutton and Reinisch Ltd.
*Chemical Description:* Sodium salt of the monosulfosuccinate of lauric isopropanolamide polyglycol ether (medium chain)
*Category:* Detergent
*Applications:* Hair shampoos
*Form:* Gel; Color: Pale yellow; Odor: Slight typical
*Properties:* Sp. gr.: ≈ 1.0; Visc.: Medium
*Composition:* 50% active ingred.; other components: water
*Reaction:* Neutral
*Containers:* Lacquered drums

**Condanol SCK 2040.** Dutton and Reinisch Ltd.
*Chemical Description:* Detergent combination
*Category:* Detergent raw material, anionic
*Applications:* Preparation of liquid cleaning agents, shampoos
*Form:* Clear viscous liquid; Color: Yellow
*Properties:* Visc.: 600 cp; Biodegradable
*Composition:* 39% solids content

**Condanol TLS 40.** Dutton and Reinisch Ltd.
*Chemical Description:* Triethanolamine lauryl sulfate
*Category:* Detergent
*Applications:* Shampoo base
*Form:* Liquid; Color: Pale yellow; Odor: Slight typical
*Properties:* Sp. gr.: ≈ 1.0; Visc.: Low
*Composition:* 45% active ingred.; other components: water
*Reaction:* Neutral
*Containers:* Lacquered drums

**Condanol TMS.** Dutton and Reinisch Ltd.
*Chemical Description:* Disodium alkyl sulfosuccinamide
*Category:* Emulsifier, wetting agent, foam-additive; anionic
*Applications:* Foam-additive for natural and synthetic lattices; in carpet backing, foam mattresses, furniture foam blocks; emulsion polymerization
*Form:* Soft creamy paste or liquid
*Properties:* Stability: Stable in alkaline and acidic media; Biodegradable
*Composition:* 35% solids content

**Condanol TS 25.** Dutton and Reinisch Ltd.
*Chemical Description:* Sulfosuc-

cinate-base detergent combination
*Category:* Detergent, anionic
*Applications:* Concentrate for foam cleaners for coarse-fibered textiles (carpets, upholstery, car-seat covers)
*Form:* Liquid; Color: Light yellow
*Properties:* Clear Pt.: 8 C; Biodegradable
*Composition:* 25% solids content

**Condasol Sulfonic Acid K.** Dutton and Reinisch Ltd.
*Chemical Description:* Straight chain dodecylbenzene sulfonic acid
*Category:* Anionic surfactant
*Applications:* Dishwashing, washing and cleaning formulations
*Form:* Viscous liquid; Color: Brown
*Properties:* 90% biodegradable
*Composition:* 97% sulfonic acid; 1% sulfuric acid; 2% unsulfated matter

**Condasol T 65.** Rewo Chem.
*Chemical Description:* p-toluene sulfonic acid, 65%
*Category:* Catalyst, electrolyte, intermediate
*Applications:* Catalyst for preparation of esters, for acid hydrolysis, polymerization, dehydration, alkylation reactions; electrolyte; intermediate product in preparation of cresol
*Form:* Clear, pale liquid; Odor: Characteristic
*Composition:* 65% sulfonic acid; 1.5% sulfuric acid; 0.2% toluene; 0.3% sulfone
*Warnings and Caution:* Strong acid; rubber gloves and goggles should be worn

**Condasol TKS 90/F.** Dutton and Reinisch Ltd.
*Chemical Description:* Triammonium dodecylbenzenesulfonate
*Category:* Raw material for detergent formulations; anionic
*Applications:* Raw material for production of concentrated cleansing agents, steam-jet and high pressure cleaners, defatting agents, car shampoos; chain greasing; cleaning steel ball bearings
*Form:* Viscous liquid; Color: Dark amber
*Properties:* Biodegradable
*Composition:* 85% active ingred.

**Condensate 4476.** Proctor Chemical Company, Inc.
*Chemical Description:* Amine condensate of fatty acid
*Category:* Wetting agent, detergent, high foamer
*Applications:* Substitute for 100% coconut amine condensate, condensate PC
*Form:* Liquid; Color: 6–8 (Gardner)
*Properties:* Sp. gr.: 1.020
*Composition:* 88–90% active

**Condensate PC.** Proctor Chemical Company, Inc.
*Chemical Description:* Coconut-diethanolamine condensation product
*Category:* Wetting and rewetting agent, Detergent, high foamer, thickener, emulsifier
*Applications:* Excellent detergen-

cy on fabrics, scouring piece goods and yarns after naphthol dyeing to remove excess dyestuff adhering to the surface of the fiber, liquid dishwashing formulations, thickener even at concentrations of 5% to 10%, oil soluble emulsifier for preparing self-emulsifiable oil systems, formulating sanitary supply chemicals
*Form:* Liquid; Color: Amber
*Composition:* 100% active ingred.
*Containers:* 480 lb drum

**Corax.** Diamond Shamrock
*Chemcial Description:* Wax
*Applications:* Rubber lubricant for O-rings
*Form:* Solid
*Composition:* 100% active ingred.

**Corona Lanolin.** Croda Chem. Ltd.
*Chemical Description:* Anhydrous lanolin
*Category:* Conditioning emollient, moisturizing agent, co-solvent, plasticizer, W/O emulsifier, superfatting agent, wetting/dispersing agent
*Applications:* Pharmaceutical bases and salves, skin and hair care preparations, make-up cosmetics, soaps, protective creams

**Coronet Lanolin.** Croda Chem. Ltd.
*Chemical Description:* Super refined cosmetic grade lanolin
*Category and Applications:* See Corona lanolin
*Form:* Minimum odor, color and batch variation in this grade
*Properties:* Slightly higher M.P. than USP and BP standards

**Corrosion Inhibitor 500.** Alkaril Chem. Ltd.
*Category:* Corrosion inhibitor
*Applications:* Used where corrosion is caused by hydrochloric acid, organic acids and carbon dioxide; hydrochloric acid industrial cleaning products and pickling baths
*Form:* Clear liquid; Color: Deep amber
*Properties:* Solubility: Sol. in distilled water, aliphatic and aromatic solvents, 15% hydrochloric acid, 15% sulfuric acid, 30% phosphoric acid
*References:* Technical bulletin 101

**Corrosion Inhibitor 600.** Alkaril Chem. Ltd.
*Category:* Corrosion inhibitor
*Applications:* Filming type corrosion inhibitor when applied to steel; used in concentrated hydrochloric, phosphoric and sulfuric acids; concentrated mineral acid industrial cleaning products for pickling bath and drain cleaner applications
*Form:* Clear liquid; Color: Amber
*Properties:* Solubility: Sol. in hydrochloric, phosphoric and sulfuric acids; pH: 6.0–7.0
*References:* Technical bulletin 102

**CPA-Alpha.** Dixo Co. Inc.
*Chemical Description:* Anionic/nonionic detergent mixture
*Category:* Detergent, wetting agent, emulsifier
*Applications:* Detergent for dry-

cleaning solvent use to improve or accelerate soil removal; "prespotting" material in drycleaning; emulsifier for water and drycleaning solvents
*Form:* Liquid; Color: Very light yellow amber; Odor: Slight
*Properties:* Solubility: Completely sol. in drycleaning solvents
*Composition:* 100% active ingred.
*Containers:* 1, 5, 15, 30, and 55 gal containers

**Cremophor A 6.** BASF AG
*Chemical Description:* Ethoxylated saturated fatty alcohol
*Category:* Emulsifer, nonionic
*Applications:* Emulsifying agents for ointments, creams, liquid emulsions, gels and other cosmetic and pharmaceutical preparations
*Form:* Wax; Color: White
*Properties:* Ref. Index: 1.4420–1.4424 @ 60 C; Solubility: Sol. in: water, alcohol; Density: 0.896–0.906 g/cm$^3$ @ 60 C; HLB: 10–12; Sapon. Value: < 3; Acid No.: < 1; Hydroxyl No.: 115–135; Iodine Value: < 1; Stability: Stable to acids, bases and salts; Emulsifiability: Oil-in-water types of emulsion; pH: 6–7
*Composition:* 100% active ingred.
*References:* Technical data sheet M 5551 e

**Cremophor A 11.** BASF AG
*Chemical Description:* Ethoxylated saturated fatty alcohol
*Category:* Emulsifier, nonionic
*Applications:* Emulsifying agents for ointments, creams, liquid emulsions, gels and other cosmetic and pharmaceutical preparations
*Form:* Wax; Color: White
*Properties:* Ref. Index: 1.4464–1.4474 @ 60 C; Solubility: Sol. in: water, alcohol; Density: 0.964–0.968 g/cm$^3$ @ 60 C; HLB: 12–14; Sapon. Value: < 1; Acid No.: < 1; Hydroxyl No.: 70–80; Iodine Value: < 1; Stability: Stable to acids, bases and salts; Emulsifiability: Oil-in-water types of emulsion; pH: 6–7
*Composition:* 100% active ingred.
*References:* Technical Data sheet M 5551 e

**Cremophor A 25.** BASF AG
*Chemical Description:* Ethoxylated saturated fatty alcohol
*Category:* Emulsifier, nonionic
*Applications:* Emulsifying agents for ointments, creams, liquid emulsions, gels and other cosmetic and pharmaceutical preparations;
*Form:* Powder; Color: White
*Properties:* Ref. Index: 1.4512–1.4520 @ 60 C; Solubility: Sol. in: water, alcohol; Density: 1.020–1.028 g/cm$^3$ @ 60 C; HLB: 15–17; Sapon. Value: < 3; Acid No.: < 1; Hydroxyl No.: 35–45; Iodine Value: < 1; Stability: Stable to acids, bases and salts; Emulsifiability: Oil-in-water types of emulsions; pH: 5–7
*Composition:* 100% active ingred.
*References:* Technical Data sheet M 5551 e

**Cremophor EL.** BASF AG
*Chemical Description:* Fatty acid

esters of glycerolpolyglycols and polyglycols
*Category:* Solubilizer, emulsifier,, nonionic
*Applications:* Pharmaceuticals, cosmetics, veterinary medicine, and animal feeding stuffs; used for producing aqueous preparations of hydrophobic substances
*Form:* Liquid: Color: Pale yellow
*Properties:* Ref. Index: ≈ 1.471; Solubility: Sol. in water, ethyl alcohol, *n*-propyl alcohol, isopropyl alcohol, ethyl acetate, chloroform, carbon tetrachloride, trichlorethylene, benzene, toluene and xylene; Density: 1.05–1.06 ml; Visc.: 700–850 cps; HLB: 12–14; Sapon. Value: 65–70; Acid No.: < 2; Hydroxyl No.: 65–78; Iodine Value: 28–32; Stability: Stable to electrolytes, e.g., acids and salts; Emulsifiability: Fatty soluble vitamins A, D, E and K; ethereal oils or perfume oils; pH: 6–8 (10% aq.)
*Composition:* 100% active ingred.
*Warnings and Caution:* Should be stored in closed containers and protected from light
*References:* Technical data sheet M 1714 e

**Cremophor RH 40.** BASF AG
*Chemical Description:* Hydrophobic consisting of fatty acid glycerol polyglycol esters and fatty acid polyglycol esters
*Category:* Solubilizer, emulsifier, nonionic
*Applications:* Solubilization of essential oils and perfumery synthetics
*Form:* Paste; Color: White
*Properties:* Ref. Index: 1.453–1.457 @ 60 C; Solubility: Sol. in water, ethanol, isopropyl alcohol, *n*-propanol, ethyl acetate, chloroform, carbon tetrachloride, trichloroethylene, benzene, toluene and xylene; HLB: 14–16; Sapon Value: 50–60; Acid No.: < 1; Hydroxyl No.: 60–80; Iodine Value: < 1; Stability: Stable to aqueous alcohol; Emulsifiability: hydrophobic substances; pH: 6–7 (10% aq. sol'n.)
*Composition:* 100% active ingred.
*Warnings and Caution:* Drums or tanks should be kept tightly closed or microorganisms may grow
*References:* Technical data sheet M 5629 e

**Cremophor RH 60.** BASF
*Chemical Description:* Hydrophobic consisting of fatty acid glycerol polyglycol esters and fatty acid polyglycol esters
*Category:* Solubilizer, emulsifier, nonionic
*Applications:* Solubilization of essential oils and perfumery synthetics
*Form:* Paste; Color: White
*Properties:* Ref. Index: 1.453–1.457 @ 60 C; Solubility: Sol. in water, ethanol, isopropyl alcohol, *n*-propanol, ethyl acetate chloroform, carbon tetrachloride, trichloroethylene, benzene, toluene and xylene; HLB: 15–17; Sapon. Value: 40–50; Acid No.: < 1; Hydroxyl No.: 50–70; Iodine Value: < 1; Stability: Stable to aqueous alcohol; Emulsifiability:

Hydrophobic substances; pH: 6–7 (10% aq. sol'n.)
*Composition:* 100% active ingred.
*Warnings and Caution:* Drums or tanks should be kept tightly closed or microorganisms may grow
*References:* Technical data sheet M 5629 e

**Cremophor RH 410.** BASF
*Chemical Description:* Hydrophobic consisting of fatty acid glycerol polyglycol esters and fatty acid polyglycol esters
*Category:* Solubilizer, emulsifier, nonionic
*Applications:* Solubilization of essential oils and perfumery synthetics
*Form:* Liquid
*Properties:* Ref. Index: 1.457–1.462; Solubility: Sol. in water, ethanol, isopropyl alcohol, *n*-propanol, ethyl acetate, chloroform, carbon tetrachloride, trichloroethylene, benzene, toluene and xylene; Visc.: 1500–2000 cps; HLB: 14–16; Sapon. Value: 45–55; Acid No.: $< 1$; Iodine Value: $< 1$; Stability: Stable to aqueous alcohol; Emulsifiability: Hydrophobic substances; pH: 6–7 (10% aq.)
*Composition:* 90% active ingred.
*Warnings and Caution:* Drums or tanks should be kept tightly closed or microorganisms may grow
*References:* Technical data sheet M 5629 e

**Cremophor RH 455.** BASF AG
*Chemical Description:* Hydrophobic consisting of fatty acid glycerol polyglycol esters and fatty acid polyglycol esters
*Category:* Solubilizer, emulsifier, nonionic
*Applications:* Solubilization of essential oils and perfumery synthetics
*Form:* Liquid
*Properties:* Ref. Index: 1.459–1.464; Solubility: Sol. in water, ethanol, isopropyl alcohol, *n*-propanol, ethyl acetate, chloroform, carbon tetrachloride, trichloroethylene, benzene, toluene and xylene; Visc.: 1000–1500 cps; HLB: 14–16; Sapon. Value: 45–55; Acid No.: $< 1$; Iodine Value: $< 1$; Stability: Stable to aqueous alcohol; Emulsifiability: Hydrophobic substances; pH: 6–7 (10% aq.)
*Composition:* 90% active ingred.
*Warnings and Caution:* Drums or tanks should be kept tightly closed or microorganisms may grow
*References:* Technical data sheet M 5629 e

**Cremophor S 9.** BASF AG
*Chemical Description:* Ester of a polyethylene glycol with stearic acid
*Category:* Emulsifier, nonionic
*Applications:* Oil-in-water type emulsions; suspension stabilizer; thickening agent; good lubricating and anti-tack effect
*Form:* Solid; Color: Yellowish white
*Properties:* Ref. Index: ≈ 1.446 @ 60 C; Solubility: Sol. in water, ethyl alcohol, isopropyl alcohol,

*n*-propyl alcohol, acetone, ethyl acetate, chloroform, benzene, dimethyl phthalate, castor oil, cod liver oil and oleic acid; Density: ≈ 0.97 g/cm³ @ 60 C; HLB: ≈ 12; Sapon. Value: 88–98; Acid No.: 2; Hydroxyl No.: 80–105; Iodine Value: < 2; Stability: Stable to polyethylene glycols and other inert bases, electrolytes; Emulsifiability: Fatty oils, greases, vaseline, solid and liquid paraffin, beeswax, silicone oils, fatty alcohols
*Composition:* 100% active ingred.
*References:* Technical data sheet M 5542 e

**Cremophor WO-A.** BASF AG
*Chemical Description:* Fatty alcohol glycerol ether
*Category:* Emulsifier, nonionic
*Applications:* Water-in-oil emulsion-based cosmetics and pharmaceuticals
*Form:* Liquid; Color: Yellowish
*Properties:* Solubility: Sol. in mineral and native oils and fats; Sp. gr.: 0.950–0.980 g/cm³ @ 20 C; HLB: 4.8–5.8; Sapon. Value: < 5; Acid No.: < 0.1; Hydroxyl No.: 230–300; Iodine Value: 50–60; Stability: Stable to acids, bases and electrolytes
*Composition:* 100% active ingred.
*References:* Technical data sheet M 5614 e

**Crestalans.** Croda Chem.
*Chemical Description:* Derivatives of fractionated liquid lanolin and isopropyl esters
*Category:* Emollient, moisturizing agent, conditioners, plasticizers, super-fatting agents
*Applications:* Skin preparations, baby creams, lotions and oils, skin make-up, hair preparations; plasticizers for aerosol hair sprays and lacquers; super-fatting agents for nail varnish removers and similar cosmetic solvent compositions
*Form:* Clear liquid; Color: Pale yellow; Odor: Faint sterol

**Crill Series.** Croda Chem.
*Chemical Description:* Partial esters of sorbitan
*Category:* Nonionic surfactant, emulsifier, dispersant, co-solvent, wetting agents, antifoams, viscosity reducers, mold release agents, internal slip agents, antiblock agents
*Applications:* Cosmetics, food and food packaging, insecticides and herbicides, leather treatment, metal working fluids, oil slick dispersants, paints and inks, pharmaceuticals, plastics, polishes, textiles

**Crill 1.** Croda Chem.
*Chemical Description:* Sorbitan monolaurate
*Category:* Emulsifier, wetting and pigment dispersing aid, antifoam, corrosion inhibitor, lubricant, antistatic agent
*Applications:* Auxiliary emulsifier in cosmetics and pharmaceutical oil in water systems; antifoam in food and protein processing (e.g., milk, molasses, sugar and yeast); manufacture of paper board for food packaging; used in coating polyolefin films; dry cleaning

systems; leather treatment; metal and furniture polishes; pesticide emulsions; textile processing
*Form:* Clear viscous liquid; Color: Pale yellow
*Properties:* Solubility: Sol. in ethyl alcohol, oleyl alcohol, mineral oil; HLB: 8.6; Sap. Value: 158–170

**Crill 2**. Croda Chem.
*Chemical Description:* Sorbitan monopalmitate
*Category and Applications:* See Crill Series; may be used in many of the applications designated to Crill 3; leaves nongreasy residue
*Form:* Hard waxy solid; Color: Pale tan
*Properties:* Solubility: Partly sol. in propylene glycol, ethyl and oleyl alcohols, olive oil, oleic acid; M.P.: 46 C; HLB: 6.7; Sap. Value: 140–150

**Crill 3**. Croda Chem.
*Chemical Description:* Sorbitan monostearate
*Category:* Emulsifier, lubricant, antistat
*Applications:* O/W emulsions; cosmetic and pharmaceutical creams and lotions; silicone and wax polishes; insecticides, herbicides; metal cleaners; buffing compounds; textile lubricants; antistatic agents; food applications
*Form:* Hard waxy solid; Color: Pale tan; Odor: Low
*Properties:* Solubility: Partly sol. in oleyl alcohol, olive oil, oleic acid; M.P.: 51–54 C; HLB: 4.7; Sap. Value: 147–157

**Crill 4**. Croda Chem.
*Chemical Description:* Sorbitan monooleate
*Category:* Emulsifier, wetting agent, dispersant, lubricant, coupling agent, antifoam
*Applications:* Industrial W/O emulsifiers; cosmetic and pharmaceutical applications (as wetting agents and dispersants for medicaments and fillers, pigment dispersing aid in lipstick, eye makeup); aids dispersion for aerosol powder systems; aerosol polishes; insecticidal sprays; inks and surface coatings; metal working lubricants; cutting oils; textile lubricants; dry cleaning operations; food process antifoam; oil slick dispersant
*Form:* Viscous liquid; Color: Amber
*Properties:* Solubility: Sol. in ethyl and oleyl alcohols, mineral oil, IPM, olive oil, oleic acid; HLB: 4.3; Sap. Value: 147–160

**Crill 6**. Croda Chem.
*Chemical Description:* Sorbitan monoisostearate (based on a branched chain fatty acid; completely saturated ester)
*Category:* W/O emulsifier, wetting agent, pigment dispersant
*Form:* Viscous liquid; Color: Pale yellow
*Properties:* Solubility: Sol. in mineral oil, olive oil, partly sol. in oleyl alcohol, IPM; HLB: 4.7; Sap. Value: 143–153

**Crill 35**. Croda Chem.
*Chemical Description:* Sorbitan

tristearate
*Category and Applications:* See Crill 3
*Form:* Hard waxy solid; Color: Pale tan
*Properties:* Solubility: Partly sol. in oleyl alcohol, mineral oil, IPM, olive oil, oleic acid; M.P.: 48 C; HLB: 2.1; Sap. Value: 176–188

**Crill 43.** Croda Chem.
*Chemical Description:* Sorbitan sesquioleate
*Category and Applications:* See Crill 4
*Form:* Viscous liquid; Color: Amber
*Properties:* Solubility: Sol. in oleyl alcohol, mineral oil, olive oil, oleic acid; HLB: 3.7; Sap. Value: 149–160

**Crill 45.** Croda Chem.
*Chemical Description:* Sorbitan trioleate
*Category and Applications:* See Crill 4
*Form:* Viscous liquid; Color: Amber
*Properties:* Solubility: Sol. in oleyl alcohol, mineral oil, IPM, olive oil, oleic acid; HLB: 1.8; Sap. Value: 172–186

**Crillet Series.** Croda Chem.
*Chemical Description:* Polyethoxylated sorbitan esters
*Category:* Nonionic surfactants; emulsifier, wetting agent, solubilizer; often combined with a member of the Crill range in emulsification systems
*Applications:* Cosmetics, food and food packaging, household products, insecticides and herbicides, metal working fluids, paints and inks, pharmaceuticals, polishes and cleaners, polymers, textiles

**Crillet 1.** Croda Chem.
*Chemical Description:* POE 20 sorbitan monolaurate
*Category:* Solubilizer, emulsifier
*Applications:* Solubilizer for essential oils and perfumes in cosmetics; emulsifier and dispersant for pharmaceuticals (e.g., menthol, phenols); cosmetic creams and lotions; animal feedstuffs
*Form:* Clear liquid; Color: Yellow; Odor: Low
*Properties:* Solubility: Sol. in water, ethyl alcohol, oleyl alcohol, oleic acid; HLB: 16.7; Sap. Value: 40–51; Surface Tension: 38.5 dynes/cm (0.1% sol'n. @ 25 C)

**Crillet 2.** Croda Chem.
*Chemical Description:* POE 20 sorbitan monopalmitate
*Category:* Emulsifier, dispersant, penetrant, levelling agent, lubricant, antistat
*Applications:* Similar to Crillet 3 with slightly higher HLB
*Form:* Pasty liquid; Color: Yellow amber
*Properties:* Solubility: Sol. in water, ethyl alcohol, oleyl alcohol, oleic acid; HLB: 15.6; Sap. Value: 43–49; Surface Tension: 41.5 dynes/cm (0.1% @ 25 C)

**Crillet 3.** Croda Chem.
*Chemical Description:* POE 20 sorbitan monostearate

*Category:* Emulsifier, dispersant, penetrant, levelling agent, lubricant, antistat
*Applications:* As emulsifier, alone or with Crill 3, for cosmetics and pharmaceuticals; for solvents and polish ingredients; dispersant for insecticides, herbicides, cattle dyes; textile dyeing as penetrant and levelling agent; lubricant/antistat in synthetic fiber processing
*Form:* Yellow liquid gels to soft solid on cooling
*Properties:* Solubility: Sol. in ethyl and oleyl alcohol, oleic acid; partly sol. in water, isopropyl myristate, xylene, trichlorethylene; HLB: 14.9; Sap. Value: 45–55; Surface Tension: 42.5 dynes/cm (0.1% @ 25 C)

**Crillet 4.** Croda Chem.
*Chemical Description:* POE 20 sorbitan monooleate
*Category:* Emulsifier, dispersant, solubilizer, lubricant, detergent, antistat, wetting agent
*Applications:* Cosmetics and pharmaceuticals, often used with appropriate Crill; emulsifier for solvents, waxes, silicones, aerosol and nonaerosol polishes; oil slick dispersant, marine tank cleaners; synthetic animal feedstuffs, hand cleaners; insecticides, herbicides; leather degreasing; veterinary products
*Form:* Clear liquid; Color: Yellow amber; Odor: Faint characteristic fatty
*Properties:* Solubility: Sol. in water, ethyl and oleyl alcohols, oleic acid; partly sol. in isopropyl myristate, xylene; HLB:15.0; Sap. Value: 45–55; Surface Tension: 42.5 dynes/cm (0.1% @ 25 C)

**Crillet 6.** Croda Chem.
*Chemical Description:* POE 20 sorbitan monoisostearate
*Category and Applications:* See Crillet Series; Solubilizing characteristics similar to Crillet 3; low temperature liquidity and superior color, odor and chemical stability
*Form:* Clear liquid; Color: Yellow
*Properties:* Solubility: Sol. in water, ethyl and oleyl alcohols, oleic acid, xylene, trichlor-ethylene; HLB: 14.9; Sap. Value: 40–50; Surface Tension: 38.6 dynes/cm (0.1% @ 25 C)

**Crillet 11.** Croda Chem.
*Chemical Description:* POE 4 sorbitan monolaurate
*Category and Applications:* See Crillet 1
*Form:* Clear liquid; Color: Yellow
*Properties:* Solubility: Sol. in ethyl and oleyl alcohols, oleic acid, kerosene; partly sol. in water, isopropyl myristate, olive oil, xylene, trichlor-ethylene; HLB: 13.3; Sap. Value: 100–115; Surface Tension: 34.7 dynes/cm (0.1% @ 25 C)

**Crillet 31.** Croda Chem.
*Chemical Description:* POE 4 sorbitan monostearate
*Category and Applications:* See Crillet 3
*Form:* Waxy solid; Color: Cream pale yellow
*Properties:* Solubility: Sol. in oleyl

alcohol, oleic acid; HLB: 9.6; Sap. Value: 98–113; Surface Tension: 41.5 dynes/cm (0.1% @ 25 C)

**Crillet 35.** Croda Chem.
*Chemical Description:* POE 20 sorbitan tristearate
*Category and Applications:* See Crillet 3
*Form:* Waxy solid; Color: Cream/buff
*Properties:* Solubility: Sol. in ethyl and oleyl alcohols, oleic acid, trichlor-ethylene; partly sol. in water, mineral oil, isopropyl myristate, olive oil, kerosene, xylene; HLB: 10.5; Sap. Value: 88–98; Surface Tension: 42.5 dynes/cm (0.1% @ 25 C)

**Crillet 41.** Croda Chem.
*Chemical Description:* POE 5 sorbitan monooleate
*Category and Applications:* See Crillet 4
*Form:* Clear liquid; Color: Yellow amber
*Properties:* Solubility: Sol. in ethyl alcohol, oleyl alcohol, isopropyl myristate, oleic acid, kerosene, butyl stearate; HLB: 10.0; Sap. Value: 96–104; Surface Tension: 37 dynes/cm (0.1% @ 25 C)

**Crillet 45.** Croda Chem.
*Chemical Description:* POE 20 sorbitan trioleate
*Category and Applications:* See Crillet 4
*Form:* Clear viscous liquid; Color: Amber
*Properties:* Solubility: Sol. in ethyl and oleyl alcohols, isopropyl myristate, oleic acid, kerosene, trichlor-ethylene, butyl stearate; HLB: 11.0; Sap. Value: 82–95; Surface Tension: 41 dynes/cm (0.1% @ 25 C)

**Crodafos Series.** Croda Chem.
*Chemical Description:* Alkyl ether phosphates; available in acid form or neutralized with appropriate base
*Category:* Anionic surfactants, detergents, wetting agents, coupling agents, corrosion inhibitors, thickeners, foam stabilizers, lubricants, antistats
*Applications:* Detergents and cleaners (hard surface, floor, industrial, steam cleaners, dry cleaning process); metal working fluids, cutting oils, lubricating oils, soluble cutting oils; grinding fluids; anticorrosion agent; antistatic agent for various polymers; textiles (processing of synthetic fibers, fabric softening, dye baths, bleaching and scouring operations); herbicides, pesticides and fertilizers; emulsifiers for solvents and industrial specialties
*Form:* Many and varied
*Composition:* 99% min. active
*Warnings and Caution:* Treat Crodafos acid phosphates as acids; avoid eye and skin contact; flush with water in case of contact

**Crodafos 25D Series.** Croda Chem.
*Chemical Description:* Alkyl ether phosphates based on synthetic C12/C15 alcohols; available with 2, 5, and 10 moles ethylene oxide and as Crodafos 25D TEA salt

*Category and Applications:* See Crodafos Series
*Form:* Liquid to soft paste; Color: Colorless to pale yellow
*Properties:* pH: 1.8–2.5 (3% sol'n. @ 20 C)
*Composition:* 99% min. active

**Crodafos CS Series.** Croda Chem.
*Chemical Description:* Alkyl ether phosphates based on ceto stearyl alcohol; available with 2, 5, and 10 moles ethylene oxide and as Crodafos CS TEA salt
*Category and Applications:* See Crodafos Series
*Form:* Hard wax; Color: White to buff
*Properties:* pH: 1.8–2.5 (3% sol'n. @ 20 C)
*Composition:* 99% min. active

**Crodafos ID Series.** Croda Chem.
*Chemical Description:* Alkyl ether phosphates based on isodecanol; available with 2, 5, and 10 moles ethylene oxide and as Crodafos ID TEA salts
*Category and Applications:* See Crodafos Series
*Form:* Fluid to paste; Color: Colorless to pale yellow
*Properties:* pH: 1.8–2.5 (3% sol'n. @ 20 C)
*Composition:* 99% min. active

**Crodafos N Series.** Croda Chem.
*Chemical Description:* Phosphated oleyl polyether; available as Crodafos N3 based on Volpo N3 and Crodafos N10 based on Volpo N10; available as the free acids and as neutralized (salt) versions
*Category:* Emulsifier, solubilizer, detergent, antistat, anionic
*Applications:* Emulsifier esp. for use at low concentrations; cosmetics, pharmaceutical and toiletry preparations (e.g., bubble bath, shampoo, waterless hand cleaners)
*Form:* Paste
*Composition:* 100% active ingred.

**Crodafos O Series.** Croda Chem.
*Chemical Description:* Alkyl ether phosphates based on oleyl alcohol; members differ in polyether chain length with 2, 5, 10 moles of ethylene oxide respectively (Crodafos O.2 acid, Crodafos O.5 acid, Crodafos O.10 acid)
Also available neutralized by triethanolamine as Crodafos O.2 TEA salt, etc.
*Category and Applications:* See Crodafos Series
*Form:* Slightly viscous liquid to soft paste; Color: Pale yellow to amber
*Properties:* pH: 1.8–2.5 (3% sol'n. @ 20 C)
*Composition:* 99% active ingred.

**Crodafos SG.** Croda Chem.
*Chemical Description:* Phosphated ester derived from an alkoxylate of cosmetic grade cetyl alcohol
*Category:* Emulsifier, antistat, corrosion inhibitor, antigelling agent, anionic
*Applications:* Shampoos, cream rinses, soap-free cleansing products; depilatories, hair straighteners and other alkaline cosmetics;

corrosion inhibitors and antigelling agents in soluble salt aerosol antiperspirants
*Form:* Liquid
*Composition:* 100% active ingred.

**Crodafos T Series.** Croda Chem.
*Chemical Description:* Alkyl ether phosphates based on tridecanol; available with 2, 5, and 10 moles ethylene oxide and as Crodafos T TEA salts
*Category and Applications:* See Crodafos Series
*Form:* Liquid to viscous liquid; Color: Colorless to pale yellow
*Properties:* pH: 1.8–2.5 (3% sol'n. @ 20 C)
*Composition:* 99% min. active

**Crodalan AWS.** Croda Chem.
*Chemical Description:* Acetylated ethoxylated lanolin alcohol derivative
*Category:* Emollient, super-fatting agent, conditioning agent, O/W emulsifier, solubilizing agent, nonionic
*Applications:* Shampoos, foam baths, bubble baths, skin creams/lotions, make-up preparations, shaving compositions
*Form:* Liquid; Color: Golden; Odor: Faint fatty
*Composition:* 100% active ingred.

**Crodalan IPL.** Croda Chem.
*Chemical Description:* Isopropyl esters based on fractionated lanolin acids; available in liquid or solid form
*Category:* Emollient, wetting agent
*Applications:* Bath oils, skin, hair and make-up cosmetics
*Form:* Thin opalescent liquid or soft paste liquifying at skin temperature; Color: Pale yellow

**Crodalan LA.** Croda Chem.
*Chemical Description:* Acetylated lanolin alcohols
*Category:* Emollient, penetrant, plasticizer, nonionic
*Applications:* Cosmetics (hair sprays and lacquers, skin preparations, baby products, hair preparations, make-up); pharmaceuticals; as plasticizer in polymeric substances and high melting point fats
*Form:* Thin mobile liquid; Color: Clear; Odor: Odorless
*Properties:* Visc.: Low
*Composition:* 100% active ingred.

**Crodamol BS.** Croda Chem.
*Chemical Description:* Butyl stearate
*Category:* Emollient, plasticizer, wetting agent
*Applications:* Cosmetics: Emollient without oiliness in lotions and creams; plasticizer in hair sprays and nail varnish (also improves gloss); nail varnish removers; wetting agent for pigments and dyes in lipsticks
*Form:* Liquid; Color: Water white

**Crodamol CAP.** Croda Chem.
*Chemical Description:* Blend of branched chain esters
*Category:* Emollient
*Applications:* Skin care preparations as emollient with good water

repellency, reduced stickiness
*Form:* Color: Water white; Odor: Low

**Crodamol CL.** Croda Chem.
*Chemical Description:* Cetyl lactate
*Category:* Emollient
*Applications:* Emollient with good solubility in aqueous alcoholic solutions; dry emollient in make-up and skin care products; in preparations packed in polystyrene
*Form:* Solid; Color: Colorless to pale straw
*Properties:* Solubility: Good in aqueous alcoholic solutions; sol. in various glycols

**Crodamol CSL.** Croda Chem.
*Chemical Description:* Ceto stearyl lactate
*Category:* Emollient
*Applications:* Emollient, esp. in compositions packaged in styrene
*Properties:* Similar to Crodamol ML with slightly higher M.P.

**Crodamol DA.** Croda Chem.
*Chemical Description:* Diisopropyl adipate
*Category:* Super-fatting agent, solvent for cosmetics, plasticizer, emollient
*Applications:* Solvent for aromatic oils and perfumes, lipstick dyestuffs, after shave lotions, colognes, skin fresheners; super-fatting agent for aqueous alcoholic lotions and nail varnish removers; plasticizer in nail varnishes and hair sprays; skin care preparations and shampoos as emollient
*Form:* Liquid; Color: Water white; Odor: Neutral
*Properties:* Solubility: Sol. in aqueous alcoholic solutions; miscible with mineral oils, vegetable oils, synthetic esters, acetoglycerides, triacetin and silicone oils of low viscosity; Visc.: Very low

**Crodamol DO.** Croda Chem.
*Chemical Description:* Decyl oleate
*Category:* Emollient, lubricant, penetrant
*Applications:* Cosmetics and pharmaceuticals; leaves nonsticky dry film on skin; in lipsticks as a partial substitute for castor oil
*Form:* Liquid; Color: Pale straw; Odor: Mild characteristic; Taste: Bland

**Crodamol ICS.** Croda Chem.
*Chemical Description:* Isocetyl stearate
*Category:* Emollient, lubricant, spreading agent
*Applications:* Controls viscosity and crystallization and reduces rheological changes in emulsion systems; bath oils, skin creams and lotions, hair-dressing, make-up cosmetics; cosmetics packaged in styrene
*Form:* Liquid; Color: Water white
*Properties:* Visc.: Low; Liquid sets at ≈ 0 C

**Crodamol IPM.** Croda Chem.
*Chemical Description:* Isopropyl myristate
*Category:* Spreading agent, emol-

lient, co-solvent
*Applications:* Vehicle and co-solvent for cosmetic raw materials; bath oils, aerosol antiperspirants, deodorants, hygiene sprays, sunscreens, lipsticks, pre-shave preparations, skin care, make-up, and hair care cosmetics
*Form:* Liquid; Color: Water white; Odor: Odorless

**Crodamol LL.** Croda Chem.
*Chemical Description:* Lauryl lactate
*Category:* Emollient
*Applications:* Promotes spreading and wetting characteristics of cosmetics; leaves thin dry film on skin and hair; antiperspirants, after shaves, make-up, hair and skin care products
*Form:* Liquid; Color: Colorless to pale straw; Odor: Faint bland
*Properties:* Solubility: Sol. in highly polar materials such as glycols; does not attack styrene

**Crodamol ML.** Croda Chem.
*Chemical Description:* Myristyl lactate
*Category:* Emollient
*Applications:* Improves smoothness and gloss in lipsticks without oiliness; tack reducer in antiperspirants; hair care preparations
*Form:* Liquid; Color: White to pale yellow
*Properties:* Solubility: Sol. in aqueous alcoholic solutions; essentially lipophilic but also sol. in a variety of glycols; does not attack polystyrene

**Crodamol MM.** Croda Chem.
*Chemical Description:* Myristyl myristate
*Applications:* Substitute for spermaceti wax and/or beeswax in cosmetic and pharmaceutical formulations; hair creams, cold creams, cleansing creams, suppositories, lipsticks, eye make-up
*Form:* Waxy solid; Color: Creamy white

**Crodamol OP.** Croda Chem.
*Chemical Description:* 2 ethyl hexyl palmitate
*Category:* Lubricant
*Applications:* Aids prevention of insensible perspiration and skin respiration; used in make-up, skin and hair care preparations and to modify the occlusiveness of creams, ointments and oils
*Form:* Liquid; Color: Water white; Odor: Mild
*Properties:* Low order of reaction with polystyrene packaging materials

**Crodamol PC.** Croda Chem.
*Chemical Description:* Propylene glycol diester of selected coconut fatty acids
*Category:* Emollient, spreading agent
*Applications:* Cosmetic systems esp. bath oils, antiperspirants, lipsticks, skin and hair care preparations
*Form:* Liquid; Odor: Slight bland
*Properties:* Visc.: Low; Cloud Pt.: Very low

**Crodamol SS.** Croda Chem.

*Chemical Description:* Synthetic spermaceti NF; essentially a mixture of myristyl myristate and cetostearyl palmitate
*Category:* Emollient
*Applications:* Gives body and emollience to cosmetic and pharmaceutical formulations
*Form:* Crystalline solid; Color: Almost white

**Crodamol W.** Croda Chem.
*Chemical Description:* Blend of straight chain saturated esters
*Category:* Emollient and water repellent
*Applications:* Cosmetics and toiletries; melts rapidly on application to skin giving cooling effect; water repellent and nongreasy emollient film on application to skin
*Form:* Waxy solid; Color: White; Odor: Mild
*Properties:* M.P.: Just below skin temperature

**Crodasinic LS35.** Croda Chem.
*Chemical Description:* Sodium N-lauroyl sarcosinate
*Category:* Foaming agent, wetting agent, detergent, lubricant, antistat, corrosion inhibitor, bacteriostat, penetrant, anionic
*Applications:* For use in slightly acidic conditions; as inhibitor for hexokinase, used in toothpaste, mouth washes, other dental preparations; corrosion inhibitor in aqueous systems; pharmaceuticals, depilatories, shampoos, shaving preparations, toilet soaps; surgical scrubs; alkali cleaners, carpet shampoos; cutting oils; emulsion polymerization; food packaging; textile scouring and fulling of wool; household and industrial cleaners
*Form:* Clear liquid
*Properties:* Solubility: Water sol.; Stability: Stable to alkaline hydrolysis, under moderately acid conditions at normal temperatures; Biodegradable
*Composition:* 35% active ingred. (also available as 30% active)

**Crodateric C.** Croda Chem.
*Chemical Description:* Derived from coconut fatty acid
*Category:* Amphoteric surfactants
*Properties:* m.w.: 360
*Composition:* 50% active aqueous solution

**Crodateric Cy.** Croda Chem.
*Chemical Description:* Derived from caprylic acid
*Category:* Amphoteric surfactants
*Properties:* m.w.: 284
*Composition:* 50% active aqueous solution

**Crodateric L.** Croda Chem.
*Chemical Description:* Derived from lauric acid
*Category:* Amphoteric surfactants
*Properties:* m.w.: 340
*Composition:* 50% active aqueous solution

**Crodateric O.** Croda Chem.
*Chemical Description:* Derived from oleic acid
*Category:* Amphoteric surfactants
*Properties:* m.w.: 417

*Composition:* 50% active aqueous solution

**Crodateric O.100.** Croda Chem.
*Chemical Description:* Derived from oleic acid
*Category:* Amphoteric surfactants
*Properties:* m.w.: 417
*Composition:* 100% active

**Crodateric S.** Croda Chem.
*Chemical Description:* Derived from stearic acid
*Category:* Amphoteric surfactants
*Properties:* m.w.: 410
*Composition:* 50% active aqueous solution

**Crodazoline C.** Croda Chem.
*Chemical Description:* Fatty acid imidazoline
*Form:* Color: 9–10 (Gardner)
*Properties:* m.w.: 288; M.P.: 42 C; Surface Tension: 32.5 (1%/0.36%); pH: 11.5 (10% aq. disp.)
*Composition:* 90% min. imidazoline

**Crodazoline Cy.** Croda Chem.
*Chemical Description:* Fatty acid imidazoline
*Form:* Color: 10–11 (Gardner)
*Properties:* m.w.: 212; M.P.: 45 C; Surface Tension: 27.8 (1%/0.36%); pH: 12 (10% aq. disp.)
*Composition:* 90% min. imidazoline

**Crodazoline H.** Croda Chem.
*Chemical Description:* Fatty acid imidazoline
*Form:* Color: 8–9 (Gardner)
*Properties:* m.w.: 383; M.P.: 57 C; Surface Tension: 43 (1%/0.36%); pH: 11.5 (10% aq. disp.)
*Composition:* 90% min. imidazoline

**Crodazoline L.** Croda Chem.
*Chemical Description:* Fatty acid imidazoline
*Form:* Color: 4–5 (Gardner)
*Properties:* m.w.: 268; M.P.: 40 C; Surface Tension: 37.2 (1%/0.36%); pH: 11.3 (10% aq. disp.)
*Composition:* 90% min. imidazoline

**Crodazoline M.** Croda Chem.
*Chemical Description:* Fatty acid imidazoline
*Form:* Color: 9–10 (Gardner)
*Properties:* m.w.: 297; M.P.: 51 C; Surface Tension: 36.5 (1%/0.36%); pH: 11.5 (10% aq. disp.)
*Composition:* 90% min. imidazoline

**Crodazoline O.** Croda Chem.
*Chemical Description:* Fatty acid imidazoline
*Form:* Color: 6–7 (Gardner)
*Properties:* m.w.: 345; Pour Pt.: −12 C; Surface Tension: 26.5 (1%/0.36%); pH: 11.1 (10% aq. disp.)
*Composition:* 90% min. imidazoline

**Crodazoline S.** Croda Chem.
*Chemical Description:* Fatty acid imidazoline
*Form:* Color: 4–5 (Gardner)
*Properties:* m.w.: 338; M.P.: 52 C; Surface Tension: 43.2 (1%/0.36%); pH: 11.0 (10% aq. disp.)

*Composition:* 90% min. imidazoline

**Crodazoline T.** Croda Chem.
*Chemical Description:* Fatty acid imidazoline
*Form:* Color: 10–11 (Gardner)
*Properties:* m.w.: 350; Pour Pt.: –10 C; Surface Tension: 36 (1%/0.36%); pH: 11.2 (10% aq. disp.)
*Composition:* 90% min. imidazoline

**Crodazoline TW.** Croda Chem.
*Chemical Description:* Fatty acid imidazoline
*Form:* Color: 7–8 (Gardner)
*Properties:* m.w.: 338; M.P.: 49 C; Surface Tension: 38 (1%/0.36%); pH: 11.1 (10% aq. disp.)
*Composition:* 90% min. imidazoline

**Croderol G7000.** Croda Chem.
*Chemical Description:* Glycerine
*Category:* Humectant, moisturizing agent, diluent, plasticizer
*Applications:* Diluent and plasticizer for many polar materials; as humectant, moisturizer in cosmetics and pharmaceuticals (skin and hair care preparations, shaving compositions, etc.)

**Crodesta.** Croda Chem.
*Chemical Description:* Sucrose fatty acid esters based on stearic/palmitic acid; the monoester content varies in the series
*Category:* Emulsifier
*Applications:* Cosmetic emulsifier, esp. for lotion compositions; eye, lip, face make-up and baby products due to mildness to skin and eyes

**Crodet L4.** Croda Chem.
*Chemical Description:* Lauric acid POE 4 (4 moles of ethylene oxide)
*Category:* Emulsifier, wetting agent, solubilizer, dispersant, nonionic
*Applications:* O/W emulsifiers for cosmetics and pharmaceutical creams, lotions and ointments; solubilizer for perfumes or aqueous alcoholic preparations (bath essences, colognes, after shave lotions); L4 is useful as plasticizer for polymers and resins used in hair setting lotions and sprays
*Form:* Liquid; Color: Pale straw
*Properties:* Solubility: Sol. in ethyl and oleyl alcohols, oleic acid, ceto stearyl alcohol; HLB: 9.8; Sap. Value: 138–150; Stability: Stable to moderately strong acids and alkalis, to oxidation under normal conditions, in presence of many electrolytes

**Crodet L8.** Croda Chem.
*Chemical Description:* Lauric acid POE 8 (8 moles of ethylene oxide)
*Category and Applications:* See Crodet L4; L8 esp. used as solubilizer for perfumes and plasticizer for hair sprays
*Form:* Liquid; Color: Pale straw
*Properties:* Solubility: Sol. in water, ethyl and oleyl alcohols, oleic acid, ceto stearyl alcohol; HLB: 12.7; Sap. Value: 95–106; Stability: Stable to moderately strong acids and alkalis, to oxidation under normal conditions, in pres-

ence of many electrolytes

**Crodet L12.** Crodet Chem.
*Chemical Description:* Lauric acid POE 12
*Category and Applications:* See Crodet L4; L12 esp. as solubilizer for perfumes; dispersant in aqueous systems
*Form:* Liquid; Color: Pale straw
*Properties:* Solubility: Sol. in water, ethyl and oleyl alcohols, oleic acid, ceto stearyl alcohol; HLB: 14.5; Sap. Value: 72–82; Stability: Stable to moderately strong acids and alkalis, to oxidation under normal conditions, in presence of many electrolytes

**Crodet L24.** Crodet Chem.
*Chemical Description:* Lauric acid POE 24
*Category and Applications:* See Crodet L4; L24 esp. as solubilizer for perfumes
*Form:* Soft paste; Color: Off white
*Properties:* Solubility: Sol. in water, ethyl and ceto stearyl alcohols; HLB: 16.8; Sap. Value: 42–48; Stability: Stable to moderately strong acids and alkalis, to oxidation under normal conditions, in presence of many electrolytes

**Crodet L40.** Croda Chem.
*Chemical Description:* Lauric acid POE 40
*Category and Applications:* See Crodet L4
*Form:* Waxy solid; Color: Off white
*Properties:* Solubility: Sol. in water, ethyl and ceto stearyl alcohols; HLB: 17.9; Sap. Value: 26–31; Stability: Stable to moderately strong acids and alkalis, to oxidation under normal conditions, in presence of many electrolytes

**Crodet L100.** Croda Chem.
*Chemical Description:* Lauric acid POE 100
*Category and Applications:* See Crodet L4
*Form:* Waxy solid; Color: Pale yellow
*Properties:* Solubility: Sol. in water, ethyl and ceto stearyl alcohols; HLB: 19.1; Sap. Value: 11–15; Stability: Stable to moderately strong acids and alkalis, to oxidation under normal conditions, to presence of many electrolytes

**Crodet S4.** Croda Chem.
*Chemical Description:* Stearic acid POE 4 (4 moles of ethylene oxide)
*Category and Applications:* See Crodet L4; also as auxiliary surfactants to optimize HLB
*Form:* Soft paste; Color: Off white
*Properties:* Solubility: Sol. in ethyl and oleyl alcohols, oleic acid, ceto stearyl alcohol; essentially sol. in arachis oil and isoparaffinic solvent; HLB: 7.7; Sap. Value: 117–129; Stability: Stable to moderately strong acids and alkalis, to oxidation under normal conditions, and to many electrolytes

**Crodet S8.** Croda Chem.
*Chemical Description:* Stearic acid POE 8

*Category and Applications:* See Crodet L4
*Form:* Soft paste; Color: Off white
*Properties:* Solubility: Sol. in ethyl and oleyl alcohols, oleic acid, ceto stearyl alcohol; essentially sol. in arachis oil and isoparaffinic solvent; HLB: 10.8; Sap. Value: 84–94; Stability: Stable to moderately strong acids and alkalis, oxidation under normal conditions, and to many electrolytes

**Crodet S12.** Croda Chem.
*Chemical Description:* Stearic acid POE 12
*Category and Applications:* See Crodet L4
*Form:* Waxy solid; Color: Off white
*Properties:* Solubility: Sol. in ethyl and oleyl alcohols, oleic acid, ceto stearyl alcohol; essentially sol. in water; HLB: 13.4; Sap. Value: 65–75; Stability: Stable to moderately strong acids and alkalis, oxidation under normal conditions, to many electrolytes

**Crodet S24.** Croda Chem.
*Chemical Description:* Stearic acid POE 24
*Category and Applications:* See Crodet L4
*Form:* Waxy solid; Color: Off white
*Properties:* Solubility: Sol. in water, ethyl and ceto stearyl alcohols; partly sol. in arachis oil, oleyl alcohol, isopropyl myristate, oleic acid; HLB: 15.8; Sap. Value: 38–47; Stability: Stable to moderately strong acids and alkalis, oxidation under normal conditions, to many electrolytes

**Crodet S40.** Croda Chem.
*Chemical Description:* Stearic acid POE 40
*Category and Applications:* See Crodet L4
*Form:* Waxy solid; Color: Off white
*Properties:* Solubility: Sol. in water, ethyl and ceto stearyl alcohols; partly sol. in oleic acid, propylene glycol; HLB: 16.7; Sap. Value: 23–30; Stability: Stable to moderately strong acids and alkalis, oxidation under normal conditions, to many electrolytes

**Crodet S100.** Croda Chem.
*Chemical Description:* Stearic acid POE 100
*Category and Applications:* See Crodet L4
*Form:* Waxy solid; Color: Off white
*Properties:* Solubility: Sol. in water, ethyl and ceto stearyl alcohols; partly sol. in propylene glycol; HLB: 18.8; Sap. Value: 10–14; Stability: Stable to moderately strong acids and alkalis, oxidation under normal conditions, to many electrolytes

**Crodex A.** Croda Chem.
*Chemical Description:* Prepared from ceto stearyl alcohol and sodium lauryl sulfate
*Category:* Emulsifying wax BP, anionic
*Applications:* Pharmaceutical and

cosmetic uses; yields emulsions with dry emollience; sunscreens; cold creams; hair creams; depilatory; lubricating creams; insect repellent
*Form:* Waxy solid; Color: Almost white; Odor: Faint characteristic
*Properties:* Stability: Stable to weak acids and alkalis and some metal salts

**Crodex C.** Croda Chem.
*Chemical Description:* Prepared from ceto stearyl alcohol and Cetrimide BP
*Category:* Cationic emulsifying wax BPC; bacteriacides
*Applications:* Pharmaceuticals and cosmetics; germicidal creams, ointments and lotions, hair conditioning preparations, baby creams, skin care products
*Form:* Waxy solid; Color: Almost white; Odor: Faint characteristic

**Crodex N.** Croda Chem.
*Chemical Description:* Prepared from ceto stearyl alcohol and Cetomacrogol 1000 BPC
*Category:* Nonionic emulsifying wax BP; wetting agent and penetrant
*Applications:* Emulsifier for most emollient materials in cosmetics and pharmaceuticals
*Form:* Waxy solid; Color: Almost white; Odor: Faint characteristic

**Croduret 10.** Croda Chem.
*Chemical Description:* POE 10 hydrogenated castor oil
*Category:* Emulsifier, solubilizer, emollient, superfatting agent, detergent, nonionic
*Applications:* Cosmetics (shampoos, foam baths, skin care products; solubilizer for Vitamins A and D); Textiles (spin finish additives, emulsifiers, lubricants, and antistats); water soluble metal working fluids; emulsion polymerization; insecticides, herbicides; household detergent formulations
*Form:* Liquid; Color: Straw
*Properties:* Solubility: Sol. in oleyl alcohol, naphtha, methyl ethyl ketone, oleic acid, trichloroethylene; HLB: 6.3; Cloud Pt.: < 20 C; Sap. Value: 120–130; Surface Tension: 37 dynes/cm (0.1%)

**Croduret 30.** Croda Chem.
*Chemical Description:* POE 30 hydrogenated castor oil
*Category and Applications:* See Croduret 10
*Form:* Liquid; Color: Pale straw
*Properties:* Solubility: Sol. in water, ethyl and oleyl alcohols, naphtha, methyl ethyl ketone, oleic acid, trichloroethylene; HLB: 11.6; Cloud Pt.: 48 C; Sap. Value: 70–80; Surface Tension: 46 (0.1% sol'n.)

**Croduret 40.** Croda Chem.
*Chemical Description:* POE 40 hydrogenated castor oil
*Category and Applications:* See Croduret 10
*Form:* Viscous paste; Color: Off white
*Properties:* Solubility: Sol. in water, ethyl alcohol, naphtha, methyl ethyl ketone, oleic acid, trichloroethylene; HLB: 13.0; Cloud

Pt.: 62 C; Sap. Value: 60–65; Surface Tension: 46 dynes/cm (0.1% sol'n.)

**Croduret 60.** Croda Chem.
*Chemical Description:* POE 60 hydrogenated castor oil
*Category and Applications:* See Croduret 10
*Form:* Stiff paste; Color: Off white
*Properties:* Solubility: Sol. in water, ethyl alcohol, methyl ethyl ketone, oleic acid, trichloroethylene; HLB: 14.7; Cloud Pt.: 71 C; Sap. Value: 45–50; Surface Tension: 47.5 dynes/cm (0.1% sol'n.)

**Croduret 100.** Croda Chem.
*Chemical Description:* POE 100 hydrogenated castor oil
*Category and Applications:* See Croduret 10
*Form:* Waxy solid; Color: Off white
*Properties:* Solubility: Sol. in water, ethyl alcohol, methyl ethyl ketone, trichloroethylene; HLB: 16.5; Cloud Pt.: 65 C; Sap. Value: 25–35; Surface Tension: 46.2 dynes/cm (0.1% sol'n.)

**Crolactil CSL.** Croda Chemicals
*Chemical Description:* Acyl lactylates
*Category:* Emulsifier
*Applications:* Water-in-oil emulsions, ingredients of food products
*Form:* Powder; Color: Off-white
*Properties:* HLB: 5.0; Sapon. Value: 175–250; Acid No.: 50–86

**Crolactil SISL.** Croda Chemicals
*Chemical Description:* Acyl lactylates
*Category:* Emulsifier
*Applications:* Oil-in-water emulsions, ingredients of food products
*Form:* Liquid; Color: Yellow
*Properties:* HLB: 6.5; Sapon. Value: 210–280; Acid No.: 60–90

**Crolactil SSL.** Croda Chemicals
*Chemical Description:* Acyl lactylates
*Category:* Emulsifier
*Applications:* Oil-in-water emulsions, ingredients of food products
*Form:* Powder; Color: Off-white
*Properties:* HLB: 6.5; Sapon. Value: 210–280; Acid No.: 60–90

**Crotein A, C, O.** Croda Chem.
*Chemical Description:* Refined collagen hydrolysates (differing in molecular weights)
*Category:* Conditioning agent, foam stabilizer and booster, dye leveling agent
*Applications:* Substantive to hair and skin; protective action in hair care products for dyed, bleached, waved hair; hair conditioning and texturizing; foam stabilizers and boosters in shampoos; dye leveler in hair dyes and bleaches; cold waves and straighteners; hair sprays, rinses, sets; hand and body lotions, make-up, treatment creams, moisturizers; nail hardeners; depilatories
*Form:* Powder; Color and Odor: Very low
*Composition:* 100% active ingred.

**Crotein CCT.** Croda Chem.
*Chemical Description:* Triethanolamine salt of a complex protein coconut fatty acid ester
*Category:* Detergent, foaming agent
*Applications:* Skin cleansers, hair and body shampoos; reduces irritation of other detergents

**Crotein Q.** Croda Chem.
*Chemical Description:* Quaternary ammonium derivative of a hydrolyzed collagen protein
*Category:* Hair conditioning agent, cationic
*Applications:* Substantive to hair; hair conditioners and shampoos with reduction of split ends, superior gloss, improved manageability and texture
*Properties:* Solubility: Water sol.

**p-Cumyl phenol 7.5 mole ethoxylate (developmental product).** Monsanto Industrial Chem.
*Category:* Wetting agent, emulsifier
*Applications:* Emulsifier in solvent cleaners, in metal cleaning compounds
*Form:* Liquid; Color: Light yellow; Odor: Mild phenolic
*Properties:* Ref. Index: 1.5146 @ 60 C; Solubility: Sol. in water; Sp. gr.: 1.1121 @ 20 C; Visc.: 370 cps; Cloud Pt.: 48.8 C; Flash Pt.: 565 F; Pour Pt.: –1 F; Stability: Good; Surface Tension: 39.1 dynes/cm
*Composition:* 99.5% active ingred.; other components: 0.4% water max.; 0.1% ash max.
*Reaction:* Neutral
*Toxicity:* Moderate skin and eye irritant
sol.; Sp. gr.: 1.1191 @ 20 C; Visc.: 370 cps; Cloud Pt.: 94 C; Flash Pt.: 565 F; Pour Pt.: 56 F; Stability: Good; Surface Tension: 38.9 dynes/cm
*Composition:* 99.5% active ingred.; other components: 0.4% water max.; 0.1% ash max.
*Reaction:* Neutral
*Toxicity:* Mild eye irritant

**p-Cumyl phenol 10 mole ethoxylate (developmental product).** Monsanto Industrial Chem.
*Category:* Wetting agent, emulsifier
*Applications:* Emulsifier in solvent cleaners, in hard surface and general industrial cleaners
*Form:* Liquid; Color: Light yellow; Odor: Mild phenolic
*Properties:* Refractive Index: 1.4990 @ 60 C; Solubility: Sol. in water; Sp. gr.: 1.1141 @ 20 C; Visc.: 410 cps; Cloud Pt.: 74.6 C; Flash Pt.: 545 F; Pour Pt.: 26 F; Stability: Good; Surface Tension: 43.0 dynes/cm
*Composition:* 99.5% active ingred.; other components: 0.4% water max.; 0.1% ash max.
*Reaction:* Neutral
*Toxicity:* Mild eye irritant

**p-Cumyl phenol 15 mole ethoxylate (developmental product)** Monsanto Industrial Chem.
*Category:* Wetting agent, emulsifier
*Applications:* Emulsifier, in sol-

vent cleaners, in hard surface, metal and general industrial cleaners
*Form:* Liquid; Color: Light yellow; Odor: Mild phenolic
*Properties:* Refractive Index: 1.4878 @ 60 C; Solubility: Water-

**Cyclochem 326A.** Cyclo Chem. Corp.
*Chemical Description:* A synthetic replacement for natural beeswax
*Category:* Emulsifer, nonionic
*Applications:* Its plasticity, emulsion forming power and behavior with solvents are indistinguishable from the natural product; particularly suited to both w/o and o/w emulsions including hair creams, cold creams, lotions and lipsticks
*Form:* Flakes; Color: Off-white
*Properties:* M. P.: 51–54 C; Acid No.: 25–27

**Cyclochem CL.** Cyclo Chem. Corp.
*Chemical Description:* Cetyl lactate
*Category:* Penetrant
*Applications:* To promote penetration and feel, useful in imparting creaminess to lipstick and conditioning effect to cream mascara, imparts a luxurious feel to bath oils and lotions
*Form:* Solid; Color: Off-white
*Properties:* M. P.: 24 C; Sapon. Value: 180; Acid No.: 3.5 max.

**Cyclochem CP.** Cyclo Chem. Corp.
*Chemical Description:* Cetyl palmitate
*Category:* Ester
*Applications:* Ester developed for use as a substitute for spermaceti in cold creams, lipsticks and topical cosmetics
*Form:* Flakes; Color: Off-white
*Properties:* M. P.: 50 C; Sapon. Value: 114; Acid No.: 3.0 max.

**Cyclochem EGDS.** Cyclo Chem. Corp.
*Chemical Description:* Ethylene glycol distearate pure
*Category:* Thickener
*Applications:* Used in shampoos and cosmetic lotions as a thickener and to provide optimum opacity
*Form:* Flakes; Color: White
*Properties:* M. P.: 61 C; Sapon. Value: 195; Acid No.: 20.0 max.

**Cyclochem EGMS.** Cyclo Chem. Corp.
*Chemical Description:* Ethylene glycol monostearate pure
*Category:* Additive
*Applications:* Widely used to add viscosity, pearlescence and opacity to lotions and shampoos
*Form:* Flakes; Color: White
*Properties:* M. P.: 56 C; Sapon. Value: 182; Acid No.: 4.0 max.

**Cyclochem EM 324.** Cyclo Chem. Corp.
*Chemical Description:* A blend of natural fatty alcohols, fatty acids and other derivatives
*Category:* Emulsifier, anionic
*Applications:* Self-emulsifying; useful in formulating o/w hair and skin conditioning creams; tolerates up to 5% organic acids such as citric and lactic acids

*Form:* Flakes; Color: Off-white
*Properties:* M. P.: 47–50 C; Acid No.: 93–98

**Cyclochem EM 560.** Cyclo Chem. Corp.
*Chemical Description:* A blend of natural fatty alcohols and fatty esters
*Category:* Emulsifier, anionic
*Applications:* Self-emulsifying; useful for o/w creams on the acid side
*Form:* Wax; Color: Off-white
*Properties:* M. P.: 62–65 C; Acid No.: 1.0 max.

**Cyclochem GMO.** Cyclo Chem. Corp.
*Chemical Description:* Glyceryl monooleate
*Category:* Emulsifier, emollient, opacifier, lubricant
*Applications:* Effective water-in-oil emulsifier, emollient and opacifier, provides lubricity to pre- and aftershave lotions and cosmetic systems
*Form:* Liquid; Color: Amber
*Properties:* M.P.: 9 C; Sapon. Value: 170; Acid No.: 9.0 max.

**Cyclochem GMS.** Cyclo Chem. Corp.
*Chemical Description:* Glyceryl monostearate pure
*Category:* Emulsifier
*Applications:* A strongly lipophilic emulsifier, easily solubilized by PEG esters, enhances feel and appearance of formulations
*Form:* Flakes; Color: White
*Properties:* M. P.: 56 C; Sapon. Value: 170; Acid No.: 5.0 max.

**Cyclochem GMS 21.** Cyclo Chem. Corp.
*Chemical Description:* Glyceryl monostearate acid stable self emulsifying
*Category:* Electrolyte tolerant, emulsifier, emollient, anionic
*Applications:* High electrolyte tolerance, emulsifier and emollient for creams, lotions and related toiletry products
*Form:* Flakes; Color: Off-white
*Properties:* M. P.: 52 C; Sapon. Value: 157; Acid No.: 18.0 max.

**Cyclochem GMS 165.** Cyclo Chem. Corp.
*Chemical Description:* Glyceryl monostearate acid stable self emulsifying
*Category:* Emulsifier, electrolyte tolerant
*Applications:* A principle emulsifier in o/w creams and lotions, exhibits exceptional electrolyte tolerance and functions well in the acid pH range
*Form:* Flakes; Color: Off-white
*Properties:* M. P.: 55 C; Sapon. Value: 93; Acid No.: 2.0 max.

**Cyclochem GTIS.** Cyclo Chem. Corp.
*Chemical Description:* Isostearyl triglyceride
*Category:* Emollient, co-emulsifier, softener
*Applications:* A super emollient, its liquid nature allows it to be easily incorporated into creams and lotions; as a co-emulsifier, it

provides good texture in creams; promotes softness and a smooth feeling to the skin in creams and lotions
*Form:* Liquid; Color: Yellow
*Properties:* M. P.: $<$ 0 C; Sapon. Value: 192; Acid No.: 8.0 max.

**Cyclochem GTL.** Cyclo Chem. Corp.
*Chemical Description:* Lauric/myristic triglyceride
*Category:* Emollient
*Applications:* To promote emolliency and an oily feel to cream and lotions
*Form:* Liquid/paste
*Properties:* M. P.: 27 C; Sapon. Value: 245; Acid No.: 30.0 max.

**Cyclochem INEO.** Cyclo Chem. Corp.
*Chemical Description:* Isostearyl neopentanoate
*Category:* Penetrant
*Applications:* Highly acid and alkali stable ester with rapid penetration that promotes a dry feel to the skin
*Form:* Liquid; Color: Off-white
*Properties:* M. P.: $<$ 0 C; Sapon. Value: 149; Acid No.: 2.0 max.

**Cyclomide LIPA.** Cyclo Chem. Corp.
*Chemical Description:* Lauric monoisopropanolamide
*Category:* Viscosity builder, foam stabilizer
*Applications:* Builds viscosity, stabilizes foam and is mild to the skin
*Form:* Flakes; Color: Off-white
*Composition:* 100% active, $\approx$ 95% amide content

**Cyclochem LVL.** Cyclo Chem. Corp.
*Chemical Description:* Lauryl lactate
*Category:* Penetrant, emollient, pigment dispersant
*Applications:* To provide rapid non-oily penetration; important component of cake powders, lotions and creams; emollient and pigment dispersant for liquid makeups; reduces tackiness of aluminum salt anti-perspirants
*Form:* Liquid; Color: Off-white
*Properties:* M. P.: 4 C; Sapon. Value: 199; Acid No.: 3.0 max.

**Cyclochem ML.** Cyclo Chem. Corp.
*Chemical Description:* Myristyl lactate
*Category:* Melting ester, penetrant, emollient, pigment dispersant
*Applications:* Slightly higher melting ester with similar properties to Cyclochem LVL
*Form:* Liquid; Color: Off-white
*Properties:* M. P.: 12 C; Sapon. Value: 172; Acid No.: 3.0 max.

**Cyclochem MM/M.** Cyclo Chem. Corp.
*Chemical Description:* Myristyl myristate
*Category:* Fatty acid esters
*Applications:* With its melt point at body temperature, it imparts a soft, velvety feel which persists after rinsing; at low levels (under 1%) it provides infinite spreading so as to prevent the

watery feel of hand and body lotions high in water content
*Form:* Flakes; Color: Off-white
*Properties:* M. P.: 38 C; Sapon. Value: 130; Acid No.: 3.0 max.

**Cyclochem MST.** Cyclo Chem. Corp.
*Chemical Description:* Myristyl Stearate
*Category:* Fatty acid esters
*Applications:* To impart soft, velvety feel to emulsions where a stiffer emulsion with superior opacity is desired
*Form:* Flakes; Color: Off-white
*Properties:* M. P.: 45 C; Sapon. Value: 115; Acid No.: 5.0 max.

**Cyclochem PEG 400 DS.** Cyclo Chem. Corp.
*Chemical Description:* Polyethylene glycol 400 distearate
*Category:* Fatty acid esters, nonionic
*Applications:* These esters offer the formulator varied hydrophilic and lipophilic properties
*Form:* Soft solid; Color: White
*Properties:* M. P.: 33 C; Sapon. Value: 123; Acid No.: 10.0 max.

**Cyclochem PEG 400 ML.** Cyclo Chem. Corp.
*Chemical Description:* Polyethylene glycol 400 monolaurate
*Category:* Fatty acid esters, nonionic
*Applications:* These esters offer the formulator varied hydrophilic and lipophilic properties
*Form:* Liquid; Color: Light straw
*Properties:* M. P.: 10 C; Sapon. Value: 90; Acid No.: 5.0 max.

**Cyclochem PEG 400 MO.** Cyclo Chem. Corp.
*Chemical Description:* Polyethylene glycol 400 monooleate
*Category:* Fatty acid esters, nonionic
*Applications:* These esters offer the formulator varied hydrophilic and lipophilic properties
*Form:* Liquid; Color: Light straw
*Properties:* M. P.: 10 C; Sapon. Value: 85; Acid No.: 5.0 max.

**Cyclochem PETO.** Cyclo Chem. Corp.
*Chemical Description:* Pentaerythritol tetraoleate
*Category:* Fatty acid esters, nonionic
*Applications:* Unique liquid ester with good high temperature performance, imparts gloss in lipsticks and assists in the development of the 'wet look'
*Form:* Liquid; Color: Yellow
*Properties:* M. P.: $< 0$ C; Sapon. Value: 187; Acid No.: 5.0 max.

**Cyclochem PETS.** Cyclo Chem. Corp.
*Chemical Description:* Pentaerythritol tetrastearate
*Category:* Wax, nonionic
*Applications:* Synthetic, hard, high melting, neutral wax that provides high heat stability in lubricants; domestic substitute for expensive neutral waxes in polishes, coatings and various hot-melt compositions and in cosmetic products
*Form:* Flakes; Color: Off-white

*Properties:* M. P.: 57 C; Sapon. Value: 190; Acid No.: 9.0 max.

**Cyclochem PGMS.** Cyclo Chem. Corp.
*Chemical Description:* Propylene glycol monostearate, pure
*Category:* Emulsifier, occlusive agent, nonionic
*Applications:* An auxiliary emulsifier for lotions and make-up preparations, a good occlusive agent for reducing skin moisture loss
*Form:* Soft solid; Color: Off-white
*Properties:* M. P.: 36 C; Sapon. Value: 186; Acid No.: 5.0 max.

**Cyclochem POL.** Cyclo Chem. Corp.
*Chemical Description:* A blend of higher fatty alcohol and ethoxylated adducts
*Category:* Emulsifier, nonionic
*Applications:* Self-emulsifying; product of choice for creams and lotions containing high quantities of alkaline ingredients
*Form:* Wax; Color: Off-white
*Properties:* M. P.: 48–52 C; Acid No.: 2.0 max.

**Cyclochem SPS.** Cyclo Chem. Corp.
*Chemical Description:* Synthetic spermaceti
*Category:* Fatty acid esters, nonionic
*Applications:* Chemically and physically identical to the natural product; provides sheen and smooth feel to creams; gloss to stick makeup products; at 0.5% in aerosol talc, it reduces billowing and improves talc adhesion
*Form:* Flakes; Color: Off-white
*Properties:* M. P.: 46 C; Sapon. Value: 115; Acid No.: 3.0 max.

**Cyclochem SS.** Cyclo Chem. Corp.
*Chemical Description:* Stearyl stearate
*Category:* Fatty acid esters
*Applications:* A component of many natural waxes which provides a 'feel' on skin and hair that is less oily than mineral oil and less drying than isopropyl myristate
*Form:* Flakes; Color: Off-white
*Properties:* M. P.: 53 C; Sapon. Value: 110; Acid No.: 5.0 max.

**Cyclogol Cetyl Alcohol NF.** Cyclo Chem. Corp.
*Chemical Description:* *n*-hexadecanol
*Category:* Emollient carrier, viscosity builder, nonionic
*Applications:* Emollient carrier and viscosity builder for creams, lotions and other cosmetic emulsions
*Form:* Flakes; Color: Off-white
*Properties:* M. P.: 45–50 C; Acid No.: nil

**Cyclogol Cetyl-Stearyl Alcohol.** Cyclo Chem. Corp.
*Chemical Description:* *n*-hexadecanol and *n*-octadecanol
*Category:* Fatty alcohol, nonionic
*Applications:* Combines the viscosity effect of stearyl with the skin feel of cetyl in lotions and cosmetic emulsions
*Form:* Flakes; Color: Off-white
*Properties:* M. P.: 48–53 C; Acid

No.: nil

**Cyclogol EL.** Cyclo Chem. Corp.
*Chemical Description:* Ethoxylated fatty alcohol
*Category:* Solubilizer, nonionic
*Applications:* Valuable in formulating water soluble bath oils; solubilizer of perfume oils, super-fatting agents and microbicides
*Form:* Paste; Color: White
*Properties:* M. P.: 28–33 C; Acid No.: 2.0 max.

**Cyclogol LTH.** Cyclo Chem. Corp.
*Chemical Description:* Ethoxylated linear alcohol
*Category:* Detergent, dispersant, nonionic
*Applications:* Excellent for emulsifying oils, for degreasing and for engine cleaners
*Form:* Clear liquid
*Properties:* M. P.: 22–24 C; Acid No.: nil

**Cyclogol NI.** Cyclo Chem. Corp.
*Chemical Description:* A blend of higher fatty alcohol and ethoxylated adducts
*Category:* Emulsifier, nonionic
*Applications:* Useful in preparing o/w ointments, creams and lotions containing hard to solubilize additives such as thioglycolates, antihistamines and polyvalent salts
*Form:* Flakes; Color: White
*Properties:* M. P.: 49–50 C; Acid No.: 1.0 max.

**Cyclogol Stearyl Alcohol.** Cyclo Chem. Corp.
*Chemical Description:* *n*-octadecanol
*Category:* Viscosity builder, nonionic
*Applications:* Viscosity builder for creams, lotions and other cosmetic emulsions
*Form:* Flakes; Color: Off-white
*Properties:* M. P.: 56–60 C; Acid No.: nil

**Cyclomide AC 28.** Cyclo Chem. Corp.
*Chemical Description:* Fatty acid condensate
*Category:* Anti-corrosive mineral oil additive
*Applications:* Anti-corrosive mineral oil additive for cutting fluids and soluble cutting oil compositions
*Form:* Liquid
*Composition:* 100% active

**Cyclomide BWA/M.** Cyclo Chem. Corp.
*Chemical Description:* Fatty alkanolamide blends
*Category:* Emulsifier
*Applications:* Effective o/w emulsifier system for use as an engine degreaser
*Form:* Liquid
*Composition:* 100% active

**Cyclomide C212.** Cyclo Chem. Corp.
*Chemical Description:* Coconut monoethanolamide
*Category:* Additive
*Applications:* Effective additive for soap or synthetic based washing powders to promote and stabilize lather, increase detergent power and impart skin mildness

*Form:* Flakes; Color: Cream
*Composition:* 100% active, ≈ 95% amide content

**Cyclomide CODI.** Cyclo Chem. Corp.
*Chemical Description:* Cocamidopropyl dimethylamine
*Category:* Cationic emulsions
*Applications:* Base for preparing cationic emulsions for use in creams, lotions and hair rinse formulations
*Form:* Liquid; Color: Yellow
*Composition:* 100% active, ≈ 98% amide content

**Cyclomide DC212.** Cyclo Chem. Corp.
*Chemical Description:* 2:1 Coconut diethanolamide
*Category:* Viscosity booster, foaming agent, detergency
*Applications:* Used in cosmetic and laundry products to boost viscosity, foam quality and detergency
*Form:* Liquid; Color: Amber
*Composition:* 100% active

**Cyclomide DC212M.** Cyclo Chem. Corp.
*Chemical Description:* 2:1 Modified coconut diethanolamide
*Category:* Cleaners, solventless degreasers and wax strippers
*Applications:* For hard surface cleaners, solventless degreasers and wax strippers where high alkali compatibility is a necessity
*Form:* Liquid; Color: Amber
*Composition:* 100% active

**Cyclomide DC212/S.** Cyclo Chem. Corp.
*Chemical Description:* 1:1 Coconut diethanolamide
*Category:* Foam stabilizer, viscosity booster
*Applications:* Foam stabilizing and viscosity boosting when low cloud point is essential
*Form:* Liquid
*Composition:* 100% active, ≈ 85% amide content

**Cyclomide DC212/SE.** Cyclo Chem. Corp.
*Chemical Description:* 1:1 Coconut diethanolamide
*Category:* Thickening and foam stabilization
*Applications:* Maximizes thickening and foam stabilization in combination with alkyl aryl sulfonates or lauryl sulfates in shampoos and detergent formulations
*Form:* Liquid
*Composition:* 100% active, ≈ 95% amide content

**Cyclomide DL203.** Cyclo Chem. Corp.
*Chemical Description:* 2:1 Lauric diethanolamide
*Category:* Stabilizer
*Applications:* Outstanding power to stabilize lather of shampoos and liquid detergents
*Form:* Off-white liquid/White paste
*Composition:* 100% active

**Cyclomide DL203/S.** Cyclo Chem. Corp.
*Chemical Description:* 1:1 Lauric

diethanolamide
*Category:* Foam and detergency booster
*Applications:* High quality foam booster for cosmetic products, boosts detergency and contributes super-fatty effect
*Form:* Crystalline solid; Color: Off-white
*Composition:* 100% active, ≈ 95% amide content

**Cyclomide DL207/S.** Cyclo Chem. Corp.
*Chemical Description:* 1:1 Lauric/myristic diethanolamide
*Category:* Foam and viscosity booster, wetting agent
*Applications:* Maximum mildness combined with outstanding foam and viscosity boosting, fastest wetting action of series plus improved super-fatting effect
*Form:* Off-white liquid/White paste
*Composition:* 100% active, ≈ 95% amide content

**Cyclomide DO280/S.** Cyclo Chem. Corp.
*Chemical Description:* 1:1 Oleic diethanolamide
*Category:* Thickening and super-fatting agent, emulsifier
*Applications:* Excellent thickening and super-fatting agent for lotion shampoos, good auxiliary emulsifier
*Form:* Liquid
*Composition:* 100% active, ≈ 85% amide content

**Cyclomide DOTS.** Cyclo Chem. Corp.
*Chemical Description:* 1:1 Linoleic diethanolamide
*Applications:* Used as a viscosity builder, effective as an emulsifier and as a pigment dispersant
*Form:* Liquid; Color: Amber
*Composition:* 100% active, ≈ 95% amide content

**Cyclomide DP240/S.** Cyclo Chem. Corp.
*Chemical Description:* 1:1 Palm kernel diethanolamide
*Category:* Viscosity builder, foam stabilizer
*Applications:* Liquid viscosity builder and foam stabilizer for organic formulations
*Form:* Liquid
*Composition:* 100% active, ≈ 85% amide content

**Cyclomide IODI.** Cyclo Chem. Corp.
*Chemical Description:* Isostearamidopropyl dimethylamine
*Category:* Cationic emulsions
*Applications:* Base for preparing cationic emulsions, promotes substantivity and adhesion of other additives to the skin, enhances protein substantivity
*Form:* Clear liquid
*Composition:* 100% active, ≈ 98% amide content

**Cyclomide KD.** Cyclo Chem. Corp.

*Chemical Description:* 1:1 Coconut diethanolamide
*Category:* Viscosity builder, foam stabilizer
*Applications:* Superior to other 1:1 superamides as a viscosity builder while providing excellent foam stabilization
*Form:* Liquid; Color: Pale Yellow
*Composition:* 100% active, ≈ 95% amide content

**Cyclomide L203.** Cyclo Chem. Corp.
*Chemical Description:* Lauric monoethanolamide
*Category:* High viscosity and foam stabilizer
*Applications:* Promotes high viscosity with superior foam stability
*Form:* Flakes; Color: Off-white
*Composition:* 100% active, ≈ 95% amide content

**Cyclomide Pinemulse.** Cyclo Chem. Corp.
*Chemical Description:* Modified coconut polydiethanolamide
*Category:* Emulsifier, coupler
*Applications:* Energy saving emulsifier and coupler, designed for formulating extremely stable pine oil cleaners without the aid of heat
*Form:* Liquid; Color: Amber
*Composition:* 100% active

**Cyclomide S280.** Cyclo Chem. Corp.
*Chemical Description:* Stearic monoethanolamide
*Category:* Skin protectant
*Applications:* Outstanding skin protectant in toilet bars, creams, lotions and pastes
*Form:* Flakes; Color: Off-white
*Composition:* 100% active, ≈ 95% amide content

**Cyclomide SODI.** Cyclo Chem. Corp.
*Chemical Description:* Stearamidopropyl dimethylamine
*Category:* Cationic emulsions
*Applications:* Solid material for preparing cationic emulsions, unique effect produced when neutralized with water soluble acids
*Form:* Solid; Color: Off-white
*Composition:* 100% active, ≈ 98% amide content

**Cyclophos P4.** Cyclo Chem. Corp.
*Chemical Description:* Aliphatic
*Category:* Acid ester, anionic
*Applications:* Hydrophilic phosphate ester that imparts detergency, moderate foaming and good coupling properties in powdered machine dishwashing products and steam carpet cleaners; excellent surface tension depressant
*Form:* Clear thin liquid
*Composition:* 98% active

**Cyclophos PB12.** Cyclo Chem. Corp.
*Chemical Description:* Aromatic
*Category:* Acid ester, anionic
*Applications:* Developed for metal

cleaning; highly stable under prolonged use in soak tank cleaning; retains cleaning power under high soil loads
*Form:* Clear liquid
*Composition:* 98% active

**Cyclophos PE5.** Cyclo Chem Corp.
*Chemical Description:* Aromatic
*Category:* Acid ester, anionic
*Applications:* Hydrotrope that is dispersible in either oil or water systems; promotes detergency in dry cleaning, concrete cleaners, and powdered laundry formulations; effective at elevated temperatures, in solubilizing nonionic
*Form:* Clear liquid
*Composition:* 98% active

**Cyclophos PE9.** Cyclo Chem. Corp.
*Chemical Description:* Aromatic
*Category:* Acid ester, anionic
*Applications:* High in electrolyte tolerance; used to promote detergency and viscosity in heavy duty, all purpose liquid formulations and built steam cleaners
*Form:* Clear liquid
*Composition:* 99% active

**Cyclophos PE30.** Cyclo Chem. Corp.
*Chemical Description:* Aromatic
*Category:* Acid ester, anionic
*Applications:* Useful in emulsion polymerization to promote clarity and stability of cast acrylic and PVA films; latex stabilizer; for oil emulsion printing pastes; dyeing assistant
*Form:* Paste; Color: Off-white
*Composition:* 90% active

**Cyclophos PL3.** Cyclo Chem. Corp.
*Chemical Description:* Aliphatic
*Category:* Acid ester, anionic
*Applications:* An effective emulsifier for cosmetic oils, creams and waterless hand cleaners; a lubricity additive in soluble oils and a rust inhibitor; used in lubricating and rolling oils and water-based cutting fluids
*Form:* Liquid; Color: Pale yellow
*Composition:* 99% active

**Cyclophos PL6A.** Cyclo Chem. Corp.
*Chemical Description:* Aliphatic
*Category:* Acid ester, anionic
*Applications:* Compatible in concentrated electrolyte solutions; excellent coupler for nonionics in liquid alkaline systems; stable and non-coloring on flake caustic; moderate foamer
*Form:* Liquid; Color: Pale yellow
*Composition:* 98% active

**Cyclophos PL31.** Cyclo Chem. Corp.
*Chemical Description:* Aliphatic
*Category:* Acid ester, anionic
*Applications:* Used as coupling agent in extreme alkali systems; additive for the improvement of lubricity and the loadbearing properties of oil-based lubricants along with emulsification of cutting oils
*Form:* Clear liquid
*Composition:* 98% active

**Cyclophos TE12.** Cyclo Chem. Corp.
*Chemical Description:* Aromatic

*Category:* Acid ester, anionic
*Applications:* Highly water soluble phosphate ester that is an excellent emulsifier and solubilizer of nonionics in high electrolyte solutions; excellent alkali stability provides hard surface detergency for heavy duty cleaning formulations, household detergents, floor cleaning concentrates and metal cleaning products
*Form:* Liquid; Color: Amber
*Composition:* 98% active

**Cyclopol SB-5.** Cyclo Chem. Corp.
*Chemical Description:* Sulfosuccinate of ethoxylated lanolin alcohol
*Category:* Moisturizing agent, anionic
*Applications:* Ideal for shampoos, foam baths, skin cleansers, skin creams, hair sprays, dishwashing liquids
*Form:* Liquid; Color: Yellow
*Composition:* 48–50% active

**Cyclopol SBCA198.** Cyclo Chem. Corp.
*Chemical Description:* Sulfosuccinamate of mixed fatty amine
*Category:* Liquid surface active agent, anionic
*Applications:* Liquid surface active agent useful in latex systems; produces copious, very stable foam
*Form:* Liquid; Color: Yellow
*Composition:* 33–35% active

**Cyclopol SBCP.** Cyclo Chem. Corp.
*Chemical Description:* Sulfosuccinate of coconut alkanolamide
*Category:* Anionic
*Applications:* Functioning in a wide viscosity range, it promotes low irritation in mild shampoos, bubble baths and skin cleansers; compatible with anionics, nonionics, and amphoterics
*Form:* Liquid; Color: Yellow
*Composition:* 39–41% active

**Cyclopol SBDO.** Cyclo Chem. Corp.
*Chemical Description:* Dioctyl sulfosuccinate
*Category:* Reducing agent, anionic
*Applications:* Outstanding surface tension reducing agent
*Form:* Liquid
*Composition:* 68–70% active

**Cyclopol SBFA30.** Cyclo Chem. Corp.
*Chemical Description:* Sulfosuccinate of ethoxylated lauryl alcohol
*Category:* Raw material
*Applications:* Mild, non-irritating raw material for bubble baths, shampoos and general cleansing agents
*Form:* Liquid
*Composition:* 38–40% active

**Cyclopol SBG 212.** Cyclo Chem. Corp.
*Chemical Description:* Sulfosuccinate of disodium monococamide
*Category:* Base, foaming agent
*Applications:* Mild, non-irritating base for shampoos and skin products; a high foamer that is an easily handled liquid without use of heat
*Form:* Liquid; Color: Yellow
*Composition:* 29–31% active

**Cyclopol SBG 280.** Cyclo Chem. Corp.
*Chemical Description:* Sulfosuccinate of disodium mono-oleamide
*Category:* Foaming agent, anionic
*Applications:* Excellent surfactant that is non-irritating to the eyes and skin; produces good foam structure, viscosity control, and is compatible with anionic, nonionic and amphoteric surfactants
*Form:* Liquid; Color: Yellow
*Composition:* 29–31% active

**Cyclopol SBG 280/T.** Cyclo Chem. Corp.
*Chemical Description:* Sulfosuccinate of TEA-monooleamide
*Category:* Base, anti-irritant, anionic
*Applications:* High active, extremely mild, non-irritating base for skin products and shampoos; anti-irritant when combined with other surfactants
*Form:* Liquid; Color: Yellow
*Composition:* 39–41% active

**Cyclopol SBL203.** Cyclo Chem. Corp.
*Chemical Description:* Sulfosuccinate of lauric acid monoethanolamide
*Category:* Raw material, anionic
*Applications:* Raw material for carpet shampoos; when combined with sodium lauryl sulfate produces brittle, tack free residue; mild ingredient for bubble baths and shampoos
*Form:* Liquid; Color: Yellow
*Composition:* 38–40% active

**Cyclopol SBR-3.** Cyclo Chem. Corp.
*Chemical Description:* Sulfosuccinate of higher fatty acid monoethanolamide
*Category:* Skin-protective, anti-irritant surfactant
*Applications:* Eliminates or reduces skin irritation caused by other surfactants; recommended for skin cleansing compositions, dishwashing liquids, toiletries containing surfactants
*Form:* Liquid; Color: Yellow
*Composition:* 38–40% active

**Cyclopol SBU-185.** Cyclo Chem. Corp.
*Chemical Description:* Sulfosuccinate of undecylenic acid alkanolamide
*Category:* Fungicidal and bacteriostatic additive, anti-dandruff agent, anionic
*Applications:* Used in shampoos, body lotions, foot preparations and similar detergent compositions
*Form:* Liquid; Color: Yellow
*Composition:* 49–51% active

**Cycloryl 21.** Cyclo Chem. Corp.
*Chemical Description:* Sodium lauryl sulfate cosmetic grade
*Category:* Base, anionic
*Applications:* Low cost base for cream and liquid shampoos and specialty cleaners where fast solubility and free rinsing are sought
*Form:* Liquid/paste; Color: Pale Yellow
*Composition:* 28–30% active

**Cycloryl 21 LS.** Cyclo Chem. Corp.
*Chemical Description:* Sodium lauryl sulfate low salt
*Category:* Base, anionic
*Applications:* Low chloride base for use in shampoos, hand cleaners and detergents
*Form:* Liquid; Color: Pale Yellow
*Composition:* 28–30% active

**Cycloryl 21 SP.** Cyclo Chem. Corp.
*Chemical Description:* Sodium lauryl sulfate low cloud point
*Category:* Flowable process aid, anionic
*Applications:* Flowable process aid for the preparation of compounds, frothing agents, and detergents
*Form:* Liquid; Color: Yellow
*Composition:* 28–30% active

**Cycloryl ABSA.** Cyclo Chem. Corp.
*Chemical Description:* Linear alkyl aryl sulfonic acid
*Category:* Detergent, anionic
*Applications:* Hard surface cleaners, built detergents and shampoo system
*Form:* Liquid; Color: Amber
*Properties:* Biodegradable
*Composition:* 97% min. active

**Cycloryl DA.** Cyclo Chem. Corp.
*Chemical Description:* Diethanolamine lauryl sulfate
*Category:* Base, anionic
*Applications:* Base for preparation of clear, mild shampoos
*Form:* Liquid; Color: Yellow
*Composition:* 34–36% active

**Cycloryl DCA.** Cyclo Chem. Corp.
*Chemical Description:* Modified coconut polydiethanolamide
*Category:* Detergent, anionic
*Applications:* Concentrated detergent for human and pet shampoos; used to build viscosity and to solubilize difficult ingredients
*Form:* Liquid; Color: Amber
*Composition:* 98–100% active

**Cycloryl DDB40.** Cyclo Chem. Corp.
*Chemical Description:* Sodium alkyl aryl sulfonate
*Category:* Base, anionic
*Applications:* Base for medium duty detergents, cleaners and strippers
*Form:* Liquid; Color: Yellow
*Composition:* 38–40% active

**Cycloryl DDB60T.** Cyclo Chem. Corp.
*Chemical Description:* Triethanolamine alkyl aryl sulfonate
*Category:* Base, anionic
*Applications:* High active base for cleaners and shampoos
*Form:* Liquid; Color: Yellow
*Composition:* 58–60% active

**Cycloryl JB-8.** Cyclo Chem. Corp.
*Chemical Description:* Sodium lauryl ether sulfate plus special additives
*Category:* Base, anionic
*Applications:* Base for foaming bath gels and liquids
*Form:* Liquid; Color: Amber
*Composition:* 33–36% active

**Cycloryl LDC.** Cyclo Chem. Corp.
*Chemical Description:* Ammonium lauryl ether sulfate plus sulfonate and amide
*Category:* Base, anionic
*Applications:* Base for light duty detergent and liquid dishwashing compounds; car washing applications
*Form:* Liquid; Color: Amber
*Composition:* 60–62% active

**Cycloryl M1.** Cyclo Chem. Corp.
*Chemical Description:* Built pearlized sodium lauryl ether sulfate
*Category:* Base, anionic
*Applications:* Easy to use base for luxurious pearly cream shampoos having high flash foam
*Form:* Creamy liquid; Color: White
*Composition:* 34–36% active

**Cycloryl MA.** Cyclo Chem. Corp.
*Chemical Description:* Ammonium lauryl sulfate
*Category:* Base, anionic
*Applications:* Base for shampoos, bubble baths, and light duty detergents
*Form:* Liquid; Color: Pale Yellow
*Composition:* 28–30% active

**Cycloryl MA 330.** Cyclo Chem. Corp.
*Chemical Description:* Ammonium lauryl ether sulfate
*Category:* Base, degreasing agent, anionic
*Applications:* Base for shampoos and light duty detergents; grease cutting and oil solubilization to formulations
*Form:* Liquid; Color: Pale Yellow
*Composition:* 27–29% active

**Cycloryl MA 360.** Cyclo Chem. Corp.
*Chemical Description:* Ammonium lauryl ether sulfate
*Category:* Base, degreasing agent, anionic
*Applications:* High active base for compounding light duty dish detergents, fine fabric cleaners; superior grease cutting
*Form:* Liquid; Color: Pale Yellow
*Composition:* 58–60% active

**Cycloryl ME 60.** Cyclo Chem. Corp.
*Chemical Description:* Sodium myristyl ether sulfate
*Category:* Base, anionic
*Applications:* High active concentrated base for consumer or professional shampoos, foaming bath oils, bubble baths and cosmetic emulsions
*Form:* Liquid; Color: Pale Yellow
*Composition:* 58–60% active

**Cycloryl ME 361.** Cyclo Chem. Corp.
*Chemical Description:* Ammonium myristyl ether sulfate
*Category:* Base, anionic
*Applications:* High active concentrated base for acid shampoos and mild high foaming cleansing products
*Form:* Liquid; Color: Pale Yellow
*Composition:* 58–60% active

**Cycloryl MG.** Cyclo Chem. Corp.
*Chemical Description:* Magnesium

lauryl sulfate
*Category:* Base, anionic
*Applications:* Carpet, upholstery shampoo base having very low cloud point; forms tack free residue
*Form:* Liquid; Color: Pale Yellow
*Composition:* 28–30% active

**Cycloryl MS.** Cyclo Chem. Corp.
*Chemical Description:* Sodium myristyl sulfate
*Category:* Mild emulsifier effective, anionic
*Applications:* Mild emulsifier effective at very low levels; provides detergency to cleansing creams; excellent for fine fabric detergents
*Form:* Paste; Color: White
*Composition:* 25–27% active

**Cycloryl NA.** Cyclo Chem. Corp.
*Chemical Description:* Sodium lauryl ether sulfate
*Category:* Base, anionic
*Applications:* Base for shampoos and light duty detergents; superior electrolyte sensitivity, and good hard water stability to formulations
*Form:* Liquid; Color: Pale Yellow
*Composition:* 27–29% active

**Cycloryl NA-2.** Cyclo Chem. Corp.
*Chemical Description:* Sodium lauryl ether sulfate
*Category:* Base, anionic
*Applications:* Base for clear liquid and colored shampoos, bubble baths and light duty detergents
*Form:* Liquid; Color: Clear
*Composition:* 24–26% active

**Cycloryl NA 61.** Cyclo Chem. Corp.
*Chemical Description:* Sodium lauryl ether sulfate
*Category:* Base, anionic
*Applications:* High active base for high foaming detergents, dishwashes, car washes, textile scours and shampoos
*Form:* Liquid; Color: Pale Yellow
*Composition:* 58–60% active

**Cycloryl NA61 CG.** Cyclo Chem. Corp.
*Chemical Description:* Sodium lauryl ether sulfate
*Category:* Base, anionic
*Applications:* Cosmetics where color and high activity are important
*Form:* Liquid
*Composition:* 58–60% active

**Cycloryl NWC.** Cyclo Chem. Corp.
*Chemical Description:* Sodium lauryl ether sulfate plus builders
*Category:* Base, anionic
*Applications:* Base for shampoos and bubble baths
*Form:* Liquid; Color: Amber
*Properties:* Biodegradable
*Composition:* 62–64% active

**Cycloryl SET.** Cyclo Chem. Corp.
*Chemical Description:* Ethoxylated primary alcohol plus additives
*Category:* Base, anionic
*Applications:* High active base for superior heavy duty liquid laundry detergent
*Form:* Liquid
*Composition:* 79–81% active

**Cycloryl TD.** Cyclo Chem. Corp.
*Chemical Description:* Triethanolamine lauryl ether sulfate
*Category:* Base, anionic
*Applications:* Base for viscous shampoos with high salt tolerance
*Form:* Liquid; Color: Pale Yellow
*Composition:* 39–41% active

**Cycloryl TEALS.** Cyclo Chem. Corp.
*Chemical Description:* Triethanolamine lauryl sarcosinate
*Category:* Foaming agent, detergent, anionic
*Applications:* High flash foam, mild detergent for shampoos, bubble baths; performs well in hard water applications
*Form:* Liquid
*Composition:* 39–41% active

**Cycloryl TS30.** Cyclo Chem. Corp.
*Chemical Description:* Sodium lauryl sulfate plus additves
*Category:* Detergent, anionic
*Applications:* Carpet shampoo concentrate
*Form:* Liquid; Color: Amber
*Composition:* 28–30% active

**Cycloryl WAT.** Cyclo Chem. Corp.
*Chemical Description:* Triethanolamine lauryl sulfate
*Category:* Base, anionic
*Applications:* Base for light colored clear shampoos, bubble baths and other cleaning preparations where a creamy, mild lather is desired
*Form:* Liquid; Color: Pale Yellow
*Composition:* 38–40% active

**Cycloryl WAT-I.** Cyclo Chem. Corp.
*Chemical Description:* Triamine lauryl sulfate
*Category:* Base, anionic
*Applications:* Base for highly concentrated foaming toiletry products; useful where water content must be minimized
*Form:* Liquid
*Composition:* 86–90% active

**Cycloryl XL.** Cyclo Chem. Corp.
*Chemical Description:* Amphoteric modified lauryl sulfate
*Category:* Base, anionic
*Applications:* Base for providing conditioning and low irritation shampoos
*Form:* Liquid; Color: Pale Yellow
*Composition:* 36–38% active

**Cyclosheen 100.** Cyclo Chem. Corp.
*Chemical Description:* A blend of alkanolamides and opacifiers
*Category:* Emulsifier, nonionic
*Applications:* Used to impart stable pearl, to improve lathering and to increase viscosity of liquid shampoos and cleansers
*Form:* Soft wax; Color: Off-white
*Properties:* M. P.: 58–60 C; Acid No.: 2.0 max.

**Cycloteric BET C-30.** Cyclo Chem. Corp.
*Chemical Description:* Betaine amphoteric
*Category:* Foaming agent
*Applications:* Mild, high foaming surfactant; a broad range of viscosity control is provided when used with anionics; for promoting foam and detergency over a wide

pH range
*Form:* Liquid; Color: Yellow
*Properties:* Stability: Stable in acid and alkaline media; pH: 4.5–5.5
*Composition:* 29–31% active

**Cycloteric BET I-30.** Cyclo Chem. Corp.
*Chemical Description:* Betaine amphoteric
*Category:* Conditioning agent
*Applications:* Outstanding conditioning agent that provides an excellent viscosity building effect in compounded systems
*Form:* Liquid; Color: Yellow
*Properties:* pH: 5.5–6.5
*Composition:* 29–31% active

**Cycloteric BET O-30.** Cyclo Chem Corp.
*Chemical Description:* Betaine amphoteric
*Category:* Foaming, conditioning, and thickening agents
*Applications:* Foam stabilizer, conditioner, and thickening agent for hair and skin preparations
*Form:* Liquid; Color: Yellow
*Properties:* pH: 5.5–6.5
*Composition:* 29–31% active

**Cycloteric DC-SF.** Cyclo Chem. Corp.
*Chemical Description:* Dicarboxylate amphoteric
*Category:* Cleansing agent
*Applications:* High tolerance to alkali and other electrolytes
*Form:* Liquid; Color: Clear
*Properties:* pH: 8.0–9.0
*Composition:* 38–40% active

**Cycloteric MV-SF.** Cyclo Chem. Corp.
*Chemical Description:* Monocarboxylate amphoteric
*Category:* Low foaming, wetting agent
*Applications:* A powerful, low foaming wetting agent valuable in compounding highly acidic and highly alkaline cleansers where fast action and rust inhibition are sought
*Form:* Liquid; Color: Clear
*Properties:* pH: 9.0–11.0
*Composition:* 38–40% active

**Cycloton CT 100.** Cyclo Chem. Corp.
*Chemical Description:* Fatty quaternary compound combined with compatible emulsifiers
*Category:* Emulsifier, cationic
*Applications:* An excellent creme rinse/conditioner is produced when the cationic wax is diluted
*Form:* Flakes; Color: White
*Properties:* M. P.: 76–77 C; Acid No.: 1.0 max.

**Cyncal 80%.** Hilton-Davis
*Chemical Description:* Blend of alkyl dimethyl benzyl ammonium chlorides
*Category:* Antimicrobial agent, antistat, cationic
*Applications:* Disinfectant, sanitizer, alone or in combination with detergents for use in food, beverage processing plants, food service establishments, hospitals, industrial plants and farm use; fabric

antistatic agent; water treatment; good hard water tolerance
*Form:* Liquid; Color: Light yellow; Odor: Mild, pleasant
*Properties:* m.w.: 359 (of active); Solubility: Freely sol. in water, lower alcohols, ketones, glycols; Density: 7.8 lb/gal; Sp. gr.: 0.94; Stability: Crystallizes after prolonged storage at low temperature (returns to homogeneous state on warming to room temperature)
*Composition:* 80% active ingred.
*Reaction:* Compatible with nonionics, inorganic detergent builders, organic chelates, most polar solvents
*Toxicity:* Skin irritant; causes eye damage
*Warnings and Caution:* Corrosive. Goggles and gloves should be worn. Harmful or fatal if swallowed. Avoid contamination of food
*References:* Tech bulletin A-2-9/72

# D

**Dapral.** Akzo Chemie
*Chemical Description:* Phosphate derivatives
*Category:* Detergent, wetting agent, emulsifier
*Applications:* Detergent—textiles, laundering, dyeing, household and cosmetics; as wetting agent in dyeing; emulsifiers for making of cutting and soluble oils; dispersing waxes, pigments, resins
*Form:* Liquid, paste

**Darvan No. 1.** R. T. Vanderbilt Co.
*Chemical Description:* Complex polymerized organic salts of sulfonic acids of alkylaryl type
*Category:* Dispersing agent
*Applications:* General purpose dispersant esp. for good color; in adhesives, as defoaming agents in paperboard, in gasket sealers for food packaging; agricultural sprays; aids ball and tube mill grinding; cosmetic emulsions and suspensions esp. liquid make-up bases, eye make-up, calamine and talc lotions; household polishes; for barium sulfate suspensions in medical research and x-ray work; leather tanning processes
*Form:* Granular; Color: Buff
*Properties:* Solubility: 40% sol. in water; Density: 1.55 mg/m$^3$; Stability: Stable to freezing, mild acid and alkali; pH: 8.0–10.5 (1% sol'n.)

**Darvan No. 2.** R. T. Vanderbilt Co.
*Chemical Description:* Complex polymerized organic salts of sulfonic acids of alkylaryl type
*Category:* Dispersing agent, emulsifier
*Applications:* Pelleting aid in animal feed; agricultural sprays; aids in ball and tube mill grinding; O/W emulsions; cosmetic emulsions and suspensions esp. liquid make-up bases, eye make-up, calamine and talc lotions; household polishes; pharmaceutical tablets containing insolubles; in microbiology
*Form:* Powder; Color: Dark brown
*Properties:* Solubility: 25% water sol.; Density: 1.59 mg/m$^3$; Stability: Stable to freezing, mild

acid and alkali; pH: 7.0–8.5 (1% sol'n.)

**Darvan No. 7.** R. T. Vanderbilt Co.
*Chemical Description:* Polyelectrolyte dispersing agent in aqueous solution
*Category:* Dispersant
*Applications:* Nondiscoloring dispersant for preparing high solids dispersions; agricultural sprays; aids in ball and tube mill grinding; cosmetic emulsions esp. liquid make-up bases, eye make-up, calamine and talc lotions; household polishes; used by museums and universities in cleaning artifacts; in rubber and ceramic fields to disperse $TiO_2$
*Form:* Liquid; Color: Water white
*Properties:* Solubility: Very water sol.; Density: 1.16 $mg/m^3$; Stability: Stable to freezing, mild acid and alkali; pH: 9.5–10.5
*Composition:* 25% solids
*References:* Bulletin No. 118

**Daxad 11.** W.R. Grace & Co.
*Chemical Description:* Sodium salt of polymerized alkyl naphthalene sulfonic acids
*Category:* Dispersing agent
*Applications:* Agricultural chemicals; mastics, caulks and sealants; pigment slurries and dispersions; reduces slurry viscosity; emulsification properties
*Form:* Fine powder; Color: Buff
*Properties:* Density: 48 $lb/ft^3$; Surface Tension: 71 dynes/cm (1% sol'n. @ 20 C)

**Daxad 11G.** W.R. Grace & Co.
*Chemical Description:* Sodium salt of polymerized alkyl naphthalene sulfonic acids
*Category:* Dispersing agent
*Applications:* Agricultural chemicals; inks; mastics, caulks and sealants; synthetic polymers; emulsification; paper industry; polymerization; slurry viscosity reduction; wetting action toward carbon black
*Form:* Granules; Color: Buff
*Properties:* Density: 35 $lb/ft^3$; Stability: Stable to mild acids and alkalies; Surface Tension: 71 dynes/cm (1% @ 20 C)
*Composition:* 87% active ingred.; 4.5% max. moisture content; 8.5% max. inert solubles
*References:* Bulletin DX-72-1

**Daxad 11KLS.** W.R. Grace & Co.
*Chemical Description:* Potassium salt of polymerized alkyl naphthalene sulfonic acids
*Category:* Dispersant
*Applications:* Dyes, dyestuffs; inks; latex paints; pigment slurries and dispersions; latex viscosity reduction; textile treatment
*Form:* Granules; Color: Light buff
*Properties:* Density: 35 $lb/ft^3$; Surface Tension: 71 dynes/cm (1% @ 20 C)

**Daxad 13.** W.R. Grace & Co.
*Chemical Description:* Sodium salt of polymerized alkyl naphthalene sulfonic acids
*Category:* Dispersent

*Applications:* Disperses inorganic and organic pigments; paper coating dispersion; paper mill slime control
*Form:* Powder; Color: Amber
*Properties:* Solubility: Very sol. in water systems; Density: 47 lb/$ft^3$; Surface Tension: 66 dynes/cm (1% sol'n. @ 72 F); pH: 9.5 (1% sol'n. @ 72 F)

**Daxad 15.** W.R. Grace & Co.
*Chemical Description:* Sodium salt of polymerized alkyl naphthalene sulfonic acids
*Category:* Dispersant
*Applications:* Agricultural chemicals; ceramics; concrete admixtures; dyes; cements; industrial cleaners; inks; water treatment chemicals
*Form:* Powder; Color: Amber
*Properties:* Density: 47 lb/$ft^3$; Surface Tension: 70 dynes/cm (1% sol'n. @ 20 C)

**Daxad 16.** W.R. Grace & Co.
*Chemical Description:* Sodium salts of polymerized alkyl naphthalene sulfonic acids
*Category:* Dispersant
*Applications:* Agricultural chemicals; concrete admixtures; dyes; pigment slurries and dispersions; synthetic polymers; synthetic tanning agents; water treatment chemicals
*Form:* Transparent liquid; Color: Dark brown
*Properties:* Density: 10.4 lb/gal.; Stability: Stable to mild acids and alkalies; Surface Tension: 70 dynes/cm (1% sol'n. @ 20 C); pH: 9–10 (1% sol'n.)
*Composition:* 42.5% active ingred.
*References:* Bulletin DX-66-2

**Daxad 17.** W.R. Grace & Co.
*Chemical Description:* Sodium salts of polymerized alkyl naphthalene sulfonic acids
*Category:* Dispersant
*Applications:* Agricultural chemicals; ceramics; concretes and cements; dyes; industrial cleaners; inks; mastics; pigment slurries; synthetic polymers; synthetic tanning agents; water treatment chemicals
*Form:* Granules; Color: Amber
*Properties:* Density: 35 lb/$ft^3$; Surface Tension: 70 dynes/cm (1% sol'n. @ 20 C)

**Daxad 19.** W.R. Grace & Co.
*Chemical Description:* Sodium salts of polymerized alkyl naphthalene sulfonic acids
*Category:* Dispersant
*Applications:* Concretes and cements; gypsum wallboard; pigment slurries; synthetic tanning agents
*Form:* Powder; Color: Amber
*Properties:* Solubility: Very sol. in water systems; Density: 47 lb/$ft^3$; Surface Tension: 71 dynes/cm (1% sol'n. @ 20 C); pH: 9.0 (1% sol'n. @ 72 F)

**Daxad 21.** W.R. Grace & Co.
*Chemical Description:* Mono-calcium salt of polymerized aryl alkyl

sulfonic acids
*Category:* Dispersant
*Applications:* Agricultural chemicals; concretes and cements; dyes; gypsum wallboard; pigment slurries; water treatment chemicals
*Form:* Powder; Color: Brown
*Properties:* Density: 30 lb/ft$^3$; Stability: Stable to mild acids and alkalies; Surface Tension: 62 dynes/cm (1% sol'n. @ 20 C); pH: 5–8 (1% sol'n. @ 72 F)
*Composition:* 92% active ingred.
*References:* Bulletin DX-71-1

**Daxad 23.** W.R. Grace & Co.
*Chemical Description:* Sodium salts of polymerized substituted benzoid alkyl sulfonic acids
*Category:* Dispersant
*Applications:* Agricultural chemicals; concretes and cements; dyes; linoleum pastes; pigment slurries; water treatment chemicals
*Form:* Powder; Color: Dark brown
*Properties:* Density: 37 lb/ft$^3$; Stability: Stable to mild acids and alkalies; Surface Tension: 56 dynes/cm (1% sol'n. @ 20 C); pH: 7.0–8.5 (1% sol'n. @ 72 F)
*Composition:* 84.5% active min.
*References:* Bulletin DX-63-2

**Daxad 27.** W.R. Grace & Co.
*Chemical Description:* Sodium salts of polymerized substituted benzoid alkyl sulfonic acids formulated with an inorganic suspending agent
*Category:* Dispersant
*Applications:* Long lasting suspensions of larger size particles; agricultural chemicals; pigment slurries and dispersions
*Form:* Free-flowing powder; Color: Light gray-brown
*Properties:* Density: 48 lb/ft$^3$; Stability: Stable to mild acids and alkalies; Surface Tension: 62 dynes/cm (1% sol'n. @ 20 C); pH: 6–8 (30% aq. susp.)
*Composition:* 84.5% active min.
*References:* Bulletin DX-62-2

**Daxad 30.** W.R. Grace & Co.
*Chemical Description:* Sodium salt of a carboxylated polyelectrolyte
*Category:* Dispersant
*Applications:* Rapid dispersion of pigments in aqueous solutions; paint formulations; emulsion polymerization; large particle suspensions; agricultural chemicals; ceramics; industrial cleaners; inks; latex paints; mastics; water treatment chemicals
*Form:* Liquid; Color: Clear to slightly opalescent
*Properties:* Solubility: Sol. in water systems; Density: 9.6 lb/gal; Sp. gr.: 1.15; Visc.: 75 cps (Brookfield #2 spindle); F.P.: $< -5$ C; Stability: Stable over wide pH range; Surface Tension: 70 dynes/cm (1% sol'n. @ 20 C); pH: 10.0
*Composition:* 25% solids

**Daxad 30-30.** W.R. Grace & Co.
*Chemical Description:* Sodium salt of a carboxylated polyelectrolyte
*Category:* Dispersant
*Applications:* Rapid dispersion of pigments in aqueous solutions;

paint formulations; emulsion polymerization; large particle suspensions; agricultural chemicals; ceramics; industrial cleaners; inks; latex paints; mastics; water treatment chemicals
*Form:* Liquid; Color: Clear to slightly opalescent
*Properties:* Solubility: Sol. in water systems; Density: 9.9 lb/gal; Sp. gr.: 1.21; Visc.: 1 cps max. (Brookfield #2 spindle); F.P.: < −5 C; Stability: Stable over wide pH range; Surface Tension: 70 dynes/cm (1% sol'n. @ 20 C); pH: 10
*Composition:* 30% solids content

**Daxad 30S.** W.R. Grace & Co.
*Chemical Description:* Sodium salt of a carboxylated polyelectrolyte
*Category:* Dispersant
*Applications:* Rapid dispersion of pigments in aqueous solutions; agricultural chemicals; industrial cleaners; mastics; pigment slurries; water treatment chemicals
*Form:* Fine powder; Color: White
*Properties:* Solubility: Sol. in water systems; Density: 32 lb/ft$^3$; Stability: Stable over wide pH range; Surface Tension: 70 dynes/cm (1% sol'n. @ 20 C)
*Composition:* 90% active ingred.
*References:* Bulletin DX-61-2

**Daxad 31.** W.R. Grace & Co.
*Chemical Description:* Sodium salt of a carboxylated polyelectrolyte
*Category:* Dispersant
*Applications:* Latex paints and coatings; dispersion of organic and inorganic pigments
*Form:* Clear liquid; Color: Yellow
*Properties:* Solubility: Sol. in water systems; Density: 9.2 lb/gal; Sp. gr.: 1.11; Visc.: 30 cps (Brookfield #2 spindle); Surface Tension: 50 dynes/cm (1% sol'n. @ 25 C)
*Composition:* 25% solids content

**Daxad 32.** W.R. Grace & Co.
*Chemical Description:* Ammonium salt of carboxylated polyelectrolyte
*Category:* Dispersant
*Applications:* Ceramics; pigment slurries and dispersions
*Form:* Clear to slightly opalescent fluid; Color: Amber
*Properties:* Solubility: Very sol. in water systems; Density: 9.1–9.4 lb/gal; Sp. gr.: 1.11; Visc.: 75 cps (Brookfield #2 spindle); Surface Tension: 70 dynes/cm (1% sol'n. @ 20 C); pH: 7.5–9.0
*Composition:* 25% solids content
*References:* Bulletin DX-60-2

**DDBSA 99-b.** Monsanto Ind. Chem.
*Chemical Description:* Dodecylbenzene sulfonic acid
*Applications:* Used in low salt liquid detergents; general purpose neutralization reactions; spray dried products
*Properties:* m.w.: 316 avg.; Sp. gr.: 1.0;
*Composition:* 98% active ingred.
*Toxicity:* Causes burns; harmful if absorbed through skin
*Containers:* 450 lb net, 55 gal metal drums; 5000 gal tanktrucks;

8000 and 10,000 gal tankcars
*Warnings and Caution:* Do not get in eyes, on skin or clothing; avoid breathing vapor or mist; in case of contact flush thoroughly with water for 15 min; call physician

**Deceresol Surfactant CW.** American Cyanamid Co.
*Category:* Wetting agent, modified nonionic
*Applications:* Wetting of various textiles during processing; dyeing; desizing and boiling off cottons; bleaching; sizing; Kier boiling; papermaking; wet out of fiber mulch for dyeing process; desizing of starch-based sizes
*Form:* Liquid; Color: Clear
*Properties:* Solubility: Dissolves readily with stirring; Density: 8.8 lb/gal; Sp. gr.: 1.06; Stability: Excellent
*Composition:* 92% active ingred.; other components: water and 2-ethyl hexanol
*Reaction:* Modified nonionic
*Toxicity:* Severe eye irritant; drying to skin
*Warnings and Caution:* Rubber gloves and goggles should be worn
*References:* Bulletin 292

**Deceresol Surfactant NI Conc.** American Cyanamid Co.
*Category:* Detergent, emulsifier, wetting agent, nonionic
*Applications:* Reactant finishing of textiles, desizing, scouring, bleaching, dyeing, printing, carbonizing woolens; Kier boiling; making of cutting and soluble oils; dispersing waxes, pigments, resins
*Form:* Liquid; Color: Clear
*Properties:* Solubility: Dissolves readily with stirring in cold water; less sol. in water above 140 F; Density: 8.8 lb/gal; Sp. gr.: 1.05; Stability: Excellent; stable to freezing and thawing; pH: 6–7
*Composition:* 100% active ingred.
*Reaction:* Nonionic
*Toxicity:* Nontoxic in normal usage; avoid skin and eye contact; severe eye irritant and drying to skin
*Containers:* 470 lb containers
*Warnings and Caution:* Wear rubber gloves and goggles when handling
*References:* Bulletin 182

**Deceresol Surfactant OT 75%.** American Cyanamid Co.
*Category:* Wetting agent, anionic
*Applications:* Wetting out fabric; desizing; scouring and bleaching; dyeing and printing; finishing
*Form:* Viscous liquid; Color: Water white; Odor: Slight hexanol
*Properties:* Solubility: Completely as dilute solution; Density: 8.9 lb/gal; Sp. gr.: 1.082; Visc.: 172 cp; F.P.: 30 F; B.P.: ≈ 200 F; Stability: Excellent; Biodegradable in 24 h; pH: 5.6 (1% sol'n.)
*Composition:* 75% active ingred.; other components: 6–8% alcohol
*Toxicity:* Nontoxic in normal usage; severe eye irritant and drying to skin
*Containers:* 475 lb containers
*Warnings and Caution:* Wear rub-

ber gloves and goggles when handling

**Deceresol Surfactant OT Special 70%.** American Cyanamid Co.
*Category and Applications:* See Deceresol Surfactant OT 75%
*Form:* Thin liquid; Color: Colorless; Odor: Slight
*Properties:* Solubility: Dissolves with agitation in warm water; Density: 8.8 lb/gal; Sp. gr.: 1.05; Stability: Good; pH: 5.7
*Composition:* 70% active ingred.; other components: water and 2-ethyl hexanol
*Reaction:* Slightly acid
*Toxicity:* Nontoxic in normal usage; severe eye irritant; drying to skin
*Containers:* 475 lb containers
*Warnings and Caution:* Wear rubber gloves and goggles when handling
*References:* Bulletin 263

**Deceresol Surfactant P Special.** American Cyanamid Co.
*Category:* Detergent
*Applications:* Textile detergent
*Form:* Liquid; Color: Light amber
*Properties:* Solubility: Infinite in water; Density: 8.5 lb/gal; Stability: Stable to freeze/thaw
*Composition:* 18% active ingred.
*Reaction:* Neutral
*Containers:* 450 lb containers
*References:* Bulletin 183

**Defoamer 347D.** Hart Chem.
*Chemical Description:* Anionic specialty formulated product
*Category:* Defoamer
*Applications:* Pulp and paper effluent defoamer
*Form:* Paste; Color: Amber
*Properties:* Solubility: Fair emulsion
*Composition:* 53% active ingred.
*Reaction:* Acid
*Toxicity:* Skin and eye irritant
*Containers:* Drums, T/L
*Warnings and Caution:* Avoid contact with eyes and skin

**Defoamer 543.** Hart Chem.
*Chemical Description:* Nonionic specialty formulated product
*Category:* Defoamer
*Applications:* Paper machine and industrial effluent defoamer
*Form:* Liquid; Color: Yellow
*Properties:* Refractive Index: 1.4550; Solubility: Fair emulsion; Density: 0.874; Visc.: 31 cps
*Composition:* 100% active ingred.
*Reaction:* Neutral
*Toxicity:* Skin and eye irritant
*Containers:* Drums, T/L
*Warnings and Caution:* Avoid contact with skin and eyes

**Defoamer 2000 Series.** Hart Chem.
*Chemical Description:* Nonionic specialty formulated products
*Category:* Defoamer
*Applications:* Defoamers for brownstock, screen room, effluent and bleach plant
*Form:* Liquid; Color: Cream
*Properties:* Density: 0.870; Visc.: 700 cps
*Composition:* 100% active ingred.
*Toxicity:* Skin and eye irritant
*Containers:* Drums, T/T, T/C

**Defoamer 4201.** Hart Chem.
*Chemical Description:* Nonionic specialty formulated product
*Category:* Defoamer
*Applications:* Brownstock defoamer
*Form:* Liquid; Color: White
*Properties:* Solubility: Dispersible Density: 0.87; Visc.: 450 cps
*Composition:* 100% active ingred.
*Toxicity:* Skin and eye irritant
*Containers:* Drums, T/T, T/C
*Warnings and Caution:* Avoid contact with skin and eyes

**Defoamer 4203.** Hart Chem.
*Chemical Description:* Nonionic specialty formulated product
*Category:* Defoamer
*Applications:* Brown stock defoamer
*Form:* Liquid; Color: White
*Properties:* Solubility: Dispersible; Sp. gr.: 0.86; Visc.: 450 cps
*Composition:* 100% active ingred.
*Toxicity:* Skin and eye irritant
*Containers:* Drums, T/T, T/C
*Warnings and Caution:* Avoid contact with skin and eyes

**Defoamer 4205.** Hart Chem.
*Chemical Description:* Nonionic specialty formulated product
*Category:* Defoamer
*Applications:* Screen room and effluent defoamer
*Form:* Liquid; Color: White
*Properties:* Solubility: Fair emulsion; Sp. gr.: 0.87; Visc.: 150 cps
*Composition:* 100% active ingred.
*Toxicity:* Skin and eye irritant
*Containers:* Drums, T/T, T/C
*Warnings and Caution:* Avoid contact with skin and eyes

**Defoamer 4418.** Hart Chem.
*Chemical Description:* Nonionic specialty formulated product
*Category:* Defoamer
*Applications:* Board machine defoamer
*Form:* Liquid; Color: White
*Properties:* Solubility: Dispersible; Sp. gr.: 0.87; Visc.: 350 cps
*Composition:* 100% active ingred.
*Toxicity:* Skin and eye irritant
*Containers:* Drums, T/T, T/C
*Warnings and Caution:* Avoid contact with skin and eyes

**Defoamer 4440.** Hart Chem.
*Chemical Description:* Nonionic specialty formulated product
*Category:* Defoamer
*Applications:* Cold stock screen room defoamer
*Form:* Liquid; Color: White
*Properties:* Solubility: Dispersible; Sp. gr.: 0.86
*Composition:* 100% active ingred.
*Toxicity:* Skin and eye irritant
*Containers:* Drums, T/T, T/C
*Warnings and Caution:* Avoid contact with skin and eyes

**Defoamer C5B.** Hart Chem.
*Chemical Description:* Nonionic specialty formulated product
*Category:* Defoamer
*Applications:* Board machine and industrial effluent defoamer
*Form:* Liquid; Color: Cream
*Properties:* Solubility: Fair emulsion; Sp. gr.: 0.86; Visc.: 400 cps

*Composition:* 100% active ingred.
*Toxicity:* Skin and eye irritant
*Containers:* Drums, T/T, T/C
*Warnings and Caution:* Avoid contact with skin and eyes

**Defoamer HF-2.** Hart Chem.
*Chemical Description:* Nonionic specialty formulated product
*Category:* Defoamer
*Applications:* Paper machine defoamer
*Form:* Paste; Color: White
*Properties:* Solubility: Stable emulsion; Visc.: 6000 cps
*Composition:* 30% active ingred.
*Toxicity:* Skin and eye irritant
*Containers:* 45 IG Drums
*Warnings and Caution:* Avoid contact with skin and eyes

**Defoamer LH-192.** Hart Chem.
*Chemical Description:* Nonionic specialty formulated product
*Category:* Defoamer
*Applications:* Liquid phosphate rock defoamer
*Form:* Liquid; Color: Brown
*Properties:* Solubility: Dispersible; Visc.: 140 cps
*Composition:* 100% active ingred.
*Reaction:* Acid
*Toxicity:* Skin and eye irritant
*Containers:* T/L, T/T
*Warnings and Caution:* Avoid contact with skin and eyes

**Defoamer PBR-2.** Hart Chem.
*Chemical Description:* Nonionic specialty formulated product
*Category:* Defoamer
*Applications:* Control of foam problems in linerboard and corrugated mills
*Form:* Liquid; Color: White
*Properties:* Solubility: Dispersible; Sp. gr.: 0.86; Visc.: 450 cps
*Composition:* 100% active ingred.
*Toxicity:* Skin and eye irritant
*Containers:* Drums, T/L, mini bulk
*Warnings and Caution:* Avoid contact with skin and eyes

**Defoamer PM.** Hart Chem.
*Chemical Description:* Nonionic specialty formulated product
*Category:* Defoamer
*Applications:* Paper machine defoamer
*Form:* Liquid; Color: Amber
*Properties:* Refractive Index: 1.4864; Solubility: Fair emulsion; Sp. gr.: 0.88; Visc.: 40 cps
*Composition:* 100% active ingred.
*Reaction:* Neutral
*Toxicity:* Skin and eye irritant
*Containers:* Drums, T/L
*Warnings and Caution:* Avoid contact with skin and eyes

**Defoamer S.** Hart Chem.
*Chemical Description:* Nonionic silicone emulsion
*Category:* Defoamer
*Applications:* Multipurpose silicone defoaming agent for industrial and textile
*Form:* Viscous liquid; Color: White
*Properties:* Solubility: Readily dispersible; Sp. gr.: 0.988; Visc.: 1500 cps
*Reaction:* Acid
*Toxicity:* Skin and eye irritant
*Containers:* Drums, T/L
*Warnings and Caution:* Avoid contact with skin and eyes

**Delvet 68.** Diamond Shamrock
*Category:* Nonionic surfactant
*Applications:* Chlorinated paraffin flame retardant additive
*Form:* Fluid
*Composition:* 68% active ingred.

**Delvet 70.** Diamond Shamrock
*Category:* Nonionic surfactant
*Applications:* Chlorinated paraffin flame retardant additive
*Form:* Fluid
*Composition:* 68% active ingred.

**Demulfer Series.** Toho Chem.
*Category:* Demulsifier
*Applications:* Demulsifier for crude oil production
*Form:* Liquid

**Denlube 33-B.** Graden Chem.
*Chemical Description:* Ammonium stearate dispersion
*Category:* Foam rubber additive
*Applications:* Frothing aid and foam stabilizer in acrylic and SBR latex systems
*Form:* Liquid; Color: White; Odor: Ammoniacal
*Properties:* Density: 8.0 lb/gal; Stability: Unstable above 100 F when heated; pH: 10.0 ± 1.0 (2% sol'n.)
*Composition:* 30% active
*Containers:* 55 gal drums
*Warnings and Caution:* Solidifies on freezing but regains normal fluidity on thawing; should be stored at room temperature and kept tightly covered to avoid loss of ammonia

**Denlube 40.** Graden Chem.
*Chemical Description:* High density polyethylene emulsion
*Category:* Lubricant, nonionic
*Applications:* By lubricating the coating, it improves flow and leveling characteristics, making possible smooth even application; reduces cracking of the dry coating and provides roll release
*Form:* Liquid; Color: Amber
*Properties:* Density: 8.4 lb/gal; pH: 8.5–9.0 (1% sol'n.)
*Composition:* 22% active
*Containers:* 55 gal drums
*Warnings and Caution:* Freezing seriously damages the product; containers should be kept tightly closed when not in use to prevent evaporation loss

**Denlube 50 NFG.** Graden Chem.
*Chemical Description:* Scale wax emulsion
*Category:* Foam rubber additive
*Applications:* Added to latex during foaming to preserve the resilience of the foam rubber article; it lubricates the inside surfaces of the cells in the foam producing an anti-blocking effect
*Form:* Liquid; Color: White
*Composition:* 50% active
*Containers:* 55 gal drum
*Warnings and Caution:* Freezing seriously damages the product; protect from cold in transit and storage

**Denlube 1025-KA.** Graden Chem.
*Chemical Description:* A blend of higher molecular weight sulfonated

aliphatic compounds
*Category:* Antiblocking agent
*Applications:* Strip-stock lubricant during rubber processing; it also acts as a processing aid
*Form:* Paste; Color: Cream
*Properties:* Solubility: Disp. in all proportions in water at 180 F; pH: 8.5–9.0 (1% sol'n.)
*Composition:* 80% active
*Containers:* 450 lb net drums

**Denphos P-610.** Graden Chem.
*Chemical Description:* Phosphate ester
*Category:* Detergent, emulsifier, wetting agent, anionic
*Applications:* Used as the sole detergent and wetting agent because of its stability and solubility; low foaming product recommended for industrial and household cleaners
*Form:* Liquid; Color: Water-white to pale yellow; Odor: Mild
*Properties:* Solubility: Sol. in aqueous solutions containing high concentrations of alkaline builders; Sp. gr.: 1.155; pH: 1.5–2.5 (2% sol'n.)
*Composition:* 100% active
*Containers:* 500 lb net drums

**Densol 284.** Graden Chem.
*Chemical Description:* Highly sulfated fatty acid
*Category:* Stabilizer for natural and synthetic latex, anionic
*Applications:* Protects natural and synthetic latices from coagulation during high speed stirring or pumping
*Form:* Liquid; Color: Dark amber; Odor: Fatty
*Properties:* Stable to strong acids, such as dilute acetic or hydrochloric acid
*Containers:* 55 gal drums

**Densol 1010.** Graden Chem.
*Chemical Description:* Modified sodium polyacrylate
*Category:* Thickener, plasticizer, thinning compound
*Applications:* Extensively used in rug and upholstery backing adhesives; stabilizer for latex compounding systems
*Form:* Gel; Color: Light amber
*Properties:* Density: 8.5 lb/gal; pH: 10 (2% sol'n.)
*Composition:* 10% active
*Containers:* 55 gal drums

**Densol 6920.** Graden Chem.
*Chemical Description:* Synthetic replacement for sulfated castor oil
*Category:* Emulsifier, anionic
*Applications:* A good emulsifier for mineral oil as well as effective dispersant or emulsifier for colors in oil systems; prevents hard settling of pigments during storage
*Form:* Liquid
*Properties:* Sp. gr.: 0.972 @ 60 F; pH: 5.4 (10% sol'n.)
*Composition:* 75% active
*Containers:* 55 gal drums

**Densol 6930.** Graden Chem.
*Chemical Description:* Synthetic replacement for sulfated sperm oil
*Category:* Emulsifier
*Properties:* pH: 5.4 (10% sol'n.)

*Containers:* 55 gal drum

**Densol P-82.** Graden Chem.
*Chemical Description:* Sulfated oleic ester
*Category:* Wetting agent, emulsifier
*Applications:* Excellent wetting agent; rewetting agent for lath paper and corrugating medium, .009 paper; performs well in up to 1% sodium hydroxide solution; readily dissolves in water from 60 F to boiling; soluble in a number of organic solvents
*Form:* Liquid; Color: Amber
*Properties:* Density: 8.57 lb/gal; Sp. gr.: 1.0288; pH: 5.5 (2% sol'n.)
*Composition:* 63% active
*Containers:* 55 gal drums

**Denwet CM.** Graden Chem.
*Chemical Description:* Sodium dioctyl sulfosuccinate
*Category:* Wetting agent, anionic
*Applications:* Reduces surface and interfacial tension; aids in wetting, dispersing, emulsifying, penetrating and solubilizing; used in emulsion polymerization of latex, particularly polyvinylacetate
*Form:* Clear liquid
*Properties:* Density: 9.0 lb/gal; pH: 7.0 ± 0.5 (2% sol'n.)
*Composition:* 75 ± 1% active
*Containers:* 55 gal liquapak drum

**Denwet RG-7.** Graden Chem.
*Chemical Description:* Sulfonated alkyl ester
*Category:* Wetting agent, anionic
*Applications:* Reduces surface and interfacial tension; aids in wetting, dispersing, emulsifying, penetrating and solubilizing; drycleaning detergent which does not absorb on fabrics and develop rancid odors; contributes significantly to the adhesion of paint films to various substrates, especially shiny or glossy surfaces; used in oil solutions of herbicides and pesticides as a wetting and spreading agent; readily soluble in a wide range of solvents which makes it useful as a coupling agent or solubilizer for emulsifiers of lower solubility
*Form:* Clear liquid
*Properties:* Density: 9.0 lb/gal; pH: 7.0 ± 0.5 (2% sol'n.)
*Composition:* 75 ± 1% active
*Containers:* 55 gal liquapak drum

**Depuma.** American Color & Chem.
*Chemical Description:* Silicone type antifoaming agent
*Category:* Nonionic antifoaming agent
*Applications:* Textiles and general defoaming needs for water systems, in acid and alkaline liquors; skins, laundering, leather, dyeing, metallurgy, mines and minerals and papermaking industires; bleaching, dyeing, tanning, finishing, sizing, coating, electroplating processes; rotary washer; waterpaint calsomines, adhesives; insecticides, herbicides, fungicides; removing starches, wall paper; Kier boiling; fat liquoring; printing; making of bitumen emulsions
*Form:* Pourable viscous emulsion; Color: Pearly white
*Properties:* Solubility: Water dis-

persible; Density: 8.2 lb/gal; Stability: Indefinitely stable at room temperature; keep from freezing
*Composition:* Silicone
*Containers:* 450 lb drums

**Depuma OB.** American Color & Chem.
*Chemical Description:* Organic, nonsilicone type antifoam
*Category:* Nonionic antifoaming agent
*Applications:* Foam control for textiles in water systems; used in skins, leather, dyeing, metallurgy, mines and minerals, papermaking and paints industries; desizing and boiling off cottons; bleaching, dyeing, tanning, finishing, sizing, coating processes; water-paint calsomines, adhesives; insecticides, herbicides, fungicides; Kier boiling; fat liquoring; making of cutting and soluble oils; dispersing waxes, pigments, resins; printing; preparation of emulsified paints, bitumen emulsions
*Form:* Liquid; Color: White
*Properties:* Solubility: Sol. in most organic solvents; emulsifies in warm water; Stability: Stable; Emulsifiability: 100%; Biodegradable
*Composition:* 100% active ingred.
*Reaction:* Neutral, pH 7
*Containers:* 450 lb drums

**Deriphat.** General Mills Chem.
*Category:* Detergent, wetting agent, emulsifier
*Applications:* Detergents: textiles, leather, household, cosmetic (shampoos), cleaners of streets, buildings, hospitals; as wetting agent in bleaching process
*Form:* Liquid, powder, flake

**DET.** Swastik
*Chemical Description:* Alkyl aryl sulfonate sodium salt built with S.T.P.P. and other alkaline builders
*Category:* Detergent
*Applications:* Suitable for hand washing cottons, woolens, and synthetic fabrics
*Form:* Powder; Color: White
*Properties:* pH: 10.5 (1% sol'n.)

**DET Liquid.** Swastik
*Chemical Description:* Alkyl aryl sulfonate plus polyoxyethylene alcohol ether sulfate plus additives
*Category:* Cleansing, detergent
*Applications:* Liquid household cleanser, recommended for cleansing crockery, stainless steel utensils, tiles, linoleum, plastic wares
*Form:* Clear liquid
*Properties:* pH: 8.0–9.5 (1% sol'n.)

**DET-Washmatic.** Swastik
*Chemical Description:* Alkyl aryl sulfonate sodium salt
*Category:* Detergent
*Applications:* Suitable for use in washing machines
*Form:* Spray-dried powder; Color: Blue
*Properties:* pH: 10.5 (1% sol'n.)

**Det-O-Jet.** Alconox Inc.
*Category:* Detergent
*Applications:* Low sudsing deter-

gent with good wetting, emulsifying and penetrating properties for mechanical and ultrasonic washers; removing grease, heavy soils and stains from porcelain, glass, electronic parts
*Form:* Liquid
*Properties:* Solubility: Sol. in hot or cold water, soft or hard water; Biodegradable
*Composition:* Potassium hydroxide, silicate of soda and chlorine
*Reaction:* Highly alkaline
*Toxicity:* Skin and eye irritant
*Containers:* 1 gal container, 15 and 55 gal drums
*Warnings and Caution:* Avoid contact with skin, eyes, clothing; flush immediately with water in case of contact; do not ingest
*References:* Bulletin # 6

**Detergent CR.** Arol Chem.
*Chemical Description:* Fatty ethanolamine condensate
*Category:* Detergent, wetting agent, emulsifier
*Applications:* Textile scouring, emulsifying, wetting, penetrating and leveling agent; general purpose wetting, rewetting, penetrating and leveling agent; soaping-off prints; heavy-duty scouring; thickener in cosmetics; base for bubble baths, hand soaps; laundering, foods and household uses as detergent; Kier boiling; dispersing pigments, waxes, resins; printing
*Form:* Liquid; Color: Pale amber; Odor: Mild
*Properties:* Solubility: Readily sol. in water over wide temperature range; Sp. gr.: 1.01; Stability: Excellent
*Composition:* 100% active ingred.
*Reaction:* Alkaline
*Toxicity:* Avoid ingestion; otherwise considered nontoxic
*Containers:* 55 gal drums
*Warnings and Caution:* Avoid contact with eyes

**Detergent Honey.** Dixo Co. Inc.
*Category:* Detergent, emulsifier, wetting agent
*Applications:* Drycleaning detergent; emulsifier for a drycleaning solvent and water; wetting agent in drycleaning stain removal
*Form:* Liquid; Color: Yellow-amber; Odor: Sweet
*Properties:* Solubility: Sol. in common drycleaning solvents
*Containers:* 1, 5, 15, 30 and 55 gal metal containers

**Dilasoft RW.** Sandoz Ltd.
*Chemical Description:* Fatty acid derivative modified
*Category:* Softener, cationic/nonionic
*Applications:* Softening agent with wetting properties for cellulosic fibers
*Form:* Liquid

**Dilasoft TF.** Sandoz Ltd.
*Chemical Description:* Sulfonate
*Category:* Softening/filling agent, anionic
*Applications:* Softening/filling agent
*Form:* Liquid

**Dion 3-800.** Diamond Shamrock
*Chemical Description:* Mercaptan
*Applications:* Rapid-cure additive for epoxy resins
*Form:* Liquid
*Composition:* 100% active ingred.

**Dion Emulsifier 26.** Diamond Shamrock
*Category:* Emulsifier, nonionic
*Applications:* Emulsifier for epoxy resins
*Form:* Solid
*Composition:* 100% active ingred.

**Dion Emulsifier 37.** Diamond Shamrock
*Category:* Emulsifier, nonionic
*Applications:* Emulsifier for epoxy resins
*Form:* Liquid
*Composition:* 50% active ingred.

**Dispersant 3000E.** Hart Chem.
*Chemical Description:* Nonionic specialty formulated product
*Category:* Dispersant
*Applications:* Nonionic pitch dispersant
*Form:* Liquid; Color: Yellow
*Properties:* Solubility: Clear soluble; Sp. gr.: 1.00
*Composition:* 85% active ingred.
*Reaction:* Neutral
*Toxicity:* Skin and eye irritant
*Containers:* Drums, T/L, T/T, T/C
*Warnings and Caution:* Avoid contact with skin and eyes

**Dispersant 3050B.** Hart Chem.
*Chemical Description:* Nonionic specialty formulated product
*Category:* Dispersant
*Applications:* Nonionic pitch dispersant
*Form:* Liquid; Color: Yellow
*Properties:* Solubility: Dispersible; Sp. gr.: 1.03
*Composition:* 48% active ingred.
*Reaction:* Neutral
*Toxicity:* Skin and eye irritant
*Containers:* Drums, T/L, T/T, T/C
*Warnings and Caution:* Avoid contact with skin and eyes

**Dispersol ACA.** I.C.I. United States
*Category:* Dispersant
*Applications:* Dispersing agent for dyes, level dyeing assistant; dispersing waxes, pigments, resins; printing; dispersing dry powders
*Form:* Free-flowing powder; Color: Tan; Odor: Faint
*Properties:* Solubility: Water sol.; Sp. gr.: ≈ 1.0
*Composition:* 100% active ingred.
*Reaction:* Alkaline
*References:* TB 302R

**Dispersol CWLA.** I.C.I. United States
*Category:* Dyeing assistant, mildly cationic
*Applications:* Dye assistant for nylon or wool, esp. for acid dyes; afterscour
*Form:* Liquid; Color: Amber; Odor: Faint
*Properties:* Solubility: Water sol.; Sp. gr.: ≈ 1.0
*Composition:* Proprietary % active
*Reaction:* Acid
*Toxicity:* Causes eye burns
*Warnings and Caution:* Wear goggles; avoid inhaling vapors or

mists; in case of eye contact, flush immediately with water
*References:* TB 589

**Dispex A40.** Allied Colloids
*Chemical Description:* Ammonium salt of a polymeric carboxylic acid
*Category:* Deflocculant
*Applications:* Ceramic industry, glaze manufacture, stabilization and for preparation of clay slips for casting and spray-drying; sanitaryware; tableware
*Form:* Mobile liquid; Color: Pale yellow
*Properties:* Solubility: Miscible with water; Sp. gr.: 1.16; Visc.: 3.5–5.5 cs @ 15%; Stability: Good storage stability; imparts viscosity stability to products even at high temperature and under alkaline conditions; pH: 7.5–8.5
*Composition:* 40% active ingred.
*Containers:* 55 gal or 5 gal non-returnable steel drums
*References:* Data sheet GID/317.4

**Dispex N40.** Allied Colloids
*Chemical Description:* Sodium salt of a polymeric carboxylic acid
*Category and Applications:* See Dispex A40
*Form:* Mobile liquid; Color: Pale yellow
*Properties:* Solubility: Miscible with water; Sp. gr.: 1.30; Visc.: 15–25 cs @ 25%; Stability: Good storage stability; imparts viscosity stability to products even at high temperature and under alkaline conditions; pH: 7.0–7.5
*Composition:* 40% active ingred.
*Containers:* 55 gal or 5 gal non-returnable steel drums
*References:* Data sheet GID/317.4

**DMAMP.** IMC Chem. Group
*Chemical Description:* 2-dimethylamino-2-methyl-1-propanol
*Category:* Emulsifier, corrosion inhibitor, solubilizer, catalyst
*Applications:* Emulsifier for waxes in combination with an unsaturated fatty acid; vapor-phase corrosion inhibitor in steam-condensate systems; neutralizing agent in aqueous systems; solubilizer for resin systems; catalyst for production of polyurethane foams; curing agent for liquid epoxy resins; stabilizer; as a raw material for chemical synthesis
*Form:* Essentially anhydrous
*Properties:* Solubility: Sol. in water, organic solvents such as methanol, butanol, acetone, ethyl acetate, benzene, hexane, mineral oil, castor oil; Density: 7.5 lb/gal; Sp. gr.: 0.9; Visc.: A–$A_2$ (Gardner); F.P.: 19 C; B.P.: 160 C (@ 760 mm Hg); Flash Pt.: 135 F (TOC)
*Composition:* 100% active ingred.
*Reaction:* Alkaline
*Toxicity:* Causes skin and eye burns
*Containers:* 1, 5, 55 gal drums; tank cars; tank trucks; store in iron or steel
*Warnings and Caution:* Goggles, gloves, protective clothing should be worn; do not ingest; in case of contact, flush eyes or skin with plenty of water; combustible liquid
*References:* NP Series TDS No. 18

**DMAMP-80.** IMC Chem. Group
*Chemical Description:* 2-dimethylamino-2-methyl-1-propanol
*Category and Applications:* See DMAMP
*Form:* Liquid
*Properties:* Density: 7.9 lb/gal; Sp. gr.: 0.95; Visc.: A–$A_2$ (Gardner); F.P.: −20 C; B.P.: ≈ 98 C; Flash Pt.: 150 F (TOC)
*Composition:* 80% active ingred.; 20% water
*Reaction:* Highly alkaline
*Toxicity:* Causes skin and eye burns
*Containers:* 1, 5, 55 gal drums; tank cars; tank trucks; store in iron or steel
*Warnings and Caution:* Goggles, gloves, protective clothing should be worn; do not ingest; in case of contact, flush with plenty of water; combustible liquid
*References:* NP Series TDS No. 18

**Dobane 45.** Shell Chemicals
*Chemical Description:* Narrow cut linear alkylbenzene
*Category:* Detergent
*Applications:* Base material for preparation of linear alkylbenzene sulfonate; suitable for applications in powder detergent formulations
*Form:* Clear liquid
*Properties:* Ref. Index: 1.4830; Solubility: Not miscible with water; Density: 0.859 kg/l @ 20 C; Visc.: 13 cs @ 20 C; M.P.: −69 C; Flash Pt.: 172 C; Stability: Stable over a wide range of temperature; Biodegradable: 90%
*Toxicity:* Eye irritation and repeated or prolonged contact can cause skin irritation
*Containers:* Mild steel or stainless steel 182 kg drums
*References:* Data sheet 2A 116

**Dobane 055.** Shell Chemicals
*Chemical Description:* Linear alkylbenzene
*Category:* Detergent, anionic after sulfonation
*Applications:* Base material for preparation of linear alkylbenzene sulfonate; high foaming household powder detergent
*Form:* Clear liquid
*Properties:* Ref. Index: 1.4840; Solubility: Not miscible with water; Density: 0.860 kg/l @ 20 C; Visc.: 10 cs @ 20 C; M.P.: < −70 C; Flash Pt.: 150 C; Stability: Stable over a wide range of temperature; Biodegradable: 90%
*Composition:* 100% active ingred.
*Toxicity:* Eye irritation and repeated or prolonged contact can cause skin irritation
*Containers:* Mild steel or stainless steel 182 kg drums
*References:* Data sheet 2A 115

**Dobane 80.** Shell Chemicals
*Chemical Description:* Narrow cut linear alkylbenzene
*Category:* Detergent
*Applications:* Base material for preparation of linear alkylbenzene sulfonate; intended for use in a mixture with a higher molecular mass alkylbenzene when it endows the derived sulfonate with improved solubility and cold stability characteristics

*Form:* Clear liquid
*Properties:* Ref. Index: 1.4860; Solubility: Not miscible with water; Density: 0.862 kg/l @ 20 C; Visc.: 5 cs @ 20 C; M.P.: < –70 C; Stability: Stable over a wide range of temperature; Biodegradable: 90%
*Toxicity:* Eye irritation and repeated or prolonged contact can cause skin irritation
*Containers:* Mild steel or stainless steel 182 kg drums
*References:* Data sheet 2A 110

**Dobane 83.** Shell Chemicals
*Chemical Description:* Linear alkylbenzene
*Category:* Detergent, anionic after sulfonation
*Applications:* Base material for preparation of linear alkylbenzene sulfonate; light duty liquids where the sulfonate gives excellent solubility and cold stability characteristics
*Form:* Clear liquid
*Properties:* Ref. Index: 1.4855; Solubility: Note miscible with water; Density: 0.861 kg/l @ 20 C; Visc.: 7 cs @ 20 C; M.P.: < –70 C; Flash Pt.: 138 C; Stability: Stable over a wide range of temperature; Biodegradable: 90%
*Composition:* 100% active
*Toxicity:* Eye irritation and repeated or prolonged contact can cause skin irritation
*Containers:* Mild steel or stainless steel 182 kg drums
*References:* Data sheet 2A 111

**Dobane 102.** Shell Chemicals
*Chemical Description:* Narrow cut linear alkylbenzene
*Category:* Detergent
*Applications:* Base material for preparation of linear alkylbenzene sulfonate; designed for applications in premium quality dishwashing liquids where it provides optimum performance and ease of formulation
*Form:* Clear liquid
*Properties:* Ref. Index: 1.4850; Solubility: Not miscible with water; Density: 0.861 kg/l @ 20 C; Visc.: 7 cs @ 20 C; M.P.: < –70 C; Flash Pt.: 140 C; Stability: Stable over a wide range of temperature; Biodegradable: 90%
*Toxicity:* Eye irritation and repeated or prolonged contact can cause skin irritation
*Containers:* Mild steel or stainless steel 210 liter drums
*References:* Data sheet 2A 112

**Dobane 123.** Shell Chemicals
*Chemical Description:* Narrow cut linear alkylbenzene
*Category:* Detergent
*Applications:* Base material for preparation of linear alkylbenzene sulfonate; recommended for applications in household powder detergents
*Form:* Clear liquid
*Properties:* Ref. Index: 1.4835; Solubility: Note miscible with water; Density: 0.860 kg/l @ 20 C; Visc.: 11 cs @ 20 C; M.P.: < –70 C; Flash Pt.: 160 C; Stability: Stable over a wide range of temperature; Biodegradable: 90%
*Toxicity:* Eye irritation and re-

peated or prolonged contact can cause skin irritation
*Containers:* Mild steel or stainless steel 210 liter drums
*References:* Data sheet 2A 114

**Dobane 945.** Shell Chemicals
*Chemical Description:* Broad cut linear alkylbenzene
*Category:* Detergent
*Applications:* Base material for preparation of linear alkylbenzene sulfonate; high foaming household powder detergents; suitable as a biodegradable replacement for branched chain alkylbenzene
*Form:* Clear liquid
*Properties:* Ref. Index: 1.4835; Solubility: Not miscible with water; Density: 0.860 kg/l @ 20 C; Visc.: 11 cs @ 20 C; M.P.: –70 C; Flash Pt.: 142 C; Stability: Stable over a wide range of temperature; Biodegradable: 90%
*Toxicity:* Eye irritation and repeated or prolonged contact can cause skin irritation
*Containers:* Mild steel or stainless steel 182 kg drums
*References:* Data sheet 2A 117

**Dobane HP.** Shell Chemicals
*Chemical Description:* Broad cut linear alkylbenzene
*Category:* Detergent
*Applications:* Base material for preparation of linear alkylbenzene sulfonate; high foaming household powder detergents and light duty liquids
*Form:* Clear liquid
*Properties:* Ref. Index: 1.4845; Solubility: Not miscible with water; Density: 0.861 kg/l @ 20 C; Visc.: 9 cs @ 20 C; M.P.: < –70 C; Flash Pt.: 140 C; Stability: Stable over a wide range of temperature; Biodegradable: 90%
*Toxicity:* Eye irritation and repeated or prolonged contact can cause skin irritation
peated or prolonged contact can cause skin irritation
*Containers:* Mild steel or stainless steel 182 kg drums
*References:* Data sheet 2A 113

**Dobanic Acid 83.** Shell Chemicals
*Chemical Description:* Linear alkylbenzene sulfonic acid
*Category:* Detergent, anionic
*Containers:* Mild steel or stainless steel 182 kg drums
*References:* Data Sheet 2A 118

**Dobane JN.** Shell Chemicals
*Chemical Description:* Linear alkylbenzene
*Category:* Detergent, anionic after sulfonation
*Applications:* Base material for preparation of linear alkylbenzene sulfonate; preparation of all types of controlled foaming and high foaming household powders and light duty liquids
*Form:* Clear liquid
*Properties:* Ref. Index: 1.4850; Solubility: Not miscible with water; Density: 0.861 kg/l @ 20 C; Visc.: 8 cs @ 20 C; M.P.: < –70 C; Flash Pt.: 144 C; Stability: Stable over a wide range of temperature; Biodegradable: 90%
*Composition:* 100% active
*Toxicity:* Eye irritation and re-

*Applications:* A component of dishwashing and general cleaning liquids where it enables excellent cold stability characteristics
*Form:* Liquid; Color: Brown; Odor: Sulfur dioxide
*Properties:* Solubility: Miscible in water up to 40%; Density: 1.06 kg/l @ 20 C; Visc.: 1550 cs @ 20 C; M.P.: −10 to −12 C; Biodegradable: > 90%
*Composition:* 97% active
*Toxicity:* Eye irritation and repeated or prolonged contact can cause skin irritation
*Containers:* Mild steel 222 kg drums
*Warnings and Caution:* When handling product, always wear safety goggles/face shield to protect the eyes and face against splashes; wear rubber/plastic gloves, protective apron
*References:* Data sheet 2A 120

**Dobanic Acid 102.** Shell Chemicals
*Chemical Description:* Linear alkylbenzene sulfonic acid
*Category:* Detergent
*Applications:* A component for premium quality dishwashing liquids where it provides optimum performance with excellent cold stability charactertistics
*Form:* Liquid; Color: Brown; Odor: Sulfur dioxide
*Properties:* Solubility: Miscible in water up to 40%; Density: 1.06 kg/l @ 20 C; Visc.: 1800 cs @ 20 C; M.P.: −10 to −12 C; Biodegradable: > 90%
*Composition:* 97% active
*Toxicity:* Eye irritation and repeated or prolonged contact can cause skin irritation
*Containers:* Mild steel 222 kg drums
*Warnings and Caution:* When handling product, always wear safety goggles/face shield to protect the eyes and face against splashes; wear rubber/plastic gloves, protective apron
*References:* Data sheet 2A 121

**Dobanic Acid JN.** Shell Chemicals
*Chemical Description:* Linear alkylbenzene sulfonic acid
*Category:* Detergent, anionic
*Applications:* A component of all types of high foaming and controlled foaming household powders and light duty liquids
*Form:* Liquid; Color: Brown; Odor: Sulfur dioxide
*Properties:* Solubility: Miscible in water up to 40%; Density: 1.05 kg/l @ 20 C; Visc.: 1600 cs @ 20 C; M.P.: −10 to −12 C; Biodegradable: > 90%
*Composition:* 97% active
*Toxicity:* Eye irritation and repeated or prolonged contact can cause skin irritation
*Containers:* Mild steel 222 kg drums
*Warnings and Caution:* When handling product, always wear safety goggles/face shield to protect the eyes and face against splashes; wear rubber/plastic gloves, protective apron
*References:* Data sheet 2A 122

**Dobanol 23.** Shell Chemicals
*Chemical Description:* A narrow cut blend of primary alcohols of high purity
*Category:* Base material
*Applications:* Base material for preparation of alcohol ethoxylates, sulfates, ethoxysulfates
*Form:* Liquid; Odor: Mild
*Properties:* Ref. Index: 1.4420; Solubility: Not miscible with water; Density: 0.831 kg/l; Visc.: 14 cs @ 40 C; M. P.: 17–19 C; Flash Pt.: 132 C; Acid No.: $<0.1$; Hydroxyl No.: 290 ± 5; Iodine Value: 0.1; Biodegradable: $>95\%$
*Composition:* 100% active
*Toxicity:* Eye irritation and repeated or prolonged contact can cause skin irritation
*Containers:* Mild steel 168 kg drums
*References:* Data sheet 2A 131

**Dobanol 23-2.** Shell Chemicals
*Chemical Description:* Primary alcohol ethoxylate
*Category:* Detergent, nonionic
*Applications:* Intermediate for production of primary alcohol ethoxysulfates as components of shampoos, bubble baths, and other toiletries
*Form:* Clear liquid
*Properties:* Density: 0.91 kg/l @ 20 C; Visc.: 15 cs @ 40 C; M.P.: 4–5 C; HLB: 6.2; Flash Pt.: 148 C; Acid No.: 0.05–0.2; Hydroxyl No.: 199 ± 5; Biodegradable: $>95\%$
*Composition:* 100% active
*Toxicity:* Eye irritation and repeated or prolonged contact can cause skin irritation
*Containers:* Mild steel 190 kg drums
*Warnings and Caution:* Storage temperature should not exceed 50 C
*References:* Data sheet 2A 144

**Dobanol 23-6.5.** Shell Chemicals
*Chemical Description:* Primary alcohol ethoxylate
*Category:* Detergent, nonionic
*Applications:* Used in high performance clothes washing powders and in light duty liquids
*Form:* Slurry; Color: White
*Properties:* Solubility: Miscible in all proportions but high viscosity gels may be formed at mid-range concentrations: Density: 0.96 kg/l @ 40 C; Visc.: 26 cs @ 40 C; M.P.: 17–19 C; HLB: 11.9; Cloud Pt.: 43 ± 3 C (5% aq. sol'n.); Flash Pt.: 172 C; Acid No.: 0.05–0.2; Hydroxyl No.: 117 ± 5; pH: 6.8 (0.5% aq. sol'n.)
*Composition:* 100% active
*Toxicity:* Eye irritation and repeated or prolonged contact can cause skin irritation
*Containers:* Mild steel 190 kg drums
*Warnings and Caution:* Storage temperature should not exceed 50 C
*References:* Data sheet 2A 145

**Dobanol 25.** Shell Chemicals
*Chemical Description:* A blend of primary alcohols of high purity
*Category:* Base material
*Applications:* Base material for

preparation of alcohol ethoxylates, sulfates, and ethoxysulfates
*Form:* Liquid; Odor: Mild
*Properties:* Ref. Index: 1.4440; Solubility: Not miscible with water; Density: 0.831 kg/l; Visc.: 14 cs @ 40 C; M.P.: 20–22 C; Acid No.: < 0.1; Hydroxyl No.: 270 ± 5; Iodine Value: < 0.1; Biodegradable
*Composition:* 100% active
*Toxicity:* Eye irritation and repeated contact can cause skin irritation
*Containers:* Mild steel 168 kg drums
*Warnings and Caution:* Storage temperature should not exceed 45 C
*References:* Data sheet 2A 133

**Dobanol 25-3.** Shell Chemicals
*Chemical Description:* Primary alcohol ethoxylate
*Category:* Detergent, nonionic
*Applications:* Intermediate for production of primary alcohol ethoxysulfates as components of high performance dishwashing liquids
*Form:* Clear liquid
*Properties:* Density: 0.93 kg/l @ 20 C; Visc.: 18 cs @ 40 C; M.P.: 5–6 C; HLB: 7.8; Flash Pt.: 162 C; Acid No.: 0.05–0.2; Hydroxyl No.: 166 ± 6; Biodegradable: > 95%
HLB: 7.8; Flash Pt.: 162 C; Acid No.: 0.05–0.2; Hydroxyl No.: 166 ± 6; Biodegradable: > 95%
*Composition:* 100% active
*Toxicity:* Eye irritation and repeated or prolonged contact can cause skin irritation
*Containers:* Mild steel 195 kg drums
*Warnings and Caution:* Storage temperature should not exceed 50 C
*References:* Data sheet 2A 147

**Dobanol 25-3A/27 EU.** Shell Chem.
*Chemical Description:* Aqueous-ethanol sol'n. of primary alcohol ethoxysulfate (ammonia cation)
*Category:* Detergent
*Applications:* Component of higher performance light duty liquid formulations which may require ethanol and urea as solubilizing additives
*Form:* Clear liquid
*Properties:* Solubility: Miscible in all proportions; Density: 1.00 kg/l @ 20 C; Visc.: 30 cs @ 20 C; Flash Pt.: 33 C; pH: 7.0–9.0 (5% aq. sol'n.); Biodegradable: > 95%
*Composition:* 25.5–27.5% active
*Toxicity:* Eye irritation and repeated or prolonged contact can cause skin irritation
*Containers:* Mild steel 215 kg drums
*Warnings and Caution:* Storage temperature should not exceed 40 C; the vapor is inflammable; use flame-proofed equipment and ensure storage and handling areas are free of naked lights and sparks
*References:* Data sheet 2A 164

**Dobanol 25-3A/60.** Shell Chem.
*Chemical Description:* Aqueous-ethanol concentrate of primary alcohol ethoxysulfate (ammonia

cation)
*Category:* Detergent, anionic
*Applications:* Component of high active matter, high performance light duty liquid formulations
*Form:* Liquid
*Properties:* Solubility: Miscible in all proportions but high viscosity gels may be formed at mid-range concentrations; Density: 1.04 kg/l @ 20 C; Visc.: 60 cs @ 20 C; Cloud Pt.: 8 C; Clear Pt.: 10 C; Flash Pt.: 26 C; pH: 6.5–8.0 (5% aq. sol'n.); Biodegradable: > 95%
*Composition:* 59–61% active
*Toxicity:* Eye irritation and repeated or prolonged contact can cause skin irritation
*Containers:* Mild steel 218 kg drums
*Warnings and Caution:* Storage temperature should not exceed 40 C; the vapor is inflammable; use flame-proofed equipment and ensure storage and handling areas are free of naked lights and sparks
*References:* Data sheet 2A 162

**Dobanol 25-3S/27.** Shell Chemicals
*Chemical Description:* Aqueous solution of primary aclohol ethoxysulfate (sodium cation)
*Category:* Detergent, anionic
*Applications:* Component of light duty liquid detergents
*Form:* Clear liquid
*Properties:* Solubility: Miscible in all proportions; Density: 1.05 kg/l @ 20 C; Visc.: 60 cs @ 20 C; Cloud Pt.: 1 C; Clear Pt.: 14 C; pH: 6.5–8.0 (5% aq. sol'n.); Biodegradable: > 95%
*Composition:* > 26.0% active
*Toxicity:* Eye irritation and repeated or prlonged contact can cause skin irritation
*Containers:* Mild steel 218 kg drums
*Warnings and Caution:* Storage temperature should not exceed 40 C
*References:* Data sheet 2A 161

**Dobanol 25-3S/60.** Shell Chem.
*Chemical Description:* Aqueous-ethanol concentrate of primary alcohol ethoxysulfate (sodium cation)
*Category:* Detergent, anionic
*Applications:* Component of high active matter, high performance, light, duty, liquid formulations
*Form:* Liquid
*Properties:* Solubility: Miscible in all proportions but high viscosity gels may be formed at mid-range concentrations; Density: 1.05 kg/l @ 20 C; Visc.: 60 cs @ 20 C; Flash Pt.: 25 C; pH: 6.5–8.0 (5% aq. sol'n.); Biodegradable: > 95%
*Composition:* 58–60% active
*Toxicity:* Eye irritation and repeated or prolonged contact can cause skin irritation
*Containers:* Mild steel 220 kg drums
*Warnings and Caution:* Storage temperature should not exceed 40 C; the vapor is inflammable; use flame-proofed equipment and ensure storage and handling areas are free of naked lights and sparks
*References:* Data sheet 2A 163

**Dobanol 25-3S/70.** Shell Chem.
*Chemical Description:* Aqueous concentrate of primary alcohol ethoxysulfate (sodium cation)
*Category:* Detergent
*Applications:* Component of high active matter, high performance, light duty, liquid formulations
*Form:* Clear gelled liquid
*Properties:* Solubility: Miscible in all proportions but high viscosity gels may be formed at mid-range concentrations; Density: 1.08 kg/l @ 20 C; Visc.: 5000 cP @ 20 C at shear rate 10 reciprocal s; pH: 7.5–9.5 (5% aq. sol'n.); Biodegradable: > 95%
*Composition:* 68–72% active
*Toxicity:* Eye irritation and repeated or prolonged contact can cause skin irritation
*Containers:* Mild steel drums
*Warnings and Caution:* Storage temperature should not exceed 40 C
*References:* Data sheet 2A 165

**Dobanol 25-7.** Shell Chemicals
*Chemical Description:* Primary alcohol ethoxylate
*Category:* Detergent, nonionic
*Applications:* Used in high performance clothes washing powders
*Form:* Paste/solid; Color: White
*Properties:* Solubility: Miscible in all proportions but high viscosity gels may be formed at mid-range concentrations; Density: 0.97 kg/l @ 40 C; Visc.: 31 cs @ 40 C; M.P.: 21–23 C; HLB: 12.0; Cloud Pt.: 51 ± 3 C (5% aq. sol'n.); Flash Pt.: 180 C; Acid No.: 0.05–0.2; Hydroxyl No. 109 ± 5; pH: 6.8 (0.5% aq. sol'n.)
*Composition:* 100% active
*Toxicity:* Eye irritation and repeated or prolonged contact can cause skin irritation
*Containers:* Mild steel 190 kg drums
*Warnings and Caution:* Storage temperature should not exceed 50 C
*References:* Data sheet 2A 148

**Dobanol 25-9.** Shell Chemicals
*Chemical Description:* Primary alcohol ethoxylate
*Category:* Detergent, nonionic
*Applications:* Used in high performance clothes washing powders
*Form:* Solid; Color: White
*Properties:* Solubility: Miscible in all proportions but high viscosity gels may be formed at mid-range concentrations; Density: 0.98 kg/l @ 40 C; Visc.: 40 cs @ 40 C; M.P.: 25–28 C; HLB: 13.2; Cloud Pt.: 79 ± 4 (5% aq. sol'n.); Flash Pt.: 190 C; Acid No.: 0.05–0.2; Hydroxyl No.: 92 ± 4; pH: 6.8 (0.5% aq. sol'n.)
*Composition:* 100% active
*Toxicity:* Eye irritation and repeated or prolonged contact can cause skin irritation
*Containers:* Mild steel (open top) 204 kg drums
*Warnings and Caution:* Storage temperature should not exceed 50 C
*References:* Data sheet 2A 149

**Dobanol 45.** Shell Chemicals
*Chemical Description:* A narrow cut blend of primary alcohols of high purity
*Category:* Base material
*Applications:* Base material for preparation of alcohol ethoxylates, sulfates, ethoxysulfates
*Form:* Liquid; Odor: Mild
*Properties:* Solubility: Not miscible with water; Density: 0.824 kg/l @ 40 C; Visc.: 17 cs @ 40 C; M.P.: 29–31 C; Flash Pt.: 152 C; Acid No.: < 0.1; Hydroxyl No.: 255 ± 5; Iodine Value: 0.1; Biodegradable
*Composition:* 100% active ingred.
*Toxicity:* Eye irritation and repeated or prolonged contact can cause skin irritation
*Containers:* Mild steel 168 kg drums
*Warnings and Caution:* Storage temperature should not exceed 45 C
*References:* Data sheet 2A 134

**Dobanol 45-7.** Shell Chemicals
*Chemical Description:* Primary alcohol ethoxylate
*Category:* Detergent, nonionic
*Applications:* Used in high performance clothes washing powders
*Form:* Solid; Color: White
*Properties:* Solubility: Miscible in all proportions but high viscosity gels may be formed at mid-range concentrations; Density: 0.96 kg/l @ 50 C; Visc.: 33 cs @ 40 C; M.P.: 22–24 C; HLB: 11.6; Cloud Pt.: 46 ± 3 C (5% aq. sol'n.); Flash Pt.: 185 C; Acid No.: 0.05–0.2; Hydroxyl No.: 106 ± 5; pH: 6.8 (0.5% aq. sol'n.)
*Composition:* 100% active
*Toxicity:* Eye irritation and repeated or prolonged contact can cause skin irritation
*Containers:* Mild steel 195 kg drums
*Warnings and Caution:* Storage temperature should not exceed 50 C
*References:* Data sheet 2A 150

**Dobanol 45-11.** Shell Chemicals
*Chemical Description:* Primary alcohol ethoxylate
*Category:* Detergent, nonionic
*Applications:* Used in high performance clothes washing powders
*Form:* Solid; Color: White
*Properties:* Solubility: Miscible in all proportions but high viscosity gels may be formed at mid-range concentrations; Density: 0.99 kg/l @ 50 C; Visc.: 47 cs @ 40 C; M.P.: 28–31 C; HLB: 13.7; Cloud Pt.: 88 ± 4 C (5% aq. sol'n.); Flash Pt.: 200 C; Acid No.: 0.05–0.2; Hydroxyl No.: 80 ± 4; pH: 6.8 (0.5% aq. sol'n.)
*Composition:* 100% active
*Toxicity:* Eye irritation and repeated or prolonged contact can cause skin irritation
*Containers:* Mild steel 197 kg drums
*Warnings and Caution:* Storage temperature should not exceed 50 C
*References:* Data sheet 2A 151

**Dobanol 45-18.** Shell Chemicals
*Chemical Description:* Primary alcohol ethoxylate
*Category:* Detergent, nonionic
*Applications:* Used in high performance clothes washing powders
*Form:* Solid; Color: White
*Properties:* Solubility: Miscible in all proportions but high viscosity gels may be formed at mid-range concentrations; Density: 1.02 kg/l @ 50 C; Visc.: 42 cs @ 60 C; M.P.: 35–39 C; HLB: 15.7; Cloud Pt.: > 100 C (5% aq. sol'n.); Flash Pt.: 200 C; Acid No.: 0.05–0.2; Hydroxyl No.: 55 ± 4; pH: 6.8 (0.5% aq. sol'n.)
*Composition:* 100% active
*Toxicity:* Eye irritation and repeated or prolonged contact can cause skin irritation
*Containers:* Mild steel drums
*Warnings and Caution:* Storage temperature should not exceed 50 C
*References:* Data sheet 2A 152

**Dobanol 91.** Shell Chemicals
*Chemical Description:* A blend of primary alcohols of high purity
*Category:* Base material
*Applications:* Base material for preparation of alcohol ethoxylates, sulfates, ethoxysulfates
*Properties:* Ref. Index: 1.4370; Solubility: Not miscible with water; Density: 0.832 kg/l; Visc.: 9 cs @ 40 C; M.P.: –8 to –10 C; Flash Pt.: 107 C (PMCC); Acid No.: < 0.1; Hydroxyl No.: 350 ± 5; Iodine Value: 0.1; Biodegradable
*Composition:* 100% active
*Toxicity:* Eye irritation and repeated or prolonged contact can cause skin irritation
*Containers:* Mild steel 175 kg drums
*Warnings and Caution:* Storage temperature should not exceed 45 C
*References:* Data sheet 2A 130

**Dobanol 91-2.5.** Shell Chemicals
*Chemical Description:* Primary alcohol ethoxylate
*Category:* Chemical intermediate, nonionic
*Applications:* Used for the production of primary alcohol ethoxysulfates; component of emulsifier formulations
*Form:* Clear liquid
*Properties:* Density: 0.93 kg/l @ 20 C; Visc.: 11 cs @ 40 C; M. P.: –20 to –22 C; HLB: 8.1; Flash Pt.: 120 C (PMCC); Acid No.: 0.05-0.2; Hydroxyl No.: 208 ± 6; Biodegradable: > 95%
*Composition:* 100% active
*Toxicity:* Eye irritation and repeated or prolonged contact can cause skin irritation
*Containers:* Mild steel 195 kg drums
*Warnings and Caution:* Storage temperature should not exceed 50 C
*References:* Data sheet 2A 140

**Dobanol 91-5.** Shell Chemicals
*Chemical Description:* Primary alcohol ethoxylate
*Category:* Wetting agent, emulsi-

fier, nonionic
*Applications:* Component of hard surface liquid cleaners
*Form:* Clear liquid
*Properties:* Solubility: Miscible in all proportions; Density: 0.98 kg/l @ 20 C; Visc.: 18 cs @ 40 C; M. P.: 3–6 C; HLB: 11.6; Cloud Pt.: 36 ± 3 (5% aq. sol'n.); Flash Pt.: 146 C (PMCC); Acid No.: 0.05–0.2; Hydroxyl No.: 148 ± 4; pH: 6.8 (0.05% aq. sol'n.)
*Composition:* 100% active
*Toxicity:* Eye irritation and repeated or prolonged contact can cause skin irritation
*Containers:* Mild steel 200 kg drums
*Warnings and Caution:* Storage temperature should not exceed 50 C
*References:* Data sheet 2A 141

**Dobanol 91-6**. Shell Chemicals
*Chemical Description:* Primary alcohol ethoxylate
*Category:* Detergent, wetting agent, nonionic
*Applications:* Detergent and wetting agent in textile processing, degreasing, metal cleaning, pulp deresination
*Form:* Clear liquid
*Properties:* Solubility: Miscible in all proportions; Density: 0.99 kg/l @ 20 C; Visc.: 21 cs @ 40 C; M. P.: 6–9 C; HLB: 12.5; Cloud Pt.: 56 ± 3 (5% aq. sol'n.); Flash Pt.: 150 C (PMCC); Acid No.: 0.05–0.2; Hydroxyl No.: 133 ± 6; pH: 6.8 (0.5% aq. sol'n.)
*Composition:* 100% active
*Toxicity:* Eye irritation and repeated or prolonged contact can cause skin irritation
*Containers:* Mild steel 203 kg drums
*Warnings and Caution:* Storage temperature should not exceed 50 C
*References:* Data sheet 2A 142

**Dobanol 91-8**. Shell Chemicals
*Chemical Description:* Primary alcohol ethoxylate
*Category:* Detergent, nonionic
*Applications:* Component of dishwashing liquids in blends with other surfactants
*Form:* Liquid/slurry
*Properties:* Solubility: Miscible in all proportions but high viscosity gels may be formed at mid-range concentrations; Density: 1.00 kg/l @ 20 C; Visc.: 28 cs @ 40 C; M. P.: 14–17 C; HLB: 13.7; Cloud Pt.: 81.5 ± 3.5 (5% aq. sol'n.); Flash Pt.: 160 C (PMCC); Acid No.: 0.05–0.2; Hydroxyl No.: 110 ± 5; pH: 6.8 (0.5% aq. sol'n.)
*Composition:* 100% active
*Toxicity:* Eye irritation and repeated or prolonged contact can cause skin irritation
*Containers:* Mild steel 200 kg drums
*Warnings and Caution:* Storage temperature should not exceed 50 C
*References:* Data sheet 2A 143

**Dobanol 235**. Shell Chemicals
*Chemical Description:* A blend of primary alcohols of high purity
*Category:* Base material

*Applications:* Base material for preparation of alcohol ethoxylates, sulfates, ethoxysulfates
*Properties:* Ref. Index: 1.4430; Solubility: Not miscible with water; Density: 0.831 kg/l; Visc.: 14 cs @ 40 C; M.P.: 18–20 C; Flash Pt.: 134 C (PMCC); Acid No.: < 0.1; Hydroxyl No.: 280 ± 5; Iodine Value: 0.1; Biodegradable
*Toxicity:* Eye irritation and repeated or prolonged contact can cause skin irritation
*Containers:* Mild steel drums
*Warnings and Caution:* Storage temperature should not exceed 45 C
*References:* Data sheet 2A 132

**Dobanol 235-3.** Shell Chemicals
*Chemical Description:* Primary alcohol ethoxylate
*Category:* Chemical intermediate
*Applications:* Used for production of primary alcohol ethoxysulfate as component of high performance dishwashing liquids and toiletries
*Form:* Clear liquid
*Properties:* Density: 0.92 kg/l @ 20 C; Visc.: 17 cs @ 40 C; M. P.: 4–5 C; HLB: 7.8; Flash Pt.: 160 C (PMCC); Acid No.: 0.05–0.2; Hydroxyl No.: 172 ± 6; Biodegradable: > 95%
*Composition:* 100% active
*Toxicity:* Eye irritation and repeated or prolonged contact can cause skin irritation
*Containers:* Mild steel drums
*Warnings and Caution:* Storage temperature should not exceed 50 C
*References:* Data sheet 2A 146

**Doittol 891.** Diamond Shamrock
*Category:* Anionic surfactant
*Applications:* Rubber anti-blocking agent
*Form:* Liquid
*Composition:* 20% active ingred.

**Doittol K-21.** Diamond Shamrock
*Category:* Anionic surfactant
*Applications:* Rubber anti-blocking agent
*Form:* Liquid
*Composition:* 25% active ingred.

**Dow Corning 190 Surfactant.** Dow Corning
*Chemical Description:* Silicone-glycol copolymer
*Category:* Silicone surfactant
*Applications:* Used for producing flexible slab stock (urethane) foam; has high processing latitude; good efficiency
*Form:* Low-viscosity liquid; Color: 2 (Gardner)
*Properties:* Sp. gr.: 1.035; Visc.: 1500 cst; Flash Pt.: 250 F (CCC); Stability: Good in water and amine premixes
*Containers:* 40 and 441 lb containers net weight
*References:* Bulletin 22-315

**Dow Corning 193 Surfactant.** Dow Corning
*Chemical Description:* Nonhydrolyzable silicone glycol copolymer
*Applications:* For producing all

types of rigid urethane foam; to manufacture board, laminates, foam in place, spray foam and shoe soles
*Form:* Clear viscous liquid; Color: 2 (Gardner)
*Properties:* Refractive Index: 1.45; Sp. gr.: 1.07; Visc.: 465 cs; F.P.: 50 F; Flash Pt.: 300 F (COC); Pour Pt.: 52 F; Stability: Good premix stability; becomes hazy < 77 F; solidifies to soft wax < 50 F
*Containers:* 40, 441 lb containers
*References:* Bulletin 22-316A

**Dow Corning 196 Surfactant.** Dow Corning Corp.
*Chemical Description:* Silicone glycol copolymer
*Category:* Profoamer
*Applications:* Used for producing low-density, flexible-slab urethane foams; with high efficiency and excellent processing latitude
*Form:* Low viscosity liquid; Color: < 4 (Gardner)
*Properties:* Solubility: Sol. in water, polyol, fluorocarbon or water-amine streams; Sp. gr.: 1.035; Visc.: 2000 cst; Flash Pt.: 225 F (open cup)
*Toxicity:* Mild eye irritant; moderate skin irritant on prolonged contact
*Containers:* 441 lb drums; 2000 and 4000 gal tank trucks
*References:* Bulletin 22-291

**Dow Corning 197 Surfactant.** Dow Corning Corp.
*Chemical Description:* Nonhydrolyzable silicone
*Category:* Profoamer
*Applications:* Profoamer for high- and low-density rigid urethane foams with good system compatibility, low freeze point, excellent emulsifying properties
*Form:* Thin liquid; Color: Clear yellow
*Properties:* Refractive Index: 1.446; Sp. gr.: 1.05; Visc.: 300 cst; Flash Pt.: 120 F
*Toxicity:* Skin and eye irritant; prolonged inhalation of vapors may be harmful
*Containers:* 40 and 440 lb containers
*Warnings and Caution:* Combustible; keep away from heat and flame. Use with adequate ventilation. Avoid prolonged breathing of vapors and eye and skin contact
*References:* Bulletin 22-249a

**Dow Corning 198 Surfactant.** Dow Corning Corp.
*Chemical Description:* Silicone glycol copolymer
*Applications:* High processing latitude surfactant for high- and low-density flexible urethane foams, hot molding and filled foam operations
*Form:* Low viscosity liquid; Color: < 4 (Gardner)
*Properties:* Solubility: Sol. in water, polyol, fluorocarbon, or water-amine streams; Sp. gr.: 1.035; Visc.: 1500 cst; Flash Pt.: > 150F (open cup)
*Toxicity:* Slight eye irritant; prolonged skin contact may cause moderate irritation

*Containers:* 441 lb drums, 2000 and 4000 gal tank trucks
*References:* Bulletin 22-301

**Dow Corning 470A Fluid.** Dow Corning
*Chemical Description:* Silicone glycol copolymer
*Category:* Nonionic emulsifier, emulsifier additive, wetting agent, surface tension depressant
*Applications:* Used to make powders more wettable and dispersible; used to treat gels, thickeners, clays, powdered gums, dyes, pigments and water-dispersible powders; emulsifier additive; polish additive; glass polish ingredient; soap and detergent additive; cosmetic and aerosol shave ingred.
*Form:* Liquid; Color: Medium straw
*Properties:* Refractive Index: 1.45; Solubility: Water sol. @ $<$ 160 F; Sp. gr.: 1.075; Visc.: 275 cst; Flash Pt.: 500 F
*Composition:* 100% solids
*Containers:* 1 gal pails, 5, 55 gal drums
*References:* Bulletin 22-205

**Dow Corning 1248 Fluid (formerly Q4-3558).** Dow Corning Corp.
*Chemical Description:* Silicone polycarbinols (graft copolymer type)
*Applications:* Improves properties of organics such as release, abrasion resistance, lubricity, water repellency, softening and flexibility, low- and high-temperature flexibility; improves properties of urethanes
*Form:* Liquid; Color: Clear
*Properties:* Refractive Index: 1.4110; Solubility: Sol. in ethanol, isopropanol, methylene-chloride, n-hexane, toluene (@ 10%); Visc.: 160 cs; Flash Pt.: 250 F; Pour Pt.: −130 F
*Containers:* 5 and 55 gal containers
*References:* Product Information

**Dow Corning 1250 Surfactant.** Dow Corning Corp.
*Chemical Description:* Silicone in solvent (xylene)
*Category:* Silicone profoamer
*Applications:* Mechanical frothing of vinyl plastisols into a uniform fine-celled foam with better water resistance; allow foam fusing at low temperatures
*Form:* Thin liquid; Color: Very light straw
*Properties:* Sp. gr.: 1.00; Visc.: 6 cst; Flash Pt.: 85 F (open cup)
*Composition:* 50% silicone content
*Toxicity:* Skin irritant on prolonged contact; avoid prolonged breathing of vapors
*Containers:* 8, 40, 440 lb containers
*Warnings and Caution:* Solvent is combustible; keep away from heat and open flame; use with adequate ventilation. Avoid prolonged contact with skin and breathing of vapors
*References:* Bulletin 22-008c
*Patent:* U.S. 3,511,788

**Dow Corning 1251 Surfactant.** Dow

Corning Corp.
*Chemical Description:* Silicone in solvent (xylene)
*Category and Applications:* See Dow Corning 1250 Surfactant
*Form:* Thin liquid; Color: Very light straw
*Properties:* Sp. gr.: 1.00; Visc.: 4 cst; Flash Pt.: 85 F (open cup)
*Composition:* 50% silicone content
*Toxicity:* Skin irritant on prolonged contact
*Containers:* 8, 40, 440 lb containers
*Warnings and Caution:* Combustible solvent; keep away from heat and open flame; use with adequate ventilation; avoid prolonged contact with skin or breathing of vapors
*Patent:* U.S. 3,511,788

**Dow Corning 1252 Surfactant.** Dow Corning Corp.
*Chemical Description:* Silicones in solvent (texanol isobutyrate)
*Category and Applications:* See Dow Corning 1250 Surfactant
*Form:* Thin liquid; Color: Straw
*Properties:* Sp. gr.: 1.06; Visc.: 250 cs; Flash Pt.: 135 F (open cup)
*Composition:* 50% silicone content
*Toxicity:* Skin irritant on prolonged content
*Containers:* 8, 40, 440 lb containers
*Warnings and Caution:* Combustible solvent; keep away from heat and open flame; use with adequate ventilation. Avoid prolonged contact with skin or inhalation of vapors
*References:* Bulletin 22-008c
*Patent:* U.S. 3,511,788

**Dow Corning 1253 Surfactant.** Dow Corning Corp.
*Chemical Description:* Silicones in solvent (texanol isobutyrate)
*Category and Applications:* See Dow Corning 1250 Surfactant
*Form:* Thin liquid; Color: Straw
*Properties:* Sp. gr.: 1.07; Visc.: 250 cs; Flash Pt.: 135 F
*Composition:* 50% silicone content
*Toxicity:* Skin irritant on prolonged contact
*Containers:* 8, 40, 440 lb containers
*Warnings and Caution:* Combustible solvent; keep away from heat and open flame. Avoid prolonged breathing or contact with skin
*References:* Bulletin 22-008c
*Patent:* U.S. 3,511,788

**Dow Corning FF-400.** Dow Corning Corp.
*Chemical Description:* Organosiloxane
*Category:* Fiber finish
*Applications:* Textile finishes with high heat stability, high lubrication at low add-on; filament spin finishes; staple fiber finishes; knitting yarn lubricants; coning oil ingredients; texturizing finishes; additives to organic finishes; thread finishes
*Form:* Fluid; Color: Light amber

*Properties:* Solubility: Sol. in polar solvents such as isopropanol, low m.w. alcohols, glycol ethers; water-dispersible; Sp. gr.: 1.022; Visc.: 300 cs; Flash Pt.: 177 C (open cup)
*Composition:* 100% active ingred.
*References:* Bulletin 25-008

**Dow Corning FF-412.** Dow Corning Corp.
*Chemical Description:* Organopolysiloxane
*Category and Applications:* See Dow Corning FF-400
*Form:* Fluid; Color: Light amber
*Properties:* Solubility: Sol. in low m.w. alcohols, glycol ethers; dispersible in water; Sp. gr.: 1.040; Visc.: 200 cs; Flash Pt.: 177 C (open cup)
*Composition:* 100% active ingred.
*References:* Bulletin 25-008

**Dow Corning FF-414.** Dow Corning Corp.
*Chemical Description:* Organopolysiloxanes
*Category and Applications:* See Dow Corning FF-400
*Form:* Fluid; Color: Light amber
*Properties:* Solubility: Sol. in water, alcohols, glycol ether; Sp. gr.: 1.070; Visc.: 465 cs; Flash Pt.: 177 C (open cup)
*Composition:* 100% active ingred.
*References:* Bulletin 25-008

**Dow Corning Q2-8026 Fluid.** Dow Corning Corp.
*Chemical Description:* Silicone polycarbinols (graft copolymer type)
*Applications:* See Dow Corning 1248 Fluid
*Form:* Liquid; Color: Clear straw
*Properties:* Refractive Index; 1.45; Solubility: Sol. in water, ethanol, isopropanol, methylene-chloride, toluene; dispersible in n-hexane (@ 10%); Visc.: 465 cs; Flash Pt.: 400 F; Pour Pt.: 60 F
*Toxicity:* Slight eye irritant on direct contact
*Containers:* 5 and 55 gal containers

**Dow Corning Q4-3557 Fluid.** Dow Corning Corp.
*Chemical Description:* Silicone polycarbinols (graft copolymer type)
*Applications:* See Dow Corning 1248 Fluid
*Form:* Liquid; Color: Clear straw
*Properties:* Refractive Index: 1.43; Solubility: Sol. in ethanol, isopropanol, methylene-chloride, n-hexane, toluene; dispersible in mineral oil (@ 10%); Visc.: 40 cs; Flash Pt.: > 200 F; Pour Pt.: −80 F
*Containers:* 5 and 55 gal containers

**Dow Corning Q4-3667 Fluid.** Dow Corning Corp.
*Chemical Description:* Silicone polycarbinols (ABA type)
*Applications:* See Dow Corning 1248 Fluid
*Form:* Liquid; Color: Clear straw
*Properties:* Refractive Index: 1.44; Solubility: Sol. in ethanol, isopropanol, methylene-chloride, toluene; dispersible in water, n-hexane

(@ 10%); Visc.: 320 cs; Flash Pt.: 285 F; Pour Pt.: 65 F
*Containers:* 5 and 55 gal containers

**Dowfax 2A1.** Dow Corning Corp.
*Chemical Description:* Sodium dodecyl diphenyl ether disulfonate
*Category:* Detergent, emulsifier, wetting agent, solubilizer
*Applications:* Emulsion polymerization, dyeing assistant; primary emulsifier for SB & ABS polymers; leveling agent for acid dyeing of nylon; phenolic solubilizer for disinfectants; as detergent for dyeing, mines and minerals, cleaners of street, buildings, hospitals; wetting agent in electroplating; preparation of emulsified paints, bitumen emulsions
*Form:* Liquid; Color: Amber
*Properties:* Sp. gr.: 1.16; Visc.: 145 cps @ 0.10%; Surface Tension: 31 dynes/cm @ 0.10%
*Composition:* 45% active ingred.
*Toxicity:* Causes eye burns; may cause skin irritation; toxic to fish
*Containers:* 55 gal drum and bulk

**Dowfax 2A0.** Dow
*Chemical Description:* Dodecyl diphenyl-ether disulfonic acid
*Category:* Intermediate
*Applications:* Used to prepare non-sodium salts of Dowfax 2A1 for special applications
*Form:* Liquid; Color: Black
*Properties:* Sp. gr.: 1.11; Visc.: 510 cps; Surface Tension: 30 dynes/cm @ 0.10%
*Composition:* 40% active ingred.
*Reaction:* Acid
*Toxicity:* Irritant to skin, hair, fingers, eyes
*Containers:* 55 gal drum and bulk
*Warnings and Caution:* Strong acid—causes burns, toxic to fish

**Dowfax 3B2.** Dow Corning Corp.
*Chemical Description:* Sodium n-decyl diphenyl-ether disulfonate
*Category:* Detergent, wetting agent, emulsifier, solubilizer
*Applications:* Pesticide spray adjuvant, esp. for methane arsonate herbicides; phenolic solubilization for disinfectants; detergent in strong alkaline/acid cleaners, cement, transport and industrial cleaners; preparation of emulsified paints
*Form:* Liquid; Color: Red-brown
*Properties:* Sp. gr.: 1.17; Visc.: 120 cps; Surface Tension: 38 dynes/cm @ 0.10%
*Composition:* 45% active ingred.
*Toxicity:* Causes eye burns; may cause skin irritation; toxic to fish
*Containers:* 22 gal drum and bulk

**Dowfax 9N6.** Dow Chem. Europe
*Category:* Detergent, emulsifier, wetting agent, dispersing agent
*Applications:* Household wax and polishes; solvent based cleaners; emulsifier and wetting agent for pesticides; emulsifier for metal cleaners, drycleaning "soap," household-type wax and polish coemulsifier; detergent in metallurgy and household products; dispersing waxes, pigments, resins
*Form:* Liquid; Color: Colorless
*Properties:* Refractive Index:

1.486; Solubility: Water insol.; Sp. gr.: 1.04; Visc.: 220 cst; Stability: Good
*Composition:* 100% active ingred.
*Reaction:* Neutral
*Toxicity:* Slightly irritating to eyes
*Containers:* Bulk, drums
*Warnings and Caution:* Avoid eye contact and ingestion

**Dowfax 9N9.** Dow Chem. Europe
*Category:* Detergent, emulsifier, wetting agent
*Applications:* Household liquid cleaners; industrial detergent; leather and laundering industries; emulsifier for latex paints; textile processing aid; paper industry processing aid; fat liquoring; preparation of emulsified paints, bitumen emulsions; wetting agent for desizing and boiling off cottons, carbonizing woolens, abatement of dust, wetting pulp in beater, insecticides
*Form:* Liquid; Color: Slightly yellow
*Properties:* Refractive Index: 1.480; Solubility: Water sol.; Sp. gr.: 1.06; Visc.: 270 cst; Stability: Stable; Surface Tension: 33 dynes/cm
*Composition:* 100% active ingred.
*Reaction:* Neutral
*Toxicity:* Slightly irritating to eyes
*Containers:* Bulk, drrums
*Warnings and Caution:* Avoid eye contact and ingestion

**Dowfax 9N30.** Dow Chem. Europe
*Category:* Detergent, emulsifier, wetting agent
*Applications:* Heavy duty detergents; detergent and wetting agent in system involving near boiling or boiling temperatures and/or high concentrations of electrolytes; dispersing agent; preparation of emulsified paints, bitumen emulsions
*Form:* Solid; Color: White
*Properties:* Ref. Index: 1.470; Solubility: Water sol.; Sp. gr.: 1.05; M.P.: 49 C; Stability: Stable; Surface Tension: 42 dynes/cm
*Composition:* 100% active ingred.
*Reaction:* Neutral
*Toxicity:* Slight eye irritant
*Containers:* Bulk, drums
*Warnings and Caution:* Avoid eye contact and ingestion

**Dowfax 10-5.** Dow Chem. Europe
*Category:* Detergent, wetting agent, dispersing agent
*Applications:* Liquid detergents for household and industrial use; dry cleaning formulations; textiles, leather, laundering applications as detergents; desizing and boiling off cottons, removal of spots, abatement of dust, insecticides as wetting agent
*Form:* Liquid; Color: Colorless
*Properties:* Refractive Index: 1.439; Solubility: Water sol.; Sp. gr.: 0.97; Visc.: 29 cst; M.P.: 5 C; Stability: Stable; Surface Tension: 28 dynes/cm
*Composition:* 100% active ingred.
*Reaction:* Neutral
*Toxicity:* Slight eye irritant
*Containers:* Bulk, drums
*Warnings and Caution:* Avoid eye

contact and ingestion

**Dowfax 10-6.** Dow Chem. Europe
*Category:* Detergent, wetting agent
*Applications:* Liquid detergents for household and industrial use; dry cleaning formulations; textiles, leather, laundering uses as detergents; desizing and boiling off cottons, removal of spots, abatement of dust, insecticides as wetting agents
*Form:* Liquid; Color: Colorless
*Properties:* Refractive Index: 1.443; Solubility: Water sol.; Sp. gr.: 0.98; Visc.: 50 cst; M.P.: 8 C; Stability: Stable; Surface Tension: 28 dynes/cm
*Composition:* 100% active ingred.
*Reaction:* Neutral
*Toxicity:* Slight eye irritant
*Containers:* Bulk, drums
*Warnings and Caution:* Avoid eye contact and ingestion

**Dowfax 12-3.** Dow Chem. Europe
*Category:* Detergent, emulsifier, raw material for ethersulfates
*Applications:* Raw material for cosmetic grade ethersulfates for shampoos and bubble bath; quench oil additive; automotive engine cleaner; mineral oil emulsions
*Form:* Liquid; Color: Colorless
*Properties:* Refractive Index: 1.439; Solubility: Water insol.; Sp. gr.: 0.93; Visc.: 30 cst; Stability: Stable
*Composition:* 100% active ingred.
*Reaction:* Neutral
*Toxicity:* Slight eye irritant
*Containers:* Bulk, drums
*Warnings and Caution:* Avoid eye contact and ingestion

**Dowfax 14-3.** Dow Chem. Europe
*Category:* Detergent, emulsifier
*Applications:* Raw material for detergent grade ethersulfates for liquid detergents, quench oil additive, automotive engine cleaner, mineral oil emulsions
*Form:* Liquid; Color: Colorless
*Properties:* Refractive Index: 1.438; Solubility: Water insol.; Sp. gr.: 0.925; Visc.: 35 cst; Stability: Stable
*Composition:* 100% active ingred.
*Reaction:* Neutral
*Toxicity:* Slight eye irritant
*Containers:* Bulk, drums
*Warnings and Caution:* Avoid eye contact and ingestion

**Dowfax 14-12.** Dow Chem. Europe
*Category:* Detergent, wetting agent
*Applications:* Heavy duty detergents; as detergent and wetting agent in systems involving boiling or near boiling temperatures and/or high concentrations of electrolytes; as detergents in textiles, laundering, household applications
*Form:* Paste; Color: Slightly yellow
*Properties:* Refractive Index: 1.449; Solubility: Water sol.; Sp. gr.: 0.98; M.P.: 43 C; Stability: Stable; Surface Tension: 32 dynes/cm
*Composition:* 100% active ingred.
*Reaction:* Neutral
*Toxicity:* Slight eye irritant

*Containers:* Bulk, drums
*Warnings and Caution:* Avoid eye contact and ingestion

**Dowfax 18-11.** Dow Chem. Europe
*Category:* Detergent, wetting agent, emulsifier
*Applications:* Heavy duty detergents; formulations of low foaming heavy duty detergent for textiles, laundering and household uses; emulsifier for cosmetics and pharmaceuticals
*Form:* Solid; Color: White
*Properties:* Refractive Index: 1.450; Solubility: Water sol.; Sp. gr.: 0.97; M.P.: 33 C; Stability: Stable; Surface Tension: 36 dynes/cm
*Composition:* 100% active ingred.
*Reaction:* Neutral
*Toxicity:* Slight eye irritant
*Containers:* Bulk, drums
*Warnings and Caution:* Avoid eye contact and ingestion

**Dowfax 18-25.** Dow Chem. Europe
*Category:* Detergent
*Applications:* Heavy duty detergents; detergent and wetting agent in systems involving near boiling or boiling temperatures and/or high concentrations of electrolytes; detergents for textiles, laundering and household uses
*Form:* Solid; Color: White
*Properties:* Refractive Index: 1.453; Solubility: Water sol.; Sp. gr.: 1.01; M.P.: 50 C; Stability: Stable; Surface Tension: 45 dynes/cm
*Reaction:* Neutral
*Toxicity:* Slight eye irritant
*Containers:* Bulk, drums
*Warnings and Caution:* Avoid eye contact and ingestion

**Dowfax 63N10.** Dow Chem. Europe
*Category:* Detergent, defoamer, demulsifier
*Applications:* As detergent for dishwashing machines; heavy duty liquid detergents; defoamer for anionic detergent slurries, for breaking oil emulsions; as detergent for textiles, laundering, foods, household uses; dispersing waxes, pigments, resins
*Form:* Liquid; Color: Colorless
*Properties:* Solubility: Slightly water soluble; Sp. gr.: 1.02; Visc.: 340 cst; Stability: Stable; Surface Tension: 42 dynes/cm
*Composition:* 100% active ingred.
*Reaction:* Neutral
*Toxicity:* Slight eye irritant
*Containers:* Bulk, drums
*Warnings and Caution:* Avoid eye contact and ingestion

**Dowfax 70N14D.** Dow Chem. Europe
*Category:* Demulsifier
*Applications:* Crude oil demulsifier; break of crude oil emulsions
*Form:* Liquid; Color: Slightly yellow
*Properties:* Solubility: Slightly sol. in water; dispersible in water; Sp. gr.: 0.95; Visc.: 25 cst; F.P.: −50 C; B.P.: 64 C; Stability: Stable
*Composition:* 65% active ingred.
*Reaction:* Neutral
*Toxicity:* To be determined; eye irritant

*Containers:* Bulk
*Warnings and Caution:* Fire hazard; vapor-air mixture can be explosive

**Dresinate 81 Liquid Soap.** Hercules
*Chemical Description:* Modified aqueous solution of sodium soap
*Category:* Emulsifier ingredient
*Applications:* Emulsifier ingred. and/or stabilizer in solvent cleaners and soluble oils for metal working, disinfectants, oil-well drilling muds, drawing and grinding compounds; as plasticizing agents in specialty coatings
*Form:* Liquid; Color: Pale
*Properties:* Solubility: Dilutable with water and aqueous solutions of alcohols and glycols; Density: 8.7 lb/gal @ 21 C; Visc.: 5.7 poises @ 140 F
*Composition:* 87% solids content
*Containers:* Bung-type and open-head drums (465 lb net wt)
*References:* PD No. 7246

**Dresinate 90 Liquid Soap.** Hercules
*Chemical Description:* Modified aqueous solution of potassium soap
*Category and Applications:* See Dresinate 81
*Form:* Liquid; Color: Pale
*Properties:* Solubility: Dilutable with water and aqueous solutions of alcohols and glycols; Density: 9.0 lb/gal @ 21 C; Visc.: 14.7 poises @ 140 F
*Composition:* 90% solids content
*References:* PD No. 7246

**Dresinate 91 Liquid Soap.** Hercules
*Chemical Description:* Modified aqueous solution of potassium soap
*Category and Applications:* See Dresinate 81
*Form:* Liquid; Color: Pale
*Properties:* Solubility: Dilutable with water and aqueous solutions of alcohols and glycols; Density: 8.8 lb/gal @ 21 C; Visc.: 5.3 poises @ 140 F
*Composition:* 88% solids content
*References:* PD No. 7246

**Dresinate 92 Liquid Soap.** Hercules
*Chemical Description:* Modified aqueous solution of potassium soap
*Category and Applications:* See Dresinate 81
*Form:* Liquid; Color: Dark
*Properties:* Solubility: Dilutable with water and aqueous solutions of alcohols and glycols; Density: 9.0 lb/gal @ 21 C; Visc.: 9.2 poises @ 140 F
*Composition:* 88% solids
*References:* PD No. 7246

**Dresinate 93 Liquid Soap.** Hercules
*Chemical Description:* Modified aqueous solution of potassium soap
*Category and Applications:* See Dresinate 81
*Form:* Liquid; Color: Dark
*Properties:* Solubility: Dilutable with water and aqueous solutions of alcohols and glycols; Density: 9.5 lb/gal @ 21 C; Visc.: 7.3 poises @ 140 F
*Composition:* 87% solids content
*References:* PD No. 7246

**Dresinate 94 Liquid Soap.** Hercules
*Chemical Description:* Modified aqueous solution of potassium soap
*Category and Applications:* See Dresinate 81
*Form:* Liquid; Color: Pale
*Properties:* Solubility: Dilutable with water and aqueous solutions of alcohols and glycols; Density: 9.3 lb/gal @ 21 C; Visc.: 12.4 poises @ 140 F
*Composition:* 87% solids
*References:* PD No. 7246

**Dresinate 95 Liquid Soap.** Hercules
*Chemical Description:* Modified aqueous solution of potassium soap
*Category and Applications:* See Dresinate 81
*Form:* Liquid; Color: Dark
*Properties:* Solubility: Dilutable with water and aqueous solutions of alcohols and glycols; Density: 9.1 lb/gal; Visc.: 4.5 poises @ 140 F
*Composition:* 88% solids
*References:* PD No. 7246

**Dresinate 214 Paste Soap.** Hercules
*Chemical Description:* Potassium soap of a pale modified rosin
*Category:* Emulsifier, wetting agent, foaming agent, dispersant
*Applications:* Used in the manufacture of adhesives, polymer emulsions, synthetic latices; as sizing agents; in foaming and flotation operations; for pigment wetting and dispersing
*Form:* Paste
*Properties:* Solubility: Dilutable with water and aqueous solutions of alcohols and glycols; Density: 9.22 lb/gal @ 60 C; Visc.: 23 poises @ 140 F; Stability: Tends to separate or stratify on prolonged standing
*Composition:* 80% solids content
*Containers:* Open-head metal drums (475 lb net wt)
*References:* PD No. 7247-1

**Dresinate 515 Paste Soap.** Hercules
*Chemical Description:* Potassium soap of a pale modified rosin
*Category and Applications:* See Dresinate 214; Dresinate 515 is a nonstratifying soap recommended to replace Dresinate 214
*Form:* Paste
*Properties:* Solubility: Dilutable with water and aqueous solutions of alcohols and glycols; Density: 9.24 lb/gal @ 60 C; Visc.: 64 poises @ 140 F
*Composition:* 80% solids
*Containers:* Open-head metal drums (475 lb net wt)
*References:* PD No. 7247-1

**Dresinate 731 Paste Soap.** Hercules
*Chemical Description:* Sodium soap of a pale modified rosin
*Category and Applications:* See Dresinate 214
*Form:* Paste
*Properties:* Solubility: Dilutable with water and aqueous solutions of alcohols and glycols; Density: 9.1 lb/gal @ 60 C; Visc.: 11 poises @ 140 F
*Composition:* 70% solids
*Containers:* Open-head metal drums (475 lb net wt)

*References:* PD No. 7247-1

**Dresinate TX Dry Powder Soap.** Hercules
*Chemical Description:* Sodium soap of a tall oil
*Category and Applications:* See Dresinate X
*Form:* Free-flowing powder; Color: Dark brown
*Properties:* Solubility: Sol. in water and aqueous solutions of alcohols and glycols; Density: 26 lb/ft$^3$
*Composition:* 96% min. solids
*Containers:* Multiwall moisture vapor-resistant paper bags (50 lb net wt)
*References:* PD No. 7248-1

**Dresinate X Dry Powder Soap.** Hercules
*Chemical Description:* Sodium soap of a pale rosin
*Category:* Emulsifier, detergent aid, dispersant, stabilizer
*Applications:* As ingredients of highly built alkaline metal cleaners and other industrial cleaners, paintbrush cleaners, driveway cleaners, household cleaners, scouring powders; as dispersants for pigments, in the production of resinated colors; as stabilizers for synthetic rubber latices, conditioners for sulfur dusts; as emulsifiers for vegetable and mineral oils, pine oil and asphalt
*Form:* Free-flowing powder; Color: Pale cream
*Properties:* Solubility: Sol. in water and aqueous solutions of alcohols and glycols; Density: 26 lb/ft$^3$
*Composition:* 95% min. solids
*Containers:* Multiwall moisture vapor-resistant paper bags (50 lb net wt)
*References:* PD No. 7248-1

**Dresinate XX Dry Powder Soap.** Hercules
*Chemical Description:* Sodium soap of a dark rosin
*Category and Applications:* See Dresinate X
*Form:* Free-flowing powder; Color: Light brown
*Properties:* Solubility: Sol. in water and aqueous solutions of alcohols and glycols; Density: 28 lb/ft$^3$
*Composition:* 96% min. solids
*Containers:* Multiwall moisture vapor-resistant paper bags (50 lb net wt)
*References:* PD No. 7248-1

**Drew BRX-354.** Drew Produtos Quimicos Ltda.
*Chemical Description:* Coco amide propyl amine oxide
*Category:* Detergent, nonionic
*Applications:* Bubble bath and acid shampoos formulations
*Form:* Liquid
*Composition:* 30% active

**Drewclad 2.** PVO International
*Category:* Mothproofing agent
*Applications:* Permanent treatment for wool; for spray-on application in acid systems to raw stock dyed wool

*Form:* Liquid
*Properties:* Stability: Very stable to strong acids; Emulsifiability: Self-emulsifiable
*Composition:* 20% active ingred.

**Drewclad 33.** PVO International
*Category:* Mothproofing agent
*Applications:* Permanent treatment for wool; for use in the dye bath; resistant to extraction by laundering or drycleaning
*Form:* Liquid
*Composition:* 20% active ingred.

**Drewclad C/F.** PVO International
*Category:* Mothproofing agent
*Applications:* Permanent treatment for wool; stable and resistant to extraction
*Form:* Liquid

**Drewfax 220.** Drew Produtos Quimicos Ltda.
*Chemical Description:* Sodium alkil suf.
*Category:* Wetting agent, emulsifier, anionic
*Applications:* Wetting agent and emulsifier for polymerization
*Form:* Liquid
*Composition:* 29–30% active

**Drewfax 250.** Drew Produtos Quimicos Ltda.
*Chemical Description:* Sodium alkil suf.
*Category:* Wetting agent, anionic
*Applications:* Wetting agent for pesticides
*Form:* Powder
*Composition:* 90% active

**Drewfax 502.** Drew Produtos Quimicos Ltda.
*Chemical Description:* Sulfosuccinate derivative
*Category:* Wetting agent, emulsifier, anionic and nonionic
*Applications:* Wetting agent and emulsifier for polymerization and flowable pesticides
*Form:* Liquid

**Drewfax 528.** Drew Produtos Quimicos Ltda.
*Chemical Description:* Dioctyl ester of sodium sulfosuccinate
*Category:* Wetting agent, emulsifier, anionic
*Applications:* Wetting agent and emulsifier for polymerization
*Form:* Paste
*Composition:* 70% active

**Drewfax 539.** Drew Produtos Quimicos Ltda.
*Chemical Description:* Sulfosuccinate derivative
*Category:* Wetting agent, emulsifier, anionic and nonionic
*Applications:* Wetting agent and emulsifier for polymerization and flowable pesticides
*Form:* Liquid

**Drewfax 549.** Drew Produtos Quimicos Ltda.
*Chemical Description:* Sulfosuccinate derivative
*Category:* Wetting agent, emulsifier, anionic
*Applications:* Wetting agent and emulsifier for polymerization and flowable pesticides
*Form:* Liquid

**Drewfax 550.** Drew Produtos Quimicos Ltda.
*Chemical Description:* Sulfosuccinate derivative
*Category:* Wetting agent, emulsifier, anionic
*Applications:* Wetting agent and emulsifier for polymerization and flowable pesticides
*Form:* Liquid

**Drewfax 560.** Drew Produtos Quimicos Ltda.
*Chemical Description:* Sulfosuccinate derivative
*Category:* Wetting agent, emulsifier, anionic
*Applications:* Wetting agent and emulsifier for polymerization and flowable pesticides
*Form:* Powder

**Drewfax 577.** Drew Produtos Quimicos Ltda.
*Chemical Description:* Sulfosuccinate derivative
*Category:* Wetting agent, emulsifier, anionic
*Applications:* Wetting agent and emulsifier for polymerization and flowable pesticides
*Form:* Powder

**Drewlate 30.** PVO International
*Chemical Description:* Acetylated tartrated monoglyceride of vegetable fat
*Category:* Food emulsifier
*Form:* Syrupy liquid; Color: Amber

**Drewlene 10.** PVO International
*Chemical Description:* Propylene glycol monostearate
*Category:* Food emulsifier
*Applications:* Used to increase whipping ability and aeration in cake batters, icings and toppings
*Form:* Solid, flake or bead; Color: Yellow
*Properties:* M.P.: 110 F; HLB: 3.4; Sapon. Value: 170–190

**Drewmulse 3-1-O.** PVO International
*Chemical Description:* Triglycerol monooleate
*Category:* Nonethoxylated emulsifiers, solubilizers, dispersants, nonionic
*Applications:* W/O and O/W emulsions; emollient creams and lotions; clear emulsions; pharmaceutical vehicles; cosmetics; FDA approved for internal use (degraded by the body to glycerine and fatty acid)
*Form:* Liquid; Color: 8 (Gardner)
*Properties:* Solubility: Sol. in isopropanol, peanut oil, mineral oil; dispersible in water; HLB: 7.0; Sap. Value: 125–150
*References:* PVO brochure

**Drewmulse 3-1-S.** PVO International
*Chemical Description:* Triglycerol monostearate
*Category and Applications:* See Drewmulse 3-1-O
*Form:* Solid; Color: 10 (Gardner)
*Properties:* Solubility: Dispersible in water, isopropanol, propylene glycol, peanut oil, mineral oil; HLB: 7.0; Sap. Value: 120–140
*References:* PVO brochures

**Drewmulse 6-2-S.** PVO International
*Chemical Description:* Hexaglycerol distearate
*Category and Applications:* See Drewmulse 3-1-O
*Form:* Solid; Color: 13 (Gardner)
*Properties:* Solubility: Sol. hot in isopropanol, peanut and mineral oils; disp. in propylene glycol; HLB: 8.0; Sap. Value: 105–125
*References:* PVO brochures

**Drewmulse 10.** PVO International
*Chemical Description:* Glycerol mono fat from soya oil
*Category:* Lipophilic emulsifier
*Applications:* Various food industry uses as emulsifier, dispersing aid, antistaling agent; improves stability and shelf life, chewing or flow properties; aids aeration; dough conditioners; antistick agent
*Form:* Plastic; Color: Cream
*Properties:* M.P.: 118 F; HLB: 2.8; Sap. Value: 170–190
*Composition:* 40% alpha mono
*References:* PVO brochures

**Drewmulse 10-4-O.** PVO International
*Chemical Description:* Decaglycerol tetraoleate
*Category and Applications:* See Drewmulse 3-1-O
*Form:* Liquid; Color: 8 (Gardner)
*Properties:* Solubility: Sol. in isopropanol, peanut and mineral oils; disp. in water; HLB: 6.0; Sap. Value: 125–145
*References:* PVO brochures

**Drewmulse 10-8-O.** PVO International
*Chemical Description:* Decaglycerol octaoleate
*Category and Applications:* See Drewmulse 3-1-O
*Form:* Liquid; Color: 8 (Gardner)
*Properties:* Solubility: Sol. in peanut and mineral oils; HLB: 4.0; Sap. Value: 155–175
*References:* PVO brochures

**Drewmulse 10-10-O.** PVO International
*Chemical Description:* Decaglycerol decaoleate
*Category and Applications:* See Drewmulse 3-1-O
*Form:* Liquid; Color: 8 (Gardner)
*Properties:* Solubility: Sol. in peanut and mineral oils; HLB: 3.0; Sap. Value: 165–185
*References:* PVO brochures

**Drewmulse 10-10-S.** PVO International
*Chemical Description:* Decaglycerol decastearate
*Category and Applications:* See Drewmulse 3-1-O
*Form:* Solid; Color: 8 (Gardner)
*Properties:* Solubility: Sol. hot in peanut and mineral oils; slightly sol. in propylene glycol; HLB: 3.0; Sap. Value: 160–180
*References:* PVO brochures

**Drewmulse 15.** PVO International
*Chemical Description:* Glycerol mono cottonseed oil
*Category and Applications:* See Drewmulse 10
*Form:* Soft solid; Color: Cream

*Properties:* M.P.: 110 F
*Composition:* 40% alpha mono
*References:* PVO brochures

**Drewmulse 20/200/V.** PVO International
*Chemical Description:* Glycerol monostearate
*Category and Applications:* See Drewmulse 10
*Form:* Beads; Color: Cream
*Properties:* M.P.: 140 F
*Composition:* 40% alpha mono
*References:* PVO brochures

**Drewmulse 21.** PVO International
*Chemical Description:* Glycerol monostearate from soya oil
*Category and Applications:* See Drewmulse 10
*Form:* Bead or flake; Color: Cream
*Properties:* HLB: 8.4; Sap. Value: 160–180
*Composition:* 30% alpha mono
*References:* PVO brochures

**Drewmulse 30/900.** PVO International
*Chemical Description:* Glycerol monostearate
*Category and Applications:* See Drewmulse 10
*Form:* Beads; Color: Cream
*Properties:* M.P.: 138 F
*Composition:* 52% alpha mono
*References:* PVO brochures

**Drewmulse 50.** PVO International
*Chemical Description:* Glycerol mono fat
*Category and Applications:* See Drewmulse 10
*Form:* Cream; Color: Cream
*Composition:* 52% alpha mono
*References:* PVO brochures

**Drewmulse 55.** PVO International
*Chemical Description:* Glycerol mono shortening from tallow and soya
*Category and Applications:* See Drewmulse 10
*Form:* Plastic; Color: Cream
*Properties:* Tallow: M.P.: 130–135 F; HLB: 3.4; Sap. Value: 165–185; Soya: M.P.: 116–122 F; HLB: 2.4
*Composition:* 52% alpha mono
*References:* PVO brochures

**Drewmulse 70.** Drew Produtos Quimicos Ltda.
*Chemical Description:* Blend of glycerol monoesters
*Category:* Emulsifier, nonionic
*Applications:* Food emulsifier, baking additives, margarine, and shortening emulsifiers
*Form:* Liquid
*Composition:* 13.5–19% active

**Drewmulse 75.** Drew Produtos Quimicos Ltda.
*Chemical Description:* Sorbitan ester

*Category:* Emulsifier, nonionic
*Applications:* See Drewmulse 70
*Form:* Paste
*Composition:* 100% active

**Drewmulse 85.** PVO International
*Chemical Description:* Glycerol monooleate from tallow and soya
*Category and Applications:* See Drewmulse 10
*Form:* Semiliquid; Color: Light yellow
*Properties:* HLB: 3.4; Sap. Value: 160–180
*Composition:* 40% alpha mono
*References:* PVO brochure

**Drewmulse 200.** PVO International
*Chemical Description:* Glycerol monostearate from soya and tallow
*Category and Applications:* See Drewmulse 10
*Form:* Beads; Color: Cream
*Properties:* M.P.: 135–145 F; HLB: 2.8; Sap. Value: 154–170 (soya), 163–177 (tallow)
*Composition:* 40% alpha mono
*References:* PVO brochures

**Drewmulse 300.** PVO International
*Chemical Description:* Glycerol monostearate (80%) and polysorbate (20%) blend from soya and tallow oils
*Category and Applications:* See Drewmulse 10
*Form:* Beads; Color: Cream
*Properties:* M.P.: ≈ 135 F; HLB: 4.3
*Composition:* 30% alpha mono
*References:* PVO brochures

**Drewmulse 365.** Drew Productos Quimicos Ltda.
*Chemical Description:* Glycerol monostearate (67%) and polysorbate 65 (33%) blend from soya and tallow oils
*Category and Applications:* See Drewmulse 10
*Form:* Beads; Color: Cream
*Properties:* M.P.: ≈ 136 F; HLB: 5.3
*Composition:* 25–30% alpha mono
*References:* PVO brochures

**Drewmulse 700.** PVO International
*Chemical Description:* Glycerol monostearate (80%) and polysorbate 80 (20%) blend from soya and tallow oils
*Category and Applications:* See Drewmulse 10
*Form:* Beads; Color: Cream
*Properties:* M.P.: ≈ 135 F; HLB: 5.2
*Composition:* 30% alpha mono
*References:* PVO brochures

**Drewmulse 900.** PVO International
*Chemical Description:* Glycerol monostearate
*Category and Applications:* See Drewmulse 10
*Form:* Beads; Color: Cream
*Properties:* M.P.: ≈ 138 F; HLB: 3.8; Sap. Value: 160–180
*Composition:* 52% alpha mono
*References:* PVO brochures

**Drewmulse 1128.** PVO International
*Chemical Description:* Glycerol monostearate and polyoxyethylene propylene glycol stearate

*Category and Applications:* See Drewmulse CNO
*Form:* Flake; Color: 2 (Gardner)
*Properties:* HLB: 8.4; Stability: Acid stable; Emulsifiability: Self-emulsifying
*Composition:* 20% mono

**Drewmulse 1129.** PVO International
*Chemical Description:* Glycerol monostearate and polyoxyethylene stearate
*Category and Applications:* See Drewmulse CNO
*Form:* Flake
*Properties:* HLB: 11.0; Stability: Acid stable; Emulsifiability: Self-emulsifying
*Composition:* 20% mono

**Drewmulse CNO.** PVO International
*Chemical Description:* Glycerol monococate
*Category:* Emulsifier, stabilizer, opacifier, viscosity builder
*Applications:* Emulsifier, stabilizer and viscosity builder in cosmetics and pharmaceuticals; opacifier; bases for creams and lotions; cream rinses; hair conditioners
*Form:* Solid; Color: 4 (Gardner)
*Properties:* HLB: 3.8
*Composition:* 50% mono

**Drewmulse DGMS.** Drew Produtos Quimicos Ltda.
*Chemical Description:* Diethylene glycol monostearate
*Category:* Emulsifier, opacifier and pearlescence agents, nonionic
*Applications:* Emulsifier for W/O and O/W systems; opacifier and pearlescence agents for shampoos, additives for pigment milling
*Form:* Flakes
*Composition:* 85% active

**Drewmulse EGDS.** Drew Produtos Quimicos Ltda.
*Chemical Description:* Ethylene glycol distearate
*Category and Applications:* See Drewmulse DGMS
*Form:* Flakes
*Composition:* 48% active

**Drewmulse EGMS.** Drew Produtos Quimicos Ltda.
*Chemical Description:* Ethylene glycol monostearate
*Category and Applications:* See Drewmulse DGMS
*Form:* Flakes
*Composition:* 48% active

**Drewmulse GMC-8.** PVO International
*Chemical Description:* Glycerol monocaprylate-caprate
*Category and Applications:* See Drewmulse CNO; also for food applications (see Drewmulse 10)
*Form:* Liquid; Color: 3 (Gardner)
*Properties:* HLB: 8.3
*Composition:* 75% mono

**Drewmulse GMO.** PVO International
*Chemical Description:* Glycerol monooleate
*Category and Applications:* See

Drewmulse CNO
*Form:* Liquid; Color: 6 (Gardner)
*Properties:* HLB: 3.4
*Composition:* 40% mono

**Drewmulse GMRO.** Drew Produtos Quimicos Ltda.
*Chemical Description:* Glycerol monoricinoleate
*Category and Applications:* See Drewmulse DGMS
*Form:* Liquid

**Drewmulse GMS.** Drew Produtos Quimicos Ltda.
*Chemical Description:* Glycerol monostearate
*Category and Applications:* See Drewmulse DGMS
*Form:* Flakes
*Composition:* 40–45% active

**Drewmulse HM 100.** PVO International
*Chemical Description:* Glycerol monostearate and polyoxyethylene stearate
*Category and Applications:* See Drewmulse CNO; self-emulsifying and acid stable; esp. effective in antiperspirant formulations
*Form:* Flake; Color: 2 (Gardner)
*Properties:* HLB: 8.4
*Composition:* 24% mono

**Drewmulse M.** Drew Produtos Quimicos Ltda.
*Chemical Description:* Blend of glycerol monoesters
*Category:* Emulsifier, nonionic
*Applications:* See Drewmulse 70
*Form:* Paste
*Composition:* 32% active

**Drewmulse PGML.** PVO International
*Chemical Description:* Propylene glycol monolaurate
*Category:* Chemical specialty
*Form:* Liquid

**Drewmulse PGMS.** Drew Produtos Quimicos Ltda.
*Chemical Description:* Propylene glycol monostearate
*Category and Applications:* See Drewmulse DGMS
*Form:* Flakes
*Composition:* 48% active

**Drewmulse POE-SML.** PVO International
*Chemical Description:* Polyoxyethylene 20 sorbitan monolaurate
*Category:* Nonionic emulsifiers, solubilizers, dispersants, wetting agents, detergents, viscosity control agents
*Applications:* W/O emulsifiers for cosmetics and pharmaceuticals; emollient creams and lotions; shave creams; fragrance and flavor solubilizers; shampoos; topical ointments; vitamins; germicides in aqueous systems
*Form:* Liquid
*Properties:* HLB: 16.7; Sap. Value: 40–51

**Drewmulse POE-SMO.** PVO International
*Chemical Description:* Polyoxyethylene 20 sorbitan monooleate
*Category and Applications:* See

Drewmulse POE-SML
*Form:* Liquid
*Properties:* HLB: 15.0; Sap. Value: 45–55

**Drewmulse POE-SMS.** PVO International
*Chemical Description:* Polyoxyethylene 20 sorbitan monostearate
*Category and Applications:* See Drewmulse POE-SML
*Form:* Solid
*Properties:* HLB: 14.9; Sap. Value: 45–55

**Drewmulse POE-STS.** PVO International
*Chemical Description:* Polyoxyethylene 20 sorbitan tristearate
*Category and Applications:* See Drewmulse POE-SML
*Form:* Solid
*Properties:* HLB: 10.5; Sap. Value: 88–98

**Drewmulse PNO.** Drew Produtos Quimicos Ltda.
*Chemical Description:* Blend of glycerol monoesters
*Category:* Emulsifier, nonionic
*Applications:* See Drewmulse 70
*Form:* Paste
*Composition:* 38–42% active

**Drewmulse SML.** PVO International
*Chemical Description:* Sorbitan monolaurate
*Category and Applications:* See Drewmulse POE-SML
*Form:* Liquid
*Properties:* HLB: 8.6; Sap. Value: 159–171

**Drewmulse SMO.** PVO International
*Chemical Description:* Sorbitan monooleate
*Category and Applications:* See Drewmulse POE-SML
*Form:* Liquid
*Properties:* HLB: 4.3; Sap. Value: 144–156

**Drewmulse SMS.** PVO International
*Chemical Description:* Sorbitan monostearate
*Category and Applications:* See Drewmulse POE-SML
*Form:* Solid
*Properties:* HLB: 4.7; Sap. Value: 147–157

**Drewmulse STS.** PVO International
*Chemical Description:* Sorbitan tristearate
*Category and Applications:* See Drewmulse POE-SML
*Form:* Solid
*Properties:* HLB: 2.1; Sap. Value: 170–190

**Drewmulse TP.** PVO International
*Chemical Description:* Glycerol monostearate (triple pressed)
*Category and Applications:* See Drewmulse CNO
*Form:* Flake; Color: 4 (Gardner)
*Properties:* HLB: 3.8
*Composition:* 40% mono

**Drewmulse V.** PVO International
*Chemical Description:* Glycerol monostearate (vegetable)

*Category and Applications:* See Drewmulse CNO
*Form:* Bead; Color: 6 (Gardner)
*Properties:* HLB: 3.8
*Composition:* 40% mono

**Drewmulse V-SE.** PVO International
*Chemical Description:* Glycerol monostearate (self-emulsifying)
*Category and Applications:* See Drewmule CNO
*Form:* Bead; Color: 6 (Gardner)
*Properties:* HLB: 8.4
*Composition:* 30% mono

**Drewplast 017.** PVO International
*Category:* Plastic additive, antistatic and slip agent
*Applications:* Antistatic and slip agent for PVC and polyolefin resin; esp. for PVC food wrap; FDA listed GRAS
*Properties:* Stability: Excellent heat stability

**Drewplast 029.** PVO International
*Category:* Plastic additive, antistatic agent
*Applications:* Antistatic agent for polyolefins and polystyrene; food grade, accepted by FDA as direct food additive

**Drewplast 030.** PVO International
*Category:* Plastic additive, antifogging agent, lubricant
*Applications:* Antifogging agent for PVC meat and produce wrap; internal lubricant
*Properties:* Stability: Excellent heat stability

**Drewplast 050.** Drew Produtos Quimicos Ltda.
*Chemical Description:* Glycerol ester
*Category:* Plasticizer, nonionic
*Form:* Paste

**Drewplast 051.** PVO International
*Category:* Plastic additive, lubricant
*Applications:* Antistatic internal lubricant with antifogging properties in PVC and polyolefins; FDA listed GRAS
*Properties:* Stability: Excellent heat stability

**Drewplast 100.** PVO International
*Category:* Plastic additive, antifogging agent
*Applications:* Economical antifogging agent for PVC food wrap; improves machinability of film or sheet in automatic wrapping machines; FDA approved

**Drewplast 111.** Drew Produtos Quimicos Ltda.
*Chemical Description:* Fatty ester
*Category:* Plasticizer, nonionic
*Form:* Liquid

**Drewplast 175.** Drew Produtos Quimicos Ltda.
*Chemical Description:* Fatty ester
*Category:* Plasticizer, nonionic
*Form:* Liquid

**Drewpol 3-1-O.** PVO International
*Chemical Description:* Triglycerol monooleate
*Category:* Food emulsifier

*Applications:* The Drewpols have various functions as food additives such as solubilizer for flavors and colors; improves dispersibility, shelf life, texture, volume, emulsion stability; provides gloss in confectionary coatings; imparts aeration, dryness; foam control
*Form:* Liquid; Color: Amber
*Properties:* HLB: 7.0; Sap. Value: 120–160

**Drewpol 3-1-S.** PVO International
*Chemical Description:* Triglycerol monostearate
*Category and Applications:* See Drewpol 3-1-O
*Form:* Beads; Color: Amber
*Properties:* HLB: 7.0; Sap. Value: 120–140

**Drewpol 3-1-SH.** PVO International
*Chemical Description:* Triglycerol mono shortening (Drewmulse 8731-S)
*Category and Applications:* See Drewpol 3-1-O
*Form:* Thick liquid; Color: Amber
*Properties:* HLB: 7.4; Sap. Value: 110–140

**Drewpol 6-2-S.** PVO International
*Chemical Description:* Hexaglycerol distearate
*Category and Applications:* See Drewpol 3-1-O
*Form:* Beads; Color: Amber
*Properties:* HLB: 4.0; Sap. Value: 100–130

**Drewpol 10-4-O.** PVO International
*Chemical Description:* Decaglycerol tetraoleate
*Category and Applications:* See Drewpol 3-1-O
*Form:* Liquid; Color: Amber
*Properties:* HLB: 6.0; Sap. Value: 125–145

**Drewpol 10-8-O.** PVO International
*Chemical Description:* Decaglycerol octaoleate
*Category and Applications:* See Drewpol 3-1-O
*Form:* Liquid; Color: Amber
*Properties:* HLB: 4.0; Sap. Value: 150–180

**Drewpol 10-10-O.** PVO International
*Chemical Description:* Decaglycerol decaoleate
*Category and Applications:* See Drewpol 3-1-O
*Form:* Liquid; Color: Amber
*Properties:* HLB: 3.0; Sap. Value: 160–190

**Drewpol 10-10-S.** PVO International
*Chemical Description:* Decaglycerol decastearate
*Category and Applications:* See Drewpol 3-1-O
*Form:* Solid; Color: Tan
*Properties:* HLB: 3.0; Sap. Value: 150–190

**Drewpon 40.** Drew Produtos Quimicos Ltda.
*Chemical Description:* Triethanolamine lauryl sulfate
*Category:* Emulsifier, detergent, wetting agent, anionic

*Applications:* Liquid and emulsion shampoos, bubble baths, rug shampoo, household cleaning products, emulsion polymerization, wetting agents, pigment dispersants, dishwashing detergents
*Form:* Liquid
*Composition:* 39–42% active

**Drewpon 40LS.** Drew Produtos Quimicos Ltda.
*Chemical Description:* Triethanolamine lauryl sulfate
*Category:* Emulsifier, detergent, wetting agent, anionic
*Applications:* See Drewpon 40
*Form:* Liquid
*Composition:* 38–40% active

**Drewpon 100.** Drew Produtos Quimicos Ltda.
*Chemical Description:* Sodium lauryl sulfate
*Category:* Detergent, wetting agent, anionic
*Applications:* Used as detergent, wetting dispersant for tooth pastes, shampoos, bubble baths, pesticides
*Form:* Powder
*Composition:* 90% active

**Drewpon AC-50.** Drew Produtos Quimicos Ltda.
*Chemical Description:* Amphoteric imidazoline
*Category:* Detergent, amphoteric
*Applications:* Detergent for non-irritating shampoos
*Form:* Liquid
*Composition:* 38% active

**Drewpon AS.** Drew Produtos Quimicos Ltda.
*Chemical Description:* Sodium lauryl sulfate
*Category:* Detergent, anionic
*Applications:* Detergent for non-irritating shampoos
*Form:* Paste
*Composition:* 28–30% active

**Drewpon CN.** Drew Produtos Quimicos Ltda.
*Chemical Description:* Sodium lauryl sulfate
*Category:* Detergent, anionic
*Applications:* Detergent for non-irritating shampoos
*Form:* Liquid/paste
*Composition:* 28-30% active

**Drewpon EBO.** Drew Produtos Quimicos Ltda.
*Chemical Description:* Sodium lauryl ether sulfate
*Category:* Detergent, anionic
*Applications:* Detergent for non-irritating shampoos
*Form:* Liquid
*Composition:* 26–28% active

**Drewpon ECM.** Drew Produtos Quimicos Ltda.
*Chemical Description:* Monoethanolamine lauryl ether sulfate
*Category:* Detergent, anionic
*Applications:* Detergent for non-irritating shampoos
*Form:* Liquid
*Composition:* 30–32% active

**Drewpon EDT.** Drew Produtos Quimicos Ltda.
*Chemical Description:* Triethanol-

amine lauryl ether sulfate
*Category:* Detergent, anionic
*Applications:* Detergent for non-irritating shampoos
*Form:* Liquid
*Composition:* 34–36% active

**Drewpon EG.** Drew Produtos Quimicos Ltda.
*Chemical Description:* Sodium lauryl ether sulfate
*Category:* Detergent, anionic
*Applications:* Used for pearlescent shampoos
*Form:* Liquid
*Composition:* 22% active

**Drewpon EKZ.** Drew Produtos Quimicos Ltda.
*Chemical Description:* Ammonium lauryl ether sulfate
*Category:* Detergent, anionic
*Applications:* Detergent for non-irritating shampoos
*Form:* Liquid
*Composition:* 23–25% active

**Drewpon ELB.** Drew Produtos Quimicos Ltda.
*Chemical Description:* Sodium lauryl ether sulfate
*Category:* Detergent, anionic
*Applications:* Detergent for non-irritating shampoos
*Form:* 27–28% active

**Drewpon ESG.** Drew Produtos Quimicos Ltda.
*Chemical Description:* Magnesium lauryl ether sulfate
*Category:* Detergent, anionic
*Applications:* Detergent for non-irritating shampoos
*Form:* Liquid
*Composition:* 26–28% active

**Drewpon ETH.** Drew Produtos Quimicos Ltda.
*Chemical Description:* Sodium lauryl ether sulfate
*Category:* Detergent, anionic
*Applications:* Detergent for non-irritating shampoos
*Form:* Liquid
*Composition:* 26–28% active

**Drewpon EVE.** Drew Produtos Quimicos Ltda.
*Chemical Description:* Sodium lauryl ether sulfate
*Category:* Detergent, anionic
*Applications:* Detergent for non-irritating shampoos
*Form:* Liquid
*Composition:* 26–28% active

**Drewpon MEA.** Drew Produtos Quimicos Ltda.
*Chemical Description:* Monoethanolamine lauryl sulfate
*Category:* Detergent, anionic
*Applications:* Detergent for non-irritating shampoos
*Form:* Liquid
*Composition:* 32–34% active

**Drewpon MG.** Drew Produtos Quimicos Ltda.
*Chemical Description:* Magnesium lauryl sulfate
*Category:* Detergent, anionic
*Applications:* Detergent for non-irritating shampoos
*Form:* Liquid
*Composition:* 28–30% active

**Drewpon NH.** Drew Produtos Quimicos Ltda.
*Chemical Description:* Ammonium lauryl sulfate
*Category:* Detergent, anionic
*Applications:* Detergent for non-irritating shampoos
*Form:* Liquid
*Composition:* 27–29% active

**Drewpon NHAC.** Drew Produtos Quimicos Ltda.
*Chemical Description:* Ammonium lauryl sulfate
*Category:* Detergent, anionic
*Applications:* Detergent for non-irritating shampoos
*Form:* Liquid
*Composition:* 37.5% active

**Drewpone 60.** PVO International
*Chemical Description:* Polyoxyethylene 20 sorbitan monostearate
*Category:* Food emulsifier
*Applications:* Hydrophilic food emulsifiers generally used with more lipophilic emulsifiers; the Drewpones have various food applications including as foaming agents, dough conditioners; to improve volume, aeration, texture, gloss, dispersiblity (for color and flavor, in vitamin products), dryness, stability
*Form:* Liquid to soft solid; Color: Yellow
*Properties:* HLB: 14.9; Sap. Value: 45–55
*Composition:* 65–69.5% ETO

**Drewpone 65.** PVO International
*Chemical Description:* Polyoxyethylene 20 sorbitan tristearate
*Category and Applications:* See Drewpone 60
*Form:* Soft solid; Color: Amber
*Properties:* HLB: 10.5; Sap. Value: 88–98
*Composition:* 46–50% ETO

**Drewpone 80.** PVO International
*Chemical Description:* Polyoxyethylene 20 sorbitan monooleate
*Category and Applications:* See Drewpone 60
*Form:* Liquid; Color: Yellow
*Properties:* HLB: 15.0; Sap. Value: 45–55
*Composition:* 65–69.5% ETO

**Drewsorb 60.** PVO International
*Chemical Description:* Sorbitan monostearate
*Category:* Food emulsifier
*Applications:* Used by itself or with hydrophilic emulsifiers in food applications
*Form:* Beads; Color: Cream
*Properties:* Solubility: Dispersible in water; HLB: 4.7; Sap. Value: 147–157

**Druspin 1223.** PVO International
*Chemical Description:* Ester
*Category:* Textile lubricant
*Applications:* Lubricant to minimize breakage in worsted carding; used for ordinary filet wire in the card clothing
*Form:* Liquid
*Properties:* Visc.: Low; Stability: Good
*Composition:* 100% active ingred.

**Druspin 1400.** PVO International
*Chemical Description:* Ester
*Category:* Textile lubricant
*Applications:* Lubricant to minimize breakage in worsted carding; used for metallic wire in the card clothing
*Form:* Liquid
*Properties:* Visc.: Low; Stability: Good
*Composition:* 100% active ingred.

**Druspin LO-VIS.** PVO International
*Chemical Description:* Blended product
*Category:* Textile lubricant, antistat, emulsifier
*Applications:* Low viscosity precarding lubricant for worsted processing; esp. when nonionic scouring in absence of added electrolyte results in static charged wool
*Form:* Liquid
*Properties:* Visc.: Low
*Composition:* 100% active ingred.

**Druspin Supreme.** PVO International
*Chemical Description:* Blend of mineral oil, emulsifiers, lubricants, antistats
*Category:* Emulsifier, lubricant, antistatic agent
*Applications:* Antistatic worsted oil with excellent emulsion stability; imparts good "spinability," lubrication, antistatic protection
*Form:* Liquid
*Composition:* 100% active ingred.

**Drustat 75.** PVO International
*Chemical Description:* Complex quarternized fatty amide
*Category:* Cationic antistatic agent
*Applications:* Antistatic agent for all synthetic fibers, imparting softness, lubricity
*Form:* Liquid; Color: Amber
*Composition:* 100% active ingred.
*Reaction:* Compatible with silicone water repellent systems

**Drustat T.** PVO International
*Chemical Description:* Quarternized fatty amide
*Category:* Cationic antistatic agent
*Applications:* Antistatic agent on wool and all synthetic fibers (e.g., polyamide, polyester, acrylics, acetate, dynel); nondiscoloring to fabric; resistant to removal by washing
*Form:* Paste; Color: Cream
*Properties:* Solubility: Readily water dispersible
*Composition:* 50% active
*Reaction:* Compatible with aminoplast resin systems

**Dugista.** Van Schuppen
*Chemical Description:* Blend of de-oiled soybean lecithin, primary dried yeast and wheat germs
*Category:* Dietary supplement
*Applications:* Dietary supplement in correcting and preventing arterial diseases; general stimulant of activity; natural sources of choline, inositol, phosphatides, B-vitamins, vitamin E, amino acids and minerals
*Form:* Granular; Color: Light yel-

low; Odor and Taste: Bland
*Composition:* 65–70% phosphatides; 12% min. protein; 12–15% carbohydrates; 3–4% vegetable oil; 2.5–3% phosphorus; 2–2.5% nitrogen; 2.5–3% tricalciumphosphate
*References:* Product Information No. 62

**Dulectin Granular.** Van Schuppen
*Chemical Description:* De-oiled soybean lecithin (complex mixture of acetone-insoluble phosphatides from soybeans)
*Category:* Dietary supplement
*Applications:* Dietary supplement in correcting and preventing heart and arterial diseases; generally as a stimulant of activity
*Form:* Granular; Color: Light yellow-brown; Odor and Taste: Bland
*Properties:* Solubility: Dispersible in water; sol. in animal, mineral and vegetable oils and fats, ether and various hydrocarbons
*Composition:* 97% min. acetone-insoluble phosphatides
*References:* Product information No. 61A

**Dulectin Powdered.** Van Schuppen
*Chemical Description:* De-oiled soybean lecithin (complex mixture of acetone-insoluble phosphatides from soybeans)
*Category:* Food additive, dietary supplement
*Applications:* In margarine, macaroni and baking industries as colloidal, interfacial, emulsifying, softening and antioxidant agents; in dried full cream milk powder; as a dietary supplement in correcting and preventing heart and arterial diseases; activity stimulant
*Form:* Dry, free-flowing powder; Color: Light yellow; Odor and Taste: Bland
*Properties:* Solubility: Dispersible in water; sol. in animal, mineral and vegetable oils and fats, ether and various hydrocarbons
*Composition:* 97% min. acetone-insoluble phosphatides
*References:* Product information No. 61 B

**Duofol T.** Hart Chem.
*Chemical Description:* Anionic sulfated ester
*Category:* Wetting agent
*Applications:* Sulfonated ester used for wetting, dispersing and finishing
*Form:* Liquid; Color: Amber
*Properties:* Solubility: Clear sol. 5% water; Sp. gr.: 1.068; Visc.: 70 cps
*Composition:* 55% active ingred.
*Reaction:* Neutral
*Toxicity:* Skin and eye irritant
*Containers:* Drums, T/L
*Warnings and Caution:* Avoid contact with skin and eyes

**Duomac C.** Akzo Chemie
*Chemical Description:* Coco-1,3-diaminopropane diacetate
*Category:* Corrosion inhibitor, bactericide in drilling fluids
*Properties:* Solubility: 100% in isopropanol; Sparingly sol. in water @ 60 C; 100% in $C_2H_5OH$ @

30 C; Sp. gr.: 0.85 @ 25 C; M. P.: 20–24 C
*Composition:* 89% active
*Containers:* Net 172 kg open head drum
*Warnings and Caution:* Avoid contact with strong oxidizing agents, acids and bases

**Duomac T.** Armak
*Chemical Description:* N-tallow-1,3-propanediamine diacetate
*Category and Applications:* The Duomac diamine acetate salts are water-soluble derivatives of the corresponding diamines; used when a water-soluble cationic reagent is required for pigment flushing, froth flotation of minerals, flocculation
*Form:* Paste; Color: Brown; Odor: Vinegar
*Properties:* m.w.: 480; Solubility: Sol. in water, chloroform, ethanol, isopropanol; Sp. gr.: 0.892@ 90 C; M.P.: 181 F; B.P.: > 300 C; Flash Pt.: 335 F (COC)
*Composition:* 85% Duomeen T as the acetate
*Toxicity:* Corrosive to skin and eyes; causes severe burns on prolonged contact
*Warnings and Caution:* Protective goggles and gloves must be worn
*References:* Bulletin 76-19

**Duomeen C.** Armak
*Chemical Description:* N-coco-1,3-propanediamine
*Category:* Organic bifunctional bases, corrosion inhibitors, fuel oil additives, flotation agents, cationic
*Applications:* Form strongly bonded films on surfaces of metals, textiles, plastics; as corrosion inhibitors, multi-purpose gasoline and fuel oil additives, flotation agents, epoxy curing agents; in preparation of amine salts of some herbicides
*Form:* Liquid; Color: Dark amber; Odor: Ammonia
*Properties:* m.w.: 276; Solubility: Sol. in naphtha, mineral oil, isopropanol; Sp. gr.: 0.836; M.P.: 71 F; B.P.: > 300 C; Flash Pt.: 295 F (COC)
*Composition:* 43% min. primary amine; 43% min. secondary amine
*Reaction:* Readily neutralized to form either water- or oil-soluble mono- or diacid salts
*Toxicity:* Corrosive to skin and eyes; causes severe burns on prolonged contact
*Warnings and Caution:* Protective goggles and gloves must be worn
*References:* Bulletin 76-19

**Duomeen C Special.** Armak
*Chemical Description:* N-coco-1,3-propanediamine
*Category and Applications:* See Duomeen C
*Form:* Liquid; Color: Yellow; Odor: Ammonia
*Properties:* m.w.: 276; Solubility: Sol. in naphtha, mineral oil, isopropanol; Sp. gr.: 0.836; M.P.: 80 F; B.P.: > 300 C; Flash Pt.: 295 F (COC)
*Composition:* 45% min. primary amine; 45% min. secondary amine
*Reaction:* Readily neutralized to

form either water- or oil-soluble mono- or diacid salts
*Toxicity:* Corrosive to skin and eyes; causes severe burns on prolonged contact
*Warnings and Caution:* Protective goggles and gloves must be worn
*References:* Bulletin 76-19

**Duomeen L-11.** Armak
*Chemical Description:* N-s-alkyl-1,3-propanediamine
*Category and Applications:* See Duomeen C
*Form:* Liquid; Color: Yellow; Odor: Ammonia
*Properties:* m.w.: 300; Solubility: Sol. in mineral oil, isopropanol; Sp. gr.: 0.841; M.P.: –32 F; B.P.: > 300 C; Flash Pt.: 226 F (COC)
*Toxicity:* Corrosive to skin and eyes; causes severe burns on prolonged contact
*Warnings and Caution:* Protective goggles and gloves must be worn
*References:* Bulletin 76-19

**Duomeen L-15.** Armak
*Chemical Description:* N-s-alkyl-1,3-propanediamine
*Category and Applications:* See Duomeen C
*Form:* Liquid; Color: Yellow; Odor: Ammonia
*Properties:* m.w.: 383; Solubility: Sol. in mineral oil, isopropanol; Sp. gr.: 0.842; M.P.: 44 F; B.P.: > 300 C; Flash Pt.: 266 F (COC)
*Toxicity:* Corrosive to skin and eyes; causes severe burns on prolonged contact
*Warnings and Caution:* Protective goggles and gloves must be worn
*References:* Bulletin 76-19

**Duomeen O.** Armak
*Chemical Description:* N-oleyl-1,3-propanediamine
*Category and Applications:* See Duomeen C
*Form:* Paste; Color: Yellow; Odor: Ammonia
*Properties:* m.w.: 350; Solubility: Sol. in naphtha @ 35 C, isopropanol; Sp. gr.: 0.841; M.P.: 59 F; B.P.: > 300 C; Flash Pt.: 370 F (COC)
*Composition:* 43% min. primary amine; 43% min. secondary amine
*Reaction:* Readily neutralized to form either water- or oil-soluble mono- or diacid salts
*Toxicity:* Corrosive to skin and eyes; causes severe burns on prolonged contact
*Warnings and Caution:* Protective goggles and gloves must be worn
*References:* Bulletin 76-19

**Duomeen T.** Armak
*Chemical Description:* N-tallow-1,3-propanediamine
*Category and Applications:* See Duomeen C; also as textile finishing agent, as dispersant for inorganic pigments in paints, as larvicidal oil additive
*Form:* Paste; Color: Dark amber; Odor: Ammonia
*Properties:* m.w.: 350; Solubility: Sol. in naphtha @ 35 C, isopropanol; Sp. gr.: 0.841; M.P.: 115 F; B.P.: > 300 C; Flash Pt.: 370 F (COC)
*Composition:* 43% min. primary

amine; 43% min. secondary amine
*Reaction:* Readily neutralized to form either water- or oil-soluble mono- or diacid salts
*Toxicity:* Corrosive to skin and eyes; causes severe burns on prolonged contact
*Warnings and Caution:* Protective goggles and gloves must be worn
*References:* Bulletin 76-19

**Duoquad T-50.** Akzo Chemie
*Chemical Description:* N, N, N′, N′, N′ - pentamethyl-*n*-tallow-1,3-propanadiammonium dichloride
*Category:* Detergent, corrosion inhibitor, metal cleaner, cationic
*Form:* Liquid; Color: Max. 13 (Gardner)
*Properties:* Sp. gr.: 0.90; F.P.: −20 C; Flash Pt.: 13 C; pH: 6–8
*Composition:* 48–52% active
*Toxicity:* Skin irritant, severe eye irritant
*Containers:* Net 180 kg drums
*Warnings and Caution:* Avoid contact with strong oxidizing agents and overexposure

**Duomeen T Special.** Armak
*Chemical Description:* N-tallow-1,3-propanediamine
*Category and Applications:* See Duomeen C
*Form:* Paste; Color: Dark amber; Odor: Ammonia
*Properties:* m.w.: 330; Solubility: Sol. in naphtha, isopropanol; Sp. gr.: 0.841; M.P.: 115 F; B.P.: > 300 C; Flash Pt.: 370 F (COC)
*Composition:* 43% min. primary amine; 43% min. secondary amine
*Reaction:* Readily neutralized to form either water- or oil-soluble mono- or diacid salts
*Toxicity:* Corrosive to skin and eyes; causes severe burns on prolonged contact
*Warnings and Caution:* Protective goggles and gloves must be worn
*References:* Bulletin 76-19

**Duomeen TDO.** Armak
*Chemical Description:* N-tallow-1,3-propanediamine dioleate
*Category and Applications:* See Duomeen C; also in metal treatment as a film and boundary lubricant, a metal drawing additive, corrosion inhibitor; in paint industry as a dispersant for inorganic and organic pigments; added to primer paints to insure adhesion to metallic surfaces and displace moisture
*Form:* Semisolid; Color: Dark; Odor: Ammonia
*Properties:* m.w.: 924; Solubility: Sol. in naphtha, mineral oil, isopropanol; Sp. gr.: 0.865 @ 65 C; M.P.: 77 F; B.P.: > 300 C; Flash Pt.: 455 F (COC)
*Composition:* 33% Duomeen T
*Toxicity:* Corrosive to skin and eyes; causes severe burns on prolonged contact
*Warnings and Caution:* Protective goggles and gloves must be worn
*References:* Bulletin 76-19

**Duomeen TX.** Akzo Chemie
*Chemical Description:* Tallow-1,3-diaminopropane
*Category:* Emulsifier, base material

*Applications:* Base material for ethloxylated products; emulsifier for bitumen emulsions
*Form:* Solid; Color: 7 (Gardner); Odor: Ammoniacal
*Properties:* Solubility: Insoluble in water @ 25 C; Sp. gr.: 0.84; M.P.: 46–55 C; Flash Pt.: 180 C; Iodine Value: 30–40
*Composition:* 92% min. active
*Toxicity:* Skin irritant, severe eye irritant
*Containers:* Net 165 kg open head steel drums
*Warnings and Caution:* Avoid contact with strong oxidizing agents and strong acids

**Du Pont Retarder LAN.** Du Pont
*Chemical Description:* Alkyl trimethylammonium halide
*Category:* Dye retardant, stripping agent, emulsifier, cationic
*Applications:* Dye retarding and stripping agent with azoic dyes; esp. in application of cationic dyes to acrylic fibers
*Form:* Liquid; Color: Dark red-brown
*Properties:* Solubility: Miscible with water; Stability: Stable to acids, alkalis and in storage; Surface Tension: 43.4 dynes/cm (0.1% sol'n.)
*Composition:* 20% active ingred.
*References:* Data Sheet No. 20

**Duponol 80.** Du Pont
*Chemical Description:* Sodium n-octyl sulfate technical
*Category:* Wetting agent, dispersant, penetrant, softening agent, emulsifier, anionic
*Applications:* Textiles, leather and metal processes such as mineral dyeing, leather tanning, soldering fluxes, metal pickling and cleaning compositions; highly surface active in electrolyte solutions of moderate to high concentrations
*Form:* Liquid; Color: Light yellow
*Properties:* m.w.: 232; Solubility: Sol. in presence of many multivalent ions; Visc.: 30 cps @ 27 C; Surface Tension: 51.7 dynes/cm (0.1% sol'n.)
*Composition:* 33–35% active
*References:* DS No. 9

**Duponol C.** Du Pont
*Chemical Description:* Sodium lauryl sulfate USP
*Category:* Emulsifier, dispersant, detergent, wetting agent, anionic
*Applications:* Cosmetic, dental and medical preparations; O/W emulsions; emollient creams, vanishing creams, hand lotions, brushless shaving creams, cold creams
*Form:* Spray-dried beads; Color: White; Odor: Mildly fatty
*Properties:* Solubility: 10–15 g in 100 ml water @ 25 C; Density: 3.3 lb/gal; Stability: Stable to alkali and in normal storage; hydrolyzes in strong acid solutions
*Composition:* 90–96% sodium lauryl sulfate
*Containers:* 51-gal Leverpak (150 lb net wt)
*Warnings and Caution:* Store in well-closed containers; corrosive to most metals

**Duponol D Paste.** Du Pont
*Chemical Description:* Sodium lauryl oleyl sulfate technical
*Category:* Detergent, emulsifier, wetting agent, dispersant, anionic
*Applications:* Textile industry for heavy duty scouring and dyeing; dispersing agent for dyes; scouring of shearlings; leather degreasing; scouring and conditioning of paper mill felts; in metal cleaning compositions
*Form:* Paste; Color: Light yellow
*Properties:* m.w.: 328; Surface Tension: 30.5 dynes/cm (0.1% sol'n.)
*Composition:* 38% active ingred.
*References:* DS No. 8

**Duponol EP.** Du Pont
*Chemical Description:* Diethanolamine lauryl sulfate (technical)
*Category:* Foaming agent, detergent, wetting agent, emulsifier, anionic
*Applications:* Cosmetics esp. clear liquid shampoos and pearlescent liquid cream lotion shampoos; as detergent in textile scouring esp. for fabrics with "Lycra" spandex yarn
*Form:* Slightly viscous liquid; Color: Very pale golden; Odor: Bland and clean
*Properties:* Solubility: Sol. in water in all proportions; in polar solvents with some electrolyte precipitation; Density: 8.4 lb/gal; Visc.: 50–150 cps @ 27 C; Cloud Pt.: 5 C; Stability: Stable to freezing, heat (150 F), uv light, aging
*Composition:* 33–36% active
*Warnings and Caution:* Some corrosion on iron, steel, aluminum
*References:* D-2955

**Duponol G.** Du Pont
*Chemical Description:* Fatty alcohol amine sulfate
*Category:* Emulsifier, wetting agent, anionic
*Applications:* Powerful oil- and wax-soluble emulsifier; wetting agent in water and aqueous electrolyte solutions; at low concentrations can be added to flammable solvents to increase the electrical conductance and reduce static-caused fire hazards
*Form:* Viscous translucent paste; Color: Yellow; Odor: Mild, fatty, penetrating
*Properties:* Solubility: Sol. in ethyl alcohol, ethyl acetate, Stoddard solvent, perchlorethylene, Cellosolve, toluene, mineral oil; at least 50% sol. in water; Density: 8.2 lb/gal; Stability: Stable to alkalis and in normal storage; hydrolyzes in strong acid; Surface Tension: 29.6 dynes/cm (0.1% sol'n.)
*Composition:* 92% active ingred.
*References:* Data sheet No. 12

**Duponol LS Paste.** Du Pont
*Chemical Description:* Sodium oleyl sulfate technical
*Category:* Scouring agent, rewetting agent, anionic
*Applications:* Scouring agent in textile and leather industries; as rewetting agent for paper; in washing paper mill felts
*Form:* Paste; Color: Light tan

*Properties:* m.w.: 355; Surface Tension: 37.8 dynes/cm (0.1% sol'n.)
*Composition:* 26% active ingred.
*References:* Data Sheet No. 13

**Duponol ME Dry Surface Active Agent.** Du Pont
*Chemical Description:* Sodium lauryl sulfate, technical (manufactured from synthetic fatty alcohols)
*Category:* Detergent, wetting agent, foaming agent, emulsifier, anionic
*Applications:* Detergent esp. effective for greasy soils; general industrial applications
*Form:* Powder; Color: White to medium cream
*Properties:* m.w.: 302; Solubility: 30% sol. in water; active ingred. sol. in methanol and ethanol; Density: 3.3 lb/gal; Cloud Pt.: 70 F (1% aqueous sol'n.); Stability: Stable to alkalis and in normal storage; hydrolyzes in acids; Surface Tension: 35 dynes/cm (0.1% sol'n.)
*Composition:* 90–96% active
*References:* Data Sheet No. 28

**Duponol QC.** Du Pont
*Chemical Description:* Based on sodium lauryl sulfate
*Category:* Shampoo base, anionic
*Applications:* Highly effective for clear liquid, liquid cream lotion and paste shampoos with excellent foaming properties
*Form:* Slightly viscous liquid; Color: Essentially water white; Odor: Bland, clean; characteristic lauryl alcohol sulfate
*Properties:* Solubility: Sol. in water in all proportions, in polar solvents with some electrolyte precipitation; Visc.: < 500 cps @ 80 F; Cloud Pt.: ≈ 65 F; Stability: Stable to freezing, uv light, aging; stable to heat 150 F for several hours
*Composition:* 29–30% active
*References:* Brochure D-3077

**Duponol RA.** Du Pont
*Chemical Description:* Fortified sodium ether-alcohol sulfate
*Category:* Detergent, dispersant, emulsifier, penetrating and leveling agent, wetting agent, anionic
*Applications:* Textile processing as dispersing, emulsifying, penetrating and leveling agents; as wetting agent in aqueous solutions at elevated temperatures; in scouring and dyeing of fabrics and yarns of natural and man-made fibers
*Form:* Clear liquid; Color: Light amber
*Properties:* Solubility: Miscible with water in all proportions; Density: 8.7 lb/gal; Sp. gr.: 1.04; Visc.: 178 cps (Brookfield @ 80 F); Cloud Pt.: 68 F; Flash Pt.: > 212 F (TOC); Stability: Stable to alkalis and in normal storage; stable to low concentrations in acid media; Surface Tension: 33 dynes/cm (0.1% sol'n.)
*References:* Brochure D-5205

**Duponol WA Dry Surface Active Agent.** Du Pont
*Chemical Description:* Sodium lauryl sulfate surfactant

*Category:* Detergent, emulsifier, wetting agent, penetrant, anionic
*Applications:* Detergent esp. effective on greasy soils; for dry mixing with alkaline detergent builders and general industrial applications; reduces surface tension of aqueous systems; produces dense, stable foam
*Form:* Powder; Color: White
*Properties:* Solubility: Sol. up to 40% product conc. in water; the active ingred. is sol. in highly polar solvents (e.g., methanol, ethanol); Density: 2.6 lb/gal; Cloud Pt.: 37 F (1% sol'n.); Stability: Stable to alkalis, in storage; hydrolyzes in acid conditions; Surface Tension: 33 dynes/cm (0.1% sol'n.)
*Composition:* 44% sodium alkyl sulfate

**Duponol WA Paste.** Du Pont
*Chemical Description:* Alcohol sulfate product
*Category and Applications:* See Duponol WAQA

**Duponol WAQA.** Du Pont
*Chemical Description:* Alcohol sulfate product
*Category:* Shampoo base
*Applications:* Shampoo formulations esp. liquid cream lotion, solid cream or paste shampoos; controlled degreasing action, superior cleansing and foaming

**Duponol WAQE.** Du Pont
*Chemical Description:* Sodium lauryl sulfate technical
*Category:* Emulsifier, anionic
*Applications:* Emulsion polymerization systems esp. preparation of plastomeric, elastomeric latices and dry resins
*Form:* Liquid; Color: Pale yellow
*Properties:* m.w.: 302; Solubility: Sol. in water in all proportions; sol. in polar solvents with some electrolyte precipitation; Density: 8.6 lb/gal; Visc.: 500 cps max. @ 27 C; Cloud Pt.: ≈ 65 F; Stability: Stable to freezing, aging; stable to heat (150 F) for several hours
*Composition:* 29–30.5% active
*Containers:* Perma-Plex drums; tank trucks, tank cars
*References:* Brochure D-2989

**Duponol WN.** Du Pont
*Chemical Description:* Sodium octyl/decyl sulfate, technical
*Category:* Wetting and penetrating agent, anionic
*Applications:* Wetting agent and penetrant esp. in aqueous solutions of high electrolyte content; leather tanning, electroplating, oil well acidizing, dyeing of vat colors, caustic impregnation of warps, wool chlorination, acid fulling, cleaning of fur pelts
*Form:* Liquid; Color: Pale yellow
*Properties:* m.w.: 246; Visc.: 30 cps @ 27 C; Surface Tension: 52.5 dynes/cm (0.1% sol'n.)
*Composition:* 33–35% active
*References:* Data Sheet No. 19

**Duponol XL.** Du Pont
*Category:* Shampoo surfactant, anionic-amphoteric

*Applications:* Shampoo formulations with exceptional foaming power; substantive to hair and skin; controlled degreasing
*Form:* Moderately viscous liquid; Color: Golden; Odor: Faintly spicy
*Properties:* Solubility: Sol. in water in all proportions; in polar solvents with some electrolyte precipitation; Visc.: 2500–4500 cps @ 80 F; Cloud Pt.: ≈ 50 F; Clear Pt.: 50–55 F; Stability: Stable to freezing, heat (150 F); to uv light, to aging
*Composition:* 36% solids content

**Dur-Em 104K.** Durkee Ind. Foods
*Chemical Description:* Mono- and diglycerides from hydrogenated vegetable oil with BHA and BHT, citric acid
*Category:* Food emulsifier, nonionic
*Applications:* Bakery, prepared mixes, margarines, convenience foods, frozen desserts; products designated K are available in kosher form
*Form:* Plastic; Color: Cream
*Properties:* M.P.: 115–121 F
*Composition:* 100% active ingred.

**Dur-Em 107K.** Durkee Ind. Foods
*Chemical Description:* Mono- and diglycerides from hydrogenated vegetable oil with BHA, BHT and citric acid
*Category and Applications:* See Dur-Em 104K
*Form:* Bead; Color: White
*Properties:* M.P.: 140–145 F
*Composition:* 100% active ingred.
*Containers:* 200 lb drums, bulk

**Dur-Em 114K.** Durkee Ind. Foods
*Chemical Description:* Mono- and diglycerides from hydrogenated vegetable oil with BHA, BHT and citric acid
*Category and Applications:* See Dur-Em 104K
*Form:* Plastic; Color: Cream
*Properties:* M.P.: 110–120 F
*Composition:* 100% active ingred.
*Containers:* 50 lb carton, bulk

**Dur-Em 117K.** Durkee Ind. Foods
*Chemical Description:* Mono- and diglycerides from hydrogenated vegetable oil with citric acid
*Category and Applications:* See Dur-Em 104K; also in chewing gums
*Form:* Beads, flakes; Color: White
*Properties:* M.P.: 145–150 F
*Containers:* 50 lb bags, 200 lb drums, bulk

**Dur-Em 127K.** Durkee Ind. Foods
*Chemical Description:* Mono- and diglycerides from hydrogenated vegetable oil with citric acid
*Category and Applications:* Peanut butter stabilizer
*Form:* Beads; Color: White
*Properties:* M.P.: 140–145 F
*Composition:* 100% active ingred.
*Containers:* 50 lb carton, bulk

**Dur-Em 136K.** Durkee Ind. Foods
*Chemical Description:* Mono- and diglycerides from hydrogenated

vegetable oils with BHA, BHT and citric acid
*Category and Applications:* See Dur-Em 104K; also for vegetable dairy systems
*Form:* Plastic
*Composition:* 100% active ingred.

**Dur-Em 187.** Durkee Ind. Foods
*Chemical Description:* Mono- and diglycerides from hydrogenated animal fats with citric acid
*Category and Applications:* See Dur-Em 104K
*Form:* Bead or flake; Color: White
*Properties:* M.P.: 140–145 F
*Composition:* 100% active ingred.
*Containers:* Bulk only

**Dur-Em 195.** Durkee Ind. Foods
*Chemical Description:* Mono- and diglycerides from hydrogenated animal and vegetable fats with BHA, BHT, citric acid
*Category and Applications:* Dry mix systems, dog foods, prepared mixes, continuous and conventional breads, cakes and sweet doughs
*Form:* Plastic; Color: Cream
*Properties:* M.P.: 123–129 F
*Composition:* 100% active ingred.
*Containers:* Bulk only

**Dur-Em 204K.** Durkee Ind. Foods
*Chemical Description:* Mono- and diglycerides from hydrogenated vegetable oil with BHA, BHT, citric acid
*Category and Applications:* See Dur-Em 104K
*Form:* Plastic; Color: Cream
*Properties:* M.P.: 120–130 F
*Composition:* 100% active ingred.
*Containers:* 50 lb carton, bulk

**Dur-Em 207-E.** Durkee Ind. Foods
*Chemical Description:* Mono- and diglycerides with citric acid
*Category and Applications:* For use directly in continuous and conventional breads, sweet doughs; emulsifier hydrate for breads; cosmetic/pharmaceutical uses
*Form:* Powder; Color: White
*Properties:* M.P.: 140–146 F; Emulsifiability: Self-emulsifying
*Composition:* 100% active ingred.
*Containers:* 50 lb carton

**Dur-Em 207K.** Durkee Ind. Foods
*Chemical Description:* Mono- and diglycerides from hydrogenated vegetable oil with citric acid
*Category and Applications:* See Dur-Em 104K
*Form:* Bead or plastic; Color: White
*Properties:* M.P.: 140–146 F
*Composition:* 100% active ingred.
*Containers:* 200 lb drum, 50 lb bag, bulk

**Dur-Em 227K.** Durkee Ind. Foods
*Chemical Description:* Mono- and diglycerides from hydrogenated vegetable oil with citric acid
*Category and Applications:* See Dur-Em 104K
*Form:* Bead
*Composition:* 100% active ingred.

**Dur-Em 287.** Durkee Ind. Foods
*Chemical Description:* Mono- and

diglycerides from hydrogenated animal fat with citric acid
*Category and Applications:* See Dur-Em 104K
*Form:* Plastic
*Composition:* 100% active ingred.

**Dur-Em 287E.** Durkee Ind. Foods
*Chemical Description:* Mono- and diglycerides from hydrogenated animal fat with citric acid
*Category:* Nonionic food emulsifier
*Applications:* Conventional and continuous breads, sweet doughs; emulsifier hydrate for breads and nonfat cakes
*Form:* Powder; Color: White
*Properties:* M.P.: 140–146 F
*Composition:* 100% active ingred.

**Dur-Em 295.** Durkee Ind. Foods
*Chemical Description:* Mono- and diglycerides from hydrogenated animal and vegetable fats with and without BHA, BHT and citric acid
*Category and Applications:* See Dur-Em 104K
*Form:* Plastic; Color: Cream
*Properties:* M.P.: 123–129 F
*Composition:* 100% active ingred.

**Duratex.** Durkee Ind. Foods
*Chemical Description:* Partially hydrogenated cottonseed oil
*Category:* Lubricant, release agent
*Applications:* Lubricant for tablets; release agent; candy dusting and coating for hydroscopic materials
*Form:* Powder
*Properties:* M.P.: 142–148 F
*Containers:* 50 lb carton

**Durfax 60K.** Durkee Ind. Foods
*Chemical Description:* Polyoxyethylene 20 sorbitan monostearate
*Category:* Nonionic food emulsifier
*Applications:* O/W emulsions in food industry
*Form:* Viscous liquid; Color: Yellow
*Properties:* Solubility: Sol. in water, and cottonseed oil (at low levels); HLB: 14.9
*Composition:* 100% active ingred.
*Containers:* 470 lb packaging, bulk

**Durfax 65K.** Durkee Ind. Foods
*Chemical Description:* Polyoxyethylene 20 sorbitan tristearate
*Category and Applications:* See Durfax 60K
*Form:* Solid; Color: Tan
*Properties:* Solubility: Sol. at low levels in cottonseed oil; dispersible in water
*Composition:* 100% active ingred.
*Containers:* 470 lb packaging, bulk

**Durfax 80K.** Durkee Ind. Foods
*Chemical Description:* Polyoxyethylene 20 sorbitan monooleate
*Category and Applications:* See Durfax 60K
*Form:* Liquid; Color: Yellow
*Properties:* Solubility: Sol. in water and at low levels in cottonseed oils; HLB: 15.9
*Composition:* 100% active ingred.
*Containers:* 470 lb packaging, bulk

**Durfax EOM K.** Durkee Ind. Foods
*Chemical Description:* Ethoxylated mono- and diglycerides
*Category:* Food emulsifier, nonionic
*Applications:* Dough conditioner in yeast-raised baked goods; cakes, icings
*Form:* Paste
*Properties:* M.P.: 80–85 F; HLB: 13.1
*Composition:* 100% active ingred.

**Durkee 07.** Durkee Ind. Foods
*Chemical Description:* Partially hydrogenated cottonseed oil
*Category:* Food additive
*Applications:* Peanut butter stabilizer, breads, shortenings, chewing gum, caramel coatings
*Form:* Flake or Beads
*Properties:* M.P.: 142–148 F
*Containers:* 200 lb drum, 50 lb bag, bulk

**Durkee 17.** Durkee Ind. Foods
*Chemical Description:* Partially hydrogenated soybean oil
*Category:* Food additive
*Applications:* Shortenings
*Form:* Flake
*Properties:* M.P.: 152–158 F
*Containers:* 50 lb bag, bulk

**Durkee 27.** Durkee Ind. Foods
*Chemical Description:* Partially hydrogenated palm oil
*Category:* Food additive
*Applications:* Peanut butter stabilizer
*Form:* Bead
*Properties:* M.P.: 136–143 F
*Containers:* 50 lb carton, bulk

**Durlac 100WK.** Durkee Ind. Foods
*Chemical Description:* Glyceryl lactoesters of fatty acids from hydrolyzed vegetable oil
*Category:* Food emulsifier, nonionic
*Applications:* Prepared mixes, whipped toppings, other vegetable dairy systems; cakes, shortenings
*Form:* Flake
*Properties:* M.P.: 115–130 F
*Composition:* 100% active
*Containers:* 50 lb bag

**Durlac 200K.** Durkee Ind. Foods
*Chemical Description:* Glyceryl lactoesters of fatty acids from vegetable oil
*Category and Applications:* See Durlac 100WK; also for cosmetic and pharmaceutical use
*Form:* Solid
*Properties:* M.P.: 97–103 F
*Composition:* 100% active ingred.
*Containers:* 50 lb carton, bulk

**Durlac 300K.** Durkee Ind. Foods
*Chemical Description:* Lactylated fatty acid esters of glycerin and propylene glycol (PGLS)
*Category:* Food emulsifier, nonionic
*Applications:* Vegetable dairy systems, cakes, cake mixes, aerating agent, cosmetic and pharmaceutical uses
*Form:* Flake; Color: Cream
*Properties:* M.P.: 95–105 F
*Composition:* 100% active ingred.
*Containers:* 50 lb carton, bulk

**Durpeg 400MO.** Durkee Ind. Foods
*Chemical Description:* Mono- and dioleate esters with polyethylene glycol from animal and vegetable sources
*Category:* Food emulsifier, nonionic
*Applications:* In calf milk replacers, processing aid in animal feeds
*Form:* Semi-solid; Color: Amber
*Properties:* HLB: 11.1
*Composition:* 100% active ingred.
*Containers:* 450 lb packaging

**Durpro 107K.** Durkee Ind. Foods
*Chemical Description:* Propylene glycol, mono- and diesters, and mono- and diglycerides with citric acid
*Category:* Food emulsifier, nonionic
*Applications:* Cake and cake mix emulsifier
*Form:* Flake; Color: Cream
*Properties:* M.P.: 117–125; HLB: 2.5
*Composition:* 100% active ingred.
*Containers:* 50 lb carton

**Durpro 187.** Durkee Ind. Foods
*Chemical Description:* Propylene glycol, mono- and diesters, and mono- and diglycerides
*Category:* Food emulsifier, nonionic
*Applications:* Cake and cake mix emulsifier
*Form:* Flake; Color: Cream
*Properties:* M.P.: 117–125 F; HLB: 2.5
*Composition:* 100% active ingred.

**Durtan 60K.** Durkee Ind. Foods
*Chemical Description:* Sorbitan monostearate
*Category:* Food emulsifier, nonionic
*Applications:* Cakes and cake mixes; chocolate compound coatings as gloss-enhancers; vegetable dairy systems
*Form:* Beads; Color: Cream
*Properties:* M.P.: 121–127 F; HLB: 4.7
*Composition:* 100% active ingred.

**Dusoran MD.** Van Schuppen
*Chemical Description:* Lanolin alcohols, distilled
*Category:* Stabilizer, softener, emulsifier, nonionic
*Applications:* When mixed with paraffins, it is a choice emulsifier for the preparation of water-in-oil emulsions; in absorption bases, nourishing creams (moisturizing and conditioning), emollient creams and lotions, hair creams, cleansing preparations
*Form:* Waxy solid; Color: Light yellow
*Properties:* Solubility: Sol. in mineral oils, abs. alcohol, alcohol 95%, chloroform, ether, light petroleum, toluene; Sapon. Value: 8.0; Acid No.: 2.0; Hydroxyl No.: 130–150
*Composition:* 100% active
*References:* Data sheet 33

**Dusoran NG.** Van Schuppen
*Chemical Description:* Wool and lanolin alcohols
*Category:* Stabilizer, softener, emulsifier
*Applications:* When mixed with paraffins, it is a choice emulsifier

for the preparation of water-in-oil emulsions; in absorption bases, nourishing creams (moisturizing and conditioning), lotions, hair creams, cleansing preparations, lipstick bases
*Form:* Solid; Color: Tan; Odor: Faint
*Properties:* Solubility: Sol. in mineral oils, abs. alcohol, alcohol 95%, chloroform, toluene; Sapon. Value: 10; Acid No.: 3.0; Hydroxyl No.: 120–180
*References:* Data sheet 31 B

**Dusoran NO.** Van Schuppen
*Chemical Description:* Wool and lanolin alcohols
*Category:* Stabilizer, softener, emulsifier
*Applications:* When mixed with paraffins, it is a choice emulsifier for the preparation of water-in-oil emulsions; in absorption bases, nourishing creams (moisturizing and conditioning), lotions, hair creams, cleansing preparations, lipstick bases
*Form:* Solid; Color: Light brown; Odor: Faint
*Properties:* Solubility: Sol. in mineral oils, abs. alcohol, alcohol 95%, chloroform, toluene; Sapon. Value: 10; Acid No.: 2; Hydroxyl No.: 120–180
*References:* Data sheet 31 A

**Dyesperse DC.** Hart Chem.
*Chemical Description:* Anionic naphthalene formaldehyde sulfonate
*Category:* Dispersant
*Applications:* Dispersant for disperse and acid dyes
*Form:* Liquid; Color: Amber
*Properties:* Refractive Index: 1.4028; Solubility: Clear sol.; Sp. gr.: 1.138; Visc.: 10 cps
*Composition:* 33% active ingred.
*Reaction:* Neutral
*Toxicity:* Skin and eye irritant
*Containers:* Drums, T/L
*Warnings and Caution:* Avoid contact with skin and eyes

**Dymsol 12-03-A.** Diamond Shamrock
*Category:* Anionic emulsifier
*Applications:* Biodegradable primary or secondary emulsifier for emulsion polymerization
*Form:* Liquid
*Composition:* 59% active ingred.

**Dymsol 31-P.** Diamond Shamrock
*Category:* Anionic surfactant, stabilizer
*Applications:* Biodegradable mechanical and chemical stabilizer for latex
*Form:* Liquid
*Properties:* Biodegradable
*Composition:* 55% active ingred.

**Dymsol 36-A.** Diamond Shamrock
*Category:* Anionic emulsifier
*Applications:* Biodegradable emulsifier for vinyl chloride monomer
*Form:* Liquid
*Composition:* 24.5% active ingred.

**Dymsol 38-C.** Diamond Shamrock
*Category:* Anionic emulsifier
*Applications:* Biodegradable secondary emulsifier to cut down

coagulum in SBR production
*Form:* Liquid
*Composition:* 70% active ingred.

**Dymsol 2031.** Diamond Shamrock
*Category:* Anionic emulsifier
*Applications:* Biodegradable primary or secondary emulsifier for emulsion polymerization
*Form:* Liquid
*Composition:* 60% active ingred.

**Dymsol B.** Diamond Shamrock
*Category:* Nonionic surfactant
*Applications:* Rubber anti-blocking agent
*Form:* Dispersion
*Composition:* 45% active ingred.

**Dymsol L.** Diamond Shamrock
*Category:* Nonionic stabilizer
*Applications:* Biodegradable stabilizer giving good wetting, controlled viscosity and sharp coagulation point
*Form:* Liquid
*Composition:* 100% active ingred.

**Dymsol LP.** Diamond Shamrock
*Category:* Anionic emulsifier
*Applications:* Biodegradable primary emulsifier for large particle size latices
*Form:* Liquid
*Properties:* Biodegradable
*Composition:* 33% active ingred.

**Dymsol N.** Diamond Shamrock
*Category:* Nonionic stabilizer
*Applications:* Biodegradable chemical stabilizer for latex
*Form:* Liquid
*Composition:* 97.5% active ingred.

**Dymsol S.** Diamond Shamrock
*Category and Applications:* Replacement for sulfated sperm oil fatliquors; tanning leather
*Form:* Liquid; Color: Greenish brown
*Containers:* 55 gal lined openhead steel drums
*References:* Bulletin LE-55 Rev.

**Dymsol TXM.** Diamond Shamrock
*Category:* Lubricant, nonionic
*Applications:* Polyester lubricant liquid for plastics industry
*Form:* Liquid
*Composition:* 100% active ingred.

**Dymsol VD.** Diamond Shamrock
*Category:* Anionic surfactant
*Applications:* Latex thickener
*Form:* Liquid
*Composition:* 15% active ingred.

**Dyqex.** Georgia-Pacific
*Chemical Description:* Sodium lignosulfonate
*Category:* Dye dispersant extender
*Applications:* Compatible with primary dispersants and contributes the additional advantages of low viscosity in use and of exceptional redispersibility of dye powders
*Form:* Powder; Color: Light

# E

**EC-25K.** Durkee Ind. Foods
*Chemical Description:* Propylene glycol mono- and diesters of fats and fatty acids, mono- and diglycerides, partially hydrogenated soybean oil with lecithin, BHA, citric acid
*Category:* Food emulsifier, nonionic
*Applications:* Cakes, mixes; used with all-purpose shortenings for baker's cakes or with emulsified shortenings for very lean cakes
*Form:* Plastic; Color: Light straw
*Properties:* M.P.: 90–100 F; HLB: 2.6
*Composition:* 100% active ingred.
*Containers:* 50 lb carton

**EC-117MK.** Durkee Ind. Foods
*Chemical Description:* Mono- and diglycerides and lactylic esters of fatty acids
*Category:* Food emulsifier, nonionic
*Applications:* Emulsified system for freeze-thaw coffee whitener, aerating composition, foam mat drying
*Form:* Beads; Color: Cream to light tan
*Properties:* M.P.: 123–133 F; HLB: 4.6
*Composition:* 100% active ingred.
*Containers:* 50 lb carton

**Ecconol 61.** Essential Chem.
*Chemical Description:* Foam-stabilized anionic-nonionic complex of amide sulfonate
*Category:* Detergent
*Applications:* Dishwash and carwash products; high-sudsing cleaners
*Form:* Liquid; Color: Amber
*Properties:* Solubility: Water sol.; Biodegradable
*Composition:* 67% active ingred.

**Ecconol 66.** Essential Chem.
*Chemical Description:* Fatty amide condensate
*Category:* Detergent, wetting agent, thickener, emulsifier, nonionic
*Applications:* Cleaning compounds and lubricants; liquid detergents
*Form:* Liquid; Color: Amber
*Properties:* Biodegradable

*Composition:* 100% active ingred.

**Ecconol 606.** Essential Chem.
*Chemical Description:* Organic alkylaryl sulfonate
*Category:* Detergent, wetting agent, anionic
*Applications:* Cleaning compounds; detergency, foaming, wetting, surface tension reducing properties for liquid formulations
*Form:* Liquid; Color: Amber
*Properties:* Biodegradable
*Composition:* 60% active ingred.

**Ecconol 628.** Essential Chem.
*Chemical Description:* Coconut oil amide
*Category:* Detergent, wetting agent, nonionic
*Applications:* Cleaning agents, thickener, foam stabilizer
*Form:* Liquid
*Properties:* Biodegradable
*Composition:* 100% active ingred.

**Ecconol B.** Essential Chem.
*Chemical Description:* Fatty alkanolamide
*Category:* Detergent, wetting agent, emulsifier, nonionic
*Applications:* Cleaning agents, balanced water-sol. base detergent for floor and wall cleaners, heavy-duty formulations
*Form:* Liquid
*Properties:* Biodegradable
*Composition:* 90% active ingred.

**Ekaline G-80 Flakes.** Sandoz
*Category:* Detergent, wetting agent, emulsifier, dispersant
*Applications:* General purpose scouring, emulsifying and leveling agent; promotes disperse dye migration on filiment nylon carpets; scouring/dyeing of wool for removal of residual grease and leveling of dyes; afterscouring disperse dyeings and prints esp. polyester
*Form:* Flakes; Color: White
*Properties:* Solubility: 20% in water
*Composition:* 100% active ingred.
*Reaction:* Neutral

**Ekaline G-80 Liquid.** Sandoz
*Category:* Detergent, wetting agent, emulsifier, dispersant
*Applications:* General purpose scouring, emulsifying and leveling agent; promotes disperse dye migration on filement nylon carpets; scouring/dyeing of wool for removal of residual grease and leveling of dyes; afterscouring disperse dyeings and prints, esp. polyester
*Form:* Liquid; Color: Colorless
*Properties:* Solubility: Miscible with water; Sp. gr.: 1.02
*Composition:* 20% active ingred.
*Reaction:* Acid

**Elfan 200 Pulver.** Akzo Chemie
*Chemical Description:* Laurylalcoholsulfate sodium salt (basis: native fatty alcohol $C_{12}$)
*Category:* Detergent
*Applications:* Cosmetics; tooth pastes
*Form:* Powder, beads; Color: White
*Properties:* Density: 220 g/l; Stability: In storage, protect against humidity

*Composition:* 90% active ingred.
*Reaction:* Neutral
*Warnings and Caution:* Avoid contact with acids because of hydrolysis

**Elfan 240 T/S.** Akzo Chemie
*Chemical Description:* Laurylalcohol sulfate, triethanolamine salt (basis: synthetic fatty alcohol $C_{12}/C_{14}$)
*Category:* Detergent
*Applications:* Skins and cosmetics applications; shampoos
*Form:* Liquid; Color: Yellowish
*Properties:* Solubility: Sol. in water; Sp. gr.: 1.04; Visc.: 150 cps @ 20 C; Stability: No special storage precautions
*Composition:* 40% active ingred.; 2% max. unsulfated parts
*Reaction:* Neutral
*Containers:* 20 ton containers; 125 kg drums
*Warnings and Caution:* Prevent contact with acids because of hydrolysis

**Elfan 260 S.** Akzo Chemie
*Chemical Description:* Lauryl alcohol sulfate sodium salt (basis: synthetic fatty alcohol $C_{12}/C_{16}$)
*Category:* Detergent, emulsifier
*Applications:* Textiles, skins, cosmetics uses as detergents (esp. hairwashing, pasty cleaners); emulsion polymerization
*Form:* Liquid, paste; Color: Yellowish
*Properties:* Solubility: Sol. in water @ 30 C min.; Sp. gr.: 1.04; Visc.: 200 cps @ 30 C; F.P.: 15 C
*Composition:* 30% active ingred.; 1.2% max. $Na_2SO_4$; 1.2% max. unsulfated parts
*Reaction:* Neutral
*Containers:* 125 kg drums, bulk
*Warnings and Caution:* Prevent contact with acids because of hydrolysis

**Elfan 280.** Akzo Chemie
*Chemical Description:* Coco fatty alcohol sulfate sodium salt from native fatty alcohol $C_{12}/C_{18}$
*Category:* Detergent, anionic
*Applications:* General purpose heavy duty detergents for textiles, skins, leather, laundering, dyeing, metallurgy, mines and minerals, cement, papermaking, paints, varnish, foods, transport, household and cosmetic uses; hand cleaning
*Form:* Paste; Color: Yellowish
*Properties:* Solubility: 700 g/l water @ 50 C; Sp. gr.: 1.05; Visc.: 8000 cps @ 50 C; pH: 8.5 ± 1
*Composition:* 42% active ingred.; other components: 1%max. inorganic salts, 1.5% max. unsulfated parts
*Reaction:* Acid
*Containers:* 125 kg drums
*Warnings and Caution:* Prevent contact with acids because of hydrolysis

**Elfan 680.** Akzo Chemie
*Chemical Description:* Oleyl-cetyl alcohol sulfate sodium salt from oleyl-cetyl alcohol
*Category:* Detergent, anionic
*Applications:* Powder detergents, pasty cleaning agents and detergents; textile uses
*Form:* Paste; Color: Yellow-brown

*Properties:* Solubility: < 10 g/l water @ 20 C; Sp. gr.: 1.05; M.P.: 40 C; Stability: No special precautions for storage; pH: 11 ± 1
*Composition:* 50% active ingred.; other components: 3% max. $Na_2SO_4$, 1.5% max. NaCl, 7% max. unsulfated parts
*Reaction:* Alkaline
*Containers:* 125 kg drums
*Warnings and Caution:* Prevent contact with acids because of hydrolysis

**Elfan A.** Akzo Chemie
*Chemical Description:* Betain-type surfactant
*Category:* Wetting agent, emulsifier, amphoteric
*Applications:* Cosmetics
*Form:* Liquid

**Elfan A 432.** Akzo Chemie
*Chemical Description:* Amphoteric compound of the betain type from lauric-myristic acid
*Category:* Detergent, amphoteric
*Applications:* Cosmetics and disinfecting cosmetics; baby shampoos, bubble baths
*Form:* Liquid; Color: Yellowish
*Properties:* Solubility: Water-sol.; Sp. gr.: 1.04; F.P.: < 0 C; Stability: Protect against frost; Isoelectric Pt.: 6.5; pH: 5 ± 1
*Composition:* 30% active ingred.; other components: 5% inorganic salt
*Reaction:* Acid
*Containers:* 20 ton containers; 125 kg drums

**Elfan KT 550.** Akzo Chemie
*Chemical Description:* Coco-tallow fatty alcohol sulfate sodium salt, mixing ratio 50:50 from native fatty alcohol $C_{12}/C_{18}$
*Category:* Detergent, anionic
*Applications:* Heavy duty detergent in powder form, washing and cleaning pastes; textiles, skins, leather, laundering, dyeing, mines and minerals, cement, papermaking, paints, foods, transport, household and cosmetic applications
*Form:* Paste; Color: Yellowish
*Properties:* Solubility: Poor sol. in water; Sp. gr.: 1.05; Visc.: 10,000 cps @ 50 C; Stability: No special precautions for storage; pH: 8 ± 0.5
*Composition:* 42% active ingred.; other components: 1% max. $Na_2SO4$, 0.4% max. NaCl; 1.5% max. unsulfated parts
*Reaction:* Alkaline
*Containers:* 125 kg drums
*Warnings and Caution:* Prevent contact with acids because of hydrolysis

**Elfan L 310.** Akzo Chemie
*Chemical Description:* Ethylene glycol distearate
*Category:* Additive for shampoos and bubble baths, anionic
*Applications:* Additive for shampoos and bubble baths (pearlescent effect)
*Form:* Solid, flakes; Color: Light yellow
*Properties:* Solubility: Dispersible in water; Sp. gr.: 0.96; M.P.: 64–66 C; Stability: No special precau-

tions; Sapon. Value: 200 max.; OH No.: ca. 30; Acid No.: 6 max.
*Reaction:* Neutral
*Containers:* Paperbags 25 kg

**Elfan NS.** Akzo Chemie
*Chemical Description:* Lauryl ether sulfates
*Category:* Detergent, wetting agent
*Applications:* Dishwashing detergents, shampoos; textiles, laundering, dyeing, papermaking, household, cosmetics uses
*Form:* Liquid or paste
*Composition:* 30 or 70% active

**Elfan NS 242.** Akzo Chemie
*Chemical Description:* Lauryl ether sulfate sodium salt from native fatty alcohol $C_{12}/C_{14}$ with ca. 2.5 mol EO
*Category:* Detergent, wetting agent
*Applications:* Liquid shampoos, bubble baths, dishwashing liquids; textiles, skins, household and cosmetic applications and in fire-extinguishers
*Form:* Paste; Color: Yellowish
*Properties:* Solubility: 400 g/l water; Visc.: 30,000 cps (@ 20 C), 15,000 cps (@ 40 C); F.P.: $< 0$ C; Stability: Store between 0–50 C; pH: 7.5 ± 0.5
*Composition:* 70% active ingred.; other components: 0.6% max. $Na_2SO_4$, 2% max. unsulfated parts
*Reaction:* Neutral
*Containers:* 20 ton containers
*Warnings and Caution:* Prevent contact with acids because of hydrolysis

**Elfan NS 243 S.** Akzo Chemie
*Chemical Description:* Lauryl ether sulfate sodium salt from synthetic fatty alcohol $C_{12}/C_{14}$ with ca. 3 mol. EO
*Category:* Detergent, wetting agent, anionic
*Applications:* Liquid shampoos, bubble baths, dishwashing liquids; textiles, skins, household, cosmetic applications and in fire extinguishers
*Form:* Liquid; Color: Clear
*Properties:* Solubility: Water sol.; Sp. gr.: 1.04; Visc.: 50 cps @ 20 C; F.P.: $< 0$ C; Stability: Store between 0 and 50 C; pH: 7 ± 0.5
*Composition:* 28% active ingred.; other components: 0.3% max. $Na_2SO_4$; 0.8% unsulfated parts
*Reaction:* Neutral
*Containers:* 20 ton containers
*Warnings and Caution:* Prevent contact with acids because of hydrolysis

**Elfan NS 243 S Conc.** Akzo Chemie
*Chemical Description:* Lauryl ether sulfate sodium salt from synthetic fatty alcohol $C_{12}/C_{14}$ with ca. 3 mol. EO
*Category:* Detergent, wetting agent, anionic
*Applications:* Liquid shampoos, bubble baths, dishwashing liquids; textiles, skins, household and cosmetic applications and in fire extinguishers
*Form:* Paste; Color: Yellowish
*Properties:* 400 g/l water; Sp. gr.: 1.12; Visc.: 30,000 cps (@ 20 C), 15,000 cps (@ 40 C); F.P.: $< 0$ C;

Stability: Store between 0 and 50 C; pH: 8.5 ± 1
*Composition:* 70% active ingred.; other components: 0.6% max. $Na_2SO_4$, 2% max. unsulfated parts
*Reaction:* Weakly alkaline
*Containers:* 20 ton containers
*Warnings and Caution:* Prevent contact with acids because of hydrolysis

**Elfan NS 252 S.** Akzo Chemie
*Chemical Description:* Lauryl ether sulfate sodium salt from synthetic fatty alcohol $C_{12}/C_{15}$ with ca. 2.5 mol. EO
*Category:* Detergent, wetting agent, anionic
*Applications:* Liquid shampoos, bubble baths, dishwashing liquids; textiles, skins, household and cosmetic applications and in fire-extinguishers
*Form:* Liquid; Color: Yellowish
*Properties:* Solubility: Water sol.; Sp. gr.: 1.05; Visc.: 150 cps @ 20 C; F.P.: < 0 C; Stability: Store between 0 and 50 C; pH: 7 ± 0.5
*Composition:* 28% active ingred.; other components: 0.3% max. $Na_2SO_4$, 0.8% max. unsulfated parts
*Reaction:* Neutral
*Containers:* 20 ton containers
*Warnings and Caution:* Prevent contact with acids because of hydrolysis

**Elfan NS 252 S Conc.** Akzo Chemie
*Chemical Description:* Lauryl ether sulfate sodium salt from synthetic fatty alcohol $C_{12}/C_{15}$ with ca. 2.5 mol. EO
*Category:* Detergent, wetting agent, anionic
*Applications:* Liquid shampoos, bubble baths, dishwashing liquids; textiles, skins, household and cosmetic uses and in fire extinguishers
*Form:* Paste; Color: Yellowish
*Properties:* Solubility: 400 g/l water; Sp. gr.: 1.11; Visc.: 30,000 cps (@ 20 C), 15,000 cps (@ 40 C); F.P.: < 0 C; Stability: Store between 0 and 50 C; pH: 8.5 ± 1
*Composition:* 70% active ingred.; other components: 0.5% max. $Na_2SO_4$, 2% max. unsulfated parts
*Reaction:* Alkaline
*Containers:* 20 ton containers
*Warnings and Caution:* Prevent contact with acids because of hydrolysis

**Elfan NS 423 SH.** Akzo Chemie
*Chemical Description:* Lauryl ether sulfate sodium salt from synthetic fatty alcohol $C_{12}/C_{14}$ with increased amount of $C_{14}$ with ca. 3 mol. EO
*Category:* Detergent, anionic
*Applications:* Dishwashing liquids; textiles, skins, household detergents
*Form:* Liquid; Color: Yellowish
*Properties:* Solubility: Water-sol.; Visc.: 100 cps @ 20 C; F.P.: < 0 C; Stability: Store between 0 and 50 C; pH: 7.5 ± 0.5
*Composition:* 28% active ingred.; other components: 0.5% max. $Na_2SO_4$, 0.8% max. unsulfated parts
*Reaction:* Neutral
*Container:* 20 ton containers
*Warnings and Caution:* Prevent

contact with acids because of hydrolysis

**Elfan NS 423 SH Conc.** Akzo Chemie
*Chemical Description:* Lauryl ether sulfate sodium salt from synthetic fatty alcohol $C_{12}/C_{14}$ with increased amount of $C_{14}$ with ca. 3 mol. EO
*Category:* Detergent, anionic
*Applications:* Textiles, skins, household detergents esp. dishwashing liquids
*Form:* Paste; Color: Yellowish
*Properties:* Solubility: 400 g/l water; Sp. gr.: 1.10; Visc.: 30,000 cps (@ 20 C), 15,000 cps (@ 40 C); F.P.: < 0 C; Stability: Store between 0 and 50 C; pH: 8.5 ± 1
*Composition:* 70% active ingred.; other components: 0.6% max. $Na_2SO_4$, 2% max. unsulfated parts
*Reaction:* Alkaline
*Containers:* 20 ton containers
*Warnings and Caution:* Prevent contact with acids because of hydrolysis

**Elfan OS.** Akzo Chemie
*Chemical Description:* Olefin sulfonates
*Category:* Detergent, wetting agents, emulsifier
*Applications:* Dishwashing detergents, shampoos; textiles, laundering, dyeing, household and cosmetic detergents; making of cutting and soluble oils
*Form:* Liquid
*Composition:* 30–40 % actives

**Elfan OS 46.** Akzo Chemie
*Chemical Description:* Alpha-olefin sulfonate sodium salt from alpha-olefin $C_{14}/C_{16}$
*Category:* Detergent, anionic
*Applications:* Shampoos, bubble baths, dishwashing liquids, liquid and pasty cleaning agents; skins, household and cosmetic cleaners
*Form:* Liquid; Color: Yellow
*Properties:* Solubility: Water sol.; Sp. gr.: 1.05; Visc.: 250 cps @ 20 C; F.P.: < 0 C; Stability: Store between 10 and 40 C; pH: 8 ± 1
*Composition:* 37% active ingred.; other components: 1% max. $Na_2SO_4$, 1.5% max. unsulfated parts
*Reaction:* Alkaline
*Containers:* 20 ton containers

**Elfan PM 600.** Akzo Chemie
*Chemical Description:* Surfactant concentrate from fatty alcohol sulfates with increased amount of lauric alcohol and special additives for the skin
*Category:* Detergent, anionic
*Applications:* Household cleaning, light duty detergents
*Form:* Paste; Color: Light yellow
*Properties:* Solubility: Poor sol. in water; Sp. gr.: 1.06; Stability: No special precautions: pH: 8.5 ± 1
*Composition:* 60% active ingred.; other components: 2% min. alkylol-amide; 6% max. inorganic salts; 7% max. unsulfated parts
*Reaction:* Alkaline
*Containers:* 125 kg drums

**Elfan SG 35.** Akzo Chemie
*Chemical Description:* Silver gloss concentrate from ether sulfate

and ethylene glycol distearate
*Category:* Detergent, anionic
*Applications:* Cosmetics; shampoo additive with silver gloss; base for pearlescent shampoos and bubble bath
*Form:* Paste; Color: White
*Properties:* Solubility: Dispersible in water; Sp. gr.: 1.04; Visc.: 10,000 cps @ 20 C; Stability: Store between 10 and 40 C; pH: 8.5 ± 1
*Composition:* 37% active ingred.; other components: water
*Reaction:* Alkaline
*Containers:* 125 kg drums

**Elfan SP 500.** Akzo Chemie
*Chemical Description:* Surfactant concentrate from ether sulfate and alkyl benzene sulfonate with a nonionic surfactant
*Category:* Detergent, anionic
*Applications:* Household cleaners, all purpose cleaners, liquid dishwashing
*Form:* Liquid; Color: Yellowish
*Properties:* Solubility: 250 g/l water; Sp. gr.: 1.05; Visc.: 5000 cps @ 20 C; F.P.: < 0 C; Stability: Protect against frost; pH: 7 ± 0.5
*Composition:* 50% active ingred.; other components: water
*Reaction:* Neutral
*Containers:* 20 ton containers, 125 kg drums

**Elfan WA.** Akzo Chemie
*Chemical Description:* Dodecylbenzene sulfonate sodium salt from n-dodecylbenzene
*Category:* Detergent, emulsifier, anionic
*Applications:* Heavy duty detergent, dishwashing liquids; textiles and household cleaning; emulsion polymerization
*Form:* Paste; Color: Yellowish
*Properties:* Solubility: 250 g/l water @ 20 C; Sp. gr.: 1.08; F.P.: < 0 C; M.P.: 5 C; Stability: Store in tanks between 80–90 C; pH: 9 ± 1
*Composition:* 50% active ingred.; other components: 1% max. $Na_2SO_4$, 1% max. unsulfated parts
*Reaction:* Alkaline
*Containers:* 20 ton tanks, 125 kg drums

**Elfan WA Pulver.** Akzo Chemie
*Chemical Description:* Dodecylbenzene sulfonate sodium salt from n-dodecylbenzene
*Category:* Detergent
*Applications:* Heavy duty detergent for household and textiles uses
*Form:* Powder, beads; Color: Light-yellowish
*Properties:* Solubility: 100 g/l water @ 20 C; Density: 300 g/l; Stability: Hygroscopic; store dry; pH: 8–9
*Composition:* 80% active ingred.; other components: 2% max. $Na_2SO_4$, 18% $Na_2CO_3$, 1% max. unsulfated parts
*Reaction:* Alkaline
*Containers:* Paperbag 25 kg

**Elfan WA Sulfosäure.** Akzo Chemie
*Chemical Description:* Dodecylbenzene sulfonic acid from n-dodecylbenzene
*Category:* Raw material

*Applications:* Raw material for inorganic and organic neutralisates
*Form:* Liquid; Color: Dark brown; Odor: $SO_3$
*Properties:* Solubility: Water sol.; Sp. gr.: 1.06; Visc.: 500 cps @ 20 C; Flash Pt.: 150 C (Pensky-Martens, closed cup); Stability: Store preferaby at room temperature; Acid No.: 185 max.
*Composition:* 98% active ingred.; other components: 1.5% max. $H_2SO_4$, 1.8% unsulfated parts
*Reaction:* Acid
*Toxicity:* Skin and eye irritant
*Containers:* 20 ton containers
*Warnings and Caution:* Etching and corrosive to metals; severely irritating liquid

**Elfan WAT.** Akzo Chemie
*Chemical Description:* Dodecylbenzene sulfonate triethanolamine salt from n-dodecylbenzene
*Category:* Detergent, emulsifier, anionic
*Applications:* Liquid detergents for textile and household uses; emulsifier for insecticides
*Form:* Liquid; Color: Yellow-brown
*Properties:* Solubility: Water sol.; Sp. gr.: 1.06; Visc.: 1000 cps @ 20 C; F.P.: $<$ 0 C; Stability: No special precautions for storage; pH: 6 ± 0.5
*Composition:* 50% active ingred.; other components: 0.8% max. unsulfated parts
*Reaction:* Acid
*Containers:* 20 ton containers, 125 kg drums

**Elfanol.** Akzo Chemie
*Chemical Description:* Semiesters of succinic acid with ethoxylated fatty alcohols, etc.
*Category:* Detergent, wetting agent
*Applications:* Shampoos, hand cleaners; textiles, laundering, dyeing, papermaking, household and cosmetic cleaners; as wetting agents for bleaching, dyeing, coating
*Form:* Liquid, paste
*Composition:* 40–60% actives

**Elfanol 510.** Akzo Chemie
*Chemical Description:* Sulfosuccinic acid monoester of a fatty acid alkylol amide sodium salt; basis: tallow fatty acid monoethanolamide
*Category:* Detergent, anionic
*Applications:* Light duty detergents and Syndet bars; detergents for textiles, skins, leather, laundering, dyeing, metallurgy, mining, cement, papermaking, paints, foods, transport, household and cosmetic industries
*Form:* Paste; Color: Light yellow
*Properties:* Solubility: Poor water sol.; Sp. gr.: 1.06; Stability: Not stable @ pH $>$ 8 and pH $<$ 6; pH: 6.5 ± 1
*Composition:* 50% active ingred.; other components: water
*Reaction:* Acid
*Containers:* 125 kg drums

**Elfanol 616.** Akzo Chemie
*Chemical Description:* Sulfosuccinic acid monoester of an ethoxylated fatty alcohol sodium salt;

basis: synthetic fatty alcohol $C_{12}$/ $C_{14}$ with ca. 3 mol. EO
*Category:* Detergent, anionic
*Applications:* Cosmetics, shampoos, bubble and baby baths
*Form:* Liquid; Color: Clear
*Properties:* Solubility: Water sol.; Sp. gr.: 1.05; Visc.: 100 cps @ 20 C; F.P.: 0 C; Stability: Not stable at pH > 8 or pH < 6; pH: 7
*Composition:* 30% active ingred.; other components: 0.05% max. $Na_2SO_4$
*Reaction:* Neutral
*Containers:* 125 kg drums, 2 ton containers

**Elfanol 850.** Akzo Chemie
*Chemical Description:* Sulfosuccinic acid monoester of an ethoxylated fatty acid alkylolamide sodium salt; basis: ethoxylated coco fatty acid monoethanol amide
*Category:* Detergent, anionic
*Applications:* Hair shampoos, baby shampoos, bubble baths, other cosmetics
*Form:* Liquid; Color: Yellow-brown
*Properties:* Solubility: Poor water sol.; Sp. gr.: 1.06; F.P.: 14 C; Stability: Not stable at pH > 8 or pH < 6; pH: 7
*Composition:* 45% active ingred.
*Reaction:* Neutral
*Containers:* 125 kg drums

**Elfanol 883.** Akzo Chemie
*Chemical Description:* Sulfosuccinic acid diisooctylester sodium salt from 2-ethylhexanol
*Category:* Wetting agent, anionic
*Applications:* Wetting agent for technical processes
*Form:* Liquid; Color: Yellowish
*Properties:* Solubility: 15 g/l water @ 20 C; Sp. gr.: 1.08; Visc.: 50 cps @ 20 C; F.P.: < 0 C; Flash Pt.: 25 C (Pensky-Martens, closed cup); Stability: Not stable at pH > 8 or pH < 6; pH: 7
*Composition:* 63% active ingred.; other components: 6% isopropylalcohol; other: water
*Reaction:* Neutral
*Containers:* 125 kg drums
*Warnings and Caution:* Product contains isopropyl alcohol

**Elfapur.** Akzo Chemie
*Chemical Description:* Elfapur n ethoxylated alkyl phenols. Elfapur ethoxylated amines, amides, alcohols
*Category:* Detergent, wetting agent, emulsifier
*Applications:* Detergent formulations for textiles, laundering, dyeing, household, cosmetics industries; as wetting agents in bleaching, dyeing, waterproofing; as emulsifier for making of cutting and soluble oils, dispersing waxes, pigments and resins
*Form:* Liquid, paste

**Elfapur KA 45.** Akzo Chemie
*Chemical Description:* Coco fatty acid amide polyglycolether from coco fatty acid monoethanolamide
*Category:* Detergent, nonionic
*Applications:* Detergent for textiles, household esp. liquid dishwashing

*Form:* Liquid; Color: Yellowish
*Properties:* Solubility: 300 g/l water @ 20 C; Sp. gr.: 1.05; F.P.: 10–15 C; Cloud Pt.: 74–76 C (1% in water); Flash Pt.: 180 C (Pensky-Martens closed cup); Stability: No special precautions
*Composition:* 88% active ingred.; other components: 3% max. PEG
*Reaction:* Neutral
*Containers:* 125 kg drums, 20 ton containers

**Elfapur LM 25.** Akzo Chemie
*Chemical Description:* Lauryl alcohol polyglycol èther from native fatty alcohol $C_{12}/C_{14}$
*Category:* Raw material
*Applications:* Raw material for anionic surfactants esp. ethersulfates
*Form:* Liquid; Color: Yellowish
*Properties:* Solubility: Poor water sol.; Sp. gr.: 1.01; F.P.: 5 C; Cloud Pt.: 56–60 C (10% in 25% butyldiglycol); Flash Pt.: 180 C (P-M closed cup); OH No.: 187 ± 2; Stability: No special precautions
*Composition:* 97% active ingred.; other components: 3% max. PEG
*Reaction:* Neutral
*Containers:* 125 kg drums

**Elfapur LM 30 S.** Akzo Chemie
*Chemical Description:* Lauryl alcohol polyglycol ether from synthetic fatty alcohol $C_{12}/C_{14}$
*Category:* Raw material
*Applications:* Raw material for anionic surfactants esp. ethersulfates
*Form:* Liquid; Color:Yellowish
*Properties:* Solubility: Poor water sol.; F.P.: 5 C; Cloud Pt.: 60–64 C (10% in 25% butyl diglycol); OH No.: 170 ± 2; Stability: No special precautions
*Composition:* 97% active ingred.; other components: 3% max. PEG
*Reaction:* Neutral
*Containers:* 125 kg drums

**Elfapur LM 75 S.** Akzo Chemie
*Chemical Description:* Lauryl alcohol polyglycol ether from synthetic fatty alcohol $C_{12}/C_{14}$
*Category:* Detergent, nonionic
*Applications:* Detergent for textiles and household uses
*Form:* Liquid; Color: Yellowish
*Properties:* Solubility: 300 g/l water @ 20 C; Sp. gr.: 1.00; F.P.: 10 C; M.P.: 10–15 C; Cloud Pt.: 60–62 C (1% in water); Stability: No special precautions
*Composition:* 87% active ingred.; other components: 3% max. PEG, 10% water; also available water-free
*Reaction:* Neutral
*Containers:* 125 kg drums

**Elfapur LP 25 S.** Akzo Chemie
*Chemical Description:* Lauryl alcohol polyglycol ether from synthetic fatty alcohol $C_{12}/C_{15}$
*Category:* Raw material
*Applications:* Raw material for anionic surfactants esp. ethersulfates
*Form:* Liquid; Color: Yellowish
*Properties:* Solubility: Poor water sol.; Sp. gr.: 1.01; F.P.: 5 C;

Cloud Pt.: 50–54 C (10% in 25% butyl diglycol); Flash Pt.: 180 C (P-M closed cup); OH No.: 179 ± 2; Stability: No special precautions
*Composition:* 97% active ingred.; other components: 3% PEG
*Reaction:* Neutral
*Containers:* 125 kg drums

**Elfapur ML 30 SH.** Akzo Chemie
*Chemical Description:* Lauryl alcohol polyglycol ether from synthetic fatty alcohol $C_{12}/C_{14}$ with increased amount of $C_{14}$
*Category:* Raw material
*Applications:* Raw material for anionic surfactants esp. ethersulfates
*Form:* Liquid; Color: Yellowish
*Properties:* Solubility: Poor water sol.; Sp. gr.: 1.01; F.P.: 5 C; Cloud Pt.: 62–66 C (10% in 25% butyl diglycol); Flash Pt.: 180 C (P-M closed cup); OH No. 165 ± 2; Stability: No special precautions
*Composition:* 97% active ingred.; other components: 3% max. PEG
*Reaction:* Neutral
*Containers:* 125 kg drums

**Elfapur N 50.** Akzo Chemie
*Chemical Description:* Nonylphenol polyglycol ether
*Category:* Detergent, emulsifier, nonionic
*Applications:* Detergent for textiles, household uses
*Form:* Liquid; Color: Yellowish
*Properties:* Solubility: Poor water sol.; Sp. gr.: 1.05; F.P.: –5 C; Cloud Pt.: 60–62 C (10% in 25% butyl diglycol); Flash Pt.: 90 C (P-M closed cup); Stability: No special precautions
*Composition:* 97% active ingred.; other components: 3% max. PEG
*Reaction:* Neutral
*Containers:* 125 kg drums, 20 ton containers

**Elfapur N 70.** Akzo Chemie
*Chemical Description:* Nonylphenol polyglycol ether
*Category:* Detergent, emulsifier
*Applications:* Household and textile applications
*Form:* Liquid; Color: Yellowish; Gel at 50–70% active ingred.
*Properties:* Solubility: 300 g/l water @ 20 C; Sp. gr.: 1.05; F.P.: 0 C; Cloud Pt.: 22–24 C (1% in water); Flash Pt.: 103 C (P-M, closed cup); Stability: No special precautions
*Composition:* 97% active ingred.; other components: 3% max. PEG
*Reaction:* Neutral
*Containers:* 125 kg drums, 20 ton containers

**Elfapur N 90.** Akzo Chemie
*Chemical Description:* Nonylphenol polyglycol ether
*Category:* Detergent, emulsifier, nonionic
*Applications:* Textiles and household applications
*Form:* Liquid; Color: Yellowish; Gel at 50–70% active ingred.
*Properties:* Solubility: 300 g/l water @ 20 C; Sp. gr.: 1.06; F.P.: 15 C; Cloud Pt.: 60–62 C (1% in water); Flash Pt.: 114 C (P-M, closed cup); Stability: No special

precautions
*Composition:* 97% active ingred.; other components: 3% max. PEG
*Reaction:* Neutral
*Containers:* 125 kg drums, 20 ton containers

**Elfapur N 120.** Akzo Chemie
*Chemical Description:* Nonylphenol polyglycol ether
*Category:* Detergent, emulsifier
*Applications:* Textile and household detergents; liquid dishwashing and all purpose cleaners
*Form:* Liquid; Color: Yellowish; Gel at 50–70% active ingred.
*Properties:* Solubility: 300 g/l water @ 20 C; Sp. gr.: 1.07; M.P.: 13–18 C; Cloud Pt.: 88–90 C (1% in water); Flash Pt.: 102 C (P-M closed cup); Stability:No special precautions
*Composition:* 97% active ingred.; 3% max. PEG
*Reaction:* Neutral
*Containers:* 125 kg drums, 20 ton containers

**Elfapur N 150.** Akzo Chemie
*Chemical Description:* Nonylphenol polyglycol ether
*Category:* Detergent, emulsifier, nonionic
*Applications:* Textiles, household cleaning
*Form:* Paste; Color: Yellowish; Gel at 40–70% active ingred.
*Properties:* Solubility: 400 g/l water @ 20 C; Sp. gr.: 1.07; M.P.: 25–30 C; Cloud Pt.: 64–66 C (1% in 10% NaCl); Flash Pt.: 94 C (P-M, closed cup); Stability: No special precautions
*Composition:* 97% active ingred.; other components: 3% max. PEG
*Reaction:* Neutral
*Containers:* 20 ton containers, 125 kg drums

**Elfapur O 80.** Akzo Chemie
*Chemical Description:* Oleyl-cetyl alcohol polyglycol ether from oleyl-cetyl alcohol
*Category:* Detergent, nonionic
*Applications:* Textiles industry; foam controlled heavy duty detergent
*Form:* Liquid; Color: Yellow-brown; Gel at 15–90% active
*Properties:* Solubility: 400 g/l water @ 20 C; Sp. gr.: 1.05; M.P.: 25–30 C; Cloud Pt.: 44–46 C (1% in water); Iodine No.: 22; Stability: No special precautions
*Composition:* 97% active ingred.; other components: 3% max. PEG
*Reaction:* Neutral
*Containers:* 125 kg drums, 20 ton containers

**Elfapur O 160.** Akzo Chemie
*Chemical Description:* Oleyl-cetyl alcohol polyglycol ether
*Category:* Detergent
*Applications:* Textiles industry
*Form:* Paste; Color: Yellowish; Gel at 30–80% active ingred.
*Properties:* Solubility: 400 g/l water @ 20 C; Sp. gr.: 1.06; M.P.: 29–35 C; Cloud Pt.: 64–66 C (1% in 10% NaCl); Stability: No special precautions
*Composition:* 97% active ingred.; other components: 3% max. PEG

*Reaction:* Neutral
*Containers:* 125 kg drums, 20 ton containers

**Elfapur T 110.** Akzo Chemie
*Chemical Description:* Tallow fatty alcohol polyglycol ether
*Category:* Detergent, nonionic
*Applications:* Foam controlled heavy duty detergent powders; textile use
*Form:* Paste; Color: Yellowish; Gel at 30–80% active ingred.
*Properties:* Solubility: 300 g/l water @ 20 C; Sp. gr.: 1.05; M.P.: 35–40 C; Cloud Pt.: 77–79 C (1% in water); Flash Pt.: 125 C (P-M, closed cup); Stability: 70 C max. for containers
*Composition:* 97% active ingred.; other components: 3% max. PEG
*Reaction:* Neutral
*Containers:* 125 kg drums, 20 ton containers

**Elfapur T 250 Pulver.** Akzo Chemie
*Chemical Description:* Tallow fatty alcohol polyglycol ether
*Category:* Detergent, nonionic
*Applications:* Foam controlled heavy duty detergent, scouring powder; textiles industry
*Form:* Powder, beads; gel at active ingred. > 25%; Color: White
*Properties:* Solubility: 250 g/l water @ 20 C; Density: 300 g/l; Cloud Pt.: 77–79 C (1% in 10% NaCl); Stability: No special precautions
*Composition:* 98% active ingred.; Other components: 2% max. PEG
*Reaction:* Neutral
*Containers:* 25 kg paperbags

**Elfugin ANT.** Sandoz Ltd.
*Chemical Description:* Phosphoric ester
*Category:* Antistatic, anionic
*Applications:* Antistatic agent
*Form:* Liquid

**Elfugin KNP.** Sandoz Ltd.
*Chemical Description:* Amine derivative
*Category:* Antistatic, nonionic/cationic
*Applications:* Antistatic agent
*Form:* Liquid

**Elfugin V.** Sandoz Ltd.
*Chemical Description:* Ethylene oxide adduct
*Category:* Antistatic, amphoteric
*Applications:* Antistatic agent for synthetic fibers
*Form:* Liquid/paste

**Eltesol AX 40.** Albright & Wilson Ltd. Marchon Div.
*Chemical Description:* Ammonium xylene sulfonate
*Category:* Hydrotrope, anionic
*Applications:* In manufacture of heavy duty detergent powders to reduce viscosity of the slurry before spray-drying; as cloud point depressants in formulating light duty liquid detergents; as solubilizers and coupling agents in heavy duty detergent formulation
*Form:* Liquid; Color: Pale yellow
*Properties:* pH: 7–8.5
*Composition:* 41% active ingred.; other components: 1.5% max. inorganic sulfate; balance water
*Reaction:* Alkaline

**Eltesol PT 93.** Albright & Wilson Ltd. Marchon Div.
*Chemical Description:* Potassium toluene sulfonate
*Category and Applications:* See Eltesol AX 40
*Form:* Powder; Color: Off white
*Properties:* pH: 9–10.5
*Composition:* 93% min. active ingred.; other components: 4.5% max. inorganic sulfate; balance water
*Reaction:* Alkaline

**Eltesol PX 93.** Albright & Wilson Ltd. Marchon Div.
*Chemical Description:* Potassium xylene sulfonate
*Category and Applications:* See Eltesol AX 40
*Form:* Powder; Color: Off white
*Properties:* pH: 9–10.5
*Composition:* 93% min. active ingred.; 4.5% max. inorganic sulfate; balance water
*Reaction:* Alkaline

**Eltesol ST 34.** Albright & Wilson Ltd. Marchon Div.
*Chemical Description:* Sodium toluene sulfonate
*Category and Applications:* See Eltesol AX 40
*Form:* Liquid; Color: Pale yellow
*Properties:* pH: 7–10
*Composition:* 34% active ingred.; other components: 4.5% max. inorganic sulfate; balance water
*Reaction:* Alkaline

**Eltesol ST 90.** Albright & Wilson Ltd. Marchon Div.
*Chemical Description:* Sodium toluene sulfonate
*Category and Applications:* See Eltesol AX 40
*Form:* Powder; Color: Off white (also available as pellets)
*Properties:* pH: 9–10.5
*Composition:* 89% min. active ingred.; other components: 8.5% max. inorganic sulfate; balance water
*Reaction:* Alkaline

**Eltesol SX 30.** Albright & Wilson Ltd. Marchon Div.
*Chemical Description:* Sodium xylene sulfonate
*Category and Applications:* See Eltesol AX 40
*Form:* Liquid; Color: Pale yellow
*Properties:* pH: 7–10
*Composition:* 30% active ingred.; other components: 1.5% max. inorganic sulfate; balance water
*Reaction:* Alkaline

**Eltesol SX 93.** Albright & Wilson Ltd. Marchon Div.
*Chemical Description:* Sodium xylene sulfonate
*Category and Applications:* See Eltesol AX 40
*Form:* Powder; Color: Off white (also available in pellet form)
*Properties:* pH: 9–10.5
*Composition:* 93% min. active ingred.; other components: 4.5% max. inorganic sulfate; balance water
*Reaction:* Alkaline

**Eltesol TA65.** Albright & Wilson Ltd. Marchon Div.
*Chemical Description:* Toluene sulfonic acid
*Category:* Intermediate and catalyst
*Applications:* As hardening agents in plastics industry; as activators for nicotine insecticides; for metal descaling and in resin-bound sand castings; as catalysts in the preparation of esters, in acetylation
*Form:* Clear liquid; Color: Light amber
*Properties:* Density: 1.2 g/cc; Visc.: 9–12 cs
*Composition:* 65.5% sulfonic acid

**Eltesol TA96.** Albright & Wilson Ltd. Marchon Div.
*Chemical Description:* Toluene sulfonic acid
*Category and Applications:* See Eltesol TA65
*Form:* Thick crystalline liquid; Color: Brown
*Properties:* Density: 1.3 g/cc @ 50 C
*Composition:* 97% sulfonic acid

**Eltesol XA65.** Albright & Wilson Ltd. Marchon Div.
*Chemical Description:* Xylene sulfonic acids
*Category and Applications:* See Eltesol TA65
*Form:* Clear liquid; Color: Light amber
*Properties:* Density: 1.2 g/cc; Visc.: 9–12 cs
*Composition:* 65% sulfonic acid

**Eltesol XA/M65.** Albright & Wilson Ltd. Marchon Div.
*Chemical Description:* Xylene sulfonic acid
*Category and Applications:* See Eltesol TA65
*Form:* Slightly hazy liquid; Color: Light amber
*Properties:* Density: 1.2 g/cc; Visc.: 9–12 cs
*Composition:* 65% sulfonic acid

**Emcol 27-21.** Witco Chem.
*Chemical Description:* Blend of oil soluble sulfonate and polyoxyethylene ethers
*Category:* Emulsifier, anionic/nonionic
*Applications:* Emulsifier for Ortho-Dichlorobenzene
*Form:* Liquid
*Composition:* 100% active ingred.

**Emcol 27-21N.** Witco Chem.
*Chemical Description:* Blend of oil soluble sulfonate and polyoxyethylene ethers
*Category:* Emulsifier, anionic/nonionic
*Applications:* Emulsifier for Ortho-Dichlorobenzene
*Form:* Liquid
*Composition:* 100% active ingred.

**Emcol 64-6.** Witco Chem.
*Chemical Description:* Blend of nonionic surfactants
*Category:* Emulsifier
*Applications:* For pesticide emulsifiable concentrations
*Form:* Liquid

*Composition:* 100% active ingred.

**Emcol 64-6A.** Witco Chem.
*Chemical Description:* Blend of nonionic surfactants
*Category:* Emulsifier
*Applications:* For pesticide emulsifiable concentrations
*Form:* Liquid
*Composition:* 100% active ingred.

**Emcol 64-45D.** Witco Chem.
*Chemical Description:* Blend of anionic and nonionic surfactants
*Category:* Emulsifier
*Applications:* For pesticide emulsifiable concentrates
*Form:* Liquid
*Composition:* 100% active ingred.

**Emcol 4100M.** Witco Chem. Organics Div.
*Chemical Description:* Disodium sulfosuccinate
*Category:* Foaming agent, cleansing agent, anionic
*Applications:* Bubble bath and shampoo formulations; carpet and upholstery cleaners; textiles
*Form:* Liquid
*Properties:* Solubility: Completely water sol.

**Emcol 4161L.** Witco Chem. Organics Div.
*Chemical Description:* Disodium sulfosuccinate
*Category:* Foaming agent, detergent
*Applications:* Bubble bath and shampoo formulations; cleansing agents for cosmetics and toiletries

**Emcol 4300.** Witco Chem. Organics Div.
*Chemical Description:* Disodium sulfosuccinate
*Category:* Foaming agent, detergent
*Applications:* Bubble bath and shampoo formulations; cosmetics and toiletries cleansing agent

**Emcol 4500.** Witco Chem. Organics Div.
*Chemical Description:* Sodium dioctyl sulfosuccinate
*Category:* Wetting agent, detergent, dispersant
*Applications:* Cosmetics and toiletries; textiles
*Properties:* Solubility: Completely water sol.

**Emcol 4560.** Witco Chem. Organics Div.
*Chemical Description:* Modified sodium dioctyl sulfosuccinate
*Category:* Detergent and water solubilizer
*Applications:* Drycleaning

**Emcol 4600.** Witco Chem. Organics Div.
*Chemical Description:* Sodium bis (tridecyl) sulfosuccinate
*Category:* Detergent, foam modifier, wetting agent, dispersant
*Applications:* Drycleaning; aqueous aerosol formulations; dispersing agent for pigments in oil-based systems

**Emcol AC 62-36.** Witco Chem.
*Chemical Description:* Blend of anionic and nonionic surfactants
*Category:* Emulsifier
*Applications:* Emulsifier for arsenic based herbicides
*Form:* Liquid
*Composition:* 100% active ingred.

**Emcol AC 62-37.** Witco Chem.
*Chemical Description:* Blend of anionic and nonionic surfactants
*Category and Applications:* See Emcol AC 62-36
*Form:* Liquid
*Composition:* 100% active ingred.

**Emcol AC 62-38.** Witco Chem.
*Chemical Description:* Blend of anionic and nonionic surfactants
*Category and Applications:* See Emcol AC 62-36
*Form:* Liquid
*Composition:* 100% active ingred.

**Emcol AD 5-06.** Witco Chem.
*Chemical Description:* Blend of oil soluble sulfonate and polyoxyethylene ethers
*Category:* Emulsifier, anionic/nonionic
*Applications:* For pesticide emulsifiable concentrates
*Form:* Liquid
*Composition:* 100% active ingred.

**Emcol AD 6-89A.** Witco Chem.
*Chemical Description:* Blend of oil soluble sulfonate and polyoxyethylene ethers
*Category:* Emulsifier, anionic/nonionic
*Applications:* For pesticide emulsifiable concentrates
*Form:* Liquid
*Composition:* 100% active ingred.

**Emcol AG 4-48N.** Witco Chem.
*Chemical Description:* Blend of oil soluble sulfonate and polyoxyethylene ethers
*Category:* Emulsifier, anionic/nonionic
*Applications:* For pesticide concentrates
*Form:* Liquid
*Composition:* 100% active ingred.

**Emcol AK 16-11.** Witco Chem.
*Chemical Description:* Blend of nonionic surfactants
*Category:* Emulsifier
*Applications:* Emulsifier for the direct solubilization of DDVP in water
*Form:* Liquid
*Composition:* 100% active ingred.

**Emcol AK 16-11N.** Witco Chem.
*Chemical Description:* Blend of nonionic surfactants
*Category:* Emulsifier
*Applications:* Emulsifier for the direct solubilization of DDVP in water
*Form:* Liquid
*Composition:* 100% active ingred.

**Emcol AK 16-95.** Witco Chem.
*Chemical Description:* Blend of nonionic surfactants
*Category:* Emulsifier
*Applications:* For pesticide emulsifiable concentrates
*Form:* Liquid
*Composition:* 100% active ingred.

**Emcol AK 16-97N.** Witco Chem.
*Chemical Description:* Blend of oil soluble sulfonate and polyoxyethylene ethers
*Category:* Emulsifier, anionic/nonionic
*Applications:* For pesticide emulsifiable concentrates
*Form:* Liquid
*Composition:* 100% active ingred.

**Emcol AK 18-35.** Witco Chem.
*Chemical Description:* Blend of oil soluble sulfonate and polyoxyethylene ethers.
*Category:* Emulsifier, anionic/nonionic
*Applications:* For pesticide emulsifiable concentrates
*Form:* Liquid
*Composition:* 100% active ingred.

**Emcol AK 18-72A.** Witco Chem.
*Chemical Description:* Blend of anionic and nonionic surfactants
*Category:* Emulsifier
*Applications:* Specific emulsifier for DDVP emulsifiable concentrates
*Form:* Liquid
*Composition:* 100% active ingred.

**Emcol AL 26-43.** Witco Chem.
*Chemical Description:* Blend of anionic and nonionic surfactants
*Category:* Emulsifier
*Applications:* For pesticide emulsifiable concentrates
*Form:* Liquid
*Composition:* 100% active ingred.

**Emcol AL 69-49.** Witco Chem.
*Chemical Description:* Blend of nonionic surfactants
*Category:* Emulsifier
*Applications:* Emulsifier for pesticide emulsifiable concentrates where anionic surfactants would decompose the toxicant, e.g., for dimethoate
*Form:* Liquid
*Composition:* 100% active ingred.

**Emcol AP 08-21.** Witco Chem.
*Chemical Description:* Blend of anionic and nonionic surfactants
*Category:* Emulsifier
*Applications:* Emulsifier for emulsifiable concentrates of highly concentrated pesticide blends
*Form:* Liquid
*Composition:* 100% active ingred.

**Emcol AP 08-22.** Witco Chem.
*Chemical Description:* Blend of anionic and nonionic surfactants
*Category:* Emulsifier
*Applications:* Emulsifier for emulsifiable concentrates of highly concentrated pesticide blends
*Form:* Liquid
*Composition:* 100% active ingred.

**Emcol AP 08-22C.** Witco Chem.
*Chemical Description:* Blend of anionic and nonionic surfactants
*Category:* Emulsifier
*Applications:* Emulsifier for emulsifiable concentrates of highly concentrated pesticide blends
*Form:* Liquid
*Composition:* 100% active ingred.

**Emcol AP 08-34A.** Witco Chem.
*Chemical Description:* Blend of anionic and nonionic surfactants

*Category:* Emulsifier
*Applications:* Emulsifier for pesticide emulsifiable concentrates
*Form:* Liquid
*Composition:* 100% active ingred.

**Emcol CC-9.** Witco Chem. Organics Div.
*Chemical Description:* Polypropoxylated quaternary ammonium chloride
*Category:* Dispersant, emulsifier, conditioner, detergent, antistat, wetting agent, penetrant, lubricant, cationic
*Applications:* Cosmetica and toiletries; industrial surfactant; textile industry; antistat and detergent for cosmetic aerosols; pigment dispersant in aqueous systems
*Form:* Liquid
*Composition:* 100% conc.

**Emcol CC-36.** Witco Chem. Organics Div.
*Chemical Description:* Polypropoxylated quaternary ammonium chloride
*Category:* Dispersant, antistat, detergent, emulsifier, conditioner, wetting agent, penetrant, lubricant, cationic
*Applications:* Cosmetics and toiletries; textile industry; aerosol formulations; pigment dispersant in aq. or nonaq. systems
*Form:* Liquid
*Composition:* 100% conc.

**Emcol CC-37-18.** Witco Chem. Organics Div.
*Chemical Description:* Coco betaine
*Category:* Foaming agent and stabilizer, viscosity modifier, amphoteric
*Applications:* Foaming agent and stabilizer for cosmetics and toiletries; viscosity modifier for cationic, anionic or nonionic formulations
*Form:* Liquid

**Emcol CC-42.** Witco Chem. Organics Div.
*Chemical Description:* Polypropoxylated quaternary ammonium chlorides
*Category:* Emulsifier, conditioner, dispersant, detergent, antistat, cationic
*Applications:* Drycleaning detergent; industrial and textile surfactant; cosmetics and toiletries
*Form:* Liquid

**Emcol CC-55.** Witco Chem. Organics Div.
*Chemical Description:* Polypropoxylated quaternary ammonium acetate
*Category:* Experimental cationic surfactant
*Applications:* Industrial, cosmetics and toiletries applications

**Emcol CC-57.** Witco Chem. Organics Div.
*Chemical Description:* Polypropoxylated quaternary ammonium phosphate
*Category:* Experimental cationic surfactant; detergent, antistat
*Applications:* Drycleaning surfactant, industrial uses, cosmetics and toiletries

**Emcol CS-136, CS-143, CS-151, and CS-165.** Witco Chem.
*Chemical Description:* Organic phosphate ethers
*Category:* Detergents, emulsifiers, dispersants, solubilizers, anionic
*Applications:* Series range from highly lipophilic to highly hydrophilic for use as detergents, emulsifiers, dispersants and solubilizers for nonionic surfactants
*Form:* Liquid

**Emcol CS-1361.** Witco Chem.
*Chemical Description:* Sodium salt of complex organo-phosphate ester
*Applications:* Used for scale control
*Form:* Clear liquid
*Properties:* Solubility: Sol. in mineral oil, kerosene, heavy aromatic naphtha, xylene, perchlorethylene, ethanol, carbon tetrachloride, water; dispersible in 10% caustic soda; Sp. gr.: 1.10; Pour Pt.: 40 F; Acid No.: 5 (to pH 9.5); Surface Tension: 31.9 dynes/cm (0.05%); pH: 5 (3% aq.)
*Composition:* 10% max. moisture

**Emcol D 24-25.** Witco Chem.
*Chemical Description:* Calcium sulfonate
*Category:* Emulsifier
*Applications:* Oil-in-water emulsifier for industrial use

**Emcol E-607L.** Witco Chem. Organics Div.
*Chemical Description:* Lapyrium chloride
*Category:* Emulsifier, conditioner, emollient, foaming, cleansing and substantive agent, cationic
*Applications:* Cosmetics and toiletries; industrial applications; cosmetic aerosol formulations
*Form:* Powder
*Composition:* 90% conc.

**Emcol E-607S.** Witco Chem. Organics Div.
*Chemical Description:* Stearoyl analog of Emcol E-607L
*Category:* Emulsifier, conditioner, emollient, detergent, substantive agent, cationic
*Applications:* Cosmetics and toiletries; aerosol formulations
*Form:* Powder
*Composition:* 90% conc.

**Emcol H-400X.** Witco Chem.
*Chemical Description:* Anionic/nonionic blend
*Form:* Slightly viscous liquid; Color: Clear to dark amber
*Properties:* Sp. gr.: 1.03 @ 20 C; pH: 6–7 (3% disp.)

**Emcol H-470X.** Witco Chem.
*Chemical Description:* Blend of oil soluble sulfonate and polyoxyethylene ethers
*Category:* Emulsifier
*Applications:* Emulsifier for pesticide emulsifiable concentrates
*Form:* Liquid
*Composition:* 100% active ingred.

**Emcol N-83T.** Witco Chem.
*Chemical Description:* Blend of oil soluble sulfonate and polyoxyethylene ethers
*Category:* Emulsifier

*Applications:* Emulsifier for pesticide emulsifiable concentrates
*Form:* Liquid
*Composition:* 100% active ingred.

**Emcol N-85T.** Witco Chem.
*Chemical Description:* Blend of oil soluble sulfonates and polyoxyethylene ethers
*Category:* Emulsifier
*Applications:* Emulsifier for pesticide emulsifiable concentrates
*Form:* Liquid
*Composition:* 100% active ingred.

**Emcol N-140B.** Witco Chem.
*Chemical Description:* Blend of oil soluble sulfonates and polyoxyethylene ethers
*Category:* Emulsifier
*Applications:* Emulsifier for pesticide emulsifiable concentrates
*Form:* Hazy liquid; Color: Dark amber
*Properties:* Sp. gr.: 1.01–1.03; pH: 6–7
*Composition:* 100% active ingred.

**Emcol N-141B.** Witco Chem.
*Chemical Description:* Blend of oil soluble sulfonates and polyoxyethylene ethers
*Category:* Emulsifier
*Applications:* Emulsifier for pesticide emulsifiable concentrates
*Form:* Hazy liquid; Color: Dark amber
*Properties:* Sp. gr.: 1.01–1.03; pH: 6–7
*Composition:* 100% active ingred.

**Emcol N-300B.** Witco Chem.
*Chemical Description:* Blend of oil soluble sulfonates and polyoxyethylene ethers
*Category:* Emulsifier, anionic/nonionic
*Applications:* Emulsifier for pesticide emulsifiable concentrates bal. toward high anionic content
*Form:* Slightly viscous liquid; Color: Clear to dark amber
*Properties:* Sp. gr.: 1.03 @ 20 C; pH: 6–7
*Composition:* 100% active ingred.

**Emcol N-369.** Witco Chem.
*Chemical Description:* Blend of oil soluble sulfonates and polyoxyethylene ethers
*Category:* Emulsifier
*Applications:* Emulsifier for pesticide emulsifiable concentrates
*Form:* Liquid
*Composition:* 100% active ingred.

**Emcol N-469.** Witco Chem.
*Chemical Description:* Blend of oil soluble sulfonates and polyoxyethylene ethers
*Category:* Emulsifier
*Applications:* Emulsifier for pesticide emulsifiable concentrates
*Form:* Liquid
*Composition:* 100% active ingred.

**Emcol N-500B.** Witco Chem.
*Chemical Description:* Blend of oil soluble sulfonates and polyoxyethylene ethers
*Category:* Emulsifier, anionic/nonionic
*Applications:* Emulsifier for pesticide emulsifiable concentrates balanced toward high nonionic con-

tent
*Form:* Slightly viscous liquid; Color: Clear to dark amber
*Properties:* pH: 6–7
*Composition:* 100% active ingred.

**Emcol N-710.** Witco Chem.
*Chemical Description:* Blend of oil soluble sulfonates and polyoxyethylene ethers
*Category:* Emulsifier
*Applications:* Emulsifier for pesticide emulsifiable concentrates
*Form:* Liquid
*Composition:* 100% active ingred.

**Emcol N-712.** Witco Chem.
*Chemical Description:* Blend of oil soluble sulfonates and polyoxyethylene ethers
*Category:* Emulsifier
*Applications:* Emulsifier for pesticide emulsifiable concentrates
*Form:* Liquid
*Composition:* 100% active ingred.

**Emcol N-714.** Witco Chem.
*Chemical Description:* Blend of oil soluble sulfonates and polyoxyethylene ethers
*Category:* Emulsifier
*Applications:* Emulsifier for pesticide emulsifiable concentrates
*Form:* Liquid
*Composition:* 100% active ingred.

**Emcol P10-20 BU.** Witco Chem.
*Chemical Description:* Calcium alkylarylsulfonate
*Category:* Intermediate
*Applications:* Intermediate for agricultural emulsifiers
*Form:* Viscous liquid; Color: Amber
*Properties:* Solubility: Oil sol.; Sp. gr.: 1.01–1.03; pH: 6.0–7.5 (3% in 20% isopropanol)
*Composition:* 68–70% active

**Emcol P-1045.** Witco Chem.
*Chemical Description:* Amine salt of alkyl-aryl sulfonic acid
*Category:* Wetting agent, dispersant, anionic
*Applications:* Wetting agent in oil-based systems, e.g., dry cleaning, floor wax removal, emulsion polymerization; sludge dispersant in fuel oil
*Form:* Liquid
*Properties:* Solubility: Oil sol., water dispersible
*Composition:* 90% active ingred.

**Emcol P-1049.** Witco Chem.
*Chemical Description:* Amine salt of alkyl-aryl sulfonic acid
*Category:* Anionic surfactant
*Applications:* Exceptionally water soluble surfactant for hard-to-clean applications
*Form:* Liquid
*Properties:* Solubility: Excellent water sol.
*Composition:* 60% active ingred.

**Emcol P-1052.** Witco Chem.
*Category:* Intermediate, anionic
*Applications:* Intermediate for agricultural emulsifier
*Form:* Viscous liquid; Color: Dark brown
*Properties:* pH: 6–6.4
*Composition:* 71–72% active

**Emcol P-1059.** Witco Chem.
*Chemical Description:* Amine salt of alkyl-aryl sulfonic acid
*Category:* Anionic surfactant
*Applications:* Surfactant for O/W emulsions, cleaning and degreasing
*Form:* Liquid
*Composition:* 90% active ingred.

**Emcol P-1059B.** Witco Chem.
*Chemical Description:* Amine salt of alkyl-aryl sulfonic acid
*Category:* Anionic surfactant
*Applications:* Dry cleaning surfactant; water solubilizer
*Form:* Liquid
*Composition:* 90% active ingred.

**Emcol P-1073.** Witco Chem.
*Chemical Description:* Amine salt of alkyl-aryl sulfonic acid
*Category:* Anionic surfactant
*Applications:* For viscosity control in invert emulsion drilling mud
*Form:* Viscous cloudy liquid; Color: Dark brown
*Properties:* Solubility: Oil sol.
*Composition:* 90% active ingred.

**Emcol PS-331.** Witco Chem.
*Chemical Description:* Organic phosphate ester
*Category:* Anionic surfactant
*Applications:* Surfactant for water based systems, e.g., detergents, hydrotropes, emulsion polymerization
*Form:* Liquid

**Emcol PS-400.** Witco Chem.
*Chemical Description:* Organic phosphate ester
*Category:* Anionic surfactant
*Applications:* Surfactant for water based systems, e.g., detergents, hydrotropes, emulsion polymerization
*Form:* Liquid

**Emcol PS-415.** Witco Chem.
*Chemical Description:* Organic phosphate ester
*Form:* Liquid
*Category:* Anionic surfactant
*Applications:* Surfactant for water based systems, e.g., detergents, hydrotropes, emulsion polymerization

**Emcol TS-230.** Witco Chem.
*Chemical Description:* Organic phosphate ester
*Category:* Dispersant, lubricant, emulsifier
*Applications:* Dispersant and lubricant in oil drilling muds; in water-based fluids during drilling and post-completion well treatment; O/W emulsifier; flow control agent
*Form:* Clear liquid; Color: Amber
*Properties:* Solubility: Sol. in xylene, ethanol, carbon tetrachloride, water, 10% caustic soda; Sp. gr.: 1.22; Pour Pt.: < 50 F; Acid No.: 170 (to pH 9.5); Surface Tension: 49.6 dynes/cm (0.05%); pH: 2 (3% aq.)

**Emcol ZM-1117.** Witco Chem.
*Chemical Description:* Blend of oil soluble surfactants and polyoxyethylene ethers
*Category:* Emulsifier, anionic/nonionic

*Applications:* Emulsifier for pesticide emulsifiable concentrates
*Form:* Liquid
*Composition:* 100% active ingred.

**Emcon E.** Fanning Chem.
*Chemical Description:* Mixture of glycerides, phosphatides, and egg-derived sterols
*Category:* Emulsifier, nonionic
*Applications:* Cosmetic/pharmaceutical emulsions; excellent W/O emulsifier to form bases with water absorbing properties and add emollience to formulations; fat liquoring; leather, foods, household applications
*Form:* Liquid; Color: Yellow; Odor: Bland
*Properties:* Solubility: Mineral oil, vegetable oil, most organic solvents; Density: 9.95
*Composition:* 100% active ingred.
*Reaction:* Neutral
*Toxicity:* Slightly irritating; excellent emollient
*Containers:* Steel drums

**Emphos CS-141.** Witco Chem.
*Chemical Description:* Organic phosphate ester
*Category:* Emulsifier, lubricant, wetting agent, dispersant, solubilizer, corrosion inhibitor, antistatic agent, anionic
*Applications:* Hard-surface cleaners, emulsion polymerization, textile wet processing, dry cleaning detergents, cosmetic creams and lotions, metal degreasing, paper making, pesticide formulation, corrosion inhibitors, drilling mud fluids
*Form:* Clear viscous liquid
*Properties:* Solubility: Sol. in heavy aromatic naphtha, xylene, perchlorethylene, ethanol, carbon tetrachloride, water, 10% caustic soda; partially sol. in kerosene; Sp. gr.: 1.12; Pour Pt.: > 40 F; Acid No.: 110 (to pH 9.5); Surface Tension: 30.6 dynes/cm (0.05% sol'n.); pH: 2 (3% aq.)
*Warnings and Caution:* Acid–protect with rubber gloves and goggles while handling
*References:* Bulletin 234-CS141

**Emphos CS-143.** Witco Chem.
*Chemical Description:* Phosphoric acid ester
*Category:* Emulsifier for Kelthane
*Form:* Clear liquid
*Properties:* Acid No.: 150 mg KOH/g (to pH 9.5); pH: 2.5–3 (3% aq.)
*Reaction:* Acid

**Emphos CS-151.** Witco Chem.
*Chemical Description:* Organic phosphate ester
*Category:* Emulsifier, lubricant, wetting agent, dispersant, solubilizer, corrosion inhibitor, antistat, anionic
*Applications:* Hard surface cleaners, emulsion polymerization, textile wet processing, dry cleaning detergents, cosmetic creams and lotions, metal degreasing, paper making, pesticide formulation, corrosion inhibitors, drilling mud fluids
*Form:* Clear viscous liquid
*Properties:* Solubility: Sol. in heavy aromatic naphtha, xylene,

perchlorethylene, ethanol, carbon tetrachloride, water, 10% caustic soda; partially sol. in kerosene; Sp. gr.: 1.11; Pour Pt.: > 50 F; Acid No.: 95 (to pH 9.5); Surface Tension: 36.8 dynes/cm (0.05%); pH: 2 (3% aq.)
*Reaction:* Acid
*Warnings and Caution:* Acid; protect during handling with rubber gloves and goggles
*References:* Bulletin 234-CS151

**Emerest 2301.** Emery Industries
*Chemical Description:* Methyl oleate
*Category:* Lubricant, defoamer, floatation agent, plasticizer, nonionic
*Applications:* As components in mold release agents, defoamers, flotation agents, plasticizers for cellulosic plastics, needle lubricants, bases for industrial lubricants; when sulfated, as wetting, rewetting and dye leveling agents for textiles and leather
*Form:* Liquid; Color: < 6 (Gardner)
*Properties:* Solubility: Sol. in IPA, toluol, mineral oil; Density: 7.3 lb/gal; Visc.: 5 cs @ 38 C; F.P.: –16 C; Flash Pt.: 177 C; Sapon. Value: 195; Acid No.: 4
*Composition:* 100% active ingred.

**Emerest 2302.** Emery Industries
*Chemical Description:* Propyl oleate
*Category and Applications:* See Emerest 2301
*Form:* Liquid; Color: 7 max. (Gardner)
*Properties:* Solubility: Sol. in IPA, toluol, mineral oil; Density: 7.2 lb/gal; Visc.: 6 cs @ 38 C; F.P.: –20 C; Sapon. Value: 175; Acid No.: 6
*Composition:* 100% active ingred.

**Emerest 2314.** Emery Industries
*Chemical Description:* Isopropyl myristate
*Category:* Emollient, lubricant, nonionic
*Applications:* Cosmetic emollient; sewing thread lubricant
*Form:* Liquid; Color: 1 max. (Gardner)
*Properties:* Solubility: Sol. in IPA, toluol, mineral oil; Density: 7.1 lb/gal; Visc.: 6 cs @ 38 C; F.P.: –5 C; Flash Pt.: 152 C; Sapon. Value: 210; Acid No.: 1
*Composition:* 100% active ingred.

**Emerest 2316.** Emery Industries
*Chemical Description:* Isopropyl palmitaite (from high purity palmitic acid)
*Category:* Lubricant, emollient, nonionic
*Applications:* Lubricant for synthetic fibers; cosmetic emollient
*Form:* Liquid; Color: 1 max. (Gardner)
*Properties:* Solubility: Sol. in IPA, toluol, mineral oil; Density: 7.1 lb/gal; Visc.: 8 cs @ 38 C; F.P.: 14 C; Flash Pt.: 163 C; Sapon. Value: 188; Acid No.: 1
*Composition:* 100% active ingred.

**Emerest 2326.** Emery Industries
*Chemical Description:* Butyl stearate
*Category:* Lubricant, nonionic
*Applications:* Used in aluminum foil rolling, wire drawing; in products used in synthetic fiber spin finishing, fiber and yarn processing
*Form:* Liquid; Color: 1 max. (Gardner)
*Properties:* Solubility: Sol. in IPA, toluol, mineral oil; Density: 7.2 lb/gal; Visc.: 10 cs; F.P.: 20 C; Flash Pt.: 188 .C; Sapon. Value: 172; Acid No.: 1
*Composition:* 100% active ingred.

**Emerest 2350.** Emery Industries
*Chemical Description:* Ethylene glycol monostearate
*Category:* Opacifying and pearlescing agent, nonionic
*Applications:* Liquid cosmetic and detergent compounds as opacifier and pearlescing agent
*Composition:* 100% active ingred.

**Emerest 2355.** Emery Industries
*Chemical Description:* Ethylene glycol distearate
*Category:* Opacifier, nonionic
*Applications:* Liquid detergent and cosmetic products as opacifier
*Form:* Beads; Color: 1 (Gardner)
*Properties:* Solubility: Sol. in IPA, toluol, mineral oil; F.P.: 50 C; HLB: 2.1; Sapon. Value: 185; Acid No.: 8
*Form:* Flake; Color: 4 (Gardner)
*Properties:* Solubility: Dispersible in mineral oil; F.P.: 62 C; HLB: 1.2; Sapon. Value: 195; Acid No.: 6
*Composition:* 100% active ingred.

**Emerest 2381.** Emery Industries
*Chemical Description:* Propylene glycol monostearate
*Category:* Pearlescing and spreading agent, anionic
*Applications:* Imparts pearlescence to cosmetic and dishwashing products; as spreading agent in hand lotions
*Form:* Beads; Color: 2 (Gardner)
*Properties:* Solubility: Sol. in IPA; dispersible in water, mineral oil, toluol (hot); F.P.: 40 C; HLB: 4.0; Sapon. Value: 170; Acid No.: 20
*Composition:* 100% active ingred.

**Emerest 2400.** Emery Industries
*Chemical Description:* Glycerol monostearate
*Category:* Emulsifier, nonionic
*Applications:* In hand creams and other cosmetics; in industrial lubricants and lubricant softeners for textiles
*Form:* Beads; Color: 1 (Gardner)
*Properties:* Solubility: Sol. in IPA; sol. hot in toluol, mineral oil; F.P.: 57.5 C; Sapon. Value: 173; Acid No. 3; Iodine Value: 0.5
*Composition:* 100% active ingred.

**Emerest 2401.** Emery Industries
*Chemical Description:* Glycerol monostearate
*Category:* Emulsifier, nonionic
*Applications:* Technical grade of Emerest 2400 with similar functions

*Form:* Beads; Color: 2 (Gardner)
*Properties:* Solubility: Sol. in IPA; sol. hot in toluol, mineral oil; F.P.: 58 C; Sapon. Value: 153; Acid No.: 3; Iodine Value: 1
*Composition:* 100% active ingred.

**Emerest 2407.** Emery Industries
*Chemical Description:* Glycerol monostearate, SE
*Category:* Emulsifier, modified anionic
*Applications:* Cosmetic creams and lotions
*Form:* Beads; Color: 3 (Gardner)
*Properties:* Solubility: Sol. in IPA, sol. hot in toluol, mineral oil; F.P.: 58 C; Sapon. Value: 153; Acid No.: 20; Iodine Value: 1; Emulsifiability: Self-emulsifying
*Composition:* 100% active ingred.

**Emerest 2421.** Emery Industries
*Chemical Description:* Glycerol monooleate
*Category:* Emulsifier, nonionic
*Applications:* As a component in mold release agents, a vehicle for agricultural insecticides, an anti-icing fuel additive, a rust preventive additive for compounded oils; in textiles as a lubricant component in synthetic fiber spin finishes
*Form:* Liquid; Color: 11 (Gardner)
*Properties:* Solubility: Sol. in toluol, mineral oil; Density: 7.9 lb/gal; Visc.: 91 cs @ 38 C; F.P.: 6 C; Flash Pt.: 235 C; Sapon. Value: 170; Acid No.: 6; Iodine Value: 73
*Composition:* 100% active ingred.

**Emerest 2423.** Emery Industries
*Chemical Description:* Glycerol trioleate
*Category:* Lubricant, emulsifier, nonionic
*Applications:* Called synthetic olive oil; when emulsified, as lubricant for metals, leather, textiles; sulfated GTO used as a softener for leathers and textiles
*Form:* Liquid; Color: 3 (Gardner)
*Properties:* Solubility: Sol. in toluol, mineral oil; Density: 7.6 lb/gal; Visc.: 47 cs; F.P.: 9 C; Flash Pt.: 293 C; Sapon. Value: 197; Acid No.: 5; Iodine Value: 85
*Composition:* 100% active ingred.

**Emerest 2620.** Emery Industries
*Chemical Description:* PEG-200 monolaurate
*Category:* Emulsifier, coupling agent, defoamer, nonionic
*Applications:* Defoamer in water base coatings; viscosity depressant in vinyl plastisols; viscosity control additive in hair rinse formulations; paper softener
*Form:* Liquid; Color: 1 (Gardner)
*Properties:* Solubility: Dispersible in water, mineral oil, butyl stearate, glycerol trioleate, perchloroethylene, Stod. solvent; Density: 8.2 lb/gal; Visc.: 40 cs; HLB: 9.8; Cloud Pt.: < 25 C
*Composition:* 100% active ingred.

**Emerest 2622.** Emery Industries
*Chemical Description:* PEG-200 dilaurate
*Category:* Coemulsifier, lubricant, nonionic

*Applications:* Coemulsifier and lubricant in self-emulsifiable textile and industrial oils; mold release agent; viscosity control agent; in specialty paper coatings
*Form:* Liquid; Color: 2 (Gardner)
*Properties:* Solubility: Dispersible in water; sol. in mineral oil, butyl stearate, glycerol trioleate, perchloroethylene, Stod. solvent; Density: 8.0 lb/gal; Visc.: 35 cs; HLB: 6.8; Cloud Pt.: < 25 C
*Composition:* 100% active ingred.

**Emerest 2640.** Emery Industries
*Chemical Description:* PEG-400 monostearate
*Category:* Emulsifier, lubricant, nonionic
*Applications:* Emulsifier of oils and fats in the manufacture of fluid or paste emulsions for industrial lubricants, consumer products, textile lubricants and softeners; as thickening agent and stabilizer for starch coatings on paper; as water-dispersible paper size; lubricant for channeling wire through conduit
*Form:* Soft waxy solid; Color: < 1 (Gardner)
*Properties:* Solubility: Dispersible in water, mineral oil, butyl stearate, glycerol trioleate, perchloroethylene, Stod. solvent; M.P.: 32 C; HLB: 11.7; Cloud Pt.: < 25 C
*Composition:* 100% active ingred.

**Emerest 2642.** Emery Industries
*Chemical Description:* PEG-400 distearate
*Category:* Emulsifier, thickener, nonionic
*Applications:* Cosmetic and industrial emulsions
*Form:* Soft solid; Color: 2 (Gardner)
*Properties:* Solubility: Sol. in mineral oil, butyl stearate, glycerol trioleate, perchloroethylene, Stod. solvent; dispersible in water; M.P.: 36 C; HLB: 8.1; Cloud Pt.: < 25 C
*Composition:* 100% active ingred.

**Emerest 2646.** Emery Industries
*Chemical Description:* PEG-400 monooleate
*Category:* Emulsifier, nonionic
*Applications:* Balanced emulsifier for oil-based soluble cutting oils and other specialty industrial lubricants; industrial degreasers; viscosity stabilizer for vinyl plastisols; used in textile softeners and lubricant bases; as component in drycleaning formulations
*Form:* Liquid; Color: 2 (Gardner)
*Properties:* Solubility: Dispersible in water, butyl stearate, glycerol trioleate, perchloroethylene, Stod. solvent; Density: 8.5 lb/gal; Visc.: 100 cs; HLB: 11.7; Cloud Pt.: < 25 C
*Composition:* 100% active ingred.

**Emerest 2647.** Emery Industries
*Chemical Description:* PEG-400 sesquioleate
*Category:* Emulsifier, nonionic
*Applications:* Oil and fat emulsifier in industrial and textile lubricants

*Form:* Liquid; Color: 2 (Gardner)
*Properties:* Solubility: Sol. in mineral oil, butyl stearate, glycerol trioleate, perchloroethylene; disp. in water, Stod. solvent; Density: 8.2 lb/gal; Visc.: 75 cs; HLB: 9.2; Cloud Pt.: $<$ 25 C
*Composition:* 100% active ingred.

**Emerest 2648.** Emery Industries
*Chemical Description:* PEG-400 dioleate
*Category:* Emulsifier, solubilizer, nonionic
*Applications:* Lipophilic emulsifier and solubilizer for mineral oils, fats, and solvents; emulsifier of kerosene in agricultural and pesticide sprays; latex paints, metalworking fluids, solvents, specialty and industrial lubricants
*Composition:* 100% active ingred.

**Emerest 2650.** Emery Industries
*Chemical Description:* PEG-400 monolaurate
*Category:* Wetting agent, defoamer, leveling agent, dispersant, solubilizer, nonionic
*Applications:* Wetting agent, defoamer and leveling agent in latex paints; dispersant for pigment grinding; as solubilizer for oils and solvents; as antiblock agent in vinyls; in textile fiber processing
*Form:* Liquid; Color: 1 (Gardner)
*Properties:* Solubility: Sol. in water; disp. in butyl stearate, glycerol trioleate, perchloroethylene, Stod. solvent; Density: 8.6 lb/gal; Visc.: 100 cs; HLB: 13.1; Cloud Pt.: 33 C
*Composition:* 100% active ingred.

**Emerest 2652.** Emery Industries
*Chemical Description:* PEG-400 dilaurate
*Category:* Emulsifier, nonionic
*Applications:* Oil-soluble emulsifier, used as softener, lubricant, release agent for paper; coupler and lubricant in synthetic fiber spin finishes in textile industry
*Form:* Liquid; Color: 2 (Gardner)
*Properties:* Solubility: Sol. in mineral oil, butyl stearate, glycerol trioleate, perchloroethylene, Stod. solvent; disp. in water; Density: 8.3 lb/gal; Visc.: 65 cs; HLB: 10.4; Cloud Pt.: $<$ 25 C
*Composition:* 100% active ingred.

**Emerest 2660.** Emery Industries
*Chemical Description:* PEG-600 monooleate
*Category:* Emulsifier, nonionic
*Applications:* Water-soluble emulsifier and detergent; specialty lubricants; dye leveling agent in textiles; component in systems for the acid washing of printed circuit boards
*Form:* Liquid; Color: 2 (Gardner)
*Properties:* Solubility: Sol. in water; disp. in butyl stearate, glycerol trioleate, Stod. solvent; Density: 8.7 lb/gal; Visc.: 115 cs; HLB: 13.6; Cloud Pt.: 47 C
*Composition:* 100% active ingred.

**Emery 6705.** Emery Industries
*Chemical Description:* Phenoxyethanol
*Category:* Solvent

*Applications:* Solvent for nitrocellulose, cellulose acetate, ethylcellulose, and vinyl, alkyl and ester type resins; solvent in printing inks; in special formulations for removing industrial finishes and coatings
*Form:* Liquid; Color: 1 (Gardner)
*Properties:* Solubility: Sol. in perchloroethylene, Stod. solvent; disp. in butyl stearate; Density: 9.1 lb/gal; Visc.: 21 cs
*Composition:* 100% active ingred.

**Emery 6748.** Emery Industries
*Chemical Description:* Amphoteric betaine surfactant
*Applications:* Shampoos, hair products, bath products, skin cleaners, mild baby shampoos, other specialty cosmetics
*Form:* Liquid; Color: 4 (Gardner)
*Properties:* Solubility: Sol. in water; disp. in glycerol trioleate; Density: 8.7 lb/gal; Stability: Stable over wide pH range; high tolerance to hard water and electrolytes; high foaming
*Composition:* 30% active ingred.

**Emery 6752.** Emery Industries
*Chemical Description:* Monocarboxylate coconut imidazolinium derivative
*Category:* Amphoteric surfactant
*Applications:* Cleaning compounds and shampoos; viscosity improver and foam stabilizer in combination with anionic or nonionic surfactants; detergent in alkaline systems and steam cleaning compounds
*Form:* Liquid; Color: 9 (Gardner)
*Properties:* Solubility: Sol. in water; disp. in glycerol trioleate; Density: 8.7 lb/gal; Stability: Stable over wide pH range; high tolerance to hard water and electrolytes; isoelectric point = 7
*Composition:* 38% active ingred.

**Emery 6849.** Emery Industries
*Chemical Description:* Linear POE alcohol
*Applications:* Special purpose product used in solid built toilet bowl cleaners
*Form:* Liquid; Color: 1 (Gardner)
*Properties:* Solubility: Water sol.; Density: 8.2 lb/gal; Cloud Pt.: > 100 C
*Composition:* 65% active ingred.

**Emgard 2035.** Emery Industries
*Category:* Nonionic surfactant
*Applications:* Used in surfactant systems for soluble citrus spray oils
*Form:* Liquid; Color: 4 (Gardner)
*Properties:* Solubility: Sol. in mineral oil, glycerol trioleate; disp. in butyl stearate, perchloroethylene, Stod. solvent; Density: 8.1 lb/gal; HLB: 8.5
*Composition:* 100% active ingred.

**Emgard 2051.** Emery Industries
*Category:* Emulsifier
*Applications:* Emulsifier for aromatic weed oils
*Form:* Liquid; Color: 6 (Gardner)
*Properties:* Solubility: Sol. in water, glycerol trioleate; disp. in min-

eral oil, butyl stearate, Stod. solvent; Density: 8.9 lb/gal; HLB: 15.0
*Composition:* 90% active ingred.

**Emgard 2063.** Emery Industries
*Category:* Penetrant
*Applications:* Agricultural formulations
*Form:* Liquid; Color: 1 (Gardner)
*Properties:* Solubility: Sol. in water, butyl stearate, glycerol trioleate, Stod. solvent; Density: 8.2 lb/gal; Cloud Pt.: < 25 C
*Composition:* 100% active ingred.

**Emgard 2066.** Emery Industries
*Category:* Spreader/activator
*Applications:* Agricultural surfactant systems
*Form:* Liquid; Color: 1 (Gardner)
*Properties:* Solubility: Sol. in water, butyl stearate, glycerol; disp. hot in mineral oil, Stoddard solvent; Density: 8.7 lb/gal; Cloud Pt.: 14 C
*Composition:* 100% active ingred.

**Emgard 2067.** Emery Industries
*Category:* Spreader base
*Applications:* Contains isopropanol as a water coupling agent
*Form:* Liquid; Color: 1 (Gardner)
*Properties:* Solubility: Sol. in water, mineral oil, butyl stearate, glycerol trioleate, Stod. solvent
*Composition:* 80% active ingred.

**Emid 6500.** Emery Industries
*Chemical Description:* Coconut monoethanolamide
*Category:* Foam stabilizer, thickener, nonionic
*Applications:* Shampoos, liquid detergents, rug cleaners
*Form:* Flake; Color: 4 (Gardner)
*Properties:* Solubility: Sol. in mineral oil, butyl stearate, glycerol trioleate, perchloroethylene, Stod. solvent; disp. in water; M.P.: 71 C
*Composition:* 100% active ingred.

**Emid 6510.** Emery Industries
*Chemical Description:* Lauric super diethanolamide
*Category:* Foam booster, stabilizer, thickener in combination with anionic surfactants; nonionic
*Applications:* Shampoos, bubble baths, general purpose detergents
*Form:* Solid; Color: 2 (Gardner)
*Properties:* Solubility: Sol. in perchloroethylene, Stod. solvent; disp. in water, mineral oil, butyl stearate, glycerol trioleate; M.P.: 42 C
*Composition:* 100% active ingred.

**Emid 6511.** Emery Industries
*Chemical Description:* Lauric super diethanolamide
*Category:* Emulsifier, nonionic
*Applications:* Heavy duty, built, liquid formulations
*Form:* Solid; Color: 2 (Gardner)
*Properties:* Solubility: Sol. in perchloroethylene, Stod. solvent; disp. in water, mineral oil, butyl stearate, glycerol trioleate; M.P.: 40 C
*Composition:* 100% active ingred.

**Emid 6514.** Emery Industries
*Chemical Description:* Coco super

diethanolamide
*Category:* Foam stabilizer, thickener, detergent, nonionic
*Applications:* Liquid cleaning compounds
*Form:* Liquid; Color: 3 (Gardner)
*Properties:* Solubility: Sol. in water, butyl stearate, glycerol trioleate, perchloroethylene, Stod. solvent; disp. in mineral oil; Density: 8.2 lb/gal
*Composition:* 100% active ingred.

**Emid 6515.** Emery Industries
*Chemical Description:* Coco oil super diethanolamide
*Category:* Foam booster and stabilizer, nonionic
*Applications:* Boosts and stabilizes foam; inhibits redeposition of soils
*Form:* Liquid; Color: 3 (Gardner)
*Properties:* Solubility: Sol. in water, butyl stearate, glycerol trioleate, perchloroethylene, Stod. solvent; disp. in mineral oil; Density: 8.3 lb/gal
*Composition:* 100% active ingred.

**Emid 6533.** Emery Industries
*Chemical Description:* Modified coco diethanolamide
*Category:* Emulsifier, thickening agent, foamer
*Applications:* Emulsifier, thickening agent, moderate foamer for controlled suds detergents; fulling and scouring agent for textiles
*Form:* Liquid; Color: 4 (Gardner)
*Properties:* Solubility: Sol. in water, perchloroethylene, Stod. solvent; disp. in mineral oil, butyl stearate, glycerol trioleate; Density: 8.3 lb/gal
*Composition:* 100% active ingred.

**Emid 6534.** Emery Industries
*Chemical Description:* Modified coco diethanolamide
*Category:* Thickener, foam stabilizer, emulsifier, nonionic
*Applications:* Shampoos; fatty acid soaps in cosmetic formulations
*Form:* Liquid; Color: 4 (Gardner)
*Properties:* Solubility: Sol. in water, perchloroethylene, Stod. solvent; disp. in mineral oil, butyl stearate, glycerol trioleate; Density: 8.5 lb/gal
*Composition:* 100% active ingred.

**Emid 6538.** Emery Industries
*Chemical Description:* Modified coco diethanolamide
*Category:* Thickener, foam stabilizer, nonionic
*Applications:* Built liquid detergents
*Form:* Liquid; Color: 4 (Gardner)
*Properties:* Solubility: Sol. in water, perchloroethylene, Stod. solvent; disp. in mineral oil, butyl stearate, glycerol trioleate; Density: 8.3 lb/gal
*Composition:* 100% active ingred.

**Emid 6541.** Emery Industries
*Chemical Description:* Lauric diethanolamide
*Category:* Emulsifier, foam booster, stabilizer, nonionic
*Applications:* Shampoos, general purpose cleaners

*Form:* Paste; Color: 2 (Gardner)
*Properties:* Solubility: Sol. in water, perchloroethylene, Stod. solvent; Density: 8.3 lb/gal
*Composition:* 100% active ingred.

**Emid 6543.** Emery Industries
*Chemical Description:* Stearic-oleic diethanolamide
*Category:* Coemulsifier, nonionic
*Applications:* Hydrophobic coemulsifier for mineral oils, fatty esters, and various solvents
*Form:* Liquid; Color: 4 (Gardner)
*Properties:* Solubility: Sol. in water, perchloroethylene, Stod. solvent; disp. in mineral oil, butyl stearate, glycerol trioleate; Density: 8.5 lb/gal
*Composition:* 100% active ingred.

**Emid 6544.** Emery Industries
*Chemical Description:* Capric diethanolamide
*Category:* Foam booster, wetting agent, detergency builder, nonionic
*Applications:* Used in salt and alkali systems; dedusting agent for alkaline powders
*Properties:* Solubility: Sol. in butyl stearate, perchloroethylene, Stod. solvent; disp. in water, mineral oil, glycerol trioleate; M.P.: 30 C
*Composition:* 100% active ingred.

**Emid 6545.** Emery Industries
*Chemical Description:* Oleic diethanolamide
*Category:* Emulsifier, foam suppresant, nonionic
*Applications:* Emulsifier for mineral oils in formulating antistatic fiber processing aids and yarn lubricants; foam suppresant in dye carrier and solvent emulsions
*Form:* Liquid; Color: 7 (Gardner)
*Properties:* Solubility: Sol. in mineral oil, butyl stearate, glycerol trioleate, perchloroethylene, Stod. solvent; disp. in water; Density: 7.8 lb/gal
*Composition:* 100% active ingred.

**Emkabase.** Emkay Chem. Co.
*Chemical Description:* Naphthenic, amino sulfonate blend
*Category:* Emulsifier, anionic
*Applications:* Emulsifier for mineral oil
*Form:* Liquid

**Emkagen.** Emkay Chem. Co.
*Chemical Description:* Amide amine condensate
*Category:* Detergent
*Form:* Liquid

**Emkalane.** Emkay Chem. Co.
*Category:* Dyeing assistant
*Applications:* Dyeing assistant for wools and blends

**Emkanol.** Emkay Chem. Co.
*Category:* Mercerizing assistants

**Emkanyl.** Emkay Chem. Co.
*Category:* Dyeing assistant
*Applications:* Dyeing assistant for hosiery

**Emkapol.** Emkay Chem. Co.
*Applications:* Soap products

*Form:* Paste

**Emkapon.** Emkay Chem. Co.
*Chemical Description:* Amide sulfonates
*Category:* Detergent blend

**Emkatex.** Emkay Chem.
*Applications:* Dyestuff dispersing condensate

**Empicol AL 30.** Albright & Wilson Ltd.
*Chemical Description:* Ammonium salt of sulfated narrow cut lauryl alcohol
*Category:* Detergent, anionic
*Applications:* Shampoos, carpet shampoos
*Form:* Liquid/paste; Color: Amber
*Properties:* m.w.: 291; Visc.: 14,000 cs @ 20 C; Stability: May become pastes or hazy liquids under cold storage; becomes clear liquid on warming; pH: 6.5 ± 0.5 (5% sol'n.)
*Composition:* 29% active ingred.

**Empicol CHC 30.** Albright & Wilson Ltd. Marchon Div.
*Chemical Description:* Sodium cetyl oleyl sulfate
*Category:* Detergent, anionic
*Applications:* Multi purpose textile detergent with good wetting, dispersing and emulsifying properties; wool scouring
*Form:* Soft paste; Color: Cream
*Properties:* Solubility: Readily sol. in warm water; Stability: Chemically stable when heated under alkaline conditions and in acid solutions; pH: 8.5 min. (2% sol'n.)
*Composition:* 30% active ingred.
*Reaction:* Alkaline

**Empicol DLS.** Albright & Wilson Ltd. Marchon Div.
*Chemical Description:* Diethanolamine lauryl sulfate
*Category:* Detergent, anionic
*Applications:* Liquid and lotion shampoos; mild to hair and skin
*Form:* Clear liquid; Color: Straw
*Properties:* m.w.: 379 (of active); Density: 1.05 g/cc; Visc.: 500 cs max. @ 20 C; Cloud Pt.: < 0 C; pH: 7–8 (1% aq.)
*Composition:* 34% active ingred.

**Empicol EL.** Albright & Wilson Ltd. Marchon Div.
*Chemical Description:* Concentrated blend of anionic and nonionic surfactants
*Category:* Detergent
*Applications:* Manufacture of liquid shampoos
*Form:* Viscous liquid; Color: Amber
*Properties:* Sp. gr.: 1.025 @ 20 C; Visc.: Visc. approx. doubles at 50% sol'n.; 586 cs (conc. @ 20 C); 1143 cs (50% sol'n. @ 20 C); Cloud Pt.: –2 C
*Composition:* 50% active ingred.

**Empicol EMD.** Albright & Wilson Ltd. Marchon Div.
*Chemical Description:* Blend of natural, straight chain fatty alcohol
*Category:* Detergent
*Applications:* Very mild detergent

for shampoos, children's and baby products with good foaming, hard water performance
*Form:* Liquid; Color: Pale straw; Odor: Almost odorless
*Properties:* Solubility: Completely sol. in cold water; Density: 1.05 g/cc @ 20 C; Visc.: < 250 cs @ 20 C; Cloud Pt.: < 0 C; Stability: Resistant to hydrolysis in strong alkalis; decomposed by strong acids; pH: 6.5–7.5 (5% sol'n.); Biodegradable
*Composition:* 30% active ingred.

**Empicol ESB 3.** Albright & Wilson Ltd. Marchon Div.
*Chemical Description:* Sodium salt of sulfated ethoxylated lauryl alcohol
*Category:* Shampoo base, anionic
*Applications:* Liquid and lotion shampoo base with good foaming properties
*Form:* Liquid; Color: Colorless
*Properties:* Solubility: Water sol.; Density: 1.0; Visc.: ≈ 3000 cs @ 20 C; Cloud Pt.: 0 C; Stability: Stable to alkalis, to acids under certain conditions; pH: 6.5–7.5 (5% sol'n.)
*Composition:* 27.5% active ingred.

**Empicol ESB 3/S.** Albright & Wilson Ltd. Marchon Div.
*Chemical Description:* Sodium salt of sulfated ethoxylated fatty alcohol
*Category:* Shampoo base
*Applications:* Raw material for liquid and lotion shampoos, body shampoos and bubble baths with good foaming properties, clarity
*Form:* Liquid; Color: Colorless
*Properties:* Solubility: Excellent; Density: ≈ 1.01 g/cc; Visc.: 2600 cs @ 20 C; Cloud Pt.: < 0 C; Biodegradable
*Composition:* 25% sodium lauryl ether sulfate
*Reaction:* Neutral (5% sol'n.)

**Empicol ESB 30.** Albright & Wilson Ltd. Marchon Div.
*Chemical Description:* Sodium salt of sulfated ethoxylated lauryl alcohol
*Category:* Shampoo base
*Applications:* Liquid and lotion shampoo base with good foaming properties
*Form:* Liquid; Color: Colorless
*Properties:* m.w.: 384 (of active); Solubility: Water sol.; Density: ≈ 1.0; Visc.: ≈ 20,000 cs @ 20 C; Cloud Pt.: 0 C; Stability: Stable to alkalis, to acids under certain conditions; pH: 6.5–7.5 (5% sol'n.)
*Composition:* 27.5% active ingred.

**Empicol ESB 50.** Albright & Wilson Ltd. Marchon Div.
*Chemical Description:* Sodium salt of sulfated ethoxylated lauryl alcohol
*Category:* Shampoo base
*Applications:* Liquid and lotion shampoo base with good foaming properties
*Form:* Cloudy liquid; Color: Pale straw
*Properties:* m.w.: 384 (of active); Solubility: Slowly sol. in cold water; may tend to form a gel; Den-

sity: ≈ 1.1; Visc.: Pourable liquid; Stability: Stable to alkalis, to acids under certain conditions; pH: 6.5–7.5
*Composition:* 50% active ingred.

**Empicol LZG30.** Albright & Wilson Ltd. Marchon Div.
*Chemical Description:* Sodium salt of sulfated medium cut lauryl alcohol
*Category:* Anionic surfactant, wetting agent
*Applications:* Industrial grade anionic surfactants; wetting agents with low inorganic salt content
*Form:* Paste; Color: Cream
*Composition:* 30% active ingred.
*Reaction:* Alkaline

**Empicol LZGV.** Albright & Wilson Ltd. Marchon Div.
*Chemical Description:* Sodium salt of sulfated medium cut lauryl alcohol
*Category:* Anionic surfactant, wetting agent
*Applications:* Industrial grade anionic surfactants; wetting agents with low inorganic salt content
*Form:* Needle; Color: Cream
*Composition:* 85% active ingred.
*Reaction:* Alkaline

**Empicol MD.** Albright & Wilson Ltd. Marchon Div.
*Chemical Description:* Blend of natural, straight chain fatty alcohol
*Category:* Detergent
*Applications:* Very mild detergent for shampoos, children's and baby products with good foaming, hard water performance
*Form:* Liquid; Color: Pale straw; Odor: Almost odorless
*Properties:* Solubility: Completely sol. in cold water; Density: 1.05 g/cc @ 20 C; Visc.: < 250 cs @ 20 C; Cloud Pt.: < 0 C; Stability: Resistant to hydrolysis in strong alkalis; decomposed by strong acids; pH: 6.5–7.5; Biodegradable
*Composition:* 30% active ingred.

**Empicol ML 26.** Albright & Wilson Ltd. Marchon Div.
*Chemical Description:* Magnesium salt of sulfated narrow-cut lauryl alcohol
*Category:* Detergent, anionic
*Applications:* In manufacture of shampoos, toothpastes
*Form:* Liquid; Color: Pale straw
*Properties:* Solubility: Miscible with water; may form slightly hazy solution due to residual alkalinity; add small amount of citric acid to form clear solution; Density: ≈ 1.0 g/cc; Visc.: ≈ 150 cs
*Composition:* 26% active ingred.
*Reaction:* Alkaline

**Empicol TAS90.** Albright & Wilson Ltd. Marchon Div.
*Chemical Description:* Sodium tallow alcohol sulfate
*Category and Applications:* See Empicol TAS80V; also as high temperature fabric scouring agent; enhances luster of calendered fabrics due to wax and oil emulsifying powers; additive during sizing
*Form:* Powder; Color: Cream

*Properties:* pH: 10
*Composition:* 90% active ingred.
*Reaction:* Alkaline

**Empicol TC 34.** Albright & Wilson Ltd. Marchon Div.
*Chemical Description:* Ammonium/triethanolamine salt of sulfated narrow cut lauryl alcohol
*Category:* Detergent, anionic
*Applications:* Shampoos
*Form:* Clear liquid; Color: Amber
*Properties:* m.w.: 317; Visc.: 27,000 cs @ 20 C; Stability: May become pastes or hazy liquid under cold storage; becomes clear liquid on warming; pH: 6.5 ± 0.3 (5% sol'n.)
*Composition:* 31.5% active ingred.

**Empicol TCR.** Albright & Wilson Ltd. Marchon Div.
*Chemical Description:* Ammonium/triethanolamine salt of sulfated narrow cut lauryl alcohol
*Category:* Detergent, foaming agent, anionic
*Applications:* Shampoos, liquid detergents
*Form:* Clear liquid; Color: Golden yellow
*Properties:* m.w.: 317; Visc.: 12,500 cs @ 20 C; Stability: May become pastes or hazy liquid under cold storage; becomes clear liquid on warming; pH: 6.5 ± 0.5 (5% sol'n.)
*Composition:* 21.5% active ingred.

**Empicol TDL 75.** Albright & Wilson Ltd. Marchon Div.
*Chemical Description:* Diethanolamine/triethanolamine salt of sulfated lauryl alcohol
*Category:* Shampoo raw material
*Applications:* Raw material for shampoos, bubble baths, bath oils, etc. with excellent detergency, wetting and foaming properties, good solubility; mild to skin and scalp
*Form:* Soft paste; Color: Yellow to amber
*Properties:* m.w.: 410 (of active); Solubility: Readily sol. in water; Density: ≈ 1.1 g/cc; pH: 7.5 ± 0.2 (2% sol'n.)
*Composition:* 75% active ingred.; 6% max. ethanol; 4.5% max. free fatty material; 0.5% max. chloride ion; 2.5% max. sulfate ion

**Empicol TL 40.** Albright & Wilson Ltd. Marchon Div.
*Chemical Description:* Triethanolamine salt of sulfated narrow-cut lauryl alcohol
*Category:* Detergent
*Applications:* Manufacture of liquid and lotion shampoos
*Form:* Clear liquid; Color: Very pale yellow
*Properties:* Solubility: Miscible with water; Density: 1.025 g/cc; Visc.: 300 cs @ 20 C; Stability: Stable to acids and alkalis under certain conditions; pH: 6.8–7.1 (5% sol'n.)
*Composition:* 40% triethanolamine lauryl sulfate

**Empicol TL40/S.** Albright & Wilson Ltd. Marchon Div.
*Chemical Description:* Triethanolamine salt of sulfated fatty alcohol
*Category:* Detergent, foaming a-

gent
*Applications:* Liquid and lotion shampoo base with good detergency, foaming power, solubility and viscosity characteristics
*Form:* Clear liquid; Color: Very pale yellow
*Properties:* Density: ≈ 1.025 g/cc
*Composition:* 38% active ingred.
*Reaction:* Neutral

**Empicol TLP.** Albright & Wilson Ltd. Marchon Div.
*Chemical Description:* Triethanolamine salt of sulfated narrow-cut lauryl alcohol
*Category:* Detergent
*Applications:* Manufacture of liquid and lotion shampoos; alkylolamide acts as a foam booster and stabilizer and viscosity modifier and lends some superfatting properties
*Form:* Clear liquid; Color: Golden yellow
*Properties:* Solubility: Miscible with water; Density: 1.025 g/cc; Visc.: 1000 cs @ 20 C; pH: 6.8–7.1 (5% sol'n.)
*Composition:* 40.5% triethanolamine lauryl sulfate; 1.0% alkylolamide

**Empicol TLR.** Albright & Wilson Ltd. Marchon Div.
*Chemical Description:* Triethanolamine salt of sulfated narrow-cut lauryl alcohol
*Category:* Detergent
*Applications:* Manufacture of liquid and lotion shampoos; alkylolamide acts as a foam booster and stabilizer and viscosity modifier and lends some superfatting properties
*Form:* Clear liquid; Color: Golden yellow
*Properties:* Solubility: Miscible with water; Density: 1.025 g/cc; Visc.: 4000 cs @ 20 C; pH: 6.5–7.2 (5% sol'n.)
*Composition:* 35.5% triethanolamine lauryl sulfate; 3% min. alkylolamide

**Empicol WA.** Albright & Wilson Ltd. Marchon Div.
*Chemical Description:* Sodium salt of sulfated narrow-cut lauryl alcohol
*Category:* Industrial surfactant
*Applications:* Industrial uses, e.g., preparation of rubber foams for carpet backing
*Form:* Clear liquid; Color: Pale yellow
*Properties:* m.w.: 292 (of active); Sp. gr.: ≈ 1.04; pH: 8.5–9.5 (5% aq.)
*Composition:* 27% active ingred.; other components: 5% solubilizer; 66.5% water

**Empicol XC35.** Albright & Wilson Ltd. Marchon Div.
*Chemical Description:* Blend of anionic and nonionic components
*Category:* Shampoo base
*Applications:* Ingredient for liquid cream (lotion) shampoos; formulations should contain a preservative
*Form:* Viscous pearly liquid; Odor: Almost odorless
*Reaction:* Neutral

**Empicol XC 35/S.** Albright & Wilson Ltd. Marchon Div.
*Chemical Description:* Blend of anionic and nonionic components
*Category:* Shampoo base
*Applications:* Ingredient for pearly lotion shampoos; formulations should contain a preservative
*Form:* White liquid; Odor: Almost odorless
*Properties:* Stability: Good
*Reaction:* Neutral

**Empicol XDB.** Albright & Wilson Ltd. Marchon Div.
*Chemical Description:* Detergent complex
*Category:* Detergent
*Applications:* Detergent, foaming and conditioning agent for shampoos; substantive to hair and skin
*Form:* Clear liquid; Color: Golden
*Properties:* Visc.: ≈ 4000 cs; Cloud Pt.: < 0 C; pH: 8 ± 0.5 (5% sol'n.)
*Composition:* 30% solids content; 70% water content
*Reaction:* Alkaline

**Empicol XM 17.** Albright & Wilson Ltd.
*Category:* Detergent, anionic
*Applications:* Carpet shampoo double concentrate
*Form:* Liquid; Color: Pale amber
*Properties:* Visc.: ≈ 3000 cs @ 20 C; Cloud Pt.: 5 C; pH: 6–7 (5% sol'n.)
*Composition:* 21% anionic active ingred.; 4% foam booster

**Empicol XT 45.** Albright & Wilson Ltd. Marchon Div.
*Chemical Description:* Blend of anionic and nonionic components
*Category:* Shampoo base
*Applications:* Active ingredient for lacquer removing hair shampoo
*Form:* Mobile liquid; Color: Pale yellow
*Properties:* Sp. gr.: ≈ 1.05 @ 20 C; Visc.: 100 cs max. @ 20 C; Cloud Pt.: < 0 C; pH: 9.0–9.5 (5% aq.)
*Composition:* 26–27% active ingred.
*Reaction:* Alkaline

**Empicryl APD.** Albright & Wilson Ltd. Marchon Div.
*Chemical Description:* Aqueous solution of the disodium salt of a copolymer of maleic anhydride and diisobutylene
*Category:* Dispersant
*Applications:* Pigment dispersant for water based paints; pigment stabilizer; imparts high gloss to paints
*Form:* Clear liquid; Color: Pale yellow
*Properties:* Sp. gr.: 1.12 @ 18 C; Visc.: 16 cs; F.P.: –3 C; pH: 12.0–12.7
*Composition:* 22% min. diisobutylene disodium maleate copolymers; 3% max. sodium maleate
*Reaction:* Alkaline

**Empicryl APD/B.** Albright & Wilson Ltd. Marchon Div.
*Chemical Description:* Aqeuous solution of the disodium salt of a copolymer of maleic anhydride and diisobutylene
*Category and Applications:* See Empicryl APD

*Form and Properties:* Identical to Empicryl APD but with a lower pH (concentrate) of 10.0–10.5

**Empicryl LM.** Albright & Wilson Ltd. Marchon Div.
*Chemical Description:* Lauryl/myristyl methacrylate monomer
*Applications:* Used for polymerization to yield polymers in solution or emulsion form for surface coatings for textiles, leather, etc.; stabilized with Topanol A against premature polymerization
*Form:* Color: 100 max. (Hazen)
*Properties:* Sapon. Value: 191–218
*Composition:* 62% min. lauryl methacrylate; 22% min. myristyl methacrylate

**Empicryl TM.** Albright & Wilson Ltd. Marchon Div.
*Chemical Description:* Ceto-stearyl methacrylate monomer
*Applications:* See Empicryl LM
*Form:* Color: 200 max. (Hazen)
*Properties:* Sapon. Value: 152–178
*Composition:* 57% min. stearyl methacrylate; 25% min. cetyl methacrylate

**Empigen BAC.** Albright & Wilson Ltd. Marchon Div.
*Chemical Description:* Benzalkonium chloride solution B.P.
*Category:* Disinfectant, cationic
*Applications:* Disinfectant for dairy, food processing, restaurant, brewing and bottling industries; medical and pharmaceutical applications
*Form:* Liquid; Color: Pale yellow; Odor: Aromatic
*Properties:* Solubility: Miscible with water, alcohol and acetone
*Composition:* 50% active ingred.; 0.2% max. sulfated ash
*Reaction:* Incompatible with soap and other anionic active surfactants and with alkali hydroxides

**Empigen BB.** Albright & Wilson Ltd. Marchon Div.
*Chemical Description:* Alkyl dimethyl betaine
*Category:* Foam booster/stabilizer, wetting agent, thickening agent
*Applications:* Ingredient for shampoos and detergents, esp. baby shampoos due to mildness; static control
*Form:* Liquid; Color: Almost water white
*Properties:* Stability: Excellent acid and alkali stability
*Composition:* 30% betaine

**Empilan BQ 100.** Albright & Wilson Ltd. Marchon Div.
*Chemical Description:* Polyethylene glycol monoester of a fatty acid
*Category:* Emulsifier, wetting agent, antistat, nonionic
*Applications:* Emulsifier for insecticides, spindle oils, etc; industrial wetting agent; scouring agent; antifoam in laundry works, in glue manufacture and in paper coating mixes; ingred. in hand cleaning jellies; antistat for PVC; turbine oil additive
*Form:* Liquid; Color: Dark brown;

Odor: Faint, typical
*Properties:* m.w.: 400 (of glycol base); Solubility: Dispersible in water; yields clear solutions with ethanol, xylene, oleic acid; yields turbid solutions with white spirit, mineral oil, perchlorethylene, kerosene, naphtha; Sp. gr.: 1.017 @ 20 C; Visc.: 130 cps; M.P.: ≈ −1 C
*Composition:* 100% active ingred.

**Empilan CDE.** Albright & Wilson Ltd. Marchon Div.
*Chemical Description:* Coconut diethanolamide
*Category:* Foam stabilizer, solubilizer, detergent, nonionic
*Applications:* Shampoos, liquid detergent formulations
*Form:* Liquid/paste; Color: Golden yellow
*Composition:* 100% active ingred.

**Empilan CME.** Albright & Wilson Ltd. Marchon Div.
*Chemical Description:* Coconut monoethanolamide
*Category:* Foam booster/stabilizer; viscosity modifier, nonionic
*Applications:* Foam booster/stabilizers for liquid and powder detergents; viscosity modifier for liquid detergents, dishwashers, hair shampoos, carpet shampoos; as bases for manufacture of ethoxylated derivatives
*Form:* Waxy flake; Color: Cream
*Properties:* Solubility: Dispersible in hot water; Density: 0.4 g/cc; Stability: Stable to acids and alkalis under certain conditions
*Composition:* 94.2% active ingred.

*Reaction:* Alkaline

**Empilan CDX.** Albright & Wilson Ltd. Marchon Div.
*Chemical Description:* Coconut diethanolamide
*Category and Applications:* See Empilan CDE
*Form:* Liquid/paste; Color: Straw
*Composition:* 25% free amine

**Empilan DL 40.** Albright & Wilson Ltd. Marchon Div.
*Chemical Description:* Fatty alcohol ethoxylate
*Category:* Emulsifier, wetting, dye leveling and dispersing agents, nonionic
*Applications:* Textile dye leveling and dispersing agent for cellulosic, polyester, acrylic fibers, wool and nylon; stripping of vat dyeings using sodium bisulfite and caustic soda; dyeing and printing with azoic dyes; bleaching of textiles; desizing baths; antistat; emulsifier for fatty acids and mineral oils for fiber lubricants; wetting and stabilizing agent for synthetic rubber latices
*Form:* Clear liquid; Color: Pale yellow
*Properties:* Solubility: Miscible with water; Density: 1.046 g/cc; Cloud Pt.: > 100 C in distilled water; Stability: Yields neutral solutions stable to acids, alkalis and most electrolytes at the concentrations found in textile industry
*Composition:* 40% active ingred.

**Empilan DL 100.** Albright & Wilson Ltd. Marchon Div.

*Chemical Description:* Fatty alcohol ethoxylate
*Category and Applications:* See Empilan DL 40
*Form:* Stiff waxy solid
*Properties:* Solubility: Readily sol. in warm water; M.P.: ≈ 36 C; Cloud Pt.: > 100 C (1% in distilled water)
*Composition:* 100% active ingred.

**Empilan EGMS.** Albright & Wilson Ltd. Marchon Div.
*Chemical Description:* Ethylene glycol monostearate
*Category:* Opacifier, pearling agent, emulsifier, stabilizer
*Applications:* Opacifier/pearling agent and emulsifying/stabilizing agent in liquid cream (lotion) and cream shampoos; oil-soluble emollient in hand lotions and creams
*Form:* Hard wax-like flake; Color: White
*Properties:* Solubility: Oil-soluble; M.P.: ≈ 56 C
*Composition:* 65% diester; 30% monoester; 3.5% max. free fatty acid

**Empilan GMS LSE 32.** Albright & Wilson Ltd. Marchon Div.
*Chemical Description:* Glyceryl monostearate, self-emulsifying, food grade
*Category:* Emulsifier, nonionic
*Form:* Wax-like powder; Color: White
*Properties:* m.w.: 358 (of monoglyceride); M.P.: 55–60 C; pH: 7–9 (10% aq.)
*Composition:* 32.5% min. monoglyceride

**Empilan GMS NSE 32.** Albright & Wilson Ltd. Marchon Div.
*Chemical Description:* Glyceryl monostearate, food grade
*Category:* Nonionic emulsifier
*Form:* Wax-like powder; Color: White
*Properties:* M.P.: 54–59 C; pH: 6–8 (10% aq.)
*Composition:* 32.5% min. monoglyceride

**Empilan GMS NSE 40.** Albright & Wilson Ltd. Marchon Div.
*Chemical Description:* Glyceryl monostearate, food grade
*Category:* Nonionic emulsifier
*Form:* Wax-like powder; Color: White
*Properties:* M.P.: 53–58 C; pH: 6–8 (10% aq.)
*Composition:* 38% min. monoglyceride

**Empilan GMS SE 32.** Albright & Wilson Ltd. Marchon Div.
*Chemical Description:* Glyceryl monostearate, self-emulsifying
*Category:* Nonionic emulsifier
*Form:* Wax-like powder; Color: White
*Properties:* M.P.: 55–60 C; pH: 7–9 (10% aq.)
*Composition:* 32.5% min. monoglyceride

**Empilan GMS SE 40.** Albright & Wilson Ltd. Marchon Div.
*Chemical Description:* Glyceryl monostearate
*Category:* Nonionic emulsifier
*Form:* Wax-like powder; Color: White

*Properties:* M.P.: 56–61 C; pH: 7–9 (10% aq.)
*Composition:* 38% min. monoglyceride

**Empilan KA3.** Albright & Wilson Ltd. Marchon Div.
*Chemical Description:* Fatty alcohol ethylene oxide condensates
*Category:* Emulsifier, detergent, dispersant, wetting agent, nonionic
*Applications:* For textiles as scouring and wetting agents, as emulsifiers in self emulsifiable spinning oils, as dye leveling and dispersing agents; for metal processing as wetting agents in alkaline cleaners, electrolyte and pickling baths, as additives in cutting and rolling oils; for paper industry as wetting and deinking assistants; for paint industry as emulsifiers in acrylic/vinyl acetate emulsions; in insecticides and pesticides
*Form:* Clear liquid; Color: Pale straw (@ 20 C)
*Properties:* Solubility: Sol. in mineral oils and other organic solvents; HLB: 8.9; Cloud Pt.: Insoluble; Stability: Good; Biodegradable
*Composition:* 100% active ingred.
*Reaction:* Neutral or slightly alkaline

**Empilan KA5.** Albright & Wilson Ltd. Marchon Div.
*Chemical Description:* Fatty alcohol ethylene oxide condensate
*Category and Applications:* See Empilan KA3
*Form:* Clear liquid; Color: Pale straw (@ 20 C)
*Properties:* HLB: 11.5; Cloud Pt.: 37 C (1% aq. sol'n.); Stability: Good; Biodegradable
*Composition:* 100% active ingred.
*Reaction:* Neutral or slightly alkaline

**Empilan KA590.** Albright & Wilson Ltd. Marchon Div.
*Chemical Description:* Fatty alcohol ethylene oxide condensate
*Category and Applications:* See Empilan KA3
*Form:* Clear liquid; Color: Water white
*Properties:* Density: 9.8 lb/gal; HLB: 11.4; Cloud Pt.: 95 F; Pour Pt.: 12 F; Biodegradable
*Composition:* 90% active ingred.
*Reaction:* Neutral or slightly alkaline; compatible with wide range of anionics, cationics, acids, alkalis, detergent builders and oxidizing and bleaching agents in textile industry in normal use
*Warnings and Caution:* Avoid high concentrations Empilan KA590 with high concentration of strong oxidizing agent; instability and explosion may result

**Empilan KA880.** Albright & Wilson Ltd. Marchon Div.
*Chemical Description:* Fatty alcohol ethylene oxide condensate
*Category and Applications:* See Empilan KA3
*Form:* Clear liquid; Color: Pale straw (@ 20 C)
*Properties:* HLB: 13.6; Cloud Pt.: 81 C; Stability: Good; Biodegradable
*Composition:* 80% active ingred.

*Reaction:* Neutral or slightly alkaline

**Empilan KA1080.** Albright & Wilson Ltd. Marchon Div.
*Chemical Description:* Fatty alcohol ethylene oxide condensate
*Category and Applications:* See Empilan KA3
*Form:* Clear liquid; Color: Pale straw (@ 20 C)
*Properties:* HLB: 14.6; Cloud Pt.: 95 C; Stability: Good; Biodegradable
*Composition:* 80% active ingred.
*Reaction:* Neutral or slightly alkaline

**Empilan KB 2.** Albright & Wilson Ltd. Marchon Div.
*Chemical Description:* Ethylene oxide condensate of narrow-cut lauryl alcohol
*Category:* Emulsifier, foam booster, super-fatting agents, nonionic
*Applications:* For sulfation process to yield salts of lauryl ether sulfates and phosphates; as emulsifiers, foam boosters and superfatting agents in detergents and emulsifying systems
*Form:* Liquid; Color: Almost colorless
*Properties:* Solubility: Insol. in cold water; Density: ≈ 0.9; Visc.: 25 cs; Stability: Stable to acids and alkalis
*Composition:* 100% active ingred.

**Empilan KB 3.** Albright & Wilson Ltd. Marchon Div.
*Chemical Description:* Ethylene oxide condensates of narrow-cut lauryl alcohol
*Category and Applications:* See Empilan KB 2
*Form:* Liquid; Color: Almost colorless
*Properties:* Solubility: Insol. in cold water; Density: ≈ 0.9; Visc.: 25 cs; Stability: Stable to acids and alkalis; pH: 6–7 (5% sol'n.)
*Composition:* 100% active ingred.

**Empilan KL 6.** Albright & Wilson Ltd. Marchon Div.
*Chemical Description:* Ethylene oxide condensate of cetyl/oleyl alcohol
*Category:* Emulsifier, detergent, wetting agent
*Applications:* Emulsifier for mineral oils in cosmetics; detergents and wetting agents in textile industry
*Form:* Soft paste; Color: Cream
*Properties:* Solubility: Dispersible in cold water; Density: 0.9 g/cc; M.P.: 35 C; Stability: Stable to acids and alkalis; pH: 8–10
*Composition:* 100% active ingred.
*Reaction:* Alkaline

**Empilan KL 10.** Albright & Wilson Ltd. Marchon Div.
*Chemical Description:* Ethylene oxide condensate of cetyl/oleyl alcohol
*Category and Applications:* See Empilan KL 6
*Form:* Paste; Color: Cream
*Properties:* Solubility: Sol. in cold water; Density: 0.9 g/cc; M.P.: 40 C; Stability: Stable to acids and alkalis; pH: 8–10

*Composition:* 100% active ingred.
*Reaction:* Alkaline

**Empilan KL 20.** Albright & Wilson Ltd. Marchon Div.
*Chemical Description:* Ethylene oxide condensate of cetyl/oleyl alcohol
*Category and Applications:* See Empilan KL 6
*Form:* Soft solid; Color: Cream
*Properties:* Solubility: Sol. in cold water; Density: 0.9 g/cc; M.P.: 42 C; Stability: Stable to acids and alkalis
*Composition:* 100% active ingred.
*Reaction:* Alkaline; pH = 8–10

**Empilan KM 11.** Albright & Wilson Ltd. Marchon Div.
*Chemical Description:* Polyethanoxy alkyl ether
*Category:* Emulsifier, nonionic
*Applications:* Emulsifier of waxes and certain fatty alcohols; foam control agent in synthetic heavy duty detergents; soap additive
*Form:* Paste; Color: Colorless to pale straw
*Properties:* Setting Pt.: ≈ 30 C; Hydroxyl Value: 73
*Composition:* 100% active ingred.

**Empilan KM 20.** Albright & Wilson Ltd. Marchon Div.
*Chemical Description:* Polyethanoxy alkyl ether
*Category:* Emulsifier, nonionic
*Applications:* Emulsifier of waxes and certain fatty alcohols; foam control agent in synthetic heavy duty detergents; toilet bowl cleaners
*Form:* Waxy flake/block; Color: Colorless to pale straw
*Properties:* Setting Pt.: ≈ 38 C; Hydroxyl Value: 49

**Empilan KM 50.** Albright & Wilson Ltd. Marchon Div.
*Chemical Description:* Polyethanoxy alkyl ether
*Category and Applications:* See Empilan KM 20
*Form:* Waxy flake/block (also available in powder form); Color: Colorless to pale straw
*Properties:* Setting Pt.: ≈ 40 C; Hydroxyl Value: 23

**Empilan LDE.** Albright & Wilson Ltd. Marchon Div.
*Chemical Description:* Lauric diethanolamide
*Category and Applications:* See Empilan CDE
*Form:* Solid; Color: Straw
*Composition:* 6% max. free amine

**Empilan LIS.** Albright & Wilson Ltd. Marchon Div.
*Chemical Description:* Lauric monoisopropanolamide
*Category and Applications:* See Empilan CME
*Form:* Waxy flake; Color: Cream
*Properties:* Solubility: Dispersible in hot water; Density: 0.4 g/cc; Stability: Stable to acids and alkalis under certain conditions
*Composition:* 87.5% active ingred.

**Empilan LP2.** Albright & Wilson Ltd. Marchon Div.
*Chemical Description:* Ethylene oxide condensate of Empilan CME

*Category:* Foam and detergent booster, nonionic
*Applications:* Liquid detergent additive as foam stabilizer, detergent booster, and solubilizer
*Form:* Soft paste; Color: Creamy yellow; Odor: Faint characteristic
*Properties:* Sapon. Value: 14 max. (mg KOH/g); Stability: Stable to acids and alkalis
*Composition:* 100% active ingred.
*Reaction:* Slightly alkaline

**Empilan MAA.** Albright & Wilson Ltd. Marchon Div.
*Chemical Description:* Ethylene oxide condensate of Empilan CME
*Category and Applications:* See Empilan LP2
*Form:* Soft paste; Color: Creamy yellow; Odor: Faint characteristic
*Properties:* Sapon. Value: 15 max. (mg KOH/g); Stability: Stable to acids and alkalis
*Composition:* 100% active ingred.
*Reaction:* Slightly alkaline

**Empilan NP Series.** Albright & Wilson Ltd. Marchon Div.
*Chemical Description:* Nonyl phenol ethoxylate
*Category:* Wetting agent, detergent, emulsifier, nonionics
*Applications:* Empilan NP4: emulsifiers; when sulfated yields anionic surfactants for liquid detergents; Empilan NP9 and NP12: detergent and cleaning systems; Empilan NP15: in rubber industry for latex preparation; Empilan NP20 and NP30: high temperature alkaline detergents; other series general applications: in concrete mixing, in metal polishing preparations, as demulsifiers for crude petroleum emulsions, as solubilizers for certain insecticides
*Composition:* 100% active ingred.

**Empilan NP9.** Albright & Wilson Ltd. Marchon Div.
*Chemical Description:* Nonyl phenol ethoxylate
*Category and Applications:* See Empilan NP Series
*Form:* Soft paste/liquid; Color: Pale straw
*Properties:* Density: 1.0 g/cc; Visc.: 300 cs; Cloud Pt.: 55 C (1% in water); Hydroxyl Value: 90 (mg KOH/g)
*Composition:* 100% active ingred.
*Reaction:* Neutral or slightly alkaline

**Empimin KSN 27.** Albright & Wilson Ltd. Marchon Div.
*Chemical Description:* Sodium lauryl ether sulfate based on a narrow-cut lauryl alcohol, industrial grade
*Category:* Detergent ingredient, anionic
*Applications:* Used in liquid detergents in conjunction with sodium alkyl benzene sulfonate to yield product with mild "feel" and good foaming properties
*Form:* Clear or slightly turbid liquid; Color: Pale straw
*Properties:* Visc.: 25–100 cs; Stability: Good sol. and stability under alkaline conditions; Biodegradable
*Composition:* 27% active ingred.

**Empimin KSN 60.** Albright & Wilson Ltd. Marchon Div.
*Chemical Description:* Sodium lauryl ether sulfate based on narrow-cut lauryl alcohol; with 10% ethanol as solubilizer
*Category and Applications:* See Empimin KSN 27
*Form:* Translucent liquid; Color: Pale amber
*Properties:* Visc.: 2500–3000 cs; Stability: Stable and sol. under alkaline conditions; Biodegradable
*Composition:* 60% active ingred.

**Empimin LF.** Albright & Wilson Ltd. Marchon Div.
*Chemical Description:* Blend of foaming agents
*Category:* Foaming agent
*Applications:* Foaming agent for use with noncarboxylated SBR latices in manufacture of foamed latex carpet backing
*Form:* Clear liquid; Color: Yellow
*Properties:* Visc.: 25 cs @ 20 C; Cloud Pt.: –3 C; Stability: Good; pH: 8 (1% aq.)
*Composition:* 35% active ingred.
*Reaction:* Alkaline

**Empimin MHH.** Albright & Wilson Ltd.
*Chemical Description:* Disodium N-lauryl sulfosuccinamate
*Category:* Foaming agent, anionic
*Applications:* Foaming agent for rubber latices, suggested for high density carpet
*Form:* Liquid; Color: Pale cream
*Properties:* pH: 8 (5% aq.)
*Composition:* 40% active ingred.
*Reaction:* Alkaline

**Empimin MKK.** Albright & Wilson Ltd.
*Chemical Description:* Disodium N-stearyl sulfosuccinamate
*Category:* Foaming agent, anionic
*Applications:* Foaming agent for rubber latices
*Form:* Soft paste; Color: Pale cream
*Properties:* pH: 8 (5% aq.)
*Composition:* 35% active ingred.
*Reaction:* Alkaline

**Empimin MSS.** Albright & Wilson Ltd.
*Chemical Description:* Diammonium N-lauryl sulfosuccinamate
*Category:* Foaming agent, anionic
*Applications:* Foaming agent for rubber latices, suggested as fast foaming agent for low-density backings
*Form:* Liquid; Color: Pale cream
*Properties:* pH: 8 (5% aq.)
*Composition:* 40% active ingred.
*Reaction:* Alkaline

**Empimin MTT.** Albright & Wilson Ltd.
*Chemical Description:* Disodium N-oleyl sulfosuccinamate
*Category:* Foaming agent, anionic
*Applications:* Foaming agent for rubber latices, suggested for high density carpet and for foaming of carboxylated SBR latices
*Form:* Liquid; Color: Pale cream
*Properties:* pH: 8 (5% aq.)
*Composition:* 40% active ingred.

*Reaction:* Alkaline

**Empiphos 4KP.** Albright & Wilson Ltd. Marchon Div.
*Chemical Description:* Tetra potassium pyrophosphate
*Category:* Detergent builder
*Applications:* Builder for heavy duty liquid detergents; as clarifying agent in liquid soaps; manufacture of synthetic rubber; as a buffer in electroplating baths; as a pigment dispersant and stabilizer in emulsion paints; in high pressure boiler water treatment to reduce scale
*Properties:* Solubility: 191 g/100 g water @ 25 C; Density: 75 lb/ft$^3$
*Reaction:* Alkaline; pH = 10.3 (1% sol'n.)

**Empiwax SK.** Albright & Wilson Ltd.
*Chemical Description:* Based on cetyl/stearyl alcohol
*Category:* Emulsifier, anionic
*Applications:* Self-emulsifying wax as O/W emulsifier for pharmaceuitcal and toilet preparations and ointments
*Form:* Flake; Color: White
*Properties:* Stability: 10% emulsion in water is stable

**Empiwax SK/BP.** Albright & Wilson Ltd. Marchon Div.
*Chemical Description:* Based on cetyl/stearyl alcohol
*Category:* Emulsifier, anionic
*Applications:* Self-emulsifying wax as O/W emulsifier for pharmaceutical and toilet preparations and ointments; esp. for use in penicillin cream bases; complies with B.P. specs for emulsifying wax
*Form:* Flake; Color: White
*Properties:* Stability: 10% emulsion in water is stable

**Emsorb 2500.** Emery Industries
*Chemical Description:* Sorbitan monooleate
*Category:* Emulsifier, coupling agent, lubricant, softener, nonionic
*Applications:* Textile fibers and leather; can be used with water-sol. emulsifiers in formulating clear emulsifiable concentrates of petroleum oils and waxes, natural fats, alkyl esters; cosmetics and household products
*Form:* Liquid; Color: 8 (Gardner)
*Properties:* Solubility: Sol. in mineral oil, butyl stearate, perchloroethylene, Stod. solvent; disp. in glycerol trioleate; Density: 8.3 lb/gal; Visc.: 1000 cs; HLB: 4.6
*Composition:* 100% conc.

**Emsorb 2502.** Emery Industries
*Chemical Description:* Sorbitan sesquioleate
*Category:* Emulsifier, coupling agent, nonionic
*Applications:* Primary emulsifier for W/O systems; coemulsifier for O/W systems for mineral oil, fats, waxes; coupling agent; household aerosols; cosmetics; industrial oils; textile oils
*Form:* Liquid; Color: 6 (Gardner)
*Properties:* Solubility: Sol. in mineral oil, butyl stearate, glycerol trioleate, perchloroethylene, Stod.

solvent; water insol.; Density: 8.2 lb/gal; Visc.: 1500 cs; HLB: 3.7
*Composition:* 100% conc.

**Emsorb 2503.** Emery Industries
*Chemical Description:* Sorbitan trioleate
*Category:* Coupling agent, emulsifier, nonionic
*Applications:* Processing and softening of textiles and leather
*Form:* Liquid; Color: 7 (Gardner)
*Properties:* Solubility: Sol. in mineral oil, butyl stearate, glycerol trioleate, perchloroethylene, Stod. solvent; Density: 7.9 lb/gal; Visc.: 200 cs; HLB: 2.7
*Composition:* 100% active ingred.

**Emsorb 2505.** Emery Industries
*Chemical Description:* Sorbitan monostearate
*Category:* Emulsifier, nonionic
*Applications:* Hydrophobic emulsifier for emulsions of mineral oils, fats, waxes and silicones; coemulsifier for industrial oils, household products and cosmetics; textile lubricant; with Emsorb 6905 in paraffin wax emulsions and silicone defoamers for paper and textile processing
*Form:* Solid; Color: 4 (Gardner)
*Properties:* Solubility: Disp. in butyl stearate, glycerol trioleate, perchloroethylene; water insol.; M.P.: 54 C; HLB: 5.4
*Composition:* 100% conc.

**Emsorb 2507.** Emery Industries
*Chemical Description:* Sorbitan tristearate
*Category:* Emulsifier, nonionic
*Applications:* Hydrophobic emulsifier similar to Emsorb 2503; also as a fiber-to-metal textile lubricant
*Form:* Waxy solid; Color: 4 (Gardner)
*Properties:* Solubility: Sol. in butyl stearate, glycerol trioleate, perchloroethylene, Stod. solvent; disp. in mineral oil; water insol.; M.P.: 54 C; HLB: 2.4
*Composition:* 100% conc.

**Emsorb 2510.** Emery Industries
*Chemical Description:* Sorbitan monopalmitate
*Category:* Emulsifier, nonionic
*Applications:* Oil soluble waxy emulsifier for cosmetics and household products; fiber-to-metal lubricant for synthetic and cellulosic fibers
*Form:* Solid; Color: 8 (Gardner)
*Properties:* Solubility: Sol. in Stod. solvent; disp. in butyl stearate; M.P.: 48 C; HLB: 6.6
*Composition:* 100% conc.

**Emsorb 2515.** Emery Industries
*Chemical Description:* Sorbitan monolaurate
*Category:* Emulsifier, nonionic
*Applications:* Water dispersible emulsifier with antifoam properties for oils, fats, solvents in household specialties, industrial oils, cosmetics and emulsion polymerization
*Form:* Liquid; Color: 7 (Gardner)
*Properties:* Solubility: Sol. in mineral oil, butyl stearate, glycerol trioleate, perchloroethylene, Stod. solvent; dispersible in water; Density: 8.8 lb/gal; Visc.: 4500 cs;

HLB: 7.6
*Composition:* 100% conc.

**Emsorb 6900.** Emery Industries
*Chemical Description:* POE (20) sorbitan monooleate
*Category:* Emulsifier, dispersant, solubilizer, nonionic
*Applications:* Hydrophilic emulsifier and coemulsifier for petroleum oils, fats, solvents and waxes in cosmetics, household products, industrial lubricants, textile dye carriers; as dispersant for pigments in coatings; as solubilizer for oils and fragrances; emulsifier for aliphatic alcohols in tobacco sucker control concentrates
*Form:* Liquid; Color: 5 (Gardner)
*Properties:* Solubility: Sol. in water; disp. in glycerol trioleate; Density: 9.0 lb/gal; Visc.: 400 cs; HLB: 15.1; Cloud Pt.: > 100 C
*Composition:* 100% conc.

**Emsorb 6901.** Emery Industries
*Chemical Description:* POE (5) sorbitan monooleate
*Category:* Emulsifier, lubricant, nonionic
*Applications:* O/W emulsifier and lubricant with mineral and vegetable oils in industrial lubricants; textile lubricants and softeners
*Form:* Liquid; Color: 6 (Gardner)
*Properties:* Solubility: Sol. in glycerol trioleate, Stod. solvent; disp. in water and mineral oil; Density: 8.5 lb/gal; Visc.: 450 cs; HLB: 10.1; Cloud Pt.: < 25 C
*Composition:* 100% conc.

**Emsorb 6903.** Emery Industries
*Chemical Description:* POE (20) sorbitan trioleate
*Category:* Emulsifier, lubricant, nonionic
*Applications:* O/W emulsifier for petroleum oils, fats, waxes and alkyl esters; lubricant for metals, textiles, leather; in soluble oils for metal processing and finishing; in textile processing and finishing; glass fiber lubricants; automotive lubricant additives
*Form:* Liquid; Color: 6 (Gardner)
*Properties:* Solubility: Sol. in butyl stearate, glycerol trioleate, Stod. solvent; disp. in water, mineral oil and perchloroethylene; Density: 8.6 lb/gal; Visc.: 300 cs; HLB: 10.9; Cloud Pt.: < 25 C
*Composition:* 95% active ingred.

**Emsorb 6905.** Emery Industries
*Chemical Description:* POE (20) sorbitan monostearate
*Category:* Emulsifier, lubricant, nonionic
*Applications:* O/W emulsifier for mineral oil, fats, waxes; fiber-to-metal textile lubricant; coemulsifier with Emsorb 2505 in paraffin wax emulsions for textiles, paper and wallboard coatings; cosmetic and household products
*Form:* Waxy semisolid; Color: 3 (Gardner)
*Properties:* Solubility: Sol. in water, Stod. solvent; disp. in perchloroethylene; Density: 8.9 lb/gal; Visc.: 500 cs; HLB: 15.2; Cloud Pt.: > 100 C
*Composition:* 100% conc.

**Emsorb 6906.** Emery Industries
*Chemical Description:* POE (3) sorbitan monostearate
*Category:* Emulsifier, lubricant, nonionic
*Applications:* W/O emulsifier in household formulations; fiber-to-metal lubricant for synthetic and cellulosic fibers and yarns
*Form:* Solid; Color: 3 (Gardner)
*Properties:* Solubility: Sol. in Stod. solvent; disp. in water, mineral oil, butyl stearate, glycerol trioleate, perchloroethylene; M.P.: 39 C; HLB: 9.0; Cloud Pt.: < 25 C
*Composition:* 100% conc.

**Emsorb 6907.** Emery Industries
*Chemical Description:* POE (20) sorbitan tristearate
*Category:* Emulsifier, lubricant, nonionic
*Applications:* O/W emulsifier for petroleum oils and natural fats; textile processing and finishing compounds as lubricant and softener
*Form:* Waxy solid; Color: 5 (Gardner)
*Properties:* Solubility: Sol. in mineral oil, butyl stearate, glycerol trioleate, Stod. solvent; disp. in water, perchloroethylene; M.P.: 34 C; HLB: 11.0; Cloud Pt.: < 25 C
*Composition:* 100% conc.

**Emsorb 6910.** Emery Industries
*Chemical Description:* POE (20) sorbitan monopalmitate
*Category:* Emulsifier, lubricant, nonionic
*Applications:* O/W emulsifier and lubricant; similar functions as Emsorb 6905 with lower viscosity and greater water solubility
*Form:* Viscous liquid; Color: 6 (Gardner)
*Properties:* Solubility: Sol. in water and glycerol trioleate; disp. in perchloroethylene; Density: 9.2 lb/gal; Visc.: 600 cs; HLB: 15.8; Cloud Pt.: > 100 C
*Composition:* 100% conc.

**Emsorb 6915.** Emery Industries
*Chemical Description:* POE (20) sorbitan monolaurate
*Category:* Emulsifier, solubilizer, nonionic
*Applications:* O/W emulsifier and solubilizer of petroleum oils, solvents and fats, essential oils, fragrances; cosmetic creams and lotions; as viscosity modifier in shampoos; in textile industry as emulsifier for dye carriers, as an antistatic scrooping agent in spin finishes, as a fiber processing aid
*Form:* Liquid; Color: 6 (Gardner)
*Properties:* Solubility: High water sol.; also sol. in glycerol trioleate; disp. in perchloroethylene; Density: 9.2 lb/gal; Visc.: 400 cs; HLB: 16.5; Cloud Pt.: > 100 C
*Composition:* 100% conc.

**Emsorb 6916.** Emery Industries
*Chemical Description:* POE (4) sorbitan monolaurate
*Category:* Emulsifier
*Applications:* O/W emulsifier in emulsion polymerization of PVC
*Form:* Liquid; Color: 7 (Gardner)
*Properties:* Solubility: Disp. in

water, butyl stearate, perchloroethylene and Stod. solvent; Density: 9.0 lb/gal; Visc.: 700 cs; HLB: 12.1; Cloud Pt.: < 25 C

**Emulan A.** BASF AG
*Chemical Description:* Oleic acid oxyethylate
*Category:* Emulsifier, nonionic
*Applications:* O/W emulsifier for cutting oils, rolling oils, wiredrawing oils, formwork release oils, washable oils, cold cleaners, metal cleaning and metal polishing emulsions
*Form:* Liquid; Color: Yellowish; Odor: Characteristic
*Properties:* Solubility: Sol. in mineral oils, fatty oils, fatty acids, polar organic solvents; emulsifiable in water; Sp. gr.: ≈ 0.89 (20 C); Visc.: ≈ 120 mPa•s @ 20 C; HLB: ≈ 11; Flash Pt.: ≈ 210 C; Solidification Pt.: ≈ –20 C; Sapon. Value: ≈ 110; Acid No.: 0–1; pH: 6–7.5 (1% aq.)
*Composition:* 100% active
*Warnings and Caution:* Avoid prolonged contact with skin
*References:* Technical leaflet M 5220 e

**Emulan AF.** BASF AG
*Chemical Description:* Fatty alcohol oxyethylate
*Category:* Emulsifier, nonionic
*Applications:* Paraffin wax emulsions, mineral oil emulsions
*Form:* Soft wax; Color: White
*Properties:* Solubility: Sol. in mineral oil and polar organic media; Sp. gr.: ≈ 0.91 @ 50 C; M.P.: ≈ 42 C; HLB: ≈ 11; Flash Pt.: ≈ 190 C; Sapon. Value: Almost 0; Acid No.: Almost 0; pH: 6–7.5 (1% aq.)
*Composition:* 100% active
*Warnings and Caution:* Avoid prolonged contact with skin
*References:* Technical leaflet M 5220 e

**Emulan EL.** BASF AG
*Chemical Description:* Castor oil oxyethylate
*Category:* Emulsifier, nonionic
*Applications:* Emulsifying agent in fatty acids, fatty oils, organic solvents, e.g., plastics, resins, dyes
*Form:* Liquid; Color: Pale yellow; Odor: Characteristic
*Properties:* Solubility: Sol. in water and polar organic media; Sp. gr.: ≈ 1.06 @ 30 C; Visc.: 200–500 mPa•s @ 30 C; HLB: ≈ 14; Flash Pt.: ≈ 280 C; Solidification Pt.: 15–25 C; Sapon. Value: ≈ 60; Acid No.: 0–1; pH: 6–7.5 (1% aq.)
*Composition:* 100% active
*Warnings and Caution:* Avoid prolonged contact with skin
*References:* Technical leaflet M 5220 e

**Emulan FJ.** BASF AG
*Chemical Description:* Fatty acid alkylolamide
*Category:* Emulsifier, nonionic
*Applications:* W/O emulsifier for impregnation, lubrication, polishing and cleaning; O/W emulsifier for anticorrosive component in metal working

*Form:* Liquid; Color: Brownish; Odor: Faint
*Properties:* Solubility: Sol. in mineral oils, fatty oils and mixtures; emulsifiable in water; Sp. gr.: ≈ 0.96 @ 20 C; Visc.: ≈ 200 mPa•s @ 20 C; HLB: 3–5 (W/O type); ≈ 8.5 (O/W type); Solidification Pt.: ≈ −20 C; Sapon. Value: ≈ 150; Acid No.: ≈ 2; pH: ≈ 9 (1% aq.)
*Composition:* 100% active
*Warnings and Caution:* Avoid prolonged contact with skin
*References:* Technical leaflet M 5220 e

**Emulan FM.** BASF AG
*Chemical Description:* Triethanolamine monooleic ester
*Category:* Emulsifier, nonionic
*Applications:* W/O emulsifier for impregnation, lubrication, polishing and cleaning; O/W emulsifier for polishes; rust-preventing additive for metal working and corrosion protection
*Form:* Liquid; Color: Brown; Odor: Characteristic
*Properties:* Solubility: Sol. in mineral oils, fatty oils and mixtures; Sp. gr.: ≈ 0.95 @ 20 C; Visc.: ≈ 200 mPa•s @ 20 C; HLB: 3–5 (W/O type); ≈ 9.5 (O/W type); Flash Pt.: ≈ 240 C; Solidification Pt.: ≈ −5 C; Sapon. Value: ≈ 135; Acid No.: 0–5; pH: ≈ 10 (1% aq.)
*Composition:* 100% active
*Warnings and Caution:* Avoid prolonged contact with skin
*References:* Technical leaflet M 5220 e

**Emulan OC.** BASF AG
*Chemical Description:* Fatty alcohol oxyethylate
*Category:* Emulsifier, nonionic
*Applications:* O/W emulsifier for dry bright emulsions; stabilizing agent for emulsions and dispersions
*Form:* Soft wax; Color: White or pale yellowish
*Properties:* Solubility: Sol. in water, fatty acids, waxes, and polar organic solvents; Sp. gr.: ≈ 1.03 @ 50 C; M.P.: 35–40 C; HLB: ≈ 17; Flash Pt.: ≈ 210 C; Solidification Pt.: ≈ 34 C; Sapon. Value: Almost 0; Acid No.: Almost 0; pH: 6–7.5 (1% aq.)
*Composition:* 100% active
*Warnings and Caution:* Avoid prolonged contact with skin
*References:* Technical leaflet M 5220 e

**Emulan OG.** BASF AG
*Chemical Description:* Fatty alcohol oxyethylate
*Category:* Emulsifier, dispersant, stabilizer, nonionic
*Applications:* O/W emulsifier for fatty acids, waxes, polar organic solvents; dispersant for solid substances, in grinding dyes; stabilizer, e.g., for hydraulic and anticorrosion oils
*Form:* Waxy powder in bead form
*Properties:* Solubility: Sol. in water and polar organic media; Density: ≈ 0.6 kg/l; M.P.: 45–50 C; HLB: ≈ 17; Flash Pt.: ≈ 240 C; Sapon. Value: Almost 0; Acid No.:

Almost 0; pH: 6–7.5 (1% aq.)
*Composition:* 100% active
*Warnings and Caution:* Avoid prolonged contact with skin
*References:* Technical leaflet M 5220 e

**Emulan OSN.** BASF AG
*Chemical Description:* Fatty alcohol oxyethylate
*Category:* Emulsifier, nonionic
*Applications:* O/W emulsifier for dry bright wax emulsions
*Form:* Soft wax; Color: White or pale yellowish
*Properties:* Solubility: Sol. in water, fatty acids, waxes, polar organic solvents; Sp. gr.: ≈ 1.03 @ 50 C; M.P.: 35–40 C; HLB: ≈ 17; Flash Pt.: ≈ 230 C; Solidification Pt.: ≈ 34 C; Sapon. Value: Almost 0; Acid No.: Almost 0; pH: 6–7 (1% aq.)
*Composition:* 100% active
*Warnings and Caution:* Avoid prolonged contact with skin
*References:* Technical leaflet M 5220 e

**Emulan OU.** BASF AG
*Chemical Description:* Fatty alcohol oxyethylate
*Category:* Emulsifier, dispersant, stabilizer, nonionic
*Applications:* O/W emulsifier for fatty acids, waxes, polar organic solvents; dispersing of solid substances, in grinding processes; stabilizing of emulsions and dispersions
*Form:* Soft wax; Color: White or pale yellowish
*Properties:* Solubility: Sol. in water, fatty acids, molten greases and ester waxes, polar organic solvents; Sp. gr.: ≈ 1.02 @ 50 C; M.P.: ≈ 40 C; HLB: ≈ 17; Flash Pt.: ≈ 210 C; Solidification Pt.: ≈ 35 C; Sapon. Value: Almost 0; Acid No.: Almost 0; pH: 6–7.5 (1% aq.)
*Composition:* 100% active
*Warnings and Caution:* Avoid prolonged contact with skin
*References:* Technical leaflet M 5220 e

**Emulan P.** BASF AG
*Chemical Description:* Fatty alcohol oxyethylate
*Category:* Emulsifier, nonionic
*Applications:* Emulsions of highly refined mineral oil fractions, e.g., paraffin oil, washable oils, cold cleaners, other solvent-based cleaning agents
*Form:* Clear liquid; Color: Colorless
*Properties:* Solubility: Sol. in paraffin oil and other mineral oils; emulsifiable in water; Sp. gr.: ≈ 0.91 @ 20 C; Visc.: ≈ 400 mPa•s @ 20 C; HLB: 11–12; Flash Pt.: ≈ 150 C; Solidification Pt.: ≈ 10 C; Sapon. Value: Almost 0; Acid No.: Almost 0; pH: 6–7.5 (1% aq.)
*Composition:* 100% active
*Warnings and Caution:* Avoid prolonged contact with skin
*References:* Technical leaflet M 5220 e

**Emulan PO.** BASF AG
*Chemical Description:* Alkylphe-

nol oxyethylate
*Category:* Emulsifier, nonionic
*Applications:* Cold cleaners and other solvent-based cleaners, formwork release oils, drilling oils
*Form:* Clear liquid; Color: Colorless to pale yellowish
*Properties:* Solubility: Sol. in spindle and other mineral oils, fatty oils and polar organic solvents; emulsifiable in water; Sp. gr.: ≈ 1.04 @ 20 C; Visc.: ≈ 400 mPa•s @ 20 C; HLB: ≈ 10–12; Flash Pt.: ≈ 240 C; Solidification Pt.: ≈ −15 C; Sapon. Value: Almost 0; Acid No.: Almost 0; pH: 6–7.5 (1% aq.)
*Composition:* 100% active
*Warnings and Caution:* Avoid prolonged contact with skin
*References:* Technical leaflet M 5220 e

**Emulan SH.** BASF AG
*Chemical Description:* Nitrogenous carboxylic acid
*Category:* Salts of Emulan SH as emulsifiers, wetting agents, anticorrosive agents, anionic
*Applications:* Salts of Emulan SH as emulsifiers; as anti-corrosive agents in aqueous hydraulic liquids, anticorrosion emulsions and oils; in aq. and solvent-based cleaning agents and metal degreasers, e.g., industrial and workshop cleaners
*Form:* Clear to slightly turbid liquid; Color: Brownish; Odor; Faint
*Properties:* Solubility: In unneutralized form, sol. in mineral oils, fatty oils, organic solvents; when neutralized with alkali or amine, sol. in water; Sp. gr.: ≈ 0.95 @ 20 C; Visc.: ≈ 2500 mPa•s @ 20 C; Flash Pt.: ≈ 225 C; Solidification Pt.: ≈ −10 C; Sapon. Value: ≈ 160; Acid No.: ≈ 158; pH: ≈ 4 (1% aq.)
*Composition:* 100% active
*Warnings and Caution:* Avoid prolonged contact with skin
*References:* Technical leaflet M 5220 e

**Emulan SH 85.** BASF AG
*Chemical Description:* Nitrogenous carboxylic acid
*Category and Applications:* See Emulan SH
*Form:* Clear fluid liquid; Color: Brownish
*Properties:* Solubility: In unneutralized form, sol. in mineral oils, fatty oils, polar organic solvents; when neutralized with alkali or amine, sol. in water; Sp. gr.: ≈ 0.94 @ 20 C; Visc.: ≈ 250 mPa•s @ 20 C; Flash Pt.: ≈ 90 C; Solidification Pt.: ≈ −15 C; Sapon. Value: ≈ 135; Acid No.: ≈ 127; pH: ≈ 4 (1% aq.)
*Composition:* 85% active; 12% isobutanol; 3% water
*Warnings and Caution:* Avoid prolonged contact with skin
*References:* Technical leaflet M 5220 e

**Emulbon Series.** Toho Chem.
*Category:* Emulsifier
*Applications:* Sizing assistant, lubricant
*Form:* Liquid

*Composition:* 100% conc.

**Emulgade F.** Henkel Canada
*Chemical Description:* Blend of Dehydag Wax O and Dehydag Wax E with a nonionic emulsifying agent based on ethoxylated fatty acid
*Applications:* Self-emulsifying raw material for cosmetic creams and fluid emulsions
*Composition:* 95% fatty matter

**Emulgade F Special.** Henkel Canada
*Chemical Description:* Blend of Dehydag Was O with a nonionic emulsifying agent based on ethoxylated fatty acid
*Category:* Raw material, nonionic
*Applications:* Self-emulsifying raw material for fluid emulsions of the oil-in-water type; cosmetic creams and emulsions; raw material for processing cationics
*Composition:* 99% fatty matter

**Emulphogene BC-420.** GAF Corp.
*Chemical Description:* Tridecyloxypoly (ethyleneoxy) ethanol with relatively low ethylene oxide content
*Category:* Emulsifier, detergent, dispersant, intermediate, nonionic
*Applications:* Intermediate in manufacture of high-foaming anionic surfactants of the water-sol. sulfate ester type; oil-sol. detergent and dispersant for petroleum oils
*Form:* Cloudy, slightly viscous but pourable liquid; Color: # 2 max. (VCS @ 50 C); Odor: Mild, pleasant
*Properties:* Solubility: Insol. in water, kerosene; sol. in toluene; Sp. gr.: 0.92–0.94; Cloud Pt.: Insol.; Flash Pt.: > 200 F (P-M, closed cup); Fire Pt.: > 165 C; Pour Pt.: –7 to –1 C; Stability: Stable to sulfuric acid and alkalis; pH: 6.5–8.5 (10% sol'n.)
*Composition:* 100% active ingred.

**Emulphogene BC-610.** GAF Corp.
*Chemical Description:* Tridecyloxypoly (ethyleneoxy) ethanol
*Category:* Detergent, wetting agent, emulsifier, nonionic
*Applications:* Low-foaming detergent and wetting agent for mechanical dishwashing formulations and spray-type alkaline cleaners; in pulp and paper industry; as wool-scouring agent with degreasing action at low temperatures; emulsifier for silicones and petroleum oils
*Form:* Cloudy, slightly viscous pourable liquid; Color: # 2 max. (VCS @ 50 C); Odor: Mild, pleasant
*Properties:* Solubility: Sol. in methanol, ethanol, acetone, toluene; dispersible in water and mineral oil; Sp. gr.: 0.97–0.99; Cloud Pt.: Dispersible (1% sol'n.) in distilled water; Flash Pt.: > 200 F (P-M, closed cup); Fire Pt.: > 165 C; Pour Pt.: 4.5–7.5 C; Stability: Stable to sulfuric acid and alkalis; pH: 6–8 (10% sol'n.)
*Composition:* 100% active ingred.

**Emulphogene BC-630.** GAF Corp.
*Chemical Description:* Tridecyloxypoly (ethyleneoxy) ethanol
*Category and Applications:* See

Emulphogene BC-610; slightly more hydrophilic than BC-610
*Form:* Cloudy, slightly viscous but pourable liquid; Color: # 2 max. (VCS @ 50 C); Odor: Mild, pleasant
*Properties:* Solubility: Sol. in methanol, ethanol, acetone, toluene; partially sol. in tetrachloroethylene; Sp. gr.: ≈ 0.99; Cloud Pt.: 50–60 C (1% sol'n.); Flash Pt.: > 200 F (P-M, closed cup); Fire Pt.: > 165 C; Pour Pt.: 15–20 C; Stability: Stable to sulfuric acid, alkalis and metallic ions; pH: 6–8 (10% sol'n.)
*Composition:* 100% active ingred.

**Emulphogene BC-720.** GAF Corp.
*Chemical Description:* Tridecyloxypoly (ethyleneoxy) ethanol with relatively high ethylene oxide content
*Category:* Foam stabilizer, detergent, solubilizer, wetting and rewetting agent, emulsifier, nonionic
*Applications:* Foam stabilizer and detergent booster for alkylarylsulfonates; solubilizer for salt-free alkylarylsulfonates; light- and heavy-duty detergent formulations; textile detergent in neutral, acid and alkaline media; paper industry for cleaning paper machine felts, rewetting paper towels; wetting agent for mineral acids and corrosion inhibitors; emulsifier and solubilizer for aromatic solvents, fats, essential oils, waxes
*Form:* Opaque, viscous pourable liquid; Color: # 2 max. (VCS @ 50 C); Odor: Mild, pleasant
*Properties:* Solubility: Sol. in water, alcohol, acetone, toluene; partially sol. in tetrachloroethylene; Sp. gr.: 1.00–1.03 (@ 40 C); Cloud Pt.: 58–62 C (1% sol'n.); Flash Pt.: > 200 F (P-M, closed cup); Fire Pt.: > 165 C; Pour Pt.: ≈ 20 C; Stability: Stable to sulfuric acid, alkalis, metallic ions; pH: 6–8 (10% sol'n.)
*Composition:* 100% active ingred.

**Emulphogene BC-840.** GAF Corp.
*Chemical Description:* Tridecyloxypoly (ethyleneoxy) ethanol with relatively high ethylene oxide content
*Category:* Foam stabilizer, detergent, solubilizer, wetting agent, emulsifier, nonionic
*Applications:* Foam stabilizer and detergent booster for alkylarylsulfonates; solubilizer for salt-free alkylarylsulfonates; in light- and heavy-duty detergents; detergent and wetting agent for use with strong alkalis, high temperatures; wetting agent in high concentrations of electrolytes; stabilizer and emulsifier for synthetic latices; emulsifier and solubilizer for aromatic solvents, fats, essential oils, waxes
*Form:* Paste; Color: # 2 max. (VCS @ 50 C); Odor: Mild, pleasant
*Properties:* Solubility: Sol. in water, alcohol, acetone, toluene; partially sol. in tetrachloroethylene; Sp. gr.: 1.01–1.04 (@ 40 C); Cloud Pt.: > 95 C (1% sol'n.); Flash Pt.: > 200 F (P-M, closed cup); Fire Pt.: > 165 C; Pour Pt.: 33–40 C; Stability: Stable to sulfuric acid,

alkalis, metallic ions; pH: 6–8 (10% sol'n.)
*Composition:* 100% active ingred.

**Emulphogene TB-970.** GAF Corp.
*Chemical Description:* Linear aliphatic ethoxylate
*Category:* Detergent, dispersant, textile wet processing aid, nonionic
*Applications:* Commercial and household laundry detergents; wall and floor cleaners; dairy cleaners and sanitizers; dye leveling, scouring and other textile wet processing; automatic bowl cleaners
*Form:* Waxy solid and free-flowing flakes; Color: Off-white
*Properties:* Solubility: Sol. in water, ethanol, carbon tetrachloride, "Cellosolve," xylene (warm) (10% in 90% solvent); Cloud Pt.: > 100 C (1% in distilled water); Solidification Pt.: 50 C min.; Neutralization No.: 0.5 max.; Surface Tension: 50.8 dynes/cm (0.1% conc.)
*Composition:* 100% conc.
*References:* Data sheet 9643-027

**Emulphor EL-620.** GAF Corp.
*Chemical Description:* Polyoxyethylated vegetable oil
*Category:* Emulsifier, dispersant, antistat, lubricant, solubilizer, wetting agent, nonionic
*Applications:* In water-emulsion paints based on vinyl, alkyd or acrylic latices; in flexible and rigid urethane foams; in polyester resin emulsions as binders for glass mats and preforms; for pickle-degreasing, fatliquoring and dyeing of leather; to solubilize basic and acid dyestuffs in paper industry for beater and calender operations; in textile industry as emulsifiers, dyeing assistants, antistats for acrylic fibers, solubilizers; in liquid wax polishes for floors, furniture, automobiles, shoes, etc.; in pesticides and herbicides; magnetic coupling fluids; as emulsifiers for cosmetic and pharmaceutical preparations
*Form:* Clear liquid; Color: Light brown; Odor: Mild oily
*Properties:* Solubility: Sol. in water, acetone, carbon tetrachloride, ethanol, vegetable oil, ether, toluene, butyl Cellosolve, xylene, methanol; Density: 8.7–8.8 lb/gal; Sp. gr.: 1.04–1.05; Visc.: 600–1000 cps; Cloud Pt.: 40–45 C (1% aq. sol'n.); Flash Pt.: 291–295 C; Fire Pt.: 322–326 C; Neutralization No.: 0.5 max.; Stability: Stable to metallic ions and weak alkali; partially stable to 1% acid; Surface Tension: 41 dynes/cm (0.1% sol'n.)
*Composition:* 100% active ingred.
*References:* Booklet 9655-030

**Emulphor EL-719.** GAF Corp.
*Chemical Description:* Polyoxyethylated vegetable oil
*Category and Applications:* See Emulphor EL-620
*Form:* Clear liquid; Color: Yellow; Odor: Oily
*Properties:* Solubility: Sol. in water, acetone, carbon tetrachloride, ethanol, vegetable oil, ether, toluene, butyl Cellosolve, xylene,

methanol; Density: 8.9–9.0 lb/gal; Sp. gr.: 1.06–1.07; Visc.: 500–800 cps; Cloud Pt.: 83–88 C (1% aq. sol'n.); Flash Pt.: 275–279 C; Fire Pt.: 328–332 C; Neutralization No.: 0.5 max.; Stability: Stable to metallic ions, weak alkali; partially stable to 1% acid; Surface Tension: 38 dynes/cm (0.1% sol'n.)
*Composition:* 96% min. active ingred.
*References:* Booklet 9655=030

**Emulphor EL-980.** GAF Corp.
*Chemical Description:* Polyoxyethylated vegetable oil
*Category:* Emulsifier, antistat, lubricant, nonionic
*Applications:* Spin-finish additives esp. for high temperature use; in over-spray compositions; dyeing assistants
*Form:* Waxy solid; Color: 5 max. (VCS)
*Properties:* Solubility: Readily sol. in water; Neutralization No.: 0.5 max.
*Composition:* 100% active ingred.
*References:* Data sheet 9643-014R

**Emulphor EL-985.** GAF Corp.
*Chemical Description:* Polyoxyethylated vegetable oil
*Category and Applications:* See Emulphor EL-980
*Form:* Viscous liquid; Color: 4 max. (VCS)
*Properties:* Solubility: Readily sol. in water; Neutralization No.: 0.5 max.
*Composition:* 50% active ingred.
*References:* Data sheet 9643-014R

**Emulphor ON-870.** GAF Corp.
*Chemical Description:* Polyoxyethylated oleyl alcohol
*Category:* Emulsifier, dispersant, stabilizer, solubilizer, wetting agent, nonionic
*Applications:* To prevent premature coagulation and stabilize natural or synthetic latices in rubber-based paints, adhesives, etc.; in textile industry as emulsifiers in acid degreasing, in dyeing wool/acrylic blends, as dispersing and stabilizing agent in haphthol dyeing; in leather degreasing, dyeing and fatliquoring; as emulsifiers for cosmetic and pharmaceutical creams and lotions; as solubilizer for essential oils and drugs; as wetting agent in metal cleaners with mineral acids and corrosion inhibitors
*Form:* Solid wax; Color: White
*Properties:* Solubility: Sol. in water, xylene, ethanol, ethylene glycol, butyl Cellosolve; Sp. gr.: 1.04; Cloud Pt.: > 100 C; Flash Pt.: > 93 C (P-M, closed cup); Pour Pt.: 46 C; Solidification Pt.: 43 C; Surface Tension: 37 dynes/cm (0.1% conc.)
*Composition:* 100% active ingred.
*Toxicity:* In absence of specific tests, regard as probable eye irritant
*References:* Data sheet 9643-023

**Emulphor ON-877.** GAF Corp.
*Chemical Description:* Polyoxyethylated oleyl alcohol
*Category and Applications:* See

Emulphor ON-870
*Form:* Pourable, viscous liquid; Color: 3 max. (VCS)
*Properties:* Solubility: Sol. in water, ethanol, ethylene glycol, butyl Cellosolve; Sp. gr.: 1.06; Cloud Pt.: 95–100 C (1% sol'n.); Flash Pt.: > 93 C (P-M, closed cup); Pour Pt.: 15 C; Solidification Pt.: 9 C; Surface Tension: 40.5 dynes/cm (0.1% conc.)
*Composition:* 70% active ingred.
*References:* Data sheet 9643-023

**Emulphor VN-430.** GAF Corp.
*Chemical Description:* Polyoxyethylated oleic acid
*Category:* Emulsifier, lubricant, nonionic
*Applications:* In textile industry as lubricant for wool, in cotton and rayon finishing, as dyeing aids; in leather processing to degrease pickled skins, improve fatliquoring, increase tensile strength of chrome side leathers; as oil-sol. surfactant in cutting oils, degreasing solvents and cleaners; as W/O emulsifiers for cold creams, lotions, hair-grooming preparations and other cosmetics
*Form:* Clear liquid; Color: Dark amber
*Properties:* Solubility: Dispersible in water; sol. in xylene, white mineral oil, ethanol, butyl Cellosolve; Sp. gr.: ≈ 1.0; Cloud Pt.: < 10 C (1% sol'n.); Flash Pt.: > 93 C (P-M, closed cup); Pour Pt.: 13 C; Solidification Pt.: 10 C; Surface Tension: 32 dynes/cm (0.1% conc.); pH: 7–7.5
*Composition:* 100% active ingred.
*Reaction:* Neutral
*References:* Bulletin 9643-022

**Emulphor VT-650.** GAF Corp.
*Chemical Description:* Ethoxylated stearic acid
*Category:* Emulsifier, lubricant, nonionic
*Applications:* Emulsifier for silicone and nonsilicone defoamers, low-foaming emulsifier/lubricant for dye carriers, emulsifier additive for mineral oil, alkyl esters and triglycerides; softener and lubricant for textile processing esp. as additive in oversprays and in spin finishes; dye assistant
*Form:* Soft wax; Color: 2 (VCS); Odor: Slight acetic acid
*Properties:* Ref. Index: ≈ 1.451; Solubility: Sol. in xylene, kerosene, ethanol, carbon tetrachloride, butyl Cellosolve and perchloroethylene; Density: 8.3 lb/gal; Sp. gr.: 0.99 (@ 50 C); Visc.: ≈ 9 cs (@ 100 C); Flash Pt.: 162 F (CC); Solidification Pt.: 32 C; Sapon. Value: 83–94; Neutralization No.: 0.5 max.
*Composition:* 100% active ingred.
*References:* Data sheet 9643-030

**Emulsan O.** Reilly-Whiteman Inc.
*Chemical Description:* Polyethylene glycol 400 monooleate
*Category:* Emulsifier, nonionic
*Applications:* Same as Emulsan K but with better oil solubility
*Form:* Liquid
*Composition:* 99% active ingred.

**Emulsifier WHC.** Stepan Chem. Co.
*Category:* Emulsifier

*Applications:* Waterless hand cleaner formulations
*Form:* Liquid; Color: Light amber
*Properties:* Solubility: Dispersible in water; sol. in kerosene; Density: 8.25 lb/gal; pH: 8 (1% aq.)

**Emulso E-309.** Toho Chem.
*Category:* Emulsifier
*Applications:* Oil spill dispersant
*Form:* Liquid
*Properties:* Sp. gr.: 0.789; Visc.: 4.1 cst (@ 30 C)

**Emulton AA.** Shamtan Lubricants & Chem. Ltd.
*Chemical Description:* Fatty acid-alkanolamine condensate
*Category:* Emulsifier, detergent, thickener, nonionic
*Composition:* 100% active ingred.

**Emulton G.** Shamtan Lubricants & Chem. Ltd.
*Chemical Description:* Fatty acid-glycerol esters
*Category:* Emulsifier, stabilizer, nonionic
*Form:* Solid or liquid
*Composition:* 100% active ingred.

**Emulton PG.** Shamtan Lubricants & Chem. Ltd.
*Chemical Description:* Fatty acid-polyglycol esters
*Category:* Emulsifier, lubricant, plasticizer, wetting agent, softener, defoamer, nonionic
*Form:* Solid or liquid
*Composition:* 100% active ingred.

**Emulton PS.** Shamtan Lubricants & Chem. Ltd.
*Chemical Description:* Petroleum sulfonates from sodium, calcium, barium and ammonium salts
*Category:* Emulsifiers, rust preventives and oil additives, anionic
*Composition:* ≈ 60% active in mineral oil

**Emulvis.** C.P. Hall Co.
*Chemical Description:* Polyoxyethylene stearate
*Category:* Viscosity builder, solubility retarder, nonionic
*Applications:* Viscosity builder in cosmetics (shampoos, hair rinses, skin cleaners, etc.); solubility retarder for gum arabic, methyl cellulose, natural gums and resins, in timed release tablets and coatings, soap bars; antiblock agent for decals; evaporation control for farm ponds; viscosity improver for inks; thickener in pet shampoos; melt point control in suppositories
*Form:* Flake; Color: White to cream; Odor: Bland
*Properties:* Solubility: 50% sol. in 70% isopropyl alcohol; 3% sol. in water (greater water sol. at high temperatures); M.P.: 53–58 C; pH: 4–6 (3% aq.)
*Reaction:* Acid

**Endet.** Swastik
*Chemical Description:* Heavy duty detergent powder containing proteolytic enzyme
*Category:* Detergent
*Applications:* Household heavy duty washing powder esp. for dislodging stubborn proteinous stains
*Form:* Spray dried powder (white)

and colored enzyme beads
*Properties:* pH: 9.5 (1% sol'n. in distilled water)

**Epicol TAS80V.** Albright & Wilson Ltd. Marchon Div.
*Chemical Description:* Sodium tallow alcohol sulfate
*Category:* Detergent
*Applications:* Heavy duty detergent formulations with softening effects; controlled foam formulations; esp. useful in 50/50 blend with sodium alkyl benzene sulfonates; ore flotation agent; wetting and emulsifying agents
*Form:* Needles; Color: Cream
*Properties:* pH: 10
*Composition:* 80% active ingred.
*Reaction:* Alkaline;

**Epoxol 5-2E.** Swift Chem. Co.
*Chemical Description:* Epoxidized octyl tallate
*Category:* Stabilizer
*Applications:* High oxirane stabilizing plasticizers with low volatility; produces hardness in semi-rigid vinyl formulations; extrusion aid; minimizes air release problems in plastisols
*Form:* Color: 1 (Gardner)
*Properties:* Sp. gr.: 0.9244; Visc.: A (Gardner); Sapon. Value: 139.8; Acid No.: 0.5
*References:* Booklet 3MDP7

**Epoxol 7-4.** Swift Chem. Co.
*Chemical Description:* Epoxidized soybean oil
*Category:* Plasticizer
*Applications:* High oxirane stabilizing plasticizer; FDA approved for food packaging materials
*Form:* Color: 70 (Cobalt-platinum); Odor: Low; Taste: Low
*Properties:* Ref. Index: 1.4705; Solubility: Miscible with esters, ketones, heptane, ethyl ether, vinyl plasticizers, aromatic and chlorinated hydrocarbons; Density: 8.28 lb/gal; Sp. gr.: 0.994; Visc.: 314 cps; Flash Pt.: 590 F (COC); Pour Pt.: 25 F; Sapon. Value: 178.1; Acid No.: 0.10; Hydroxyl No.: 6.5
*Containers:* Drums, tank wagons
*References:* Booklet 3MDP7

**Epoxol 8-2B.** Swift Chem. Co.
*Chemical Description:* Epoxidized butyl esters of linseed oil fatty acids
*Category:* Plasticizer
*Applications:* High oxirane stabilizing plasticizers with low viscosity, low volatility, excellent heat and light stabilization; FDA approved for food packaging materials
*Form:* Color: 1 (Gardner)
*Properties:* Sp. gr.: 0.965; Visc.: A (Gardner); Sapon. Value: 150; Acid No.: 0.25–0.50
*References:* Booklet 3MDP7

**Epoxol 9-5.** Swift Chem. Co.
*Chemical Description:* Epoxidized linseed oil
*Category:* Plasticizer
*Applications:* High oxirane stabilizing plasticizer; FDA approved for food packaging materials
*Form:* Color: 70 (Cobalt-platinum); Odor: Low; Taste: Low
*Properties:* Ref. Index: 1.4715;

Solubility: Miscible with esters, ketones, heptane, ethyl ether, vinyl plasticizers, aromatic and chlorinated hydrocarbons; Density: 8.58 lb/gal; Sp. gr.: 1.03; Visc.: 619 cps; Flash Pt.: 595 F (COC); Pour Pt.: 30 F; Sapon. Value: 172; Acid No.: 0.12; Hydroxyl No.: 12
*References:* Booklet 3MDP7

**Equex AEM.** Procter & Gamble
*Chemical Description:* Ammonium lauryl ether sulfate with alkanolamide added
*Category:* Detergent, foamer, anionic
*Applications:* Concentrate for foaming bubble baths, low pH shampoos, etc.
*Form:* Clear liquid; Color: Light amber; Odor: Sweet, pleasant
*Properties:* Density: 8.2 lb/gal; Sp. gr.: 0.98; Visc.:60 cps (Brookfield); Cloud Pt.: ≈ 40 F; Solidification Pt.: ≈ –10 F; pH: 6.3 (as is)
*Composition:* 37.5% active ingred.; other components: 9% alkanolamide; 1.2% petroleum ether extract; 20% denatured ethyl alcohol

**Equex S.** Procter & Gamble
*Chemical Description:* Sodium lauryl sulfate, cosmetic grade
*Category:* Detergent, emulsifier, anionic
*Applications:* Shampoos, liquid cleaners, specialty products
*Form:* Liquid
*Properties:* Density: 8.7 lb/gal; Sp. gr.: 1.04; Visc.: 50–200 cps (Brookfield); Cloud Pt.: ≈ 54 F; Solidification Pt.: ≈ 35 F; pH: 7.8 (10% sol'n.)
*Composition:* 29% active ingred.

**Equex SP.** Procter & Gamble
*Chemical Description:* Sodium lauryl sulfate, polymerization grade
*Category:* Emulsifier, anionic
*Applications:* Emulsion polymerization
*Form:* Clear liquid; Color: Pale yellow; Odor: Clean, pleasant
*Properties:* Density: 8.7 lb/gal; Sp. gr.: 1.04; Visc.: ≈ 90 cps (Brookfield @ 80 F); Cloud Pt.: ≈ 55 F; Solidification Pt.: ≈ 34 F; pH: 8 (10% sol'n.)
*Composition:* 29.8% active ingred.

**Equex STM.** Procter & Gamble
*Chemical Description:* Sodium and triethanolamine lauryl sulfate blend with alkanolamide added
*Category:* Detergent, foamer, anionic
*Applications:* Concentrate for shampoos, cleaners, specialty
*Form:* Thick, honey-like fluid; Color: Opaque white to straw; Odor: Clean, pleasant
*Properties:* Density: 8.7 lb/gal; Sp. gr.: 1.04; Visc.: 9000 cps (Brookfield); Cloud Pt.: 52–55 F; Solidification Pt.: 32–34 F; pH: 8 (as is)
*Composition:* 22.5% active ingred.; other components: 7.5% alkanolamide

**Equex SW.** Procter & Gamble
*Chemical Description:* Sodium

lauryl sulfate
*Category:* Detergent, wetting agent, emulsifier, anionic
*Applications:* Cream shampoos, paste cleaners, rug shampoos
*Form:* Thick, honey-like fluid; Color: Pale yellow; Odor: Clean, pleasant
*Properties:* Density: 8.8 lb/gal; Sp. gr.: 1.05; Visc.: ≈ 40,000 cps (Brookfield @ 80 F); Cloud Pt.: ≈ 77 F; Solidification Pt.: ≈ 55 F; pH: 7.8 (10% sol'n.)
*Composition:* 28.5% active ingred.

**Equex T.** Procter & Gamble
*Chemical Description:* Triethanolamine lauryl sulfate
*Category:* Detergent, anionic
*Applications:* Shampoos
*Form:* Clear liquid; Color: Water white to pale yellow; Odor: Clean, pleasant
*Properties:* Density: 8.7 lb/gal; Sp. gr.: 1.04; Visc.: ≈ 50 cps (Brookfield @ 80 F); Cloud Pt.: ≈ 28 F; Solidification Pt.: ≈ 18 F; pH: 7.3 (10% sol'n.)
*Composition:* 40% active ingred.

**ESI-Terge 40% Coconut Oil Soap.** Emulsion Systems
*Chemical Description:* Cochin oil soap
*Category:* Detergent, anionic
*Applications:* Liquid hand soap
*Form:* Liquid; Color: Straw; Odor: Coconut
*Properties:* Solubility: Water sol.; Density: 8.5 lb/gal; Visc.: Water-like; F.P.: 320 F
*Composition:* 40% active ingred.
*Reaction:* Alkaline
*Containers:* Drums

**ESI-Terge 320.** Emulsion Systems
*Chemical Description:* Phosphated nonylphenoxy polyethoxy ethanol
*Category:* Detergent, anionic
*Applications:* Detergent, viscosity aid, self-coupling wetting agent in synthetic cleaners, hard surface cleaners, wax strippers and compounds with high caustic concentrations
*Form:* Liquid; Color: Light straw; Odor: Mild
*Properties:* Solubility: Water sol.; Density: 9.34 lb/gal; Acid No.: 100 ± 10
*Composition:* 100% active ingred.
*Containers:* Drums
*Warnings and Cautions:* Avoid skin contact
*References:* Bulletin G.P. 5

**ESI-Terge 330.** Emulsion Systems
*Chemical Description:* Phosphated glycol ester
*Category:* Detergent, wetting agent, anionic
*Applications:* Self-coupling wetting agent for high alkaline hard surface cleaner; low to intermediate viscosity detergents; steam cleaners; cement cleaners; wax strippers; oven cleaners
*Form:* Liquid; Color: Light straw; Odor: Mild
*Properties:* Solubility: Water sol.; Density: 9.18 lb/gal; Acid No.: 285 ± 25
*Composition:* 100% active ingred.
*Reaction:* Acid

*Containers:* Drums
*Warnings and Caution:* Avoid skin contact
*References:* Bulletin GP-6

**ESI-Terge B-15.** Emulsion Systems
*Chemical Description:* 2:1 coconut diethanolamide
*Category:* Detergent, wetting agent, nonionic
*Applications:* Detergent for mild all purpose cleaners; detergent foam builder and thickener for all purpose cleaners
*Form:* Liquid; Color: Amber
*Properties:* Solubility: Water sol.; Density: 8.4 lb/gal; Sp. gr.: 1.01; pH: 9.8–10.8 (1% sol'n.); Biodegradable
*Composition:* 100% active ingred.
*Reaction:* Alkaline
*Containers:* 55 gal drums
*References:* Bulletin GP-3

**ESI-Terge C-5.** Emulsion Systems
*Chemical Description:* Modified coconut diethanolamide
*Category:* Detergent, wetting agent, nonionic
*Applications:* Detergent for mild all purpose cleaners; detergent foam builder and thickener
*Form:* Liquid; Color: Amber
*Properties:* Solubility: Water sol.; Density: 8.1 lb/gal; Sp. gr.: 0.97; pH: 8.5–9.5 (1% sol'n.); Biodegradable
*Composition:* 100% active ingred.
*Reaction:* Alkaline
*Containers:* 55 gal drums
*References:* Bulletin GP-3

**ESI-Terge HA-20.** Emulsion Systems
*Chemical Description:* Modified amine condensate
*Category:* Detergent base, anionic/nonionic
*Applications:* Self-coupling detergent for high and low pH cleaners compatible with glycol ethers, high viscosity all purpose cleaners; high concentrate high alkali cleaners
*Form:* Liquid; Color: Amber
*Properties:* Solubility: Water and some solvents; Density: 8.64 lb/gal; Acid No.: 60 ± 10; pH: 7.5–8.5 (1% sol'n.)
*Composition:* 100% active ingred.
*Reaction:* Alkaline
*Containers:* 55 gal drums
*References:* Bulletin GP-7

**ESI-Terge RT-61.** Emulsion Systems
*Applications:* Grease and oil emulsifiers; detergent formulations
*Form:* Liquid; Color: Water clear
*Properties:* Density: 8 lb/gal; Visc.: 45 cps (LV #1 spindle); pH: 7.5 (1% sol'n.)
*Composition:* 90% active ingred.

**ESI-Terge S-10.** Emulsion Systems
*Chemical Description:* 1:1 coconut diethanolamide
*Category:* Detergent, emulsifier, nonionic
*Applications:* Foam stabilizer, thickener; liquid dishwashing and car washing detergent
*Form:* Liquid; Color: Light straw
*Properties:* Density: 8.15 lb/gal; Sp. gr.: 0.98; pH: 9–10; Biodegradable
*Composition:* 100% active ingred.

*Reaction:* Alkaline
*Containers:* 55 gal drums
*References:* Bulletin GP-3

**ESI-Terge T-60.** Emulsion Systems
*Chemical Description:* Triethanolamine dodecyl benzene sulfonate
*Category:* Detergent, wetting agent, foam stabilizer, anionic
*Applications:* Car washing and dishwashing detergent, synthetic hand soap
*Form:* Liquid; Color: Amber
*Properties:* Density: 9.1 lb/gal; Sp. gr.: 1.09; F.P.: 320 F; pH: 6.5–7.5 (1% sol'n.); Biodegradable
*Composition:* 60% active ingred.
*Reaction:* Neutral
*Containers:* 500 lb net wt drums
*References:* Bulletin GP-4

**Estrasan 1-A.** Reilly-Whiteman Inc.
*Chemical Description:* Methyl ricinoleate
*Category:* Lubricity aid, nonionic
*Applications:* Enhances lubricity and wettability of petroleum oils for metals
*Form:* Liquid
*Composition:* 100% active

**Estrex 1G58.** Swift Chem.
*Chemical Description:* Glycerol esters
*Category:* Lubricant
*Applications:* Ironer's wax
*Form:* Beads, flakes; Color: White
*Properties:* M.P.: 57–62 C; pH: 6–7
*Composition:* 100% active ingred.
*Containers:* Drums

**Ethoduomeen T/13.** Akzo Chemie
*Chemical Description:* N,N′,N′-tris (2-hydroxyethyl)-N-tallow-1,3 diamminopropane
*Category:* Emulsifier
*Applications:* Making of bitumen emulsions, dispersing waxes
*Form:* Liquid; Color: 18 max. (Gardner)
*Properties:* Sp. gr.: 0.95; F.P.: 20 C; M.P.: 17 C; B.P.: 150 C min.; Flash Pt.: 204 C (COC)
*Composition:* 95% min. active ingred.; other components: 2% max. primary and secondary amines; 1% max. moisture
*Toxicity:* Skin irritant, severe eye irritant
*Containers:* 200 lb bung type steel drums, net 180 kg
*Warnings and Caution:* Avoid contact with strong oxidizing agents

**Ethoduomeen T/20.** Akzo Chemie
*Chemical Description:* N,N′,N′-polyoxyethylene(10)-N-tallow-1,3-diaminopropane
*Category:* Wetting agent
*Applications:* Coating preparation on paperboard
*Form:* Liquid; Color: 18 max. (Gardner)
*Properties:* Sp. gr.: 0.99; F.P.: < 25 C; B.P.: 150 C min.; Flash Pt.: 90 C min. (P-M, closed cup)
*Composition:* 95% min. active ingred.; 2% max. primary and sec. amines; 1% max. moisture
*Toxicity:* Skin irritant, severe eye irritant
*Containers:* 200 lb bung type steel drums, net 180 kg
*Warnings and Caution:* Avoid contact with strong oxidizing agents

**Ethoduomeen T/25.** Akzo Chemie
*Chemical Description:* N,N′,N′-polyoxyethylene (15)-N-tallow-1,3-diaminopropane
*Category:* Corrosion inhibitor
*Applications:* Corrosion inhibitor in water treatment chemicals in secondary oil recovery
*Form:* Liquid; Color: 18 min. (Gardner)
*Properties:* Sp. gr.: 1.02; M.P.: 25 C max. Flash Pt.: 238 C (COC)
*Composition:* 95% active ingred.; other components: 2% max. prim. and sec. amines; 1% max. moisture
*Toxicity:* Skin irritant, severe eye irritant
*Containers:* Bung type drum, net 180 kg
*Warnings and Caution:* Avoid contact with strong oxidizing agents

**Ethoduomeen TD/13.** Armak
*Chemical Description:* Ethoxylated diamine from tallow fatty acid
*Category and Applications:* See Ethoduomeen T/13
*Form:* Clear liquid; Color: 4 max. (Gardner)
*Properties:* Sp. gr.: 0.95; Surface Tension: 34.2 dynes/cm (0.1% sol'n.)
*References:* Bulletin 73-7, 78-1

**Ethoduquad T/20.** Armak
*Chemical Description:* Polyethoxylated quaternary ammonium salt from tallow fatty acid
*Category and Applications:* See Ethoquad C/12
*Form:* Clear liquid
*Properties:* pH: 6–9 (10% aq.)
*Composition:* 72% min. active
*References:* Bulletin 73-7

**Ethofat 60/15.** Armak
*Chemical Description:* Ethylene oxide condensate of stearic acid
*Category and Applications:* See Ethofat C/15
*Form:* Soft paste; Color: 8 max. (Gardner)
*Properties:* Solubility: Sol. in acetone, isopropanol, carbon tetrachloride, dioxane, benzene; disperses in water; Sp. gr.: 1.01; Sapon. Value: 110–120; Surface Tension: 39 dynes/cm (0.1% sol'n.); pH: 6.0–7.5 (10% aq.)
*References:* Bulletin No. 73-7

**Ethofat 60/20.** Armak
*Chemical Description:* Ethylene oxide condensate of stearic acid
*Category and Applications:* See Ethofat C/15
*Form:* Clear liquid; Color: 8 max. (Gardner)
*Properties:* Solubility: Sol. in acetone (< 80 C); isopropanol (< 50 C), carbon tetrachloride (< 75 C), dioxane (< 75 C), benzene; disperses in water; Sp. gr.: 1.02; Sapon. Value: 70–80; Surface Tension: 36 dynes/cm (0.1% sol'n.); pH: 6.0–7.5 (10% aq.)
*Containers:* 55 gal unlined steel drums
*References:* Bulletin No. 73-7, 78-1

**Ethofat 60/25.** Armak Co.
*Chemical Description:* Ethylene

oxide condensate of stearic acid
*Category and Applications:* See Ethofat C/15
*Form:* Paste; Color: 11 max. (Gardner)
*Properties:* Sapon. Value: 55–65; pH: 6.0–7.5 (10% aq.)
*References:* Bulletin No. 73-7

**Ethofat 142/20.** Armak
*Chemical Description:* Ethylene oxide condensate of 4% rosin fatty acid
*Category and Applications:* See Ethofat C/15
*Form:* Clear liquid; Color: 11 max. (Gardner)
*Properties:* Sp. gr.: 1.03; Sapon. Value: 72–82; Surface Tension: 39.5 dynes/cm (0.1% sol'n.); pH: 6.0–7.5 (10% aq.)
*References:* Bulletin 73-7

**Ethofat 242/25.** Armak
*Chemical Description:* Ethylene oxide condensate of 50% rosin fatty acids
*Category and Applications:* See Ethofat C/15
*Form:* Clear liquid; Color: 12 max. (Gardner)
*Properties:* Solubility: Sol. in acetone, isopropanol, carbon tetrachloride, dioxane, benzene, water (viscous sol'n.); Sp. gr.: 1.08; Sapon. Value: 55–65; Surface Tension: 42 dynes/cm (0.1% sol'n.); pH: 6.0–7.5 (10% aq.)
*Composition:* 100% conc.
*Containers:* 55 gal unlined steel drums
*References:* Bulletin 73-7

**Ethofat C/15.** Armak
*Chemical Description:* Ethylene oxide condensate of coco fatty acid
*Category:* Emulsifier, detergent, wetting agent, dispersant, suspending agent, nonionic
*Applications:* Textile, cosmetic, agricultural chemicals, metal working, leather treating, petroleum production, paints, etc.
*Form:* Clear liquid; Color: 9 (Gardner)
*Properties:* Solubility: Sol. in acetone, isopropanol, carbon tetrachloride, dioxane, benzene; disperses in water $< 70$ C; Sp. gr.: 1.00; HLB: 10.6; Sapon. Value: 120–130; Surface Tension: 33 dynes/cm (0.1% sol'n.); pH: 6–7.5 (10% aq.)
*Composition:* 100% conc.
*References:* Bulletin No. 73-7

**Ethofat C/25.** Armak
*Chemical Description:* Ethylene oxide condensate of coco fatty acid
*Category and Applications:* See Ethofat C/15
*Form:* Paste; Color: 8 (Gardner)
*Properties:* Sapon. Value: 60–70; pH: 6.0–7.5 (10% aq.)
*References:* Bulletin No. 73-7

**Ethofat O/15.** Armak
*Chemical Description:* Ethylene oxide condensate of oleic acid
*Category and Applications:* See Ethofat C/15
*Form:* Clear liquid; Color: 9 max. (Gardner)
*Properties:* Solubility: Sol. in ace-

tone (> 25 C), isopropanol, carbon tetrachloride (cloudy), dioxane, benzene; disperses in water; Sp. gr.: 0.99; HLB: 8.6. Sapon. Value: 110–120; Surface Tension: 35 dynes/cm (0.1% sol'n.); pH: 6.0–7.5 (10% aq.)
*Composition:* 100% conc.
*References:* Bulletin No. 73-7

**Ethofat O/20.** Armak
*Chemical Description:* Ethylene oxide condensate of oleic acid
*Category and Applications:* See Ethofat C/15
*Form:* Clear liquid; Color: 9 max. (Gardner)
*Properties:* Solubility: Sol. in acetone (> 50 C), isopropanol, carbon tetrachloride, dioxane, Stoddard solvent (> 25 C), benzene, water (> 75 C); Sp. gr.: 1.03; Sapon. Value: 75–85; Surface Tension: 41 dynes/cm (0.1% sol'n.)
*Composition:* 100% conc.
*Reaction:* pH = 6.0–7.5 (10% aq. sol'n.)
*References:* Bulletin No. 73-7

**Ethomeen 18/12.** Armak
*Chemical Description:* Ethoxylated tertiary amine from stearic acid
*Category and Applications:* See Ethomeen C/12
*Form:* Solid; Color: 7 max. (Gardner)
*Properties:* Solubility: Sol. in acetone (< 25 C), isopropanol, carbon tetrachloride, benzene; insol. water; Sp. gr.: 0.96
*Composition:* 95% tertiary amine
*Toxicity:* Corrosive, skin irritant, severe eye irritant
*References:* Bulletin 73-7, 78-1

**Ethomeen 18/15.** Armak
*Chemical Description:* Ethoxylated tertiary amine from stearic acid
*Category and Applications:* See Ethomeen C/12
*Form:* Liquid to paste; Color: 8 max. (Gardner)
*Properties:* Solubility: Sol. in acetone (cloudy @ < 50 C), isopropanol (< 75 C), carbon tetrachloride (cloudy), dioxane (< 75 C), Stoddard solvent (cloudy), benzene; forms gel in water; Sp. gr.: 0.98; Surface Tension: 34 dynes/cm (0.1% sol'n.)
*Toxicity:* Corrosive, skin irritant, severe eye irritant
*References:* Bulletin 73-7, 78-1

**Ethomeen 18/20.** Armak
*Chemical Description:* Ethoxylated tertiary amine from stearic acid
*Category and Applications:* See Ethomeen C/12
*Form:* Clear liquid; Color: 8 max. (Gardner)
*Properties:* Sp. gr.: 1.02; Surface Tension: 40 dynes/cm (0.1% sol'n.)
*Toxicity:* Skin irritant, severe eye irritant
*References:* Bulletin 73-7, 78-1

**Ethomeen 18/25.** Armak Co.
*Chemical Description:* Ethoxylated tertiary amine from stearic acid

*Category and Applications:* See Ethomeen C/12
*Form:* Clear liquid; Color: 8 max. (Gardner)
*Properties:* Sp. gr.: 1.04; Surface Tension: 43 dynes/cm (0.1% sol'n.)
*References:* Bulletin 73-7, 78-1

**Ethomeen 18/60.** Armak
*Chemical Description:* Ethoxylated tertiary amine from stearic acid
*Category and Applications:* See Ethomeen C/12
*Form:* Solid; Color: White
*Properties:* Solubility: Sol. in acetone (< 100 C), isopropanol (< 100 C), carbon tetrachloride, dioxane (< 100 C), benzene (< 100 C), water; Sp. gr.: 1.12; Surface Tension: 49 dynes/cm (0.1% sol'n.)
*Toxicity:* Mild skin and eye irritant
*References:* Bulletin 73=7, 78-1

**Ethomeen C/12.** Armak
*Chemical Description:* Ethoxylated tertiary amine from coco fatty acid
*Category:* Emulsifier, dispersant, cationic
*Applications:* Textile processing
*Form:* Clear liquid; Color: 6 max. (Gardner)
*Properties:* Solubility: Sol. in acetone, isopropanol, carbon tetrachloride, Stoddard solvent, benzene; forms gel in water; Sp. gr.: 0.87
*Composition:* 95% min. tertiary amine
*Toxicity:* Skin irritant, severe eye irritant
*References:* Bulletin 73-7

**Ethomeen C/15.** Armak
*Chemical Description:* Ethoxylated tertiary amine from coco fatty acid
*Category and Applications:* See Ethomeen C/12
*Form:* Clear liquid; Color: 6 max. (Gardner)
*Properties:* Solubility: Sol. in acetone (> 25 C), isopropanol, carbon tetrachloride (> 25 C), dioxane, Stoddard solvent (> 10 C), benzene, water (cloudy); Sp. gr.: 0.98; HLB: 13.9; Surface Tension: 33 dynes/ cm (0.1% sol'n.)
*Toxicity:* Skin irritant, severe eye irritant
*References:* Bulletin 73-7, 78-1

**Ethomeen C/20.** Armak
*Chemical Description:* Ethoxylated tertiary amine from coco fatty acid
*Category and Applications:* See Ethomeen C/12
*Form:* Clear liquid; Color: 11 max. (Gardner)
*Properties:* Solubility: Sol. in acetone, isopropanol, carbon tetrachloride (> 25 C), dioxane, Stoddard solvent (> 75 C), benzene, water (cloudy); Sp. gr.: 1.02; Surface Tension: 39 dynes/cm (0.1% sol'n.)
*Toxicity:* Skin irritant, severe eye irritant
*References:* Bulletin 73-7, 78-1

**Ethomeen C/25.** Armak
*Chemical Description:* Ethoxyl-

ated tertiary amine from coco fatty acid
*Category and Applications:* See Ethomeen C/12
*Form:* Clear liquid; Color: 12 max. (Gardner)
*Properties:* Solubility: Sol. in acetone, isopropanol, carbon tetrachloride (> 25 C), dioxane, Stoddard solvent (> 80 C), benzene, water; Sp. gr.: 1.04; Surface Tension: 41 dynes/cm (0.1% sol'n.)
*Toxicity:* Skin irritant, severe eye irritant
*References:* Bulletin 73-7, 78-1

**Ethomeen O/12.** Armak
*Chemical Description:* Ethoxylated tertiary amine from oleic acid
*Category and Applications:* See Ethomeen C/12
*Form:* Clear liquid; Color: 8 max. (Gardner)
*Properties:* Sp. gr.: 0.90; Surface Tension: 31.5 dynes/cm (0.1% sol'n.)
*Toxicity:* Skin irritant, severe eye irritant
*References:* Bulletin 73-7, 78-1

**Ethomeen O/15.** Armak
*Chemical Description:* Ethoxylated tertiary amine from oleic acid
*Category and Applications:* See Ethomeen C/12
*Form:* Clear liquid; Color: 8 max. (Gardner)
*Properties:* Sp. gr.: 0.96; Surface Tension: 35.3 dynes/cm (0.1% sol'n.)
*Toxicity:* Skin irritant, severe eye irritant
*References:* Bulletin 73-7, 78-1

**Ethomeen O/25.** Armak
*Chemical Description:* Ethoxylated tertiary amines from oleic acid
*Category and Applications:* See Ethomeen C/12
*Form:* Paste; Color: 8 max. (Gardner)
*Toxicity:* Skin irritant, severe eye irritant
*References:* Bulletin 73-7, 78-1

**Ethomeen S/12.** Armak
*Chemical Description:* Ethoxylated tertiary amine from soya fatty acid
*Category and Applications:* See Ethomeen C/12
*Form:* Clear heavy liquid; Color: 14 max. (Gardner)
*Properties:* Solubility: Sol. in acetate, isopropanol, carbon tetrachloride, Stoddard solvent, benzene; insol. water; Sp. gr.: 0.91; Surface Tension: 31.4 dynes/cm (0.1% sol'n.)
*Composition:* 95% tertiary amine
*Toxicity:* Skin irritant, severe eye irritant
*References:* Bulletin 73-7, 78-1

**Ethomeen S/15.** Armak
*Chemical Description:* Ethoxylated tertiary amine from soya fatty acid
*Category and Applications:* See Ethomeen C/12
*Form:* Clear liquid; Color: 14 max. (Gardner)
*Properties:* Solubility: Sol. in ace-

tone (cloudy @ > 50 C), isopropanol, carbon tetrachloride (> 25 C), dioxane, Stoddard solvent (> 75 C), benzene; forms gel or dispersion in water; Sp. gr.: 0.95; Surface Tension: 33 dynes/cm (0.1% sol'n.)
*Toxicity:* Skin irritant, severe eye irritant
*References:* Bulletin 73-7, 78-1

**Ethomeen S/20.** Armak
*Chemical Description:* Ethoxylated tertiary amine from soya fatty acid
*Category and Applications:* See Ethomeen C/12
*Form:* Clear liquid; Color: 14 max. (Gardner)
*Properties:* Solubility: Sol. in acetone (cloudy), isopropanol, carbon tetrachloride (> 25 C), dioxane, Stoddard solvent (> 50 C), benzene, water (forms viscous sol'n.); Sp. gr.: 1.02; Surface Tension: 40 dynes/cm (0.1% sol'n.)
*Toxicity:* Skin irritant, severe eye irritant
*References:* Bulletin 73-7, 78-1

**Ethomeen S/25.** Armak
*Chemical Description:* Ethoxylated tertiary amine from soya fatty acid
*Category and Applications:* See Ethomeen C/12
*Form:* Clear liquid; Color: 18 max. (Gardner)
*Properties:* Solubility: Sol. in acetone (cloudy), isopropanol, carbon tetrachloride (> 25 C), dioxane, Stoddard solvent (> 75 C), benzene, water; Sp. gr.: 1.04; Surface Tension: 43 dynes/cm (0.1% sol'n.)
*Toxicity:* Skin irritant, severe eye irritant
*References:* Bulletin 73-7, 78-1

**Ethomeen T/12.** Armak
*Chemical Description:* Ethoxylated tertiary amine from tallow fatty acid
*Category and Applications:* See Ethomeen C/12
*Form:* Paste; Color: 8 max. (Gardner)
*Properties:* Solubility: Sol. in acetone, isopropanol, carbon tetrachloride, Stoddard solvent, benzene, water (> 50 C); Sp. gr.: 0.92
*Composition:* 95% tertiary amine
*Toxicity:* Corrosive, skin irritant, severe eye irritant
*References:* Bulletin 73-7, 78-1

**Ethomeen T/15.** Armak
*Chemical Description:* Ethoxylated tertiary amine from tallow fatty acid
*Category and Applications:* See Ethomeen C/12
*Form:* Clear liquid; Color: 8 max. (Gardner)
*Properties:* Solubility: Forms cloudy sol'ns. in acetone, isopropanol, carbon tetrachloride, dioxane, Stoddard solvent, benzene; forms gel in water @ 30–80 C; Sp. gr.: 0.97; Surface Tension: 34 dynes/cm (0.1% sol'n.)
*Toxicity:* Corrosive, skin irritant, severe eye irritant
*References:* Bulletin 73-7, 78-1

**Ethomeen T/25.** Armak
*Chemical Description:* Ethoxylated tertiary amine from tallow fatty acid
*Category and Applications:* See Ethomeen C/12
*Form:* Clear liquid; Color: 8 max. (Gardner)
*Properties:* Solubility: Forms cloudy sol'ns. in acetone, isopropanol, carbon tetrachloride, dioxane, benzene (< 100 C), water; Sp. gr.: 1.03; Surface Tension: 41 dynes/cm (0.1% sol'n.)
*Toxicity:* Skin irritant, severe eye irritant
*References:* Bulletin 73-7, 78-1

**Ethomeen TD/15.** Armak
*Chemical Description:* Ethoxylated tertiary amine from tallow fatty acid
*Category and Applications:* See Ethomeen C/12
*Form:* Clear liquid; Color: 4 max. (Gardner)
*Properties:* Sp. gr.: 0.97; Surface Tension: 35.8 dynes/cm (0.1% sol'n.)
*References:* Bulletin 73-7, 78-1

**Ethomeen TD/25.** Armak
*Chemical Description:* Ethoxylated tertiary amine from tallow fatty acid
*Category and Applications:* See Ethomeen C/12
*Form:* Clear liquid; Color: 4 max. (Gardner)
*Properties:* Sp. gr.: 1.03; Surface Tension: 43.5 dynes/cm (0.1% sol'n.)
*Toxicity:* Skin irritant, severe eye irritant
*References:* Bulletin 73-7, 78-1

**Ethomid HT/15.** Armak
*Chemical Description:* Ethoxylated amide from hyd. tallow acid
*Category:* Emulsifier, dispersant, detergent, nonionic
*Form:* Solid; Color: 10 max. (Gardner)
*Properties:* Solubility: Sol. in isopropanol (< 15 C); forms dispersions in acetone (> 25 C), carbon tetrachloride (> 25 C), dioxane, Stoddard solvent (> 30 C), benzene (> 50 C), water; Sp. gr.: 1.03; Hydroxyl No.: 95–110; Surface Tension: 37 dynes/cm (0.1% sol'n.)
*Composition:* 100% conc.
*References:* Bulletin 73-7

**Ethomid HT/60.** Armak
*Chemical Description:* Ethoxylated amide from hyd. tallow acid
*Category:* See Ethomid HT/15
*Form:* Hard solid; Color: 11 max. (Gardner)
*Properties:* Solubility: Sol. in isopropanol (< 75 C), carbon tetrachloride (< 75 C), dioxane (< 75 C), benzene (< 80 C), water (< 75 C); Sp. gr.:1.14; Hydroxyl Value: 40–50; Surface Tension: 47 dynes/cm (0.1% sol'n.)
*Reaction:* Neutral
*References:* Bulletin 73-7

**Ethomid O/15.** Armak
*Chemical Description:* Ethoxylated amide from oleic acid
*Category:* See Ethomid HT/15

*Form:* Liquid; Color: 12 max. (Gardner)
*Properties:* Solubility: Sol. in isopropanol (< 50 C), carbon tetrachloride (< 25 C), dioxane (< 25 C); forms gel in water @ 50–80 C; Sp. gr.: 1.0; Hydroxyl No.: 100–120; Surface Tension: 35 dynes/cm (0.1% sol'n.)
*Reaction:* Neutral
*References:* Bulletin 73-7

**Ethoquad 18/12.** Armak
*Chemical Description:* Polyethoxylated quaternary ammonium salt from stearic acid
*Category and Applications:* See Ethoquad C/12
*Form:* Paste; Color: 6 max. (Gardner)
*Properties:* Solubility: Sol. in acetone, isopropanol (< 25 C), benzene, water; Sp. gr.: 0.919; Surface Tension: 40.2 dynes/cm (0.1% sol'n.); pH: 7–8 (10% aq.)
*Composition:* 70% min. active
*References:* Bulletin 73-7

**Ethoquad 18/25.** Armak
*Chemical Description:* Polyethoxylated quaternary ammonium salt from stearic acid
*Category and Applications:* See Ethoquad C/12
*Form:* Clear liquid; Color: 11 max. (Gardner)
*Properties:* Solubility: Sol. in acetone, isopropanol, carbon tetrachloride, dioxane, benzene, water; Sp. gr.: 1.058; Surface Tension: 50.1 dynes/cm (0.1% sol'n.); pH: 6–9 (10% aq.)
*Composition:* 95% min. active
*References:* Bulletin 73-7

**Ethoquad C/12.** Armak
*Chemical Description:* Polyethoxylated quaternary ammonium salt from coco fatty acid
*Category:* Antistatic agents, emulsifiers, dyeing assistants, cationic
*Applications:* Textile industry, electroplating bath additives
*Form:* Clear liquid; Color: 9 max. (Gardner)
*Properties:* Solubility: Sol. in acetone, isopropanol, carbon tetrachloride, dioxane (> 15 C), Stoddard solvent (> 50 C), benzene, water; Sp. gr.: 0.969; Stability: Stable to acid and alkali; Surface Tension: 35.4 dynes/cm (0.1% sol'n.); pH: 7–8 (10% aq.)
*Composition:* 75% min. active
*References:* Bulletin 73-7

**Ethoquad C/25.** Armak
*Chemical Description:* Polyethoxylated quaternary ammonium salt from coco fatty acid
*Category and Applications:* See Ethoquad C/12
*Form:* Clear liquid; Color: 11 max. (Gardner)
*Properties:* Solubility: Sol. in acetone, isopropanol, carbon tetrachloride (> 30 C), dioxane, benzene, water; Sp. gr.: 1.071; Stability: Stable to acid and alkali; Surface Tension: 43.4 dynes/cm (0.1% sol'n.)
*Composition:* 95% min. active
*References:* Bulletin 73-7

**Ethoquad O/12.** Armak
*Chemical Description:* Polyethox-

ylated quaternary ammonium salt from oleic acid
*Category and Applications:* See Ethoquad C/12
*Form:* Clear liquid; Color: 9 max. (Gardner)
*Properties:* Solubility: Sol. in acetone, isopropanol, carbon tetrachloride, dioxane, Stoddard solvent (> 10 C), benzene, water; Sp. gr.: 0.932; Surface Tension: 40.3 dynes/cm (0.1% sol'n.); pH: 7-8 (10% aq.)
*Composition:* 72% min. active
*References:* Bulletin 73-7

**Ethoquad O/25.** Armak
*Chemical Description:* Polyethoxylated quaternary ammonium salt from oleic acid
*Category and Applications:* See Ethoquad C/12
*Form:* Clear liquid; Color: 11 max. (Gardner)
*Properties:* Solubility: Sol. in acetone, isopropanol, carbon tetrachloride, dioxane, benzene, water; Sp. gr.: 1.062; Surface Tension: 40.8 dynes/cm (0.1% sol'n.); pH: 7-9 (10% aq.)
*Composition:* 95% min. active
*References:* Bulletin 73-7

**Ethosperse G-26.** Glycol Chem.
*Chemical Description:* POE 26 glycerol
*Category and Applications:* See Ethosperse LA-4
*Form:* Liquid; Color: Pale straw
*Properties:* Solubility: Sol. in water, methanol, ethanol, acetone, ethyl acetate, toluol; Sp. gr.: 1.12 (@ 38 C); Visc.: 150 cps (@ 38 C); Acid No.: 2 max.; Hydroxyl No.: 133-142

**Ethosperse LA-4.** Glyco Chem.
*Chemical Description:* POE 4 ether lauryl alcohol
*Category:* Emulsifier, nonionic
*Applications:* Cosmetic, pharmaceutical and industrial uses
*Form:* Liquid; Color: Pale straw
*Properties:* Solubility: Sol. in methanol, ethanol, acetone, ethyl acetate. toluol, mineral oil; miscible with water, vegetable oil in certain proportions; Sp. gr.: 0.95; Visc.: 30 cps; HLB: 9.5; Acid No.: 2 max.; Hydroxyl No.: 145-160
*Composition:* 100% active ingred.

**Ethosperse LA-12.** Glyco Chem.
*Chemical Description:* POE 12 ether lauryl alcohol
*Category and Applications:* See Ethosperse LA-4
*Form:* Turbid liquid to soft paste
*Properties:* Solubility: Sol. in methanol, ethanol, acetone, ethyl acetate, toluol; miscible with water in certain proportions and in vegetable oil if hot; Sp. gr.: 1.10; Visc.: 1000 cps; HLB: 4.8; Acid No.: 2 max.; Hydroxyl No. 72-82
*Composition:* 100% active ingred.

**Ethosperse SL-20.** Glyco Chem.
*Chemical Description:* POE 20 sorbitol
*Category and Applications:* See Ethosperse LA-4
*Form:* Liquid; Color: Light yellow
*Properties:* Solubility: Sol. in water, methanol, ethanol, acetone;

Sp. gr.: 1.16; Visc.: 460 cps; HLB: 16.6; Acid No.: 0.5 max.; Hydroxyl No.: 385–430
*Composition:* 100% active ingred.

**Ethosperse TDA-6.** Glyco Chem.
*Chemical Description:* POE 6 ether tridecyl alcohol
*Category and Applications:* See Ethosperse LA-4
*Form:* Liquid; Color: Pale straw
*Properties:* Solubility: Sol. in methanol, ethanol, acetone, ethyl acetate, toluol; dispersible in water; Sp. gr.: 0.98; Visc.: 80 cps; HLB: 11.0; Acid No.: 1 max.; Hydroxyl No.: 118–133
*Composition:* 100% active ingred.

**Ethylan 20.** Lankro Chem. Ltd.
*Chemical Description:* Nonyl phenol ethoxylate
*Category:* Detergent, wetting agent, stabilizer, emulsifier, nonionic
*Applications:* Detergent, foam stabilizer and booster; pesticide formulations; essential oil solubilizer; latex stabilization; emulsification of waxes and polyester resin
*Form:* Waxy solid; Color: White; Odor: Negligible
*Properties:* Solubility: Water-sol.; Sp. gr.: 1.065 @ 40 C; Visc.: 160 cs @ 20 C; HLB: 16.0; Cloud Pt.: > 100 C (1% aq.); Flash Pt.: > 400 F (COC); Pour Pt.: 30 C; Stability: Stable to electrolytes, hard water, extremes of pH and temperature; pH: 6–8 (1% aq.)
*Composition:* 100% active ingred.
*Containers:* 200 kg net mild-steel drums or bulk
*Warnings and Caution:* Avoid prolonged contact with concentrated form; spillages may be slippery; do not mix with strong oxidizing or reducing agents (potentially explosive)
*References:* Bulletin SI:13 FM678

**Ethylan 44.** Lankro Chem. Ltd.
*Chemical Description:* Nonyl phenol ethoxylate
*Category:* Emulsifier, nonionic
*Applications:* Emulsifier for hydrocarbons, silicones, in formulations of degreasers and hand cleaning gels
*Form:* Clear liquid; Color: Straw; Odor: Negligible
*Properties:* Solubility: Water-insol.; Sp. gr.: 1.023 @ 20 C; Visc.: 340 cs @ 20 C; HLB: 9; Flash Pt.: > 400 F (COC); Pour Pt.: < 0 C; Stability: Stable to electrolytes, hard water, extremes of pH and temperature; pH: 6–8 (1% aq.)
*Composition:* 100% active ingred.
*Containers:* 200 kg net mild-steel drums or bulk
*Warnings and Caution:* Avoid prolonged contact with concentrated form; spillages may be slippery; do not mix with strong oxidizing or reducing agents (potentially explosive)
*References:* Bulletin SI:13 FM678

**Ethylan 55.** Lankro Chem. Ltd.
*Chemical Description:* Nonyl phenol ethoxylate
*Category and Applications:* See Ethylan 44
*Form:* Clear liquid; Color: Straw
*Properties:* Solubility: Water-in-

sol.; Sp. gr.: 1.035 @ 20 C; Visc.: 290 cs @ 20 C; HLB: 10.5; Flash Pt.: > 400 F (COC); Pour Pt.: < 0 C; Stability: Stable to electrolytes, hard water, extremes of pH and temperature; pH: 6–8 (1% aq.)
*Composition:* 100% active ingred.
*Containers:* 200 kg net mild-steel drums or bulk
*Warnings and Caution:* Avoid prolonged contact with concentrated form; spillages may be slippery; do not mix with strong oxidizing or reducing agents (potentially explosive)
*References:* Bulletin SI:13 FM678

**Ethylan 77.** Lankro Chem. Ltd.
*Chemical Description:* Nonyl phenol ethoxylate
*Category and Applications:* See Ethylan 44
*Form:* Clear liquid; Color: Pale straw; Odor: Negligible
*Properties:* Solubility: Water-disp.; Sp. gr.: 1.038 @ 20 C; Visc.: 360 cs @ 20 C; HLB: 10.9; Cloud Pt.: < 2 C (1% aq.); Flash Pt.: > 400 F (COC); Pour Pt.: < 0 C; Stability: Stable to electrolytes, hard water, extremes of pH and temperature; pH: 6–8 (1% aq.)
*Composition:* 100% active ingred.
*Containers:* 200 kg net mild-steel drums or bulk
*Warnings and Caution:* Avoid prolonged contact with concentrated form; spillages may be slippery; do not mix with strong oxidizing or reducing agents (potentially explosive)
*References:* Bulletin SI:13 FM678

**Ethylan 172.** Lankro Chem. Ltd.
*Chemical Description:* Cetyl, oleyl alcohol ethoxylate
*Category:* Emulsifier, nonionic
*Applications:* Emulsifying mineral oils and hydrophobic waxes
*Form:* Liquid; Color: Lt. amber
*Properties:* HLB: 7.0
*Composition:* 100% active ingred.
*Reaction:* Neutral

**Ethylan A2.** Lankro Chem. Ltd.
*Chemical Description:* PEG 200 oleate
*Category:* Emulsifier, antifoam agent, nonionic
*Applications:* Kerosene and mineral oil emulsifier, antifoam
*Form:* Liquid; Color: Lt. amber
*Properties:* Sp. gr.: 0.976; Visc.: 67 cs @ 20 C; HLB: 7.0
*Composition:* 100% active ingred.
*Reaction:* Neutral

**Ethylan A3.** Lankro Chem. Ltd.
*Chemical Description:* PEG 300 oleate
*Category:* Emulsifier, antifoam
*Applications:* Emulsifier for mineral and vegetable oils, antifoam, dispersion aid
*Form:* Liquid; Color: Lt. amber
*Properties:* Sp. gr.: 1.000; Visc.: 100 cs @ 20 C; HLB 8.9
*Composition:* 100% active ingred.
*Reaction:* Neutral

**Ethylan A4.** Lankro Chem. Ltd.
*Chemical Description:* PEG 400 oleate
*Category:* Emulsifier, antistatic, nonionic
*Applications:* Oil emulsifier, anti-

stat for plastics, dispersion aid
*Form:* Liquid; Color: Lt. amber
*Properties:* Sp. gr.: 1.017; Visc.: 130 cs @ 20 C; HLB: 10.3
*Composition:* 100% active ingred.
*Reaction:* Neutral

**Ethylan A6.** Lankro Chem. Ltd.
*Chemical Description:* PEG 600 oleate
*Category:* Dispersing agent, nonionic
*Applications:* Oil emulsifier and antistat for plastics, pigment dispersant
*Form:* Liquid; Color: Lt. amber
*Properties:* Sp. gr.: 1.037; Visc.: 160 cs @ 20 C; HLB: 12.3
*Composition:* 100% active ingred.
*Reaction:* Neutral

**Ethylan A15.** Lankro Chem. Ltd.
*Chemical Description:* Coconut diethanolamide
*Category:* Emulsifier, nonionic
*Applications:* Industrial degreasing, metal cleaning applications, emulsifying solvents and polyethylene waxes
*Form:* Clear Liquid; Color: Straw
*Properties:* Sp. gr.: 1.010; Visc.: 880 cs @ 20 C
*Composition:* 99% active ingred.
*Reaction:* Alkaline

**Ethylan BCP.** Lankro Chem. Ltd.
*Chemical Description:* Nonyl phenol ethoxylate
*Category:* Detergent, wetting agent, emulsifier, nonionic
*Applications:* Textile scouring agent, dry cleaning detergent; wetting and emulsifying agent for oils and aromatic solvents
*Form:* Clear colorless liquid; Odor: Negligible
*Properties:* Solubility: Water-sol.; Sp. gr.: 1.058 @ 20 C; Visc.: 320 cs @ 20 C; HLB: 12.9; Cloud Pt.: 50 C (1% aq.); Flash Pt.: > 400 F (COC); Pour Pt.: 0 C; Stability: Stable to electrolytes, hard water, extremes of pH and temperature; pH: 6–8 (1% aq.)
*Composition:* 100% active ingred.
*Containers:* 200 kg net mild-steel drums or bulk
*Warnings and Caution:* Avoid prolonged contact with concentrated form; spillages may be slippery; do not mix with strong oxidizing or reducing agents (potentially explosive)
*References:* Bulletin SI:13 FM678

**Ethylan BV.** Lankro Chem. Ltd.
*Chemical Description:* Nonyl phenol ethoxylate
*Category:* Foam stabilizer, emulsifier, solubilizer, nonionic
*Applications:* Foam stabilizer and booster; emulsifier of solvents and oleins, in pesticides; as emulsifier and solubilizer for essential oils and perfumes; emulsion polymerization
*Form:* Soft paste; Color: White; Odor: Negligible
*Properties:* Solubility: Water-sol.; Sp. gr.: 1.083 @ 20 C; Visc.: 380 cs @ 20 C; HLB: 14.5; Cloud Pt.: 90 C (1% aq.); Flash Pt.: > 400 F (COC); Pour Pt.: 20 C; Stability: Stable to electrolytes, hard water, extremes of pH and temperature; pH: 6–8 (1% aq.)

*Composition:* 100% active ingred.
*Containers:* 200 kg net mild-steel drums or bulk
*Warnings and Caution:* Avoid prolonged contact with concentrated form; spillages may be slippery; do not mix with strong oxidizing or reducing agents (potentially explosive)
*References:* Bulletin SI:13 FM678

**Ethylan BZA.** Lankro Chem. Ltd.
*Chemical Description:* Modified alkylphenol ethoxylate
*Category:* Detergent, nonionic
*Applications:* Low-foam nonionic with high stability on solid caustic and good food soil defoaming characteristics for mechanical dishwashing; metal cleaning
*Form:* Liquid
*Properties:* Cloud Pt.: 18 C
*Composition:* 100% active ingred.

**Ethylan C 40 AH.** Lankro Chem. Ltd.
*Chemical Description:* PEG ester of unsaturated fatty acid
*Category:* Emulsifier, nonionic
*Applications:* Emulsifier for chlorinated solvents, olein and wax, pesticides; dye leveling agent; low-foaming dispersing aid; perfumery solubilizer
*Form:* Hazy liquid; Color: Yellow
*Properties:* Sp. gr.: 1.054; Visc.: 320 cs @ 40 C; HLB: 13.5
*Composition:* 100% active ingred.
*Reaction:* Neutral

**Ethylan C 160.** Lankro Chem. Ltd.
*Chemical Description:* PEG unsaturated fatty acid ester
*Category:* Emulsifier, dispersant, solubilizer
*Applications:* Olein and wax emulsifier; perfumery; low-foaming dispersant, solubilizer
*Form:* Waxy solid; Color: Cream
*Properties:* HLB: 17.6
*Composition:* 100% active ingred.

**Ethylan CD913.** Lankro Chem. Ltd.
*Chemical Description:* Synthetic lower fraction primary alcohol ethylene oxide condensate
*Category:* Detergent, emulsifier, wetting agent, solubilizer, nonionic
*Applications:* W/O and O/W emulsification; co-emulsifier for mineral oils and waxes; emulsification of alkyd resins and paraffin hydrocarbons
*Form:* Clear liquid; Color: Colorless
*Properties:* Solubility: Water insol.; Sp. gr.: 0.936 @ 20 C; Visc.: 27 cs @ 20 C; HLB: 8.8; Flash Pt.: 290 F (COC); Pour Pt.: < 0 C; pH: 6–8 (1% aq.)
*Composition:* 97% active
*Containers:* 200 kg net mild-steel drums or bulk
*Warnings and Caution:* Avoid prolonged contact with concentrated form; spillages may be slippery; do not mix with strong oxidizing or reducing agents (potentially explosive)
*References:* Bulletin SI: 8FM678

**Ethylan CD916.** Lankro Chem. Ltd.

*Chemical Description:* Synthetic lower fraction primary alcohol ethylene oxide condensate
*Category:* Emulsifier, detergent, wetting agent, nonionic
*Applications:* O/W emulsification of paraffin waxes and hydrocarbon solvents; bases for toxicant emulsifiers and co-emulsifiers for mineral oils; scouring and wetting agents; textile antistatic agents; specialty cleaners
*Form:* Clear liquid; Color: Colorless; Odor: Characteristic fatty alcohol
*Properties:* Solubility: Water-sol.; Sp. gr.: 0.993 @ 20 C; Visc.: 53 cs @ 20 C; HLB: 12.8; Cloud Pt.: 58 C; Flash Pt.: 355 F (COC); Pour Pt.: 0 C; pH: 6–8 (1% aq.)
*Composition:* 96% active
*Containers:* 200 kg net mild-steel drums or bulk
*Warnings and Caution:* Avoid prolonged contact with concentrated form; spillages may be slippery; do not mix with strong oxidizing or reducing agents (potentially explosive)
*References:* Bulletin SI: 8FM678

**Ethylan CD919.** Lankro Chem. Ltd.
*Chemical Description:* Synthetic lower fraction primary alcohol ethylene oxide condensate
*Category:* Emulsifier, detergent, wetting agent, solubilizer, nonionic
*Applications:* O/W emulsifier; wetting agents and detergents, sudsing agents for liquid detergents; scouring and dye levelling agents; coemulsifiers for aromatic solvents, natural and synthetic waxes and toxicants; detergent sanitizers; essential oil solubilizers
*Form:* Clear liquid; Color: Colorless; Odor: Characteristic fatty alcohol
*Properties:* Solubility: Water-sol.; Sp. gr.: 1.008 @ 20 C; Visc.: 77 cs @ 20 C; HLB: 14.4; Cloud Pt.: 82 C; Flash Pt.: 385 F (COC); Pour Pt.: 7 C; pH: 6–8 (1% aq.)
*Composition:* 97% active
*Containers:* 200 kg net mild-steel drums or bulk
*Warnings and Caution:* Avoid prolonged contact with concentrated form; spillages may be slippery; do not mix with strong oxidizing or reducing agents (potentially explosive)
*References:* Bulletin SI: 8FM678

**Ethylan CD9112.** Lankro Chem. Ltd.
*Chemical Description:* Synthetic lower fraction primary alcohol ethylene oxide condensate
*Category and Applications:* See Ethylan CD919
*Form:* Waxy solid; Color: White; Odor: Characteristic fatty alcohol
*Properties:* Solubility: Water-sol.; Sp. gr.: 1.031 @ 40 C; Visc.: 56 cs @ 40 C; HLB: 15.5; Cloud Pt.: 100 C; Flash Pt.: > 400 F (COC); Pour Pt.: 23 C; pH: 6–8 (1% aq.)
*Composition:* 100% active
*Containers:* 200 kg net mild-steel drums or bulk
*Warnings and Caution:* Avoid prolonged contact with concentrated form; spillages may be slippery;

do not mix with strong oxidizing or reducing agents (potentially explosive)
*References:* Bulletin SI: 8FM678

**Ethylan CD9130.** Lankro Chem. Ltd.
*Chemical Description:* Synthetic lower fraction primary alcohol ethylene oxide condensate
*Category:* Wetting agent, emulsifier, nonionic
*Applications:* O/W emulsification; emulsifier for synethetic waxes; wetting agent at high temperature; detergent sanitizers; toilet block component
*Form:* Waxy solid; Color: White; Odor: Characteristic fatty alcohol
*Properties:* Solubility: Water-sol.; Visc.: 62 cs @ 60 C; HLB: 17.8; Cloud Pt.: > 100 C; Flash Pt.: > 400 F (COC); Pour Pt.: 45 C; pH: 6–8 (1% aq.)
*Composition:* 100% active ingred.
*Containers:* 200 kg net mild-steel drums or bulk
*Warnings and Caution:* Avoid prolonged contact with concentrated form; spillages may be slippery; do not mix with strong oxidizing or reducing agents (potentially explosive)
*References:* Bulletin SI: 8FM678

**Ethylan CF71.** Lankro Chem. Ltd.
*Chemical Description:* Coconut fatty acid ester
*Category:* Emulsifier, lubricant, nonionic
*Applications:* Oil emulsifier; emulsifier for cosmetics and pharmaceuticals; fiber lubricant
*Form:* Liquid; Color: Pale straw
*Properties:* HLB: 14.0
*Composition:* 100% active ingred.

**Ethylan CH.** Lankro Chem. Ltd.
*Chemical Description:* Ethoxylated coconut fatty acid alkylolamide
*Category:* Foam stabilizer, co-emulsifier, nonionic
*Applications:* Foam stabilizer in liquid detergents, general purpose or hard surface cleaners and in shampoos; co-emulsifier for chlorinated solvents
*Form:* Clear liquid; Color: Amber; Odor: Mild
*Properties:* Solubility: Water-sol.; Sp. gr.: 1.025 @ 20 C; Visc.: 270 cs @ 20 C; Cloud Pt.: 80 C (1% aq.); Flash Pt.: > 300 F (COC); Pour Pt.: 14 C; Stability: Good in presence of hard water and electrolytes; pH: 7.0–9.0 (1% aq.)
*Composition:* 100% active ingred.
*Reaction:* Neutral
*Warnings and Caution:* Avoid prolonged contact with concentrated form; spillages may be slippery; do not mix with strong oxidizing or reducing agents (potentially explosive)
*References:* Bulletin SI:12FM678

**Ethylan CPG 630.** Lankro Chem. Ltd.
*Chemical Description:* Modified alcohol ethoxylate
*Category:* Detergent, wetting agent, nonionic
*Applications:* Low foaming wet-

ting agent and detergent for rinse aids, machine dishwashing, metal cleaning, spray cleaning and bottle washing.
*Form:* Clear liquid; Color: Colorless; Odor: Mild
*Properties:* Solubility: Water-sol.; Sp. gr.: 0.978 @ 20 C; Visc.: 65 cs @ 20 C; Cloud Pt.: 38 C (1% aq.); Flash Pt.: > 350 F (COC); Pour Pt.: 1 C; Stability: Stable to alkalis in aq. sol'n. up to boil, to acids except in presence of strong oxidizing agents, to chlorine releasing agents; pH: 6–7.5 (1% aq.)
*Composition:* 99% active ingred.
*Reaction:* Neutral
*Containers:* 200 kg net mild steel drums or bulk
*Warnings and Caution:* Avoid prolonged contact with concentrated form; spillages may be slippery; do not mix with strong oxidizing or reducing agents (potentially explosive)
*References:* Bulletin SI:10 FM678

**Ethylan CPG 660.** Lankro Chem. Ltd.
*Chemical Description:* Modified alcohol ethoxylate
*Category and Applications:* See Ethylan CPG 630
*Form:* Clear liquid; Color: Colorless; Odor: Mild
*Properties:* Solubility: Water-sol.; Sp. gr.: 0.984 @ 20 C; Visc.: 77 cs @ 20 C; Cloud Pt.: 29 C (1% aq.); Flash Pt.: > 350 F (COC); Pour Pt.: < –10 C; Stability: Stable to alkalis in aq. sol'n. up to boil, to acids except in presence of strong oxidizing agents, to chlorine releasing agents; pH: 6–7.5 (1% aq.)
*Composition:* 99% active ingred.
*Reaction:* Neutral
*Containers:* 200 kg net mild steel drums or bulk
*Warnings and Caution:* Avoid prolonged contact with concentrated form; spillages may be slippery; do not mix with strong oxidizing or reducing agents (potentially explosive)
*References:* Bulletin SI:10 FM678

**Ethylan CPG 745.** Lankro Chem. Ltd.
*Chemical Description:* Modified alcohol ethoxylate
*Category and Applications:* See Ethylan CPG 630
*Form:* Clear liquid; Color: Colorless; Odor: Mild
*Properties:* Solubility: Water-sol.; Sp. gr.: 0.985 @ 20 C; Visc.: 84 cs @ 20 C; Cloud Pt.: 37 C (1% aq.); Flash Pt.: > 350 F (COC); Pour Pt.: –6 C; Stability: Stable to alkalis in aq. sol'n. up to boil, to acids except in presence of strong oxidizing agents, to chlorine releasing agents; pH: 6–7.5 (1% aq.)
*Composition:* 99% active ingred.
*Reaction:* Neutral
*Containers:* 200 kg net mild steel drums or bulk
*Warnings and Caution:* Avoid prolonged contact with concentrated form; spillages may be slippery; do not mix with strong oxidizing or reducing agents (potentially explosive)
*References:* Bulletin SI.10 FM678

**Ethylan CPG 945.** Lankro Chem. Ltd.
*Chemical Description:* Modified alcohol ethoxylate
*Category and Applications:* See Ethylan CPG 630
*Form:* Clear liquid; Color: Colorless; Odor: Mild
*Properties:* Solubility: Water-sol.; Sp. gr.: 1.005 @ 20 C; Visc.: 88 cs @ 20 C; Cloud Pt.: 47 C (1% aq.); Flash Pt.: > 350 F (COC); Pour Pt.: 2 C; Stability: Stable to alkalis in aq. sol'n. up to boil, to acids except in presence of strong oxidizing agents, to chlorine releasing agents; pH: 6–7.5 (1% aq.)
*Composition:* 99% active ingred.
*Reaction:* Neutral
*Containers:* 200 kg net mild steel drums or bulk
*Warnings and Caution:* Avoid prolonged contact with concentrated form; spillages may be slippery; do not mix with strong oxidizing or reducing agents (potentially explosive)
*References:* Bulletin SI:10 FM678

**Ethylan CRS.** Lankro Chem. Ltd.
*Chemical Description:* Ethoxylated coconut fatty acid alkylolamide
*Category:* Emulsifier, foam stabilizer, nonionic
*Applications:* Foam stabilizer in liquid detergent, general purpose or hard surface cleaners and in shampoos; co-emulsifier for chlorinated solvents
*Form:* Clear liquid; Color: Amber; Odor: Mild
*Properties:* Solubility: Water-sol.; Sp. gr.: 1.026 @ 20 C; Visc.: 250 cs @ 20 C; Flash Pt.: > 300 F (COC); Pour Pt.: 10 C; Stability: Good in presence of hard water and electrolytes; pH: 7–9 (1% aq.)
*Composition:* 100% active ingred.
*Reaction:* Neutral
*Warnings and Caution:* Avoid prolonged contact with concentrated form; spillages may be slippery; do not mix with strong oxidizing or reducing agents (potentially explosive)
*References:* Bulletin SI:12FM678

**Ethylan D252.** Lankro Chem. Ltd.
*Chemical Description:* Synthetic primary alcohol ethoxylate
*Category:* Emulsifier, nonionic
*Applications:* w/o and o/w emulsifier for mineral oils and waxes; emulsification of alkyd resins and paraffin hydrocarbons
*Form:* Clear liquid; Color: Colorless; Odor: Faint
*Properties:* Solubility: Oil-sol.; Sp. gr.: 0.903 @ 20 C; Visc.: 31 cs @ 20 C; HLB: 5.6; Flash Pt.: 310 F (COC); Pour Pt.: 5 C; pH: 6–8 (1% aq.)
*Composition:* 100% active ingred.
*Reaction:* Neutral
*Containers:* 200 kg net mild steel drums or bulk
*Warnings and Caution:* Avoid prolonged contact with concentrated form; spillages may be slippery; do not mix with strong oxidizing or reducing agents (potentially explosive)
*References:* Bulletin SI: 7FM678

**Ethylan D253.** Lankro Chem. Ltd.
*Chemical Description:* Synthetic primary alcohol ethoxylate
*Category and Applications:* See Ethylan D252; also as viscosity depressant for PVC plastisols; intermediate for sulfation
*Form:* Clear liquid; Color: Colorless; Odor: Faint
*Properties:* Solubility: Oil-sol.; Sp. gr.: 0.930 @ 20 C; Visc.: 36 cs @ 20 C; HLB: 7.8; Flash Pt.: 335 F (COC); Pour Pt.: 3 C; pH: 6–8 (1% aq.)
*Composition:* 100% active ingred.
*Reaction:* Neutral
*Containers:* 200 kg net mild steel drums or bulk
*Warnings and Caution:* Avoid prolonged contact with concentrated form; spillages may be slippery; do not mix with strong oxidizing or reducing agents (potentially explosive)
*References:* Bulletin SI: 7FM678

**Ethylan D254.** Lankro Chem. Ltd.
*Chemical Description:* Synthetic primary alcohol ethoxylate
*Category and Applications:* See Ethylan D252; also as dry cleaning assistant
*Form:* Clear liquid; Color: Colorless; Odor: Faint
*Properties:* Solubility: Oil-sol.; Sp. gr.: 0.948 @ 20 C; Visc.: 57 cs @ 20 C; HLB: 9.8; Flash Pt.: 360 F (COC); Pour Pt.: 5 C; pH: 6–8 (1% aq.)
*Composition:* 97% active ingred.
*Reaction:* Neutral
*Containers:* 200 kg net mild steel drums or bulk
*Warnings and Caution:* Avoid prolonged contact with concentrated form; spillages may be slippery; do not mix with strong oxidizing or reducing agents (potentially explosive)
*References:* Bulletin SI: 7FM678

**Ethylan D256.** Lankro Chem. Ltd.
*Chemical Description:* Synthetic primary alcohol ethoxylate
*Category:* Emulsifier, detergent, wetting agent, nonionic
*Applications:* o/w emulsification of paraffin waxes and hydrocarbon solvents; bases for toxicant emulsifiers and co-emulsifiers for mineral oils; scouring and wetting agents; textile antistatic agents; specialty cleaners
*Form:* Hazy liquid; Odor: Faint
*Properties:* Solubility: Water-sol.; Sp. gr.: 0.972 @ 20 C; Visc.: 72 cs @ 20 C; HLB: 11.4; Cloud Pt.: 43 C (1% aq.); Flash Pt.: 375 F (COC); Pour Pt.: 15 C; pH: 6–8 (1% aq.)
*Composition:* 100% active ingred.
*Reaction:* Neutral
*Containers:* 200 kg net mild steel drums or bulk
*Warnings and Caution:* Avoid prolonged contact with concentrated form; spillages may be slippery; do not mix with strong oxidizing or reducing agents (potentially explosive)
*References:* Bulletin SI: 7FM678

**Ethylan D257.** Lankro Chem. Ltd.
*Chemical Description:* Synthetic primary alcohol ethoxylate

*Category and Applications:* See Ethylan D256
*Form:* Hazy liquid; Odor: Faint
*Properties:* Solubility: Water-sol.; Sp. gr.: 0.974 @ 40 C; Visc.: 31 cs @ 40 C; HLB: 12.2; Cloud Pt.: 49 C (1% aq.); Flash Pt.: 385 F (COC); Pour Pt.: 21 C; pH: 6–8 (1% aq.)
*Composition:* 100% active ingred.
*Reaction:* Neutral
*Containers:* 200 kg net mild steel drums or bulk
*Warnings and Caution:* Avoid prolonged contact with concentrated form; spillages may be slippery; do not mix with strong oxidizing or reducing agents (potentially explosive)
*References:* Bulletin SI: 7FM678

**Ethylan D259.** Lankro Chem. Ltd.
*Chemical Description:* Synthetic primary alcohol ethoxylate
*Category:* Detergent, wetting agent, emulsifier, solubilizer, nonionic
*Applications:* Liquid formulations; solubilizing wax, essential oils
*Form:* Solid; Color: White
*Properties:* Solubility: Water-sol.; Sp. gr.: 0.988; Visc.: 36 cs @ 40 C; Cloud Pt.: 76 C
*Composition:* 100% active ingred.

**Ethylan D2512.** Lankro Chem. Ltd.
*Chemical Description:* Synthetic primary alcohol ethoxylate
*Category:* Emulsifier, detergent, wetting agent, solubilizer, nonionic
*Applications:* o/w emulsification; detergent, wetting and sudsing agents for liquid detergents; scouring and dye levelling agents; coemulsifiers for aromatic solvents, natural and synthetic waxes and toxicants; detergent sanitizers; essential oil solubilizers
*Form:* Waxy solid; Color: White; Odor: Faint
*Properties:* Solubility: Water-sol.; Sp. gr.: 1.001 @ 40 C; Visc.: 46 cs @ 40 C; HLB: 14.2; Cloud Pt.: 92 C (1% aq.); Flash Pt.: > 400 F (COC); Pour Pt.: 29 C; pH: 6–8 (1% aq.)
*Composition:* 100% active ingred.
*Reaction:* Neutral
*Containers:* 200 kg net mild steel drums or bulk
*Warnings and Caution:* Avoid prolonged contact with concentrated form; spillages may be slippery; do not mix with strong oxidizing or reducing agents (potentially explosive)
*References:* Bulletin SI: 7FM678

**Ethylan D2560.** Lankro Chem. Ltd.
*Chemical Description:* Synthetic primary alcohol ethoxylate
*Category:* Wetting agent, emulsifier, nonionic
*Applications:* Latex additive; toilet block component; o/w emulsification; wetting at elevated temperatures
*Form:* Flake; Color: White; Odor: Faint
*Properties:* Solubility: Water-sol.; Visc.: 120 cs @ 60 C; HLB: 18.6; Cloud Pt.: > 100 C (1% aq.); Flash Pt.: > 400 F (COC); Pour Pt.: 45 C; pH: 6–8 (1% aq.)
*Composition:* 100% active ingred.

*Reaction:* Neutral
*Containers:* 200 kg net mild steel drums or bulk
*Warnings and Caution:* Avoid prolonged contact with concentrated form; spillages may be slippery; do not mix with strong oxidizing or reducing agents (potentially explosive)
*References:* Bulletin SI: 7FM678

**Ethylan DP.** Lankro Chem. Ltd.
*Chemical Description:* Nonyl phenol ethoxylate
*Category:* Foam stabilizer, emulsifier, solubilizer, nonionic
*Applications:* Foam stabilizer and booster; emulsifier of solvents and oleins, in pesticides; as emulsifier and solubilizer for essential oils and perfumes; emulsion polymerization
*Form:* Slightly hazy, viscous liquid; Odor: Negligible
*Properties:* Solubility: Water-sol.; Sp. gr.: 1.068 @ 20 C; Visc.: 400 cs @ 20 C; HLB: 14.0; Cloud Pt.: 80 C (1% aq.); Flash Pt.: > 400 F (COC); Pour Pt.: 13 C; Stability: Stable to electrolytes, hard water, extremes of pH and temperature; pH: 6–8 (1% aq.)
*Composition:* 100% active ingred.
*Containers:* 200 kg net mild-steel drums or bulk
*Warnings and Caution:* Avoid prolonged contact with concentrated form; spillages may be slippery; do not mix with strong oxidizing or reducing agents (potentially explosive)
*References:* Bulletin SI:13 FM678

**Ethylan ENTX.** Lankro Chem.Ltd.
*Chemical Description:* Alkylphenol ethoxylate
*Category:* Emulsifier, plasticizer, nonionic
*Applications:* Mineral oil and wax emulsifier; mastic plasticizer
*Form:* Liquid; Color: Pale straw; Odor: Faint
*Properties:* Solubility: Oil-sol.; Sp. gr.: 1.006 @ 20 C; Visc.: 440 cs @ 20 C; HLB: 9.0; Flash Pt.: > 350 F (COC); Pour Pt.: 2 C; Stability: Good in presence of acids, alkalis and electrolytes; pH: 6.0–7.5 (1% aq.)
*Composition:* 100% active ingred.
*Reaction:* Neutral
*Containers:* 200 kg net mild steel drums or bulk
*Warnings and Caution:* Avoid prolonged contact with concentrated form; spillages may be slippery; do not mix with strong oxidizing or reducing agents (potentially explosive)
*References:* Bulletin SI:14FM678

**Ethylan GD.** Lankro Chem. Ltd.
*Chemical Description:* Special fatty acid diethanolamide
*Category:* Emulsifier, antistat, lubricant, nonionic
*Applications:* Emulsifier for mineral oil, cutting oils, dewatering fluids, metal degreasing agents
*Form:* Clear liquid; Color: Amber; Odor: Mild
*Properties:* Solubility: Excellent oil-sol.; Sp. gr.: 0.982 @ 20 C; Visc.: 1200 cs @ 20 C; Flash Pt.: > 300 F (COC); Pour Pt.: 10 C; Stability: Good in presence of

hard water and electrolytes; pH: 8–11 (1% aq.)
*Composition:* 75% active ingred.
*Reaction:* Alkaline
*Warnings and Caution:* Avoid prolonged contact with concentrated form; spillages may be slippery; do not mix with strong oxidizing or reducing agents (potentially explosive)
*References:* Bulletin SI:12FM678

**Ethylan GMF.** Lankro Chem. Ltd.
*Chemical Description:* Alkylphenol ethoxylate
*Category:* Wetting agent, emulsifier, nonionic
*Applications:* Emusifier for hydrocarbon solvents; emulsion polymerization
*Form:* Clear liquid; Color: Straw; Odor: Faint
*Properties:* Solubility: Water-sol.; Sp. gr.: 1.050 @ 20 C; Visc.: 330 cs @ 20 C; HLB: 12.5; Cloud Pt.: 47 C (1% aq.); Flash Pt.: > 350 F (COC); Pour Pt.: 7 C; Stability: Good in presence of acids, alkalis and electrolytes; pH: 6–7.5 (1% aq.)
*Composition:* 100% active ingred.
*Reaction:* Neutral
*Containers:* 200 kg net mild steel drums or bulk
*Warnings and Caution:* Avoid prolonged contact with concentrated form; spillages may be slippery; do not mix with strong oxidizing or reducing agents (potentially explosive)
*References:* Bulletin SI:14FM678

**Ethylan HA.** Lankro Chem. Ltd.
*Chemical Description:* Nonyl phenol ethoxylate
*Category and Applications:* See Ethylan 20
*Form:* Waxy solid; Color: White; Odor: Negligible
*Properties:* Solubility: Water-sol.; Sp. gr.: 1.064 @ 60 C; Visc.: 120 cs @ 60 C; HLB: 17.4; Cloud Pt.: > 100 C (1% aq.); Flash Pt.: > 400 F (COC); Pour Pt.: 43 C; Stability: Stable to electrolytes, hard water, extremes of pH and temperature; pH: 6–8 (1% aq.)
*Composition:* 100% active ingred.
*Containers:* 200 kg net mild-steel drums or bulk
*Warnings and Caution:* Avoid prolonged contact with concentrated form; spillages may be slippery; do not mix with strong oxidizing or reducing agents (potentially explosive)
*References:* Bulletin SI:13 FM678

**Ethylan HP.** Lankro Chem. Ltd.
*Chemical Description:* Nonyl phenol ethoxylate
*Category and Applications:* See Ethylan 20
*Form:* Waxy solid; Color: White; Odor: Negligible
*Properties:* Solubility: Water-sol.; Sp. gr.: 1.072 @ 40 C; Visc.: 78 cs @ 60 C; HLB: 16.6; Cloud Pt.: > 100 C (1% aq.); Flash Pt.: > 400 F (COC); Pour Pt.: 36 C; Stability: Stable to electrolytes, hard water, extremes of pH and temperature; pH: 6–8 (1% aq.)
*Composition:* 100% active ingred.
*Containers:* 200 kg net mild-steel drums or bulk

*Warnings and Caution:* Avoid prolonged contact with concentrated form; spillages may be slippery; do not mix with strong oxidizing or reducing agents (potentially explosive)
*References:* Bulletin SI:13 FM678

**Ethylan KELD.** Lankro Chem. Ltd.
*Chemical Description:* Modified fatty alkylolamide
*Category:* Emulsifier, nonionic
*Applications:* Wax , kerosene and olein emulsification; solvent degreasing
*Form:* Clear liquid; Color: Amber; Odor: Mild
*Properties:* Solubility: Disp. in water; Sp. gr.: 1.006 @ 20 C; Visc.: 630 cs @ 20 C; Flash Pt.: > 300 F (COC); Pour Pt.: < 0 C; Stability: Good in presence of hard water and electrolytes; pH: 8–10 (1% aq.)
*Composition:* 100% active ingred.
*Warnings and Caution:* Avoid prolonged contact with concentrated form; spillages may be slippery; do not mix with strong oxidizing or reducing agents (potentially explosive)
*References:* Bulletin SI:12FM678

**Ethylan KEO.** Lankro Chem. Ltd.
*Chemical Description:* Nonyl phenol ethoxylate
*Category and Applications:* See Ethylan BCP
*Form:* Clear colorless liquid; Odor: Negligible
*Properties:* Solubility: Water-sol.; Sp. gr.: 1.060 @ 20 C; Visc.: 340 cs @ 20 C; HLB: 13.0; Cloud Pt.: 54 C (1% aq.); Flash Pt.: > 400 F (COC); Pour Pt.: 3 C; Stability: Stable to electrolytes, hard water, extremes of pH and temperature; pH: 6–8 (1% aq.)
*Composition:* 100% active ingred.
*Containers:* 200 kg net mild-steel drums or bulk
*Warnings and Caution:* Avoid prolonged contact with concentrated form; spillages may be slippery; do not mix with strong oxidizing or reducing agents (potentially explosive)
*References:* Bulletin SI:13 FM678

**Ethylan L10.** Lankro Chem. Ltd.
*Chemical Description:* PEG 1000 laurate
*Category:* Emulsifier
*Applications:* Cosmetics and pharmaceuticals
*Form:* Solid; Color: White
*Properties:* Sp. gr.: 1.056; Visc.: 76 cs @ 40 C
*Reaction:* Neutral

**Ethylan LD.** Lankro Chem. Ltd.
*Chemical Description:* Coconut diethanolamide
*Category:* Emulsifier, foam stabilizer, antistat, nonionic
*Applications:* Foam stabilizer in liquid detergents, general purpose or hard surface cleaners and in shampoos; emulsifier in hand cleaning gels; antistat for plastics; kerosene and oil emulsifier
*Form:* Clear liquid; Color: Straw; Odor: Mild
*Properties:* Solubility: Disp. in water; Sp. gr.: 0.981 @ 20 C; Visc.: 960 cs @ 20 C; Flash Pt.: > 350 F

(COC); Pour Pt.: 15 C; Stability: Good in presence of hard water and electrolytes; pH: 8.5–10.5 (1% aq.)
*Composition:* 90% active ingred.
*Warnings and Caution:* Avoid prolonged contact with concentrated form; spillages may be slippery; do not mix with strong oxidizing or reducing agents (potentially explosive)
*References:* Bulletin SI:12FM678

**Ethylan LDS.** Lankro Chem. Ltd.
*Chemical Description:* Coconut diethanolamide
*Category:* Foam stabilizer, emulsifier, nonionic
*Applications:* Foam stabilizer in liquid detergents with cold temperature stability, general purpose or hard surface cleaners and in shampoos; kerosene and oil emulsifier
*Form:* Clear liquid; Color: Straw; Odor: Mild
*Properties:* Solubility: Disp. in water; Sp. gr.: 0.990 @ 20 C; Visc.: 810 cs @ 20 C; Flash Pt.: > 350 F (COC); Pour Pt.: 10 C; Stability: Good in presence of hard water and electrolytes; pH: 8.5–10.5 (1% aq.)
*Composition:* 87% active ingred.
*Warnings and Caution:* Avoid prolonged contact with concentrated form; spillages may be slippery; do not mix with strong oxidizing or reducing agents (potentially explosive)
*References:* Bulletin SI:12FM678

**Ethylan LM.** Lankro Chem. Ltd.
*Chemical Description:* Coconut monoalkanolamide
*Category:* Emulsifier, antistat, foam stabilizer, nonionic
*Applications:* Additive for hard surface cleaners; foam stabilizer for liquid detergents
*Form:* Flake; Color: Pale straw; Odor: Mild
*Properties:* Solubility: Water-insol.; Sp. gr.: 0.909 @ 80 C; Visc.: 40 cs @ 80 C; Flash Pt.: > 350 F (COC); Pour Pt.: 63 C; Stability: Good in presence of hard water and electrolytes; pH: 8.5–10.5 (1% aq.)
*Composition:* 94% active ingred.
*Warnings and Caution:* Avoid prolonged contact with concentrated form; spillages may be slippery; do not mix with strong oxidizing or reducing agents (potentially explosive)
*References:* Bulletin SI:12FM678

**Ethylan LM2.** Lankro Chem. Ltd.
*Chemical Description:* Ethoxylated coconut fatty acid alkylolamide
*Category:* Foam stabilizer, emulsifier, nonionic
*Applications:* Detergent formulations
*Form:* Soft paste; Color: Off-white; Odor: Mild
*Properties:* Solubility: Water-sol.; Sp. gr.: 0.963 @ 40 C; Visc.: 108 cs @ 40 C; Flash Pt.: > 300 F (COC); Pour Pt.: 2 C; Stability: Good in presence of hard water and electrolytes; pH: 7.0–9.0 (1% aq.)
*Composition:* 100% active ingred.

*Warnings and Caution:* Avoid prolonged contact with concentrated form; spillages may be slippery; do not mix with strong oxidizing or reducing agents (potentially explosive)
*References:* Bulletin SI:12FM678

**Ethylan MLD.** Lankro Chem. Ltd.
*Chemical Description:* Lauric diethanolamide
*Category:* Foam booster and stabilizer, antistat, nonionic
*Applications:* Foam booster and stabilizer in toiletries and detergent formulations (e.g., shampoos, bubble baths); antistat for plastics
*Form:* Waxy flake; Color: Pale cream; Odor: Negligible
*Properties:* Solubility: Disp. in water; Visc.: 130 cs @ 60 C; Flash Pt.: > 300 F (COC); Pour Pt.: 40 C ; Stability: Good in presence of hard water and electrolytes; pH: 8–9 (1% aq.)
*Composition:* 95% active ingred.
*Reaction:* Alkaline
*Warnings and Caution:* Avoid prolonged contact with concentrated form; spillages may be slippery; do not mix with strong oxidizing or reducing agents (potentially explosive)
*References:* Bulletin SI:12FM678

**Ethylan N30.** Lankro Chem. Ltd.
*Chemical Description:* Nonyl phenol ethoxylate
*Category and Applications:* See Ethylan 20
*Form:* Waxy solid; Color: White; Odor: Negligible
*Properties:* Solubility: Water-sol.; Sp. gr.: 1.064 @ 60 C; Visc.: 90 cs @ 60 C; HLB: 17.0; Cloud Pt.: > 100 C (1% aq.); Flash Pt.: > 400 F (COC); Pour Pt.: 39 C; Stability: Stable to electrolytes, hard water, extremes of pH and temperature; pH: 6–8 (1% aq.)
*Composition:* 100% active ingred.
*Containers:* 200 kg net mild-steel drums or bulk
*Warnings and Caution:* Avoid prolonged contact with concentrated form; spillages may be slippery; do not mix with strong oxidizing or reducing agents (potentially explosive)
*References:* Bulletin SI:13 FM678

**Ethylan N50.** Lankro Chem. Ltd.
*Chemical Description:* Nonyl phenol ethoxylate
*Category and Applications:* See Ethylan 20
*Form:* Waxy solid; Color: White; Odor: Negligible
*Properties:* Solubility: Water-sol.; Sp. gr.: 1.073 @ 60 C; Visc.: 135 cs @ 60 C; HLB: 18.2; Cloud Pt.: > 100 C (1% aq.); Flash Pt.: > 400 F (COC); Pour Pt.: 43 C; Stability: Stable to electrolytes, hard water, extremes of pH and temperature; pH: 6–8 (1% aq.)
*Composition:* 100% active ingred.
*Containers:* 200 kg net mild-steel drums or bulk
*Warnings and Caution:* Avoid prolonged contact with concentrated form; spillages may be slippery; do not mix with strong oxidizing or reducing agents (potentially explosive)
*References:* Bulletin SI:13 FM678

**Ethylan N92.** Lankro Chem. Ltd.
*Chemical Description:* Nonyl phenol ethoxylate
*Category and Applications:* See Ethylan 20
*Form:* Waxy solid; Color: White; Odor: Negligible
*Properties:* Solubility: Water-sol.; Sp. gr.: 1.078 @ 60 C; Visc.: 340 cs @ 60 C; HLB: 19.0; Cloud Pt.: > 100 C (1% aq.); Flash Pt.: > 400 F (COC); Pour Pt.: 50 C; Stability: Stable to electrolytes, hard water, extremes of pH and temperature; pH: 6–8 (1% aq.)
*Composition:* 100% active ingred.
*Containers:* 200 kg net mild-steel drums or bulk
*Warnings and Caution:* Avoid prolonged contact with concentrated form; spillages may be slippery; do not mix with strong oxidizing or reducing agents (potentially explosive)
*References:* Bulletin SI:13 FM678

**Ethylan NP 1.** Lankro Chem. Ltd.
*Chemical Description:* Nonyl phenol ethoxylate
*Category:* Defoamer, nonionic
*Applications:* Defoaming applications, oil emulsifier
*Form:* Clear liquid; Color: Straw
*Properties:* Solubility: Oil-sol.; Sp. gr.: 0.987 @ 20 C; Visc.: 650 cs @ 20 C; HLB: 4.5; Flash Pt.: 300 F (COC); Pour Pt.: < 0 C; Stability: Stable to electrolytes, hard water, and extremes of pH and temperature; pH: 6–8 (1% aq.)
*Composition:* 100% active ingred.
*Containers:* 200 kg net mild-steel drums or bulk
*Warnings and Caution:* Avoid prolonged contact with concentrated form; spillages may be slippery; do not mix with strong oxidizing or reducing agents (potentially explosive)
*References:* Bulletin SI:13 FM678

**Ethylan PQ.** Lankro Chem. Ltd.
*Chemical Description:* Modified alkylphenol ethoxylate
*Category:* Solubilizer, nonionic
*Applications:* Manufacutre of iodophor
*Form:* Water-white liquid (clear @ 30 C); Odor: Faint
*Properties:* Solubility: Water-sol.; Sp. gr.: 1.045 @ 40 C; Visc.: 110 cs @ 40 C; HLB: 14.2; Cloud Pt.: 83 C (1% aq.); Flash Pt.: > 350 F (COC); Pour Pt.: 15 C; Stability: Good in presence of acids, alkalis and electrolytes pH: 6.0–7.5 (1% aq.)
*Composition:* 98% active ingred.
*Reaction:* Neutral
*Containers:* 200 kg net mild steel drums or bulk
*Warnings and Caution:* Avoid prolonged contact with concentrated form; spillages may be slippery; do not mix with strong oxidizing or reducing agents (potentially explosive)
*References:* Bulletin SI:14 FM678

**Ethylan OE.** Lankro Chem. Ltd.
*Chemical Description:* Cetyl/oleyl alcohol ethoxylate
*Category:* Emulsifier, solubilizer, stabilizer, leveling agent, nonionic
*Applications:* Emulsifier for fatty acids, alcohols and waxes; essen-

tial oil solubilizer; latex stabilizer; in the manufacture of polishes; emulsion polymerization aid; dye leveling agent
*Form:* Waxy solid; Color: Cream; Odor: Negligible
*Properties:* Solubility: Water-sol.; Sp. gr.: 1.009 @ 40 C; Visc.: 67 cs @ 40 C; HLB: 14; Cloud Pt.: 90 C (1% aq.); Flash Pt.: > 350 F (COC); Pour Pt.: 31 C; Stability: Good in presence of acids, alkalis and electrolytes; pH: 6–8 (1% aq.)
*Composition:* 100% active ingred.
*Reaction:* Neutral
*Containers:* 200 kg net mild steel drums or bulk
*Warnings and Caution:* Avoid prolonged contact with concentrated form; spillages may be slippery; do not mix with strong oxidizing or reducing agents (potentially explosive)
*References:* Bulletin SI:9FM678

**Ethylan R.** Lankro Chem. Ltd.
*Chemical Description:* Cetyl/oleyl alcohol ethoxylate
*Category and Applications:* See Ethylan OE
*Form:* Waxy solid; Color: Cream; Odor: Negligible
*Properties:* Solubility: Water-sol.; Sp. gr.: 1.026 @ 40 C; Visc.: 150 cs @ 40 C; HLB: 15.4; Cloud Pt.: > 100 C (1% aq.); Flash Pt.: > 350 F (COC); Pour Pt.: 36 C; Stability: Good in presence of acids, alkalis and electrolytes; pH: 6–8 (1% aq.)
*Composition:* 100% active ingred.
*Reaction:* Neutral
*Containers:* 200 kg net mild steel drums or bulk
*Warnings and Caution:* Avoid prolonged contact with concentrated form; spillages may be slippery; do not mix with strong oxidizing or reducing agents (potentially explosive)
*References:* Bulletin SI:9FM678

**Ethylan TC.** Lankro Chem. Ltd.
*Chemical Description:* Coconut fatty amine ethoxylate
*Category:* Wetting agent, nonionic
*Applications:* Wetting agents in acid or alkaline formulations for metal cleaning and stripping of surface coatings; fiber antistat; cosmetics
*Form:* Liquid; Color: Dark amber; Odor: Mild fatty amine
*Properties:* Solubility: Water-sol.; Sp. gr.: 1.040 @ 20 C; Visc.: 140 cs @ 20 C; HLB: 15; Flash Pt.: > 300 F (COC); Pour Pt.: < 0 C; Stability: Good in presence of acids, alkalis and high electrolyte concentrations; pH: 8.5–10.0 (1% aq.)
*Composition:* 100% active ingred.
*Reaction:* Alkaline
*Containers:* 200 kg net mild steel drums or bulk
*Warnings and Caution:* Avoid prolonged contact with concentrated form; spillages may be slippery; do not mix with strong oxidizing or reducing agents (potentially explosive)
*References:* Bulletin SI:15FM678

**Ethylan TCO.** Lankro Chem. Ltd.
*Chemical Description:* Complex

amine oxide
*Category:* Stabilizer, foam booster, emulsifier, nonionic
*Applications:* Foam and suds stabilizer in light and heavy duty liquid detergents; cosmetic formulations; foam stabilizer in no-gel carpet backing and latex foam odor control agent
*Form:* Clear to hazy liquid; Color: Straw; Odor: Mild fatty amine
*Properties:* Solubility: Water-sol.; Sp. gr.: 1.008 @ 20 C; Visc.: 159 cs @ 20 C; Flash Pt.: > 300 F (COC); Pour Pt.: 0 C; Stability: Good in presence of acids, alkalis, electrolytes; pH: 6–8 (1% aq.)
*Composition:* 40% active ingred.
*Reaction:* Neutral
*Containers:* 200 kg net mild steel drums or bulk
*Warnings and Caution:* Avoid prolonged contact with concentrated form; spillages may be slippery; do not mix with strong oxidizing or reducing agents (potentially explosive)
*References:* Bulletin SI:15 FM678

**Ethylan TF.** Lankro Chem. Ltd.
*Chemical Description:* Ethoxylated coconut fatty amine
*Category:* Emulsifier, antistat, nonionic
*Applications:* Oil emulsifier with anticorrosive properties; antistat for synthetic fibers, with polystyrene; formulation of wax polishes
*Form:* Liquid; Color: Light amber; Odor: Mild fatty amine
*Properties:* Solubility: Disp. in water; Sp. gr.: 0.912 @ 20 C; Visc.; 140 cs @ 20 C; HLB: 6; Flash Pt.: > 300 F (COC); Pour Pt.: < 0 C; Stability: Good in presence of acids, alkalis and electrolytes; pH: 8.5–10.0 (1% aq.)
*Composition:* 100% active ingred.
*Toxicity:* Slightly toxic; skin and eye irritant
*Containers:* 200 kg net mild steel drums or bulk
*Warnings and Caution:* Avoid prolonged contact with concentrated form; spillages may be slippery; do not mix with strong oxidizing or reducing agents (potentially explosive)
*References:* Bulletin SI:15 FM678

**Ethylan TH-2.** Lankro Chem. Ltd.
*Chemical Description:* Hydrogenated fatty amine ethoxylate
*Category:* Emulsifier, anticorrosive agent, antistat, nonionic
*Applications:* Oil emulsifier with anticorrosive properties; antistat for synthetic fibers and polystyrene
*Form:* Solid; Color: Pale brown; Odor: Mild fatty amine
*Properties:* Solubility: Disp. in water; Sp. gr.: 0.878 @ 60 C; Visc.: 33 cs @ 60 C; HLB: 5; Flash Pt.: > 300 F (COC); Pour Pt.: 40 C; Stability: Good in presence of acids, alkalis and electrolytes; pH: 8.5–10.0 (1% aq.)
*Composition:* 100% active ingred.
*Toxicity:* Slightly toxic; skin and eye irritant
*Containers:* 200 kg net mild steel drums or bulk
*Warnings and Caution:* Avoid prolonged contact with concentrated form; spillages may be slippery;

do not mix with strong oxidizing or reducing agents (potentially explosive)
*References:* Bulletin SI:15FM678

**Ethylan TN-10.** Lankro Chem. Ltd.
*Chemical Description:* Coconut fatty amine ethoxylate
*Category:* Emulsifier, nonionic
*Applications:* Cosmetics manufacture, hair-conditioner formulations; oil emulsifier with anticorrosive properties; antistat for synthetic fibers; formulation of wax polishes
*Form:* Liquid; Color: Dark amber; Odor: Mild fatty amine
*Properties:* Solubility: Water-sol.; Sp. gr.: 1.015 @ 20 C; Visc.: 174 cs @ 20 C; HLB: 14; Flash Pt.: > 300 F (COC); Pour Pt.: < 0 C; Stability: Good in acids, alkalis and electrolytes; pH: 8.5–10.0 (1% aq.)
*Composition:* 100% active ingred.
*Containers:* 200 kg net mild steel drums or bulk
*Warnings and Caution:* Avoid prolonged contact with concentrated form; spillages may be slippery; do not mix with strong oxidizing or reducing agents (potentially explosive)
*References:* Bulletin SI:15FM678

**Ethylan TT-15.** Lankro Chem. Ltd.
*Chemical Description:* Tallow amine ethoxylate
*Category and Applications:* See Ethylan TC
*Form:* Paste; Color: Pale brown; Odor: Mild fatty amine
*Properties:* Solubility: Water-sol.; Sp. gr.: 1.030 @ 20 C; Visc.: 252 cs @ 20 C; HLB: 14; Flash Pt.: > 300 F (COC); Pour Pt.: 0 C; Stability: Good in presence of acids, alkalis and electrolytes; pH: 8.5–10.0 (1% aq.)
*Composition:* 100% active ingred.
*Containers:* 200 kg net mild steel drums or bulk
*Warnings and Caution:* Avoid prolonged contact with concentrated form; spillages may be slippery; do not mix with strong oxidizing or reducing agents (potentially explosive)
*References:* Bulletin SI:15FM678

**Ethylan TU.** Lankro Chem. Ltd.
*Chemical Description:* Nonyl phenol ethoxylate
*Category:* Detergent, wetting and emulsifying agent, nonionic
*Applications:* Textile scouring agent, dry cleaning detergent; wetting and emulsifying agent for oils and aromatic solvents
*Form:* Clear liquid; Color: Pale straw; Odor: Negligible
*Properties:* Solubility: Water-sol.; Sp. gr.: 1.055 @ 20 C; Visc.: 340 cs @ 20 C; HLB: 12.2; Cloud Pt.: 32 C (1% aq.); Flash Pt.: > 400 F (COC); Pour Pt.: < 0 C; Stability: Stable to electrolytes, hard water, extremes of pH and temperature; pH: 6–8 (1% aq.)
*Composition:* 100% active ingred.
*Containers:* 200 kg net mild-steel drums or bulk
*Warnings and Caution:* Avoid prolonged contact with concentrated form; spillages may be slippery;

do not mix with strong oxidizing or reducing agents (potentially explosive)
*References:* Bulletin SI:13 FM678

**Ethylan VPK.** Lankro Chem. Ltd.
*Chemical Description:* PEG unsaturated fatty acid ester
*Category:* Emulsifier, lubricant, nonionic
*Applications:* Textile oil emulsifier, fiber lubricant
*Form:* Clear liquid; Color: Pale straw
*Properties:* Sp. gr.: 1.043; Visc.: 120 cs @ 20 C; HLB: 12.4
*Composition:* 100% active ingred.
*Reaction:* Neutral

**Etocas 10.** Croda Chem. Ltd.
*Chemical Description:* POE 10 castor oil
*Category:* Emulsifier, solubilizer, lubricant, emollient, superfatting agent, antistat, softener, detergents, nonionic
*Applications:* Cosmetics (perfume and essential oil solubilizers; shampoos, foam baths; skin care preparations); textile surfactants (lubricants, antistats, spin finish additives, softeners, tufting aids, dye leveling agents for synthetic fiber processing); water soluble metal working fluids; emulsion polymerization; detergents and emulsifiers in insecticides, herbicides and household products
*Form:* Liquid; Color: Pale yellow
*Properties:* Solubility: Sol. in ethyl alcohol, oleyl alcohol, naphtha, methyl ethyl ketone, oleic acid, trichloroethylene; part sol. in butyl stearate, arachis oil; disp. in water; HLB: 6.3; Cloud Pt.: $<$ 20 C (1% in 10% brine); Sapon. Value: 120–130; Acid No.: 1.0 max.; Hydroxyl No.: 120–140; Iodine Value: 55–60; Surface Tension: Insol. (0.1 deionized water); pH: 6.0–7.5
*Composition:* 97% active ingred.

**Etocas 35.** Croda Chem. Ltd.
*Chemical Description:* POE 35 castor oil
*Category and Applications:* See Etocas 10
*Form:* Liquid; Color: Pale yellow
*Properties:* Solubility: Sol. in water, ethyl and oleyl alcohols, naphtha, methyl ethyl ketone, oleic acid, trichloroethylene; part sol. in butyl stearate and arachis oil; HLB: 12.5; Cloud Pt.: 35–40 C; Sapon. Value: 62–72; Acid No.: 1.0 max.; Hydroxyl Value: 75–90; Iodine Value: 30–35; Surface Tension: 41.5 dynes/cm (0.1 deionized water); pH: 6.0–7.5

**Etocas 40.** Croda Chem. Ltd.
*Chemical Description:* POE 40 castor oil
*Category and Applications:* See Etocas 10
*Form:* Liquid; Color: Pale yellow
*Properties:* Solubility: Sol. in water, ethyl and oleyl alcohols, naphtha, methyl ethyl ketone, oleic acid, trichloroethylene; part. sol. in butyl stearate; HLB: 13.0; Cloud Pt.: 50 C; Sapon. Value: 60–65; Acid No.: 1.0 max.; Hy-

droxyl No.: 60–75; Iodine Value: 28–32; pH: 6.0–7.5
*Composition:* 97% active ingred.

**Etocas 60.** Croda Chem. Ltd.
*Chemical Description:* POE 60 castor oil
*Category and Applications:* See Etocas 10
*Form:* Soft paste; Color: Pale yellow
*Properties:* Solubility: Sol. in water, ethyl alcohol, naphtha, methyl ethyl ketone, oleic acid, trichloroethylene; part sol. in oleyl alcohol, butyl stearate; HLB: 14.7; Cloud Pt.: 60 C; Sapon. Value: 45–50; Acid No.: 1.0 max.; Hydroxyl No.: 50–60; Iodine Value: 22–26; Surface Tension: 43.2 dynes/cm (0.1 deionized water); pH: 6.0–7.5

**Etocas 100.** Croda Chem. Ltd.
*Chemical Description:* POE 100 castor oil
*Category and Applications:* See Etocas 10
*Form:* Waxy solid; Color: Pale yellow
*Properties:* Solubility: Sol. in water, ethyl alcohol, naphtha, methyl ethyl ketone, oleic acid, trichloroethylene; part sol. in oleyl alcohol; HLB: 16.5; Cloud Pt.: 66 C; Sapon. Value: 25–35; Acid No.: 1.0 max.; Hydroxyl Value: 35–45; Iodine Value: 14–18; Surface Tension: 41.6 dynes/cm (0.1 deionized water); pH: 6.0–7.5
*Composition:* 97% active ingred.

**Eureka 102.** Atlas Refinery, Inc.
*Chemical Description:* Sulfated castor oil
*Category:* Emulsifier, detergent, anionic
*Applications:* Topping oil for suede leather; grinding aid in pigment dispersions; plasticizer in finish coatings
*Form:* Liquid; Color: Light
*Properties:* pH: 7.6
*Composition:* 72% active ingred.
*Reaction:* Neutral

**Eureka 392.** Atlas Refinery, Inc.
*Chemical Description:* Sulfated fatty acid from vegetable oil
*Category:* Emulsifier, anionic
*Applications:* Carrier for refined oils; base for solvent fatliquor systems
*Form:* Liquid
*Properties:* pH: 5.6
*Composition:* 75% active ingred.

**Eureka 400.** Atlas Refinery, Inc.
*Chemical Description:* Sulfonated oil based on refined sperm oil
*Category:* Emulsifier, lubricant, softener, anionic
*Applications:* Leather processing
*Form:* Liquid; Color: Light
*Properties:* Stability: Good over entire pH range; pH: 4.5
*Composition:* 87% active ingred.

**Eureka E-2.** Atlas Refinery, Inc.
*Chemical Description:* Cationic derivative of a fatty acid diamine condensate
*Category:* Emulsifier, lubricant, softener, cationic
*Applications:* Emulsifier for raw

oils (e.g., neatsfoot, mineral)
*Form:* Heavy viscous liquid
*Properties:* Density: 8.32 lb/gal @ 60 F; Sp. gr.: 0.999 @ 60 F; pH: 6–6.5 (10% sol'n.)
*Composition:* 100% active ingred.

**Exp. Surfactant XD-8390.** Dow
*Category:* Emulsifier, wetting agent
*Applications:* Emulsion polymerization; primary emulsifier for SB and ABS polymers; provides stability and dispersibility for flowable pesticide formulations
*Form:* Liquid; Color: Red-brown
*Properties:* Sp. gr.: 1.11 @ 20 C; Visc.: 10 cps
*Composition:* 35% active ingred.
*Toxicity:* Acute oral, skin and eye irritant
*Containers:* 55 gal drums, bulk
*Warnings and Caution:* Burns the eyes; may cause skin irritation; toxic to fish

# F

**Flexricin 9.** NL Industries, Ind. Chem. Div.
*Chemical Description:* Propylene glycol monoricinoleate
*Category:* Wetting agent, solvent, plasticizier, nonionic
*Applications:* Wetting agent, dye solvent, wax plasticizer; textile, household and cosmetic applications; rewetting dried skins
*Form:* Liquid; Color: 2+ (Gardner)
*Properties:* Solubility: Sol. in toluene, butyl acetate, MEK, ethanol; Sp. gr.: 0.96; Visc.: 3 stokes
*Composition:* 100% active ingred.
*Containers:* 55 gal drums
*Warnings and Caution:* Keep containers closed; avoid contact with oxidizing materials

**Flexricin 13.** NL Industries, Ind. Chem. Div.
*Chemical Description:* Glyceryl monoricinoleate
*Category:* Wetting agent, plasticizer, release agent, antifoam agent, nonionic
*Applications:* Wetting agent, wax plasticizer, plasticizer and mold release agent for rubber polymers, antifoam agent; textile, household and cosmetic applications; rewetting dried skins
*Form:* Liquid; Color: 2 min. (Gardner)
*Properties:* Solubility: Sol. in toluene, butyl acetate, MEK, ethanol; Sp. gr.: 0.985; Visc.: 8.8 stokes
*Composition:* 100% active ingred.
*Containers:* 55 gal drums
*Warnings and Caution:* Keep containers closed; avoid contact with oxidizing materials

**Flexricin 15.** NL Industries, Ind. Chem. Div.
*Chemical Description:* Ethylene glycol monoricinoleate
*Category:* Wetting agent, plasticizer, nonionic
*Applications:* Textiles, household and cosmetic uses; rewetting dried skins
*Form:* Liquid; Color: 4 (Gardner)
*Properties:* Solubility: Sol. in toluene, butyl acetate, MEK, ethanol; Sp. gr.: 0.965; Visc.: 3.9 stokes
*Containers:* 55 gal drums
*Warnings and Caution:* Keep con-

tainers closed; avoid contact with oxidizing materials

**Fluilan.** Croda Chem. Ltd.
*Chemical Description:* Pure liquid lanolin from anhydrous lanolin BP/USP
*Category:* Conditioning agent, emollient, moisturizer, penetrant, superfatting agent, dispersant, emulsifier, nonionic
*Applications:* Skin care cosmetics, make-up
*Form:* Viscous liquid; Color: Golden; Odor: Faint characteristic lanolin
*Composition:* 100% active ingred.

**Fluilanol.** Croda Chem. Ltd.
*Chemical Description:* Liquid lanolin
*Category:* Emulsifier, emollient, dispersant, wetting agent, superfatting agent, nonionic
*Applications:* Make-up cosmetics, bath oils; improves emulsion stability, dispersing and spreading properties; wetting agent for pigments in lipophilic media; superfatting agent for soaps and detergents
*Form:* Liquid; Color: Pale; Odor: Faint sterol
*Properties:* Self-emulsifying
*Composition:* 100% active ingred.

**Fluorad FC-93.** 3M Co.
*Chemical Description:* Ammonium perfluoroalkyl sulfonate
*Category:* Wetting agent, anionic
*Applications:* Wetting agent in etching solutions in manufacture of semi-conductor devices; active in acid and alkaline media
*Form:* Clear liquid
*Properties:* Solubility: Sol. in mixtures of alcohol and water; only slightly sol. in water; Sp. gr.: 1.1; Flash Pt.: 81 F (PMCC); Stability: Excellent chemical and thermal stability; pH: 7 (1% aq.)
*Composition:* 25% active ingred.; other components: 73% water, 27% isopropyl alcohol
*Reaction:* Neutral
*Warnings and Caution:* Flammable
*References:* Bulletin Y-ITD93 (79.05)BPH

**Fluorad FC-95.** 3M Co.
*Chemical Description:* Potassium perfluoroalkyl sulfonate
*Category:* Wetting and foaming agent, anionic
*Applications:* Chrome plating baths; conversion coatings; metal and plastic etchants; acid pickling baths; alkaline cleaning systems; for systems requiring low surface tension and dense foam
*Form:* Free-flowing powder
*Properties:* Solubility: Low sol. in water and most inorganic solvents; recommended with co-solvent such as DGMBE; M.P.: Decomposes @ 390 C; Stability: Stable to strongly oxidizing and reducing media; chemically and thermally stable; Surface Tension: 22 dynes/cm (0.1 g/100 g water); pH: 7–8 (0.1% aq.)
*Composition:* 100% active ingred.
*Warnings and Caution:* Avoid eye contact, prolonged breathing; use with adequate ventilation
*References:* Bulletin Y-ISTDI

(75-.2)NPI

**Fluorad FC-98.** 3M Co.
*Chemical Description:* Potassium perfluoroalkyl sulfonate
*Category and Applications:* See Fluorad FC-95; FC-98 is slightly less surface active and produces less stable and less dense foam than FC-95
*Form:* Free-flowing powder
*Properties:* Solubility: Low sol. in water and most inorganic solvents; recommended with co-solvent such as DGMBE; M.P.: Decomposes @ 390 C; Stability: Stable to acid and oxidizing systems; chemically and thermally stable; Surface Tension: 40 dynes/cm (0.1 g/100 g water); pH: 6–8 (0.1% aq.)
*Composition:* 100% active ingred.
*Warnings and Caution:* Avoid eye contact, prolonged breathing; use with adequate ventilation
*References:* Bulletin Y-ISTDI (75-.2)NPI

**Fluorad FC-99.** 3M Co.
*Chemical Description:* Amine perfluoroalkyl sulfonate sol'n. in water
*Category:* Wetting and foaming agent, anionic
*Applications:* Chrome plating baths, metal and plastic etchants, acid cleaning and pickling baths
*Form:* Clear liquid; Color: Yellow to amber
*Properties:* Solubility: Water miscible; Sp. gr.: 1.1; Stability: Chemically and thermally stable; active in acidic systems; Surface Tension: 22 dynes/cm (0.4 g/100 g dist. water); pH: 6–7 (as is)
*Composition:* 25% active ingred.; other components: water
*Toxicity:* Considered minimally irritating
*References:* Bulletin Y-ITD99(79.03)R1

**Fluorad FC-120.** 3M Co.
*Chemical Description:* Anionic fluorochemical
*Category:* Wetting and leveling agent, anionic
*Applications:* Aqueous coating systems; clear polishes based on alkali soluble polymers
*Form:* Low viscosity liquid
*Properties:* Sp. gr.: 1.06; Flash Pt.: $>$38 C (TCC); pH: 8.5–9.5
*Composition:* 25% active ingred. in 37.5% ethylene glycol monobutyl ether and 37.5% water
*Toxicity:* Minor eye and skin irritant
*Warnings and Caution:* Avoid prolonged breathing of vapor; goggles and gloves recommended
*References:* Bulletin Y-ITD120 (68.05)R1

**Fluorad FC-126.** 3M Co.
*Chemical Description:* Anionic fluorochemical
*Category:* Anionic surfactant
*Form:* Free-flowing powder
*Applications:* Excellent surface activity in alkaline media
*Properties:* Solubility: Sol. in water, acetone, methyl alcohol; M.P.: Decomposes @ 120 C; Surface Tension: 44 dynes/cm (0.001 conc.); pH: 4–6

*Composition:* 100% active ingred.

**Fluorad FC-128.** 3M Co.
*Chemical Description:* Potassium fluorinated alkyl carboxylate
*Category:* Wetting, spreading and leveling agent, anionic
*Applications:* Excellent leveling in alkaline media; good foaming; water-based coatings; alkaline cleaners; leak detection solutions
*Form:* Powder
*Properties:* Solubility: Forms gel in water; Surface Tension: 17 dynes/cm (0.1 g/100 g dist. water); pH: 7–8 (0.1% aq.)
*Composition:* 100% active ingred.
*Warnings and Caution:* Avoid contact with skin, eyes; avoid prolonged breathing of vapors
*References:* Bulletin Y-ISTDJ (75-.2)NPI

**Fluorad FC-129.** 3M Co.
*Chemical Description:* Potassium fluorinated alkyl carboxylate
*Category:* Wetting and leveling agent, anionic
*Applications:* Floor polish, alkaline cleaners
*Form:* Liquid; Color: Amber
*Properties:* Solubility: Limited sol. in aq. acids and bases; stronger sol. in organic polar liquids (enhanced by solvents in FC-129); Sp. gr.: 1.3; Flash Pt.: 115 F (PMCC); Surface Tension: 17 dynes/cm (0.2 g/100 g dist. water); pH: 7–8 (0.1% solids in aq. sol'n.)
*Composition:* 50% active ingred.; other components: 28% water, 18% butyl Cellosolve, 4% ethyl alcohol
*Warnings and Caution:* Flammable; keep tightly closed; avoid contact with skin, eyes; avoid prolonged breathing
*References:* Bulletin Y-ITD129 (29.05)R1

**Fluorad FC-134.** 3M Co.
*Chemical Description:* Fluorinated alkyl quaternary ammonium iodides
*Category:* Wetting, spreading and leveling agent, cationic
*Applications:* Floor polishes, photographic development, corrosion inhibitors, leak detector sol'ns.; excellent activity in acid, basic and neutral media
*Form:* Powder
*Properties:* Solubility: Forms gel in water; Surface Tension: 17 dynes/cm (0.1 g/100 g dist. water); pH: 3–4 (0.1% aq.)
*Composition:* 100% active ingred.
*Warnings and Caution:* Avoid prolonged breathing, skin and eye contact

**Fluorad FC-135.** 3M Co.
*Chemical Description:* Fluorinated alkyl quaternary ammonium iodides
*Category:* Wetting and leveling agent, cationic
*Applications:* Cleaning and polishing products; substantive to many metal surfaces
*Form:* Liquid; Color: Dark amber
*Properties:* Solubility: Low sol. in most nonpolar organic solvents; limited sol. in highly polar solvents, aq. acids and bases; Sp. gr.: 1.2; Flash Pt.: 69 F (PMCC); Sur-

face Tension: 17 dynes/cm (0.2 g/100 g dist. water); pH: 3–4 (0.1% solids in aq. sol'n.)
*Composition:* 50% active solids; 34% isopropyl alcohol; 16% water
*Toxicity:* Slight skin and eye irritant
*Warnings and Caution:* Flammable; keep from heat, sparks and flame; protect skin and eyes from contact
*References:* Bulletin Y-ITD135 (59.025)R1

**Fluorad FC-143.** 3M Co.
*Chemical Description:* Ammonium perfluoroalkyl carboxylate
*Category:* Emulsifier, wetting agent, anionic
*Applications:* Fluropolymer emulsions; wetting agent in alkaline media
*Form:* Free-flowing powder
*Properties:* Solubility: Sol. in water, methyl alcohol, acetone; Density: 0.6–0.7 g/cc; M.P.: Decomposes @ 130 C; Stability: Excellent chemical and thermal stability and in acid, alkaline, oxidizing and reducing media; Surface Tension: 44 dynes/cm (0.1 g/100 g dist. water); pH: 4–6 (0.5 aq. sol'n.)
*Composition:* 100% active ingred.
*Warnings and Caution:* Avoid prolonged breathing of dust or fumes; use with adequate ventilation
*References:* Bulletin Y-ISTDH (65.2)R

**Fluorad FC-170-C.** 3M Co.
*Chemical Description:* Fluorinated alkyl polyoxethylene ethanol
*Category:* Wetting, spreading and leveling agent, nonionic
*Applications:* Floor polish formulations; good activity in water and acid solutions
*Form:* Liquid; Color: Amber
*Properties:* Solubility: Sol. in hydrochloric acid, methyl alcohol, ethyl alcohol, isopropyl alcohol; Sp. gr.: 1.32; Cloud Pt.: $< 0$ C; Flash Pt.: $> 220$ F (PMCC); Surface Tension: 20 dynes/cm (0.1 g/100 g dist. water); pH: 7–8 (1% aq.)
*Composition:* 95% active ingred.
*References:* Bulletin Y-ITD-170C (38.05)R1

**Fluorad FC-171.** 3M Co.
*Chemical Description:* Fluorinated alkyl alkoxylate
*Category:* Wetting, spreading and leveling agent, nonionic
*Applications:* Floor polishes, cleaners, high solids coatings and water reducible coatings
*Form:* Liquid; Color: Amber
*Properties:* Solubility: 20% sol. in butyl Cellosolve, Cellosolve acetate, toluene, methyl ethyl ketone, dimethyl formamide, isopropyl alcohol, dibutyl phthalate; forms cloudy sol'n. in water; Sp. gr.: 1.4; Flash Pt.: $> 220$ F (PMCC); Pour Pt.: $\approx 20$ F; Surface Tension: 20 dynes/cm (0.1 g/100 g dist. water)
*Composition:* 100% active ingred.
*Toxicity:* Minor eye irritant
*References:* Bulletin Y-ITD171

**Fluorad FC-430.** 3M Co.
*Chemical Description:* Fluorinated alkyl ester
*Category:* Wetting, leveling,

spreading and flow control agents, nonionic
*Applications:* Water-based coatings and polymer systems; solvent-based coatings; soldering systems
*Form:* Viscous liquid; Color: Straw to amber
*Properties:* Ref. Index: 1.445; Solubility: > 20 g of solid/100 ml of solvent for: dist. water, Cellosolve acetate, methyl ethyl ketone, toluene, dimethylformamide; Sp. gr.: 1.15; Visc.: 15,000 cp (Brookfield spindle # 3 @ 6 rpm); Flash Pt.: > 300 F (TOC); Surface Tension: 30.3 dynes/cm (0.1 g/100 ml dist. water)
*Composition:* 100% active ingred.
*Warnings and Caution:* Avoid prolonged breathing of vapors and eye and skin contact
*References:* Bulletin Y-ITD-430-1 (114-.2)NPI

**Fluorad FC-431.** 3M Co.
*Chemical Description:* Fluorinated alkyl ester
*Category and Applications:* See Fluorad FC-430
*Form:* Thin liquid; Color: Straw to amber
*Properties:* Ref. Index: 1.406; Solubility: > 20 g of solid/100 ml of: butyl Cellosolve, Cellosolve acetate, methyl ethyl ketone, toluene, Solvesso 100, dimethylformamide, dibutyl phthalate; Sp. gr.: 1.05; Visc.: 200 cp (Brookfield spindle # 1 @ 60 rpm); Flash Pt.: 40 F (TOC); Surface Tension: 36.6 dynes/cm (0.1 g/100 ml dist. water)
*Composition:* 50% active solids in ethyl acetate
*Warnings and Caution:* Flammable solvent; keep away from heat, sparks or flame; avoid prolonged breathing and skin and eye contact
*References:* Bulletin Y-ITD-430 (114-.2)NPI

**Fluorad FC-432.** 3M Co.
*Chemical Description:* Fluorinated alkyl esters
*Category:* Nonionic surfactant
*Applications:* Active in low polarity organic solvents; forms hydrocarbon liquids; solvent-based coatings; oil-based insecticides, grease removers, penetrating oils
*Form:* Low viscosity liquid
*Properties:* Ref. Index: 1.40; Solubility: Sol. in perchloroethylene, toluene, benzene, heptane; Sp. gr.: 0.78; Flash Pt.: 19.4 F (TCC); Surface Tension: 20.5 dynes/cm (2 g/100 g ethyl acetate)
*Composition:* 25% active in heptane
*Warnings and Caution:* Flammable solvent; keep from heat, sparks or flame; avoid prolonged breathing and skin contact
*References:* Bulletin Y-ISTDG (65.2)R

**Fluorad FC-740.** 3M Co.
*Chemical Description:* Fluorinated alkyl ester
*Category:* Well stimulation additive, nonionic
*Applications:* Well stimulation additive active in organic solvents; produces stable foams in low po-

larity hydrocarbon liquids
*Form:* Liquid; Color: Amber
*Properties:* Sp. gr.: 1.01; Visc.: 100 cps; Flash Pt.: > 100 F (PM-CC); Pour Pt.: 10 F; Surface Tension: 24.1 dynes/cm (0.2% FC-740 in kerosene @ 22 C)
*Composition:* 50% solids in S.C. (Central Solvents and Chemicals Co.) Solvent 150
*Toxicity:* Mild skin and eye irritant
*Warnings and Caution:* Keep away from heat, sparks or flame; avoid prolonged skin contact

**Fluorad FC-750.** 3M Co.
*Chemical Description:* Fluorinated alkyl quaternary ammonium iodides
*Category:* Well stimulation additive, cationic
*Applications:* Oil and gas well stimulation fluids; active in acid, neutral and alkaline media; offers lowest surface tension for aq. acids; enhances activity of organic corrosion inhibitors
*Form:* Liquid; Color: Dark
*Properties:* Sp. gr.: 1.2; Flash Pt.: 53 F (PMCC); Pour Pt.: 32–40 F; Surface Tension: 17.5 dynes/cm (100 ppm in deionized water)
*Composition:* 50% solids in 2/1 isopropanol/water
*Toxicity:* Minor skin and eye irritant
*Warnings and Caution:* Keep away from flames; protect eyes and skin from contact
*References:* Bulletin Y-ITD750 (99.05)II

**Fluorad FC-760.** 3M Co.
*Chemical Description:* Fluorinated alkyl alkoxylates
*Category:* Well stimulation additive, nonionic
*Applications:* Oil and gas well stimulation fluids; useful at low levels to reduce surface tension of fluids used for acidizing, fracturing and water flooding; active in acid, neutral or alkaline sol'ns.
*Form:* Liquid; Color: Amber
*Properties:* Sp. gr.: 1.4; Flash Pt.: > 220 F (PMCC); Pour Pt.: 20 F; Surface Tension: 19 dynes/cm (100 ppm in deionized water)
*Toxicity:* Minimal eye irritant
*References:* Bulletin Y-ITD760 (99.05)II

**Foamaster 206-A.** Diamond Shamrock Europe
*Chemical Description:* Nonsilicone type, based on chemically modified oils and fats
*Category:* Defoamer
*Applications:* Thermosol dyeing, exhaust dyeing, printing for textile industry
*Form:* Liquid
*Properties:* Solubility: Readily dispersible in water to form emulsions
*Composition:* 100% active ingred.

**Foamaster 267-A.** Diamond Shamrock Europe
*Chemical Description:* Nonsilicone type, based on chemically modified oils and fats
*Category:* Defoamer, nonionic
*Applications:* In textile industry

for thermosol and exhaust dyeing, printing; defoamer for most coatings based on synthetic latices
*Form:* Liquid
*Properties:* Solubility: Readily dispersible in water to form emulsions
*Composition:* 100% active ingred.

**Foamaster 335-A.** Diamond Shamrock Europe
*Chemical Description:* Nonsilicone type based on chemically modified oils and fats
*Category:* Defoamer
*Applications:* Textile industry for thermosol and exhaust dyeing, printing
*Form:* Liquid
*Properties:* Solubility: Readily dispersible in water to form emulsions
*Composition:* 100% active ingred.

**Foamaster 1407-50.** Diamond Shamrock Europe
*Category:* Defoamer, nonionic
*Form:* Fluidized paste
*Properties:* Solubility: Excellent water dispersibility

**Foamaster 1719-A.** Diamond Shamrock Europe
*Category:* Defoamer, nonionic
*Applications:* Soluble defoamer for use in gelatine
*Form:* Liquid
*Composition:* 50% active ingred.

**Foamaster 8034.** Diamond Shamrock Europe
*Category:* Defoamer, nonionic
*Applications:* Paints based on fine particle size acrylic emulsions; suggested for coating applications where standard defoamers are not effective; foam control in manufacture and in finished adhesives; rubber and latex latices
*Form:* Liquid
*Composition:* 100% active ingred.

**Foamaster B.** Diamond Shamrock Europe Ltd.
*Category:* Defoamer, nonionic
*Applications:* Predispersible defoamer for latex systems (paints, synthetic rubber and adhesives)
*Form:* Liquid
*Composition:* 100% active ingred.

**Foamaster DD-72.** Diamond Shamrock Europe Ltd.
*Category:* Defoamer, nonionic
*Applications:* Emulsion paints, wallpaper coatings, water-based printing inks)
*Form:* Liquid
*Composition:* 100% active ingred.

**Foamaster DF-177-F.** Diamond Shamrock Europe
*Category:* Foam suppressor, nonionic
*Applications:* Easily dispersible foam suppressor for coating and adhesive formulations
*Form:* Liquid
*Composition:* 100% active ingred.

**Foamaster DF-178.** Diamond Shamrock Europe
*Category:* Foam suppressor, nonionic
*Applications:* Coating and adhe-

sive formulations
*Form:* Liquid
*Composition:* 100% active ingred.

**Foamaster DF-198-L.** Diamond Shamrock Europe
*Category:* Defoamer, nonionic
*Applications:* Defoamer for paints based on fine particle size acrylic emulsions
*Form:* Liquid
*Composition:* 100% active ingred.

**Foamaster DNH-1.** Diamond Shamrock Europe
*Category:* Defoamer, nonionic
*Applications:* Defoamer for emulsion paints and polymers
*Form:* Liquid
*Composition:* 100% active ingred.

**Foamaster G.** Diamond Shamrock Europe
*Category:* Defoamer, nonionic
*Applications:* Dispersible latex stripping defoamer
*Form:* Liquid
*Composition:* 100% active ingred.

**Foamaster JMY.** Diamond Shamrock Europe
*Category:* Foam control agent, nonionic
*Applications:* Defoamers for adhesives, rubber and latex latices
*Form:* Liquid
*Composition:* 100% active ingred.

**Foamaster KF-99.** Diamond Shamrock Europe
*Chemical Description:* Nonsilicone types, based on chemically modified oils and fats
*Category:* Defoamer
*Applications:* Thermosol dyeing, exhaust dyeing, printing for textile industry
*Form:* Liquid
*Properties:* Solubility: Readily dispersible in water to form emulsions
*Composition:* 100% active ingred.

**Foamaster NDW.** Diamond Shamrock Europe
*Category:* Defoamer, nonionic
*Applications:* Defoamer for water-based paints, adhesive and cement screeds; coatings based on starch, protein or synthetic latex adhesives; rubber and latex latices
*Form:* Liquid
*Composition:* 100% active ingred.

**Foamaster NS-20.** Diamond Shamrock Europe
*Category:* Defoamer, nonionic
*Applications:* Silicone-free defoamer for coating or paper stock, air knife coatings
*Form:* Liquid
*Composition:* 100% active ingred.

**Foamaster NXZ.** Diamond Shamrock Europe
*Category:* Defoamer, nonionic
*Applications:* Defoamer for synthetic latices, paints adhesives; blade and roll coatings; emulsifiable latex stripping defoamer; for foam control in manufacture and in finished adhesive
*Form:* Liquid
*Composition:* 100% active ingred.

**Foamaster P.** Diamond Shamrock Europe
*Category:* Defoamer, nonionic
*Applications:* Foam control in polyvinyl alcohol systems (gives clear film) in adhesive industry
*Form:* Liquid
*Composition:* 35% active ingred.

**Foamaster PD-1.** Diamond Shamrock Europe
*Category:* Defoamer, nonionic
*Applications:* Defoamer for dry mixes, adhesives and cement screeds, joint cements, spackles, etc.
*Form:* Powder
*Composition:* 67% active ingred.

**Foamaster TDB.** Diamond Shamrock Europe
*Category:* Defoamer, nonionic
*Applications:* Foam control agent in dry adhesives, joint cements, spackles; rubber and latex latices
*Form:* Liquid
*Composition:* 100% active ingred.

**Foamaster TDB-1.** Diamond Shamrock Europe
*Category:* Defoamer, nonionic
*Applications:* Defoamer for water-soluble resins
*Form:* Liquid
*Composition:* 100% active ingred.

**Foamaster V.** Diamond Shamrock Europe
*Category:* Defoamer, nonionic
*Applications:* Dispersible latex stripping defoamer
*Form:* Liquid
*Composition:* 100% active ingred.

**Foamaster VL.** Diamond Shamrock Europe
*Category:* Defoamer, nonionic
*Applications:* Emulsifiable latex monomer stripping defoamer; latex and emulsion paints
*Form:* Liquid
*Composition:* 100% active ingred.

www.ingramcontent.com/pod-product-compliance
Ingram Content Group UK Ltd.
Pitfield, Milton Keynes, MK11 3LW, UK
UKHW020128290926
15471UKWH00003B/350

9 780820 600413